INTERNAL COMBUSTION ENGINE MODELING

INTERNAL COMBUSTION ENGINE MODELING

J. I. Ramos

Department of Mechanical Engineering
Carnegie-Mellon University
Pittsburgh, Pennsylvania

HEMISPHERE PUBLISHING CORPORATION

A member of the Taylor & Francis Group

New York Washington Philadelphia London

INTERNAL COMBUSTION ENGINE MODELING

1 2 3 4 5 6 7 8 9 0 EBEB 8 9 8 7 6 5 4 3 2 1 0 9

This book was set in Times Roman by EPS Group. The editors were Sandra Tamburrino and Evelyn Pettit. Cover design by Deborah Eubanks Riffe. Edwards Brothers, Inc., was the printer and binder.

Library of Congress Cataloging-in-Publication Data

Ramos, J. I. (Juan Ignacio)
Internal combustion engine modeling / J. I. Ramos.
p. cm.
Includes bibliographies and index.
1. Internal combustion engines—Mathematical models. I. Title.
TJ759.R36 1989 88-28757
621.43—dc19 CIP
ISBN 0-89116-157-0

To Mrs. Shirley L. Hamlin for her dedication and excellent typing.

To Dr. Nicholas S. Winowich, who has helped me in many ways.

To the Schocks (Harold Jr., Pat, Harold III, and Stanley), who helped me to recover and see light at the end of a tunnel.

CONTENTS

PREFACE

The modeling of reciprocating and rotary engines is a multidisciplinary subject that involves chemical thermodynamics, fluid mechanics, turbulence, heat transfer, combustion, and numerical methods. This book presents a comprehensive and somewhat detailed introduction to mathematical models of internal combustion engines. It is intended for researchers in fluid mechanics, combustion, turbulence, and heat transfer and for graduate and undergraduate students in thermal sciences.

Internal combustion engines are the main power plants of transportation systems and are responsible for a substantial fraction of fuel consumption. The scarcity of oil resources and the ever-increasing standards on air pollution and emissions have dictated the need for improved, more efficient, and less polluting internal combustion engines. Improvements on engine design have been achieved by traditional methods based on extensive experience. The advent of computers and the possibility of performing "numerical" experiments may provide a new way of designing internal combustion engines. In fact, a stronger interaction between engine modelers, designers, and experimenters may result in improved engine designs in the not-too-distant future.

The selection and use of a mathematical model of engine combustion depend to a considerable extent on the phenomena that researchers wish to resolve and on available computational resources. Preliminary estimates of thermal engine efficiency and power output can be obtained from air standard cycles (Chapter 1), which neglect gas exchange processes, heat transfer to cylinder walls, and variations of specific heats with temperature. Better estimates can be obtained from equilibrium charts, which provide the temperature and composition of the combustion products for adiabatic processes at constant volume (Otto cycle) or constant pressure (Diesel cycle). However, both air standard cycles and equilibrium charts overestimate the maximum cylinder pressure and temperature and the engine power output, because they neglect heat transfer losses and simulate combustion as a heat addition process that occurs at constant pressure or at constant volume.

More detailed mathematical models of internal combustion engines account for the temporal or temporal and spatial variations of the charge pressure, temperature and composition, the heat transfer losses, the gas exchange processes, the flows in the manifolds and around the intake and exhaust valves, and so on. In Chapters 2, 3, and 4, mathematical models of rotary, reciprocating compression-ignition (diesel), and reciprocating spark-ignition engines are presented. These models can be classified into two main groups: thermodynamic and dimensional models, depending on the temporal and temporal-spatial resolution of the flow field. Thermodynamic models, in turn, can be classified into two subgroups: single-zone and multizone models, according to the number of zones in which the cylinder charge is divided. Single-zone models may be interpreted as an extension of the air standard cycles presented in Chapter 1 in that the pressure, temperature, and composition are assumed uniform throughout the combustion chamber. However, these models consider heat transfer losses to the cylinder walls and heat addition as functions of time. Single-zone models can be used as predictive tools if the heat release rate or the mass burning rate is specified as a function of time. The heat release rate can be deduced from experimentally determined pressure-volume diagrams and depends on the residual mass fraction, volumetric efficiency, engine load, combustion chamber geometry, and so on. Single-zone models also can be used to analyze the heat release rate in internal combustion engines from pressure-volume diagrams. Single-zone models neglect the flame propagation process and, in some instances, have been used to predict and improve engine performance and combustion chamber design.

Multizone models of homogeneous-charge engines consider the flame propagation process through the cylinder charge, divided into three zones: the burnt gases, the quench layer, and unburnt gases. The burnt and unburnt gases are separated by a flame front, the velocity of which can be specified as a function of the cylinder pressure and temperature, rpm, turbulence intensity, and length scales. The flame speed or the mass burning rate also can be calculated from phenomenological models that account for the entrainment and combustion of the fresh mixture. The entrainment of the fresh gases into the flame front and combustion depend on the engine turbulence intensity and length scales, which can be calculated from turbulence models. Turbulence models are discussed in Chapter 2 and are extensions of those developed for statistically stationary, incompressible, thin shear flows. The flow fields in internal combustion engines, however, are compressible, involve complex recirculation zones characterized by strong streamline curvature, and are not statistically stationary. It is impossible with present computers to resolve all the time and length scales of the flows in internal combustion engines; turbulence models are required. The first issue that arises in the development of these models is averaging. Long-time-averaging techniques (Chapter 2) are strictly valid for statistically stationary flows, where the averaging time is much shorter than the characteristic unsteadiness time of the mean flow. In engines turbulence relaxation times are of the same order as cycle times, and long-time-averaging methods are not strictly applicable.

The modeling of combustion in turbulent, chemically reacting flows also is discussed in Chapter 2, where combustion regimes are classified according to the values of the turbulence intensity, laminar flame speed, Reynolds number, mixing and reaction times, laminar flame thickness, Kolmogorov eddy size, and integral length scale of the turbulence. The combustion models presented in Chapter 2 are classified into three main groups: mean-reaction-rate models, mixing-controlled models, and hybrid models. In mean-reaction-rate models the reaction time is much longer than the turbulent mixing time, whereas the opposite is true in mixing-controlled models. The limitations of turbulent combustion models are also discussed in Chapter 2, and the chapter also presents some sample results of two- and three-dimensional flow fields in rotary (Wankel) engine models.

In Chapter 3 mathematical models of diesel or compression-ignition engines are presented. These include single-zone, multizone, and dimensional models. Single-zone models of diesel engine combustion are analogous to those presented in Chapter 2 and account for the variations of the equivalence ratio due to fuel injection. Multizone models divide the cylinder charge into several elements, the evolution of which is followed as a function of time accounting for air entrainment, liquid fuel vaporization, mixing, and combustion. Chapter 3 also contains a detailed introduction to liquid fuel breakup and atomization, spray modeling, droplet-turbulence interactions, and thin and thick sprays and a discussion on diesel engine emissions (soot and carbonaceous particulates) and noise.

Chapter 4 contains a comprehensive description of thermodynamic models of spark-ignition engines, including single-zone and multizone models. However, multidimensional models are only mentioned briefly in Chapter 4 because these models are discussed at some length in Chapter 2. The multizone models discussed in Chapters 2 and 4 have been classified into three groups, depending on the specification of the mass burning rate, the turbulent flame speed, and the calculation of the mass burning rate. The limitations of these models are discussed, and comparisons with available experimental data are presented. In particular, it is shown that multizone models that use a specified burning speed do not account for the effects of the turbulence length scale on the turbulent flame speed. Chapter 4 also contains an introduction to ignition, engine knock, and unburnt hydrocarbon, nitric oxide, and carbon monoxide emissions.

In-cylinder processes are analyzed in Chapters 1, 2, 3, and 4. A brief introduction to gas exchange processes and two-stroke engines is presented in Chapter 5, where mathematical models for analyzing flows in engine manifolds and around valves are presented. The gas exchange processes, the flows in the manifolds, and the flows around the intake and exhaust valves play a major role in determining the volumetric and scavenging efficiency in four- and two-stroke engines, respectively, and in establishing the flow field within the engine cylinder. Multidimensional models of gas exchange processes are based on the equations presented in Chapter 2 and are only discussed briefly in Chapter 5.

Numerous topics are described in the book, thus reflecting the multidisciplinary nature of internal combustion engine modeling. Portions of the book may be of interest to researchers and students working on turbulence, combustion, turbulent combustion, and turbulent two-phase flows. In particular, the book includes detailed descriptions of the modeling of turbulence, combustion, two-phase flows, and droplet-turbulence interactions. However, it does not include the numerical methods that are needed to solve the flow field in internal combustion engines and models of engine components, such as carburetors, compressors, turbines, catalytic converters, intercoolers, mufflers, and air filters. The modeling of these components can be found in some of the references presented in this book.

Portions of the book can be used in courses on internal combustion engines (Chapters 1–5), turbulence (Chapter 2), combustion (Chapter 2), two-phase flows (Chapter 3), and air pollution (Chapters 3 and 4).

In preparing the book, I have tried to present a balanced account of thermodynamic and dimensional models, used numerous references and papers, and tried to minimize repetitions and duplications. Thus, for example, nitric oxide, carbon monoxide, and unburnt hydrocarbon emissions are described in Chapter 4, whereas soot, smoke, and particulates are discussed in Chapter 3. Experimental results are discussed throughout the text and particularly in Chapters 3 and 4. In some cases the predictions of reciprocating and rotary engine models are compared qualitatively and quantitatively with experimental data.

I thank those colleagues throughout the world who have provided me with information and figures included in the book. I am indebted particularly to Dr. Rodney J. Tabaczynski of Ford Motor Company, Dr. Peter J. O'Rourke of Los Alamos National Laboratory, Profesor Tom I-P. Shih of the University of Florida in Gainesville, Dr. Charles K. Westbrook of Lawrence Livermore National Laboratory, Dr. R. M. Green of Sandia National Laboratories, Drs. T.-W. Kuo and K.-J. Wu of General Motors Research Laboratories, Professor Harold J. Schock of Michigan State University, Professor Ahmed F. Ghoniem of the Massachusetts Institute of Technology, Dr. H. Lee Nguyen of NASA Lewis Research Center, and Dr. Mark H. Carpenter of NASA Langley Research Center. I am also grateful to Professors William A. Sirignano and Frediano V. Bracco for introducing me to the topics described in the book and to Mr. Rangarajan Pitchumani, who read and provided useful comments on the whole manuscript. Special thanks are due to Mrs. Shirley L. Hamlin for her excellent typing, patience, and diligence in a seemingly endless task. Her dedication and help are most appreciated. Finally, I acknowledge the NASA Lewis Research Center and U.S.–Spain Joint Committee for Scientific and Technological Cooperation for supporting my work on engine modeling.

J. I. Ramos

INTERNAL COMBUSTION ENGINE MODELING

CHAPTER
ONE

AIR STANDARD CYCLES AND EQUILIBRIUM CHARTS

1.1 INTRODUCTION

The objective of this chapter is to present the air standard cycles (Section 1.2) and equilibrium calculations (Section 1.3). Air standard cycles serve as an introduction to the more detailed and accurate models of internal combustion engines presented in Chapters 2, 3, and 4 and provide insight into some of the important parameters that affect the performance of reciprocating and rotary engines. However, air standard cycles neglect the heat transfer to the cylinder walls, replace the combustion process by a heat addition process that occurs at constant volume in the Otto cycle and at constant pressure in the diesel cycle, do not consider the gas exchange processes, and assume that the cylinder charge is a calorically perfect gas (i.e., the specific heats at constant pressure and volume are assumed constant).

Equilibrium calculations at constant pressure or constant volume provide the temperature and composition of the combustion products and can be used to estimate the performance and thermal efficiency of homogeneous-charge and direct-injection stratified-charge engines. However, these calculations neglect the heat transfer to the cylinder walls and do not consider the fuel vaporization, mixing and combustion, and finite rate chemical kinetics phenomena.

Air standard cycles and equilibrium calculations overestimate the performance, the maximum pressure and temperature, and the thermal efficiency of reciprocating and rotary engines and are based on the first law of thermodynamics. Analyses based on the second law of thermodynamics provide insights into the distribution of engine thermodynamic losses and irreversibilities and may indicate possible engine modifications to improve the fuel economy. Methods based on the second law of thermodynamics are presented in Section 1.4.

1.2 AIR STANDARD CYCLES

Air standard cycles are a useful introduction to the most detailed thermodynamic models presented in the next three chapters. They assume that the mixture charge is pure air with constant specific heats at constant pressure and volume, neglect the intake and exhaust processes (i.e., a fixed mass of air is the working fluid throughout the engine cycle), replace combustion by a heat transfer process from an external source, and complete the cycle by heat transfer to the surroundings until the air pressure and temperature correspond to the initial conditions [1].

In the next sections the efficiencies of the Otto and diesel cycles are analyzed.

1.2.1 Otto Cycle

In the Otto cycle the air at initial pressure p_1 and temperature T_1 is compressed isentropically until top dead center (TDC) (point 2 in Fig. 1.1) so that

$$p_2 = p_1 \left(\frac{V_1}{V_2}\right)^{\gamma} \qquad T_2 = T_1 \left(\frac{V_1}{V_2}\right)^{\gamma-1} \tag{1}$$

where $\gamma = C_p/C_v$ is the ratio of specific heats, C_p is the specific heat at constant pressure, and C_v is the specific heat at constant volume.

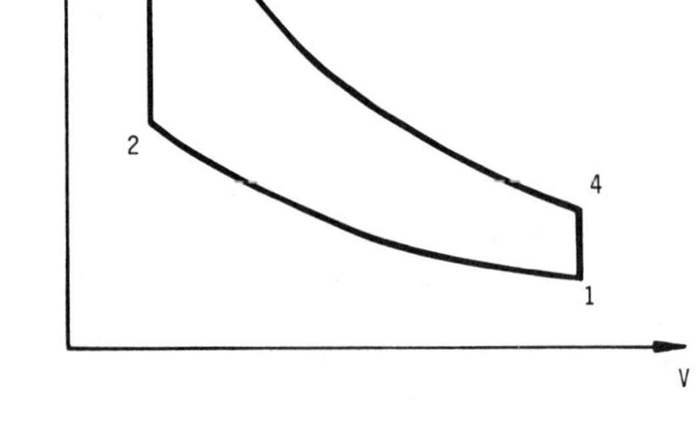

Figure 1.1 Otto cycle.

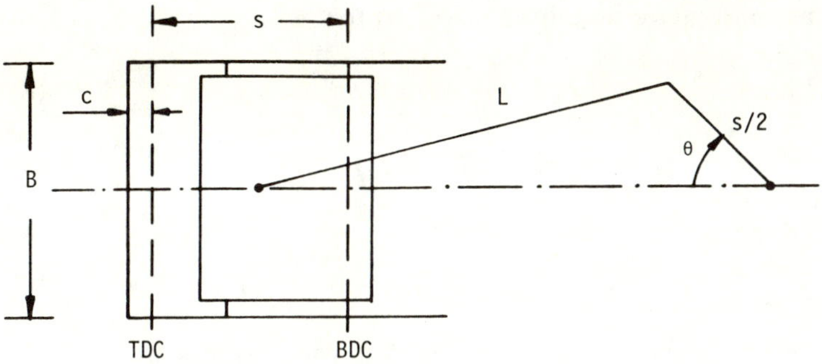

Figure 1.2 Schematic of a reciprocating internal combustion engine. B, cylinder bore; c, clearance; s, stroke; L, connecting rod length; θ, crank shaft angle; TDC, top dead center; BDC, bottom dead center.

The volumes V_1 and V_2 are given by the following expressions (Fig. 1.2):

$$V_1 = \pi B^2 \frac{s + c}{4} \qquad V_2 = \pi B^2 \frac{c}{4} \tag{2}$$

where B is the cylinder bore, c is the clearance, and s is the stroke.

The compression ratio is defined as

$$\mathrm{CR} = \frac{V_1}{V_2} = \frac{s + c}{c} \tag{3}$$

Thus, Eq. (1) can be written as

$$p_2 = p_1 \mathrm{CR}^{\gamma} \qquad T_2 = T_1 \mathrm{CR}^{\gamma - 1} \tag{4}$$

Combustion is simulated as a constant-volume heat addition process from 2 to 3, so that

$$T_3 = p_3 \frac{T_2}{p_2} = \frac{p_3}{p_2} T_1 \mathrm{CR}^{\gamma - 1} \tag{5}$$

and the heat added from 2 to 3 is

$$Q_{v_{2\to 3}} = mC_v(T_3 - T_2) \tag{6}$$

where m is the mass of air contained in the cylinder.

In the expansion stroke (from 3 to 4), the air is expanded isentropically so that

$$p_4 = p_3 \left(\frac{V_2}{V_1}\right)^{\gamma} = p_3 \mathrm{CR}^{-\gamma} \qquad T_4 = T_3 \left(\frac{V_2}{V_1}\right)^{\gamma - 1} = T_3 \mathrm{CR}^{1-\gamma} \tag{7}$$

Heat is lost at constant volume from 4 to 1 so that

$$T_4 = \frac{T_1 p_4}{p_1} \tag{8}$$

$$Q_{v_{4\to 1}} = m\, C_v\, (T_1 - T_4) \tag{9}$$

The work W done in the cycle is

$$W = Q_{v_{2\to 3}} + Q_{v_{4\to 1}} = m\, C_v\, (T_3 - T_2) + m\, C_v\, (T_1 - T_4) \tag{10}$$

and the Otto cycle efficiency η_O is given by the following expression:

$$\eta_O = \frac{W}{Q_{v_{2\to 3}}} = 1 + \frac{T_1 - T_4}{T_3 - T_2} \tag{11}$$

Substitution of Eqs. (4), (5), (7), and (8) into Eq. (11) yields

$$\eta_O = 1 - \mathrm{CR}^{1-\gamma} \tag{12}$$

Equation (12) shows that the Otto cycle efficiency is independent of the heat transferred to the cycle.

1.2.2 Diesel Cycle

The diesel cycle, shown in Fig. 1.3, consists of isentropic compression (1 to 2), heat addition at constant pressure (2 to 3), isentropic expansion (3 to 4), and heat transfer at constant volume (4 to 1).

The pressure and temperature at 2 are given by the following expressions:

$$p_2 = p_1\, \mathrm{CR}^{\gamma} \qquad T_2 = T_1\, \mathrm{CR}^{\gamma-1} \tag{13}$$

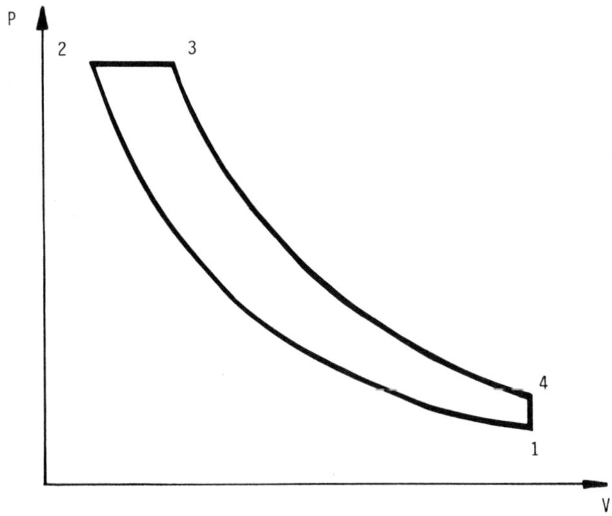

Figure 1.3 Diesel cycle.

The heat addition at constant pressure from 2 to 3 yields

$$Q_{p_{2\rightarrow 3}} = m\, C_p\, (T_3 - T_2) \tag{14}$$

where

$$T_3 = T_2 \frac{V_3}{V_2} = T_1\, \mathrm{CR}^{\gamma-1} \left(\frac{V_3}{V_2}\right) = \beta\, T_1\, \mathrm{CR}^{\gamma-1} \tag{15}$$

and $\beta = V_3/V_2$.

The isentropic expansion from 3 to 4 yields

$$T_4 = T_3 \left(\frac{V_3}{V_1}\right)^{\gamma-1} = T_1\, \beta^{\gamma} \qquad p_4 = p_3 \left(\frac{V_3}{V_1}\right)^{\gamma} = p_1\, \beta^{\gamma} \tag{16}$$

where T_4 can be calculated from the heat lost at constant volume (4 to 1) as

$$T_4 = T_1 \frac{p_4}{p_1} = T_1\, \beta^{\gamma} \tag{17}$$

The work W done in the cycle is

$$W = Q_{p_{2\rightarrow 3}} + Q_{v_{4\rightarrow 1}} = m\, C_p\, (T_3 - T_2) + m\, C_v\, (T_1 - T_4) \tag{18}$$

The diesel engine efficiency is given by the following expression:

$$\eta_D = \frac{W}{Q_{p_{2\rightarrow 3}}} = 1 + \frac{T_1 - T_4}{\gamma(T_3 - T_2)} \tag{19}$$

Substitution of Eqs. (13), (16), and (17) into Eq. (19) yields

$$\eta_D = 1 - f(\beta,\gamma)\, \mathrm{CR}^{1-\gamma} \tag{20}$$

where

$$f(\beta,\gamma) = \frac{\beta^{\gamma} - 1}{\gamma(\beta - 1)} \tag{21}$$

is greater than unity.

Equations (20) and (21) indicate that the efficiency of the diesel cycle depends on the heat transfer ($\beta > 1$). Equations (12) and (20) show that for the same compression ratio, the Otto cycle is more efficient than the diesel cycle. However, for the same maximum pressure and work, the diesel cycle is more efficient than the Otto cycle.

The efficiencies of the Otto and diesel cycles given by Eqs. (12) and (20) provide extremely optimistic numerical estimates of the engine work because (1) the heat transfer losses during compression and expansion have been neglected; (2) the charge in reciprocating engines is not pure air, and its specific heats are a function of the cylinder pressure, temperature, fuel-air ratio, and residual mass fraction; (3) combustion does not occur at constant presure (diesel cycle) or constant volume (Otto cyle); (4) in real engines the gas exchange processes modify the pressure-volume diagrams shown in Figs. 1.1 and 1.3, and reduce the power output, and the heat rejection processes do not occur at constant volume [2]; (5) in four-stroke engines, there are pumping losses that occur when the intake and exhaust valves are open; (6) combustion does not occur instantaneously; and (7) the efficiencies given by Eqs. (12) and (20) do not account for frictional forces.

Real gases have variable specific heats that depend on the fuel-air ratio, temperature, and pressure. For fuel-air ratios near stoichiometry, dissociation is important, and the variation of the specific heats with the pressure and temperature yields smaller pressure, temperature, engine efficiency, and power output.

The gas exchange process is most important in determining the temperature and pressure of the trapped gases in the cylinder during the compression stroke.

The Otto and diesel engine cycles can be extended to establish some basic relationships between heat release, cycle efficiency, and maximum pressure and to construct an equivalent air standard cycle to analyze engine test data [3]. However, these extensions are based on the unrealistic assumption that the heat added to the engine cylinder occurs at constant volume or at constant pressure, and they neglect the heat transfer losses to the cylinder walls.

1.3 EQUILIBRIUM CHARTS

The Otto and diesel cycle efficiencies presented in the previous section are based on the assumption that combustion is a heat addition process at constant volume (Otto cycle) or constant pressure (diesel cycle). In real engines chemical reactions occur in the compression stroke, and the mixture composition changes. In diesel engines the liquid fuel injected into the cylinder vaporizes with a consequent change in the composition of the gas mixture. Furthermore, the mass fraction of residual gases f mixes with the mass fraction $(1\text{-}f)$ of the incoming charge (i.e., the fluid entering the cylinder) to form the mixture. The mass fraction of the residual gases depends on the compression ratio, throttle opening, etc. [2, 4].

The composition of the combustion products can be calculated assuming that (1) the compression stroke is isentropic, (2) the mixture has a uniform composition, (3) there are no heat losses, (4) combustion occurs at constant volume (Otto cycle) or constant pressure (diesel cycle), and (5) the combustion products are in chemical equilibrium.

In order to calculate the temperature and composition of the equilibrium combustion products, consider as an example the reaction of isooctane in air:

$$C_8H_{18} + 4(O_2 + 3.76\,N_2) \rightarrow aCO_2 + bCO + cH_2O + dH_2 + eO_2 + fOH + gH + hO + lC(s) + qN_2 + mN + nNO \quad (22)$$

where the coefficients $a, b, \ldots, n$ denote the number of moles of the combustion products. The products of isooctane-air combustion may also include CH_4, NH_3, NO^+, e^-, etc., and depend on the temperature, pressure, and mixture stoichiometry. For lean mixtures ($\phi < 1$) there is an excess of oxygen to convert all the carbon and hydrogen to CO_2 and H_2O, and the main combustion products are CO_2, H_2O, O_2, and N_2. At high temperatures the dissociation of CO_2 and H_2O produces CO, O_2, H_2, O, H, and OH, and the dissociation of O_2 and H_2 yields H and O. At still higher temperatures N_2 and O_2 react and produce NO.

Equation (22) shows that there are 13 unknowns: the number of moles of the combustion products and the temperature. The mass balance equations for each atom in the system (C, H, O, and N) yield four equations:

$$8 = a + b + l \tag{23}$$

$$18 = 2c + 2d + f + g \tag{24}$$

$$8 = 2a + b + c + 2e + f + h + n \tag{25}$$

$$14.04 = 2q + m + n \tag{26}$$

If the reaction occurs at constant pressure, the first law of thermodynamics states that

$$\Delta U = \Delta Q - p\,\Delta V \qquad \Delta \mathrm{H} = \Delta Q + V\,\Delta p = \Delta Q \tag{27}$$

where U is the internal energy, ΔQ is the heat added to the system at pressure p, and V is the volume of the system.

For an adiabatic process Eq. (27) yields

$$\Delta \mathrm{H} = 0 \qquad \mathrm{H}_{\text{reactants}} = \mathrm{H}_{\text{products}} \tag{28}$$

where

$$\mathrm{H}_{\text{reactants}} = \sum_{\text{reac}} n_j h_j \qquad \mathrm{H}_{\text{products}} = \sum_{\text{prod}} n_j h_j \tag{29}$$

$$h_j = h_j^0 + \int_{T_0}^{T} C_{p_j}\,dT \tag{30}$$

where n_j is the number of moles, h_j is the specific enthalpy of species j, T_0 is a reference temperature, T is the temperature, h_j^0 is the specific enthalpy of formation of species j at T_0, and C_p is the specific heat at constant pressure. The enthalpy of formation is the heat required to form a substance from its elements in their standard states at the same temperature and pressure.

Equation (28) can be written from the temperature T_2 of the reactants to the temperature T_3 of the combustion products as

$$\sum_{\text{reac}} n_j \left(h_j^0 + \int_{T_0}^{T_2} C_{p_j}\,dT \right) = \sum_{\text{prod}} n_j \left(h_j^0 + \int_{T_0}^{T_3} C_{p_j}\,dT \right) \tag{31}$$

from which the temperature of the combustion products can be calculated, provided that the composition of the products is known.

Equations (23)–(26) and (31) form a system of five equations with 13 unknowns. Therefore, eight more equations are needed to calculate the temperature and composition of the combustion products. These equations can be obtained from eight equilibrium reactions such as, for example,

$$C + O_2 \rightleftarrows CO_2 \tag{32}$$

$$H_2 + \frac{1}{2} O_2 \rightleftarrows H_2O \tag{33}$$

$$C + \frac{1}{2} O_2 \rightleftarrows CO \tag{34}$$

$$\frac{1}{2} H_2 + \frac{1}{2} O_2 \rightleftarrows OH \tag{35}$$

$$\frac{1}{2} O_2 + \frac{1}{2} N_2 \rightleftarrows NO \tag{36}$$

$$\frac{1}{2} O_2 \rightleftarrows O \tag{37}$$

$$\frac{1}{2} H_2 \rightleftarrows H \tag{38}$$

$$\frac{1}{2} N_2 \rightleftarrows N \tag{39}$$

The equilibrium reactions are always written for the formation of one mole of the substances, and their equilibrium constants can be calculated from the law of mass action as follows. For an arbitrary equilibrium reaction such as

$$aA + bB \rightleftarrows cC + dD \tag{40}$$

the equilibrium constant at constant pressure K_p for ideal gases is given by the following expression:

$$K_p(T) = \frac{p_C^c p_D^d}{p_A^a p_B^b} = \exp\left(-\frac{\Delta F^\circ}{\tilde{R}T}\right) \tag{41}$$

where p_A, p_B, p_C, and p_D are the partial pressures of A, B, C, and D; $\tilde{R}$ is the universal gas constant; F is the Gibbs free energy defined as $F = U - TS$; S is the entropy; and ΔF° is the standard-state Gibbs free energy change.

The partial pressure of species i can be written as

$$p_i = p \frac{n_i}{\sum n_i} \tag{42}$$

where p is the mixture pressure.

Substitution of Eq. (42) into Eq. (41) yields

$$K_p = \frac{n_C^c \, n_D^d}{n_A^a \, n_B^b} \left(\frac{p}{\Sigma \, n_i} \right)^{c+d-a-b} \tag{43}$$

Equations (23)–(26), (31), and (43) form a system of algebraic equations from which T_3, a, b, c, d, e, f, g, h, l, q, m, and n can be calculated.

The values of the equilibrium constants at constant pressure and the values of the specific heats at constant pressure of many species have been collected in the JANAF tables [5].

The NASA equilibrium program [6] solves Eqs. (23)–(26), (31), and (43) by minimizing the Gibbs and Helmholtz free energies. Meintjes and Morgan [7] developed a procedure that algebraically combines the equilibrium equations, eliminates some unknowns, and scales the simplified equations to obtain a set of simultaneous nonlinear equations that are easy to solve algebraically. Their procedure combines the Gauss-Seidel and Newton methods. The Gauss-Seidel method is fast but may be unreliable, particularly when the matrices are not diagonally dominant. On the other hand, Newton's method is reliable but slow, and it has a quadratic rate of convergence when the initial guess is close enough to the solution.

For chemical reactions occurring in constant-volume processes such as those shown in Fig. 1.1, the first law of thermodynamics applied to an adiabatic process states that

$$\Delta U = \Delta Q - p \, \Delta V = 0 \qquad U_{\text{reactants}} = U_{\text{products}} \tag{44}$$

and for ideal gases

$$h = u + pv = u + RT \tag{45}$$

where h is the specific enthalpy, u is the specific internal energy, v is the specific volume, and R is the gas constant (i.e., $R = \tilde{R}/M$, where M is the molecular mass).

If the reference temperature is taken as $T_0 = 0$, Eq. (45) implies that

$$h^0 = u^0 \tag{46}$$

and

$$u = u^0 + \int_{T_0}^{T} C_v \, dT = h^0 + \int_{T_0}^{T} C_v \, dT \tag{47}$$

Thus, for adiabatic processes at constant volume, the energy equation [Eq. (31)] can be written as

$$\sum_{\text{prod}} n_j \left(h_j^0 + \int_{T_0}^{T_3} C_{v_j} \, dT \right) = \sum_{\text{reac}} n_j \left(h_j^0 + \int_{T_0}^{T_2} C_{v_j} \, dT \right) \tag{48}$$

which, with Eqs. (23)–(26) and (43), can be used to calculate the temperature and composition of the equilibrium combustion products.

The solution of Eqs. (23)–(26), (31), and (43) or Eqs. (23)–(26), (43), and (48) can be expressed in graphical form as a function of the pressure and temperature [4]. However, these graphical data provide only an approximation to the actual combustion processes in internal combustion engines because the heat transfer losses to the cylinder walls have been neglected, and combustion was assumed to occur at constant pressure (diesel cycle) or constant volume (Otto cycle). More accurate and detailed models of rotary and reciprocating engine combustion processes are presented in the next three chapters, which also include heat transfer models and finite rate chemistry phenomena.

1.4 SECOND LAW ANALYSES

The air standard cycles and equilibrium calculations presented in the two previous sections make use of only the first law of thermodynamics and mass balance equations. Recently, there has been a great interest in assessing automotive engine models in terms of the second law of thermodynamics. Second law analyses provide additional insights concerning the distribution of engine thermodynamic losses and may indicate possible engine modifications to improve fuel economy. These analyses may include turbocharging, charge air cooling, turbocompounding, insulating techniques, etc., and are based on the concept of availability [8].

The availability of a system in a given state is defined as the maximum useful work that can be obtained from the combination of the system and the surrounding atmosphere as the system goes from that state to equilibration with the atmosphere while exchanging heat only with the atmosphere [9, 10]. Availability is a property of the system, but its value depends not only on the properties of the system but also on the properties of the atmosphere.

The availability of a given system can be calculated by applying the first and second laws of thermodynamics. However, it is not a conserved quantity [11] but is related to the concept of irreversibility [12, 13]. Second law analyses determine the level of thermodynamic irreversibility of various engine processes.

Let us consider the mass, energy, and entropy balance equations in an open control volume of mean specific enthalpy h and mean specific entropy s. This control volume corresponds to a zero-dimensional model of the combustion chamber of an internal combustion engine. The mass influx $\dot{m}_i$ into the control volume may correspond to the mass exchanges through the inlet and exhaust valves, prechamber, crevices between the piston, piston rings, and cylinder wall, and fuel injected into the combustion chamber.

The mass, energy (first law), and entropy (second law) balance equations can be written as [14]

$$\dot{m} = \dot{m}_i \tag{49}$$

$$\dot{E} = \dot{m}_i h_i - \dot{Q}_w - p\dot{V} \tag{50}$$

$$\dot{S} = \dot{m}_i s_i - \frac{1}{T_w}\dot{Q}_w + \dot{S}_{\text{irr}} \tag{51}$$

where m, E, and S denote the mass, internal energy, and entropy of the combustion chamber; $\dot{Q}_w$ denotes the heat losses through the combustion chamber walls; p is the pressure; V is the volume; T_w is the wall temperature; $\dot{S}_{irr}$ (>0) is the rate of entropy generation by irreversibilities inside the control volume; and the dots denote differentiation with respect to time.

The values of $\dot{E}$ and $\dot{S}$ are related to the specific internal energy and entropy as follows:

$$\dot{E} = m\dot{e} + \dot{m}e \qquad \dot{S} = m\dot{s} + \dot{m}s \tag{52}$$

where

$$e = \frac{E}{m} \tag{53}$$

is the specific internal energy.

If the mixture in the combustion chamber behaves as a calorically perfect gas,

$$pV = mRT \tag{54}$$

$$e = e_0 + C_v(T - T_0) \tag{55}$$

$$s = s_0 + C_p \ln\left(\frac{T}{T_0}\right) - R \ln\left(\frac{p}{p_0}\right) \tag{56}$$

where C_p and C_v are the specific heats at constant pressure and volume, respectively, and e_0 and s_0 are the specific internal energy and specific entropy at pressure p_0 and temperature T_0.

Differentiation of Eqs. (54)–(56) with respect to time and substitution into Eqs. (50) and (51) yields

$$\dot{m} = \left\{\dot{m}_i h_i - \dot{Q}_w - p\dot{V} - m\left[\dot{e}_0 + C_v \frac{pV}{mR}\frac{\dot{T}}{T} + \dot{C}_v\left(\frac{pV}{mR} - T_0\right)\right]\right\} \times [e_0 + C_v(T - T_0)]^{-1} \tag{57}$$

$$\dot{S}_{irr} = \frac{\dot{Q}_w}{T_w} + \dot{m}\left(s_0 + C_p \ln\frac{T}{T_0} - R\ln\frac{p}{p_0}\right) + m\left(\dot{s}_0 + C_p\frac{\dot{T}}{T} + \dot{C}_p \ln\frac{pV}{mRT_0} - R\frac{\dot{p}}{p} - \dot{R}\ln\frac{p}{p_0}\right) - \dot{m}s_i \tag{58}$$

As a consequence of the second law of thermodynamics, $\dot{S}_{irr}$ must be nonnegative at every instant of time. This requirement provides a useful check on the consistency of any model.

Equations (50) and (51) can be integrated from t to $t + \Delta t$ to yield [15, 16]

$$E(t + \Delta t) - E(t) = H_i - Q_w - W \tag{59}$$

$$S(t + \Delta t) - S(t) = S_i - S_w + \Delta S_{irr} \tag{60}$$

where

$$H_i = \int_t^{t+\Delta t} \dot{m}_i h_i \, dt \qquad Q_w = \int_t^{t+\Delta t} \dot{Q}_w \, dt \tag{61}$$

$$W = \int_t^{t+\Delta t} p\dot{V} \, dt \qquad S_i = \int_t^{t+\Delta t} \dot{m}_i s_i \, dt \tag{62}$$

$$S_w = \int_t^{t+\Delta t} \frac{\dot{Q}_w}{T_w} \, dT \qquad \Delta S_{\text{irr}} = \int_t^{t+\Delta t} \dot{S}_{\text{irr}} \, dt \tag{63}$$

and Eq. (60) can be used to check the consistency of a particular engine model because $\Delta S_{\text{irr}} \geq 0$.

The values of C_p, C_v, R, T, e_0, and s_0 in Eqs. (57) and (58) depend on the thermodynamic model used to study combustion in reciprocating and rotary engines. These values depend on the mixture composition, pressure, and temperature and on the number of zones into which the charge is divided (see Chapters 2, 3, and 4).

The analysis presented in this section can also be applied in multidimensional engine models where the control volumes can be considered as those around the grid points. Second law analyses are very useful for choosing an engine model that is in agreement with the second law of thermodynamics. They can also be used to gain insight into the engine performance and design. Myers [17] used a second law analysis technique to show that the fuel consumption decreases with increases in the engine compression ratio.

REFERENCES

1. Benson, R. S., and Whitehouse, N. D., *Internal Combustion Engines*, Pergamon Press, New York, 1979.
2. Taylor, C. F., *The Internal Combustion Engine in Theory and Practice*, 2nd ed., vols. 1 and 2, MIT Press, Cambridge, Mass., 1985.
3. Hsu, B. D., Heat Release, Cycle Efficiency and Maximum Cylinder Pressure in Diesel Engine—The Use of an Extended Air Cycle Analysis, SAE International Off-Highway and Powerplant Congress and Exposition, Milwaukee, Wis., SAE Paper No. 841054, 1984.
4. Obert, E. F., *Internal Combustion Engines and Air Pollution*, Intext, New York, 1973.
5. Stull, D. R., and Prophet, H., *JANAF Thermochemical Tables*, 2nd ed., National Bureau of Standards, Washington, D.C., 1971.
6. Gordon, S., and McBride, B. J., Computer Program for Calculation of Complex Chemical Equilibrium Compositions, Rocket Performance, Incident and Reflected Shocks, and Chapman-Jouguet Detonations, NASA SP-273, National Aeronautics and Space Administration, Washington, D.C., 1971.
7. Meintjes, K., and Morgan, A. P., Performance of Algorithms for Calculating the Equilibrium Composition of a Mixture of Gases, *Journal of Computational Physics*, vol. 60, pp. 219–234, 1985.
8. Primus, R. J., A Second Law Approach to Exhaust Gas System Optimization, SAE International Congress and Exposition, Detroit, Mich., SAE Paper No. 840033, 1984.
9. Flynn, P. F., Hoag, K. L., Kamel, M. M., and Primus, R. J., A New Perspective on Diesel

Engine Evaluation Based on Second Law Analysis, SAE International Congress and Exposition, Detroit, Mich., SAE Paper No. 840032, 1984.
10. Kestin, J., Availability: The Concept and Associated Terminology, *Energy*, vol. 5, pp. 679–692, 1980.
11. Primus, R. J., Hoag, K. L., Flynn, P. F., and Brands, M. C., An Appraisal of Advanced Engine Concepts Using Second Law Analysis Techniques, I. Mech. E./SAE Conference Proceedings 1984, Paper No. C440/84, pp. 73–87, I. Mech. E., London, 1984.
12. Bejan, A., Second Law Analysis in Heat Transfer, *Energy*, vol. 5, pp. 721–732, 1980.
13. Bejan, A., *Entropy Generation through Heat and Fluid Flow*, Wiley, New York, 1982.
14. Brzustowski, T. A., Toward a Second-Law Taxonomy of Combustion Processes, *Energy*, vol. 5, pp. 743–755, 1980.
15. Bedran, E. C., and Beretta, G. P., General Thermodynamic Analysis for Engine Combustion Modelling, SAE International Congress and Exposition, Detroit, Mich., SAE Paper No. 850205, 1985.
16. Matthews, R. D., Beckel, S. A., and Shizki, M., A New Technique for Thermodynamic Engine Modelling, *AIAA Journal of Energy*, vol. 7, pp. 667–675, 1983.
17. Myers, P., The Art of Choosing a Model, SAE International Congress and Exposition, Detroit, Mich., SAE Paper No. 850341, 1985.

CHAPTER

TWO

MATHEMATICAL MODELS OF ROTARY ENGINES

2.1 INTRODUCTION

The performance of Wankel engines has been analyzed for a number of years at NSU in West Germany, Toyo Kogyo Co., Ltd. (now Mazda Motor Corporation) [1, 2], Curtiss-Wright Corporation [3, 4], Mazda [5], Toyota [6], Yammar Diesel Engine Co. [7], John Deere [8], and Outboard Marine Corporation [9]. These studies have shown that (1) misfire or partial burning tends to occur in rotary engines at light loads due to an excess of exhaust gas recirculation (EGR) [6]; (2) improvements in power output, fuel economy, and noise can be obtained with gas seals, thermally controlled cooling systems, and turbocharging [5]; and (3) direct-injection stratified-charge (DISC) rotary engines have a great potential as efficient, lightweight, compact, and economical power plants over a range of operating conditions [3, 4].

The DISC rotary engine has a small size, low weight, high specific power density, and multifuel capability [10]; it is particularly advantageous in areas where availability of aviation fuel is limited [11, 12]. In addition, rotary engines have fewer moving parts than reciprocating engines. However, their major problems are caused by leakage through the rotor apex seals and housing (or side) seals.

Table 2.1 summarizes current state-of-the-art performance for turbine, rotary, and piston engines that have been developed for aircraft applications. This table shows that the fuel consumption of rotary engines is slightly higher than that of piston engines; however, it is substantially lower than that of gas turbines.

Table 2.1 Engine comparison

	Piston engine	Gas turbine	Rotary engine (two rotors)
Sea-level performance (kW/hp)	263–350	263–350	263–350
Performance at 6 km (kW/hp)	263–350	150–200	263–350
Jet fuel	No	Yes	Yes
Weight	260	135	127
Fuel consumption (g/kW/hr)			
Rated	243	395	274
Cruise	225	335	244

In terms of power-to-weight ratio, the rotary engine has a higher power density than the piston engine but a lower power density than the gas turbine [12, 13]. Because of its unique geometry, the DISC rotary engine can be designed for operation with a variety of fuels while still maintaining a high power density [14–16]. It should be pointed out, however, that the performance of piston engines shown in Table 2.1 corresponds to engines developed more than 30 years ago [17] and that piston engines for aircraft applications do not seem to have been improved since then. Improvements in such engines may result in better performance than that shown in Table 2.1.

Jones et al. [14] showed that the steady-state specific fuel consumption of rotary engines may be equal to or better than that of representative automotive diesel engines. Loyd [18] showed that stratified-charge rotary engines are efficient, lightweight, compact, and economical power plants capable of a broad-range operation. Cichanowicz and Sawyer [19] extended the lean operating limit of a rotary engine by means of hydrogen-enriched fuel and observed a reduction in NO_x (nitric oxide) and CO (carbon monoxide) emissions and an increase in unburnt HC (hydrocarbon) emissions. Shimamura and Tadokoro [20] showed that catalysts can improve the fuel economy of rotary engines. King [21] analyzed the durability of natural-gas-fueled rotary engines and showed that the wear rates of the rotor apex seals and trochoid (housing) are low. Sierens and Verdonck [22] also analyzed the performance of a natural-gas-fueled stratified-charged rotary engine. Eberle and Klomp [23], Knoll et al. [24], and Orlandea et al. [25] analyzed engine seals and showed that potential improvements in the rotary engine can be achieved by reducing apex- and side-seal leakage.

The National Aeronautics and Space Administration Lewis Research Center conducted a research program aimed at studying the possible application of rotary engines to small aircraft [12, 13]. The work included the use of holographic interferometry and helium to visualize the small- and large-scale flow structures in a motored Wankel engine as a function of the engine speed and rotor position [26], the development of a real-time indicated mean effective pressure (IMEP) system to analyze the cycle efficiency of a rotary engine [27, 28], finite element structural analyses, the development of a trigger mechanism to synchronize cinematography and holographic techniques with light source timings and shaft encoders [29], and thermodynamic models that have been used with finite element methods to obtain the temperature distributions in the combustion chamber, housing, rotor, and seals of a low-heat-rejection rotary engine [30].

Flame photography and time-resolved exhaust sampling have been used to analyze the effects of the fuel-air ratio, engine load, residual gas, and engine speed on combustion and HC emissions in rotary engines [31, 32]. These studies have shown that the major sources of HC emissions are flame extinction in the trailing portion of the combustion chamber for lean air-fuel ratios and apex seal leakage for rich air-fuel mixtures. Burley et al. [33] found that the major sources of unburnt HC are apex-seal leakage, cold walls, residuals, and surface-to-volume ratio; minor sources were found to be the crevice volumes, spark plug, turbulence, intake mixture, and lubricant on the combustion chamber walls. Ferguson et al. [34] showed that most of the NO_x is exhausted with the leading gases while most of the HC is exhausted with the trailing gases. This indicates that in the combustion chamber the highest temperatures occur near the rotor leading apex and the lowest temperatures occur near the rotor trailing apex. Ito et al. [35] obtained photographs of rotary engine combustion and showed that combustion is poor at the trailing edge of the chamber due to high velocities and heat transfer losses. They also showed that stratification results in more uniform and intense combustion than in conventional rotary engines, that leading flame propagation toward the rotor leading apex is enhanced by the rotor-induced velocity field, and that toward the rotor trailing apex it is diminished by the rotor-induced velocity field.

Mathematical models have been developed to enhance understanding of the flow field, fuel-air mixing, and combustion in rotary engines. These models are based on those developed for spark-ignition and diesel engines and can be classified into two groups: thermodynamic (also called zero-dimensional) and dimensional models. Thermodynamic models can in turn be divided into single- and multizone models, whereas dimensional models can be divided into one- and multidimensional models. Single-zone models consider that the mixture composition, pressure, and temperature are uniform throughout the combustion chamber. These models can be used in heat release analyses if an experimental pressure diagram is used as input or as predictive tools if the heat release rate is specified. However, the assumption of uniform temperature throughout the combustion chamber is unrealistic. This assumption is relaxed in multizone models, where the combustion chamber mixture is divided into three regions: the unburnt and burnt gas zones and the quench layer. Multizone models require specification of the flame front geometry and flame speed, which depend on engine-operating conditions, combustion chamber geometry, turbulence, etc. Because of their simplicity, single- and multizone models can account for leakage through seals and flows into crevices.

Single- and multizone thermodynamic models of rotary engine combustion have not reached the same sophistication as those developed for spark-ignition and diesel engines. Furthermore, thermodynamic models of rotary engine combustion use heat transfer correlations that have been developed for reciprocating engines. These correlations may not be applicable to rotary engines due to the large velocities induced by the rotor, the large surface-to-volume ratio of the combustion chambers of rotary engines, the flame extinction at the rotor trailing apex, and the flame enhancement at the rotor leading apex. In addition, most of the multizone models developed to analyze rotary engine combustion do not account for the flame distortion caused by the rotor-induced velocity field. Therefore, the assumption of spherical flame propagation that is used in these models cannot be justified physically and experimentally.

Dimensional models are based on the time-averaged conservation equations of mass, linear momentum, energy, and species, and these models account for turbulence. These models provide qualitative and quantitative information about the flow field (velocity, temperature, turbulence, and species mass fractions) at certain times and locations within the combustion chamber, and they account for the spatial and temporal variations of the flow field within the combustion chamber.

In this chapter mathematical models of rotary engines are presented. The assumptions and limitations of these models are stated, and representative results are shown. The lack of experimental data does not permit verification of the numerical predictions of multidimensional models, and the presentation of results is limited to discussions on the plausibility of the predictions.

This chapter also includes some of the turbulence and combustion models that have been used to analyze the flow fields in reciprocating and rotary engines. The limitations of these models are stated, and comparisons between predictions and available experimental data are presented in order to emphasize that these models require further refinement.

The turbulence and combustion models discussed in this chapter have been used with some success to predict the flow fields in motored reciprocating engines and the spray penetration in enginelike environments. However, improvements are necessary to accurately predict turbulent chemically reacting flows, particularly those in reciprocating and rotary engines.

The rotary engine models presented in this chapter are based on those developed for spark-ignition and compression-ignition engines. However, the geometry of and flow fields in rotary engines are so different from those in reciprocating engines that modifications to the models may have to be introduced after comparing the numerical predictions with the (as-yet-unavailable) experimental data.

In the next section the rotary engine geometry is described. This section is followed by those on single- and multizone models and on multidimensional models.

2.2 ROTARY ENGINE GEOMETRY

Rotary engines have a trochoidal shape [1, 36–38], which may be generated from an epitrochoid by maintaining a constant distance between the two. This

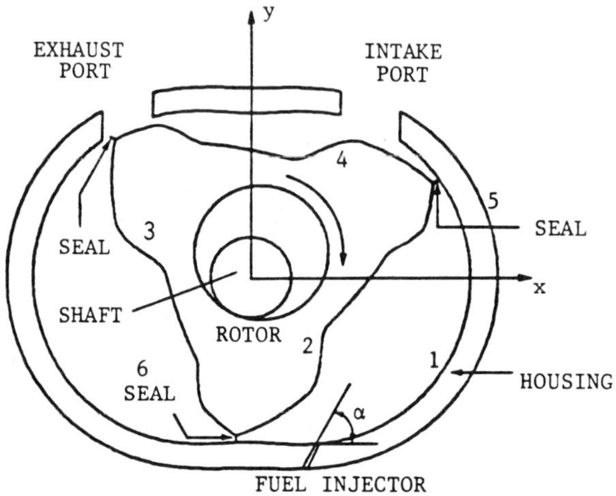

Figure 2.1 Schematic of a rotary engine.

distance corresponds to the seal head radius. Figure 2.1 shows a section of a rotary engine provided with pockets in the rotor and peripheral intake and exhaust ports. The apex seals are fixed to the rotor by retaining pins and springs. Side seals are also attached to the rotor to minimize leakage (Fig. 2.2). The NSU Wankel engine has an intake port located on the periphery of the housing, the Curtiss-Wright engine uses side ports, and the Toyo Kogyo (now Mazda Motor Corporation) engines use dual side ports [38].

Table 2.2 shows the geometry of a rotary engine, which will be used in the section on two-dimensional models. Other engine geometries can be found in references 1, 36, and 37.

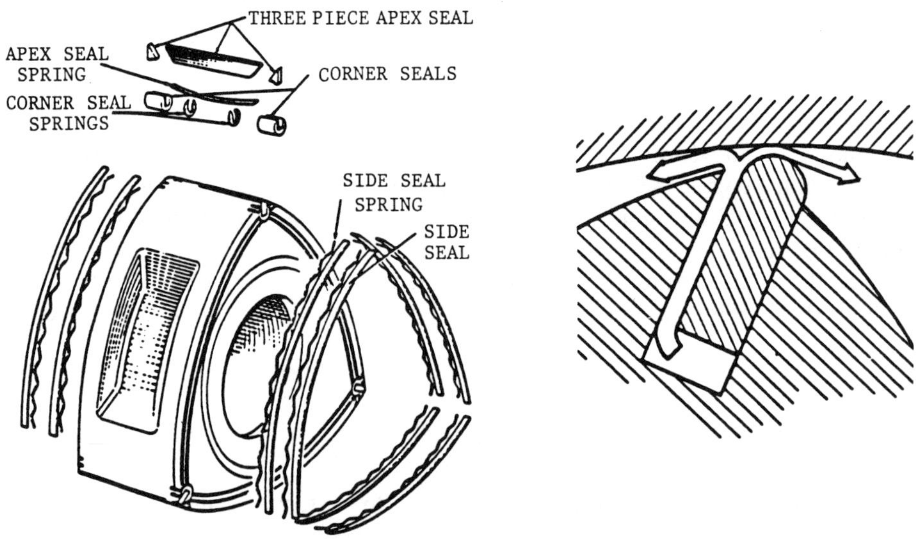

Figure 2.2 Schematic of seals and flows from crevices [40].

Table 2.2 Geometries of the rotor and housing[a]

$$x_H = E\cos(3A) + (R+C)\cos A$$

$$y_H = E\sin(3A) + (R+C)\sin A$$

$$x_R = E\sin\theta + x_0\cos\left(\frac{\theta}{3}\right) + y_0\sin\left(\frac{\theta}{3}\right)$$

$$y_R = E\cos\theta + x_0\cos\left(\frac{\theta}{3}\right) - y_0\sin\left(\frac{\theta}{3}\right)$$

$$x_0 = x_1\cos\left(\frac{\pi}{6}\right) + y_1\sin\left(\frac{\pi}{6}\right)$$

$$y_0 = y_1\cos\left(\frac{\pi}{6}\right) + x_1\sin\left(\frac{\pi}{6}\right)$$

$$x_1 = R\cos(2\beta) - \left(\frac{3E^2}{R}\right)\sin(6\beta)\sin(2\beta) + 2E\left[1 - \left(\frac{3E}{R}\right)^2\sin^2(3\beta)\right]^{1/2}\cos(3\beta)\cos(2\beta) - x_2$$

$$y_1 = R\sin(2\beta) - \left(\frac{3E^2}{R}\right)\sin(6\beta)\sin(2\beta) + 2E\left[1 - \left(\frac{3E}{R}\right)^2\sin^2(3\beta)\right]^{1/2}\cos(3\beta)\cos(2\beta) - y_2$$

$$x_2 = Q\cos(2\beta) \qquad y_2 = Q\sin(2\beta)$$

$$Q = 0 \quad \text{if } \beta_1 \le \beta \le \beta_T \quad \text{or} \quad \beta_L \le \beta \le \beta_4$$

$$Q = 0.5Q_1\left[1 - \cos\left(\frac{\beta_1 - \beta}{\beta_1 - \beta_2}\right)\right] \quad \text{if } \beta_2 \le \beta \le \beta_1$$

$$Q = Q_2 + (Q_1 - Q_2)\frac{\beta - \beta_3}{\beta_2 - \beta_3} \quad \text{if } \beta_3 \le \beta \le \beta_2$$

$$Q = 0.5Q_2\left[1 - \cos\left(\frac{\beta - \beta_4}{\beta_3 - \beta_4}\right)\right] \quad \text{if } \beta_4 \le \beta \le \beta_3$$

$$\beta_1 = \beta_L + 0.875(\beta_T - \beta_L) \qquad \beta_2 = \beta_L + 0.75(\beta_T - \beta_L)$$

$$\beta_3 = \beta_L + 0.4375(\beta_T - \beta_L) \qquad \beta_4 = \beta_L + 0.25(\beta_T - \beta_L)$$

[a] $0 \le A \le 2\pi$; $E = 0.0154$ m; $R = 0.1064$ m; $C = 0.004$ m is the distance between the rotor and housing at the leading and trailing apexes; $Q_1 = 0.015$ m; $Q_2 = 0.0198$ m; and the values of β_L and β_T depend on the rotor surfaces shown in Fig. 2.1 as follows: For surface 2: $\beta_L = \pi/6 \le \beta \le \beta_T = \pi/2$. For surface 3: $\beta_L = 5\pi/6 \le \beta \le \beta_T = 7\pi/6$. For surface 4: $\beta_L = 3\pi/2 \le \beta \le \beta_T = 11\pi/6$.

2.3 THERMODYNAMIC MODELS

In this section we consider models based on the first law of thermodynamics; these models have been used to analyze the performance and emissions of spark-ignition and compression-ignition rotary engines. Thermodynamic models can be classified into two groups: single- and multizone models. In single-zone models the mixture composition, pressure, and temperature of the combustion chamber are assumed to be uniform, and the chemical heat release is either specified (predictive analysis) or calculated from measured pressure diagrams (heat release analysis); in multizone models the mixture in the combustion chamber is divided into two or three regions: the unburnt and burnt regions and the quench layer. Multizone models require specification of the burning velocity and flame front geometry. Because of its simplicity, thermodynamic models can account for complex chemical reactions, leakage, flows into crevices, and heat transfer. The multidimensional models described in subsequent sections are based on simplified reaction mechanisms and neglect leakage through the side seals, leakage through the rotor trailing and leading apexes, and flows into crevices.

In the next sections single-zone models for spark-ignition and DISC engines are described.

2.3.1 Single-Zone Models of Spark-Ignition Engines

The first law of thermodynamics can be applied to the combustion chamber of a rotary engine and written as [39]

$$\frac{dE}{d\theta} = -p\frac{dV}{d\theta} - \frac{dQ}{d\theta} - h_{CR}\frac{dm_{CR}}{d\theta} - h_l\frac{dm_l}{d\theta} + h_{\text{inj}}\frac{dm_{\text{inj}}}{d\theta} \tag{1}$$

where p is the pressure; V is the volume; E is the internal energy; θ is the shaft angle, which is related to the shaft velocity ω through $\theta = \omega t$; t is time; $dQ/d\theta$ represents the heat transfer losses; and the third, fourth, and fifth terms on the right-hand side are, respectively, the enthalpy losses due to flows into crevices CR and due to leakage l and the enthalpy flux due to fuel injection. The fifth term on the right-hand side is identically equal to zero in the absence of fuel injection but has been kept in the equation in order to model DISC engines in Section 2.3.2.

Figure 2.3 shows a schematic of a rotary engine and indicates the energy transfers and mass flows across the combustion chamber boundaries. The mixture in the combustion chamber of a rotary engine consists of unburnt and burnt gases, and the internal energy can be written as

$$m = m_u + m_b \tag{2a}$$

$$E = m_u e_u + m_b e_b \tag{2b}$$

$$e = e^0 + \int_{T_0}^{T} C_v\, dT \tag{2c}$$

where m is the mass contained in the combustion chamber, e is the specific internal energy, e^0 is the energy of formation at the reference temperature T_0, C_v is the specific heat at constant volume, and the subscripts u and b denote unburnt and burnt gases, respectively. The mass in the combustion chamber $m = m_u + m_b$ decreases due to leakage and flow into crevices and increases because of the fuel injection, that is,

$$\frac{dm}{d\theta} = \frac{dm_{\text{inj}}}{d\theta} - \frac{dm_{CR}}{d\theta} - \frac{dm_l}{d\theta} = \frac{dm_{\text{inj}}}{d\theta} - \frac{dm^*_{CR}}{d\theta} \tag{3}$$

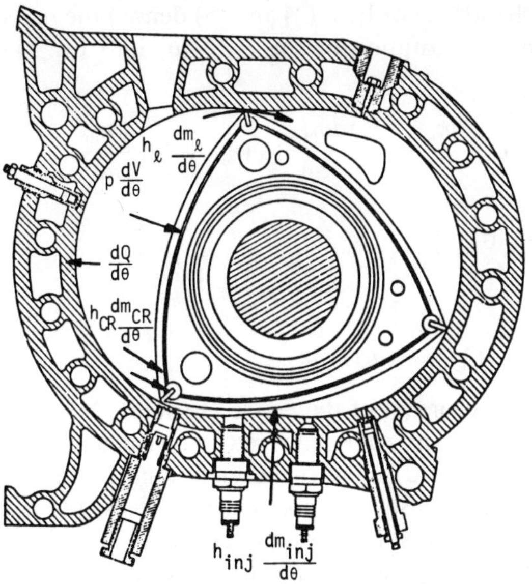

Figure 2.3 Schematic of Wankel engine indicating energy transfers and mass flows across the combustion chamber boundaries [40].

where the subscripts inj, CR, and l denote injection, crevice, and leakage, and m_{CR}^* denotes the combined effects of leakage and flows into crevices.

In Eqs. (2) and (3) we have accounted for burnt and unburnt gases in the combustion chamber in order to develop a general thermodynamic model. Equations (2) and (3) will be used in Section 2.3.2, on multizone models. In single-zone models the mixture composition, temperature, and pressure are uniform throughout the combustion chamber, and there is no distinction between the unburnt and burnt gases. We have made such a distinction in Eqs. (2) and (3) in order to develop a general formulation that is also valid for multizone models. Furthermore, such a distinction yields the chemical heat release as a function of the specific internal energies of formation of the burnt and unburnt gases and as a function of the mass burning rate. However, the heat release rate associated with combustion can be considered as an additional (heat addition) term in the first law of thermodynamics [cf. Eq. (1)].

The mass of the burnt and unburnt gases can be calculated from the following equations:

$$\frac{dm_b}{d\theta} = -\frac{dm_{b,CR}^*}{d\theta} + \frac{dm_{b,R}}{d\theta} \tag{4}$$

$$\frac{dm_u}{d\theta} = -\frac{dm_{u,CR}^*}{d\theta} + \frac{dm_{u,R}}{d\theta} + \frac{dm_{\text{inj}}}{d\theta} \tag{5}$$

$$\frac{dm_{u,R}}{d\theta} = -\frac{dm_{b,R}}{d\theta} = -\frac{dm_c}{d\theta} \tag{6}$$

where the first terms on the right-hand side of Eqs. (4) and (5) represent the mass losses of the burnt and unburnt gases due to leakage and flows into crevices and the second terms on the right-hand side of Eqs. (4) and (5) denote the mass changes due to chemical reactions. Substitution of Eqs. (2)–(6) into Eq. (1) yields

$$m\bar{c}_v \frac{dT}{d\theta} + p\frac{dV}{d\theta} + \frac{dQ}{d\theta} - (e_b - e_u)\frac{dm_{u,R}}{d\theta} + h'\frac{dm^*_{CR}}{d\theta} - (h_{\text{inj}} - e_u)\frac{dm_{\text{inj}}}{d\theta}$$
$$- e_b\frac{dm^*_{b,CR}}{d\theta} - e_u\frac{dm^*_{u,CR}}{d\theta} = 0 \tag{7}$$

where

$$m\bar{c}_v = m_u(C_v)_u + m_b(C_v)_b \tag{8}$$

and h' is the enthalpy of the flows into or out of the crevice regions.

The mass losses due to flows into crevices and leakage can be calculated by assuming (1) a constant leakage area A, (2) quasi-one-dimensional isentropic leakage flow, (3) constant-volume and constant-temperature crevices, (4) the crevice pressure equal to the chamber pressure, (5) crevices modeled as apex seals only, (6) leakage composition and temperature equal to crevice composition and temperature, and (7) burnt and unburnt gases perfectly mixed in the crevice [40]. Under these conditions

$$m_{CR} = \frac{pV_{CR}}{RT_{CR}} \tag{9a}$$

$$\frac{dm_{CR}}{d\theta} = \frac{m_{CR}}{p}\frac{dp}{d\theta} \tag{9b}$$

where the variations of the gas constant with the mixture composition, pressure, and temperature have been neglected. The mass flow rates due to leakage are given by the following expressions:

$$\frac{dm_l}{dt} = c_d A \frac{p_0}{RT_0}(\gamma RT_0)^{1/2}\left\{\frac{2}{\gamma - 1}\left[\left(\frac{p_s}{p_0}\right)^{2/\gamma} - \left(\frac{p_s}{p_0}\right)^{(\gamma+1)/\gamma}\right]\right\}^{1/2} \tag{10}$$

or if the flow is choked,

$$\frac{p_0}{p_s} > \left(\frac{2}{\gamma + 1}\right)^{\gamma/(\gamma-1)} \tag{11a}$$

$$\frac{dm_l}{dt} = c_d A \frac{p_0}{RT_0}(\gamma RT_0)^{1/2}\left(\frac{2}{\gamma + 1}\right)^{(\gamma+1)/2(\gamma-1)} \tag{11b}$$

where A is the leakage area, c_d is a discharge coefficient, γ is the ratio of specific heats, p_0 and T_0 are the stagnation pressure and temperature, and p_s is the static pressure at the restriction.

If the pressure in the combustion chamber under consideration is higher than those of the adjacent chambers, then $dm_l/dt > 0$ and the value of h' that appears in Eq. (7) is the specific enthalpy of the mixture contained in the combustion chamber under consideration. However, if $dm_l/dt < 0$, the value of h' corresponds to those of the adjacent chambers.

The difference between the internal energies of the burnt and unburnt gases can be written as

$$e_b - e_u = (e_b - e_b^0) + (e_b^0 - e_u^0) - (e_u - e_u^0) \tag{12}$$

and the chemical heat release due to combustion is

$$\frac{dQ_{CH}}{d\theta} = (e_b^0 - e_u^0)\frac{dm_{u,R}}{d\theta} \tag{13}$$

Equations (7), (9*b*), (10) or (11*b*), and (13) can be used to calculate the mean temperature of the cylinder charge once the geometry of the combustion chamber $dV/d\theta$, the heat release rate $dQ_{CH}/d\theta$, and the heat transfer losses $dQ/d\theta$ are specified.

Note that in single-zone thermodynamic models there is no distinction between the burnt and unburnt gases. Therefore, Eq. (7) can be written as

$$m\bar{c}_v\frac{dT}{d\theta} + p\frac{dV}{d\theta} + \frac{dQ}{d\theta} - \frac{dQ_{CH}}{d\theta} + (h' - e)\frac{dm_{CR}^*}{d\theta} - (h_{\text{inj}} - e)\frac{dm_{\text{inj}}}{d\theta} = 0 \tag{14}$$

where e is the specific internal energy of the mixture in the combustion chamber. The value of e can be calculated from Eq. (2*b*).

In rotary engines the leakage through the seals and the flows into crevices are very important, and the term $dm_{CR}^*/d\theta$ in Eq. (14) is not negligible. However, this term may be negligible in spark-ignition and diesel (or compression-ignition) engines. Equation (14) is also valid for reciprocating engines.

2.3.1.1 Heat transfer correlations. The heat transfer losses $dQ/d\theta$ that appear in Eq. (14) can be calculated from Nusselt-Reynolds number correlations as, for example [41],

$$\text{Nu} = \frac{hL}{\lambda} = 0.037\ \text{Re}^{0.8}\ \text{Pr}^{0.3} \tag{15}$$

where Nu, Re, and Pr are the Nusselt, Reynolds, and Prandtl numbers; h is the film transfer coefficient; L is a characteristic dimension; and λ is the gas thermal conductivity. The velocity required to evaluate the Reynolds number can be calculated by means of Woschni's correlation [42], which accounts for the mean rotor speed v_R and the increase in velocity due to combustion v_c, that is,

$$v_R = \frac{\pi}{3}R_R\omega \tag{16a}$$

$$v_c = c_c\frac{V_D T_1}{V_1 p_1}(p_{\text{fir}} - p_{\text{mot}}) \tag{16b}$$

where V_D is the displaced volume; R_R is the rotor radius; ω is the angular velocity of the shaft; T_1, p_1, and V_1 are the pressure, temperature, and volume at the beginning of combustion; the subscripts fir and mot denote firing and motored conditions, respectively; and $c_c = 0.00324$. The heat transfer rate to the walls is

$$\frac{dQ}{dt} = hA_s (T - T_w) \tag{17}$$

where

$$h = C_1[131 R_R^{-0.2} p^{0.8} T^{-0.53} (2.28v_R + C_2v_c)^{0.8}] \tag{18}$$

C_1 and C_2 are constants that must be determined by comparisons between the numerical predictions and experimental data, A_s is the total surface area of the combustion chamber, and T_w is the wall temperature.

Equation (16*b*) was derived for reciprocating engines; its validity in rotary engines has not been established.

Atesmen [43] used McAdams' correlation to analyze one-dimensional quasi-steady-state heat transfer in rotary engines. He also used the following correlation:

$$h = \frac{0.023\lambda}{D} \mathrm{Re}_D^{0.8} \mathrm{Pr}^{0.4} \tag{19}$$

where D is the hydraulic diameter of the combustion chamber.

He also used a modified Annand's empirical formula [44] given by

$$h = \frac{a\lambda}{r} \mathrm{Re}_r^b + c \frac{T^4 - T_w^4}{T - T_w} \tag{20}$$

where $a = 0.25$, $b = 0.7$, and $c = 1.30 \times 10^{-10}$ [Btu/(hr)(ft²)(°R⁴)] were determined numerically; r is the rotor radius; and the velocity used to evaluate the Reynolds number is $v = \omega r$.

Lee et al. [45] obtained the following experimental heat transfer correlations in a rotary engine:

$$\mathrm{Nu} = 0.3311 \left(\frac{T}{T_f}\right)^{4.552} \mathrm{Re}^{0.7367} \quad \text{unburnt gases} \tag{21}$$

$$\mathrm{Nu} = 13.865 \left(\frac{T}{T_f}\right)^{-3.28} \mathrm{Re}^{0.3906} \quad \text{burnt gases} \tag{22}$$

where $T_f = (T + T_w)/2$ is the film temperature, T is the bulk temperature (°F), and the Reynolds number was evaluated at the film temperature with a velocity

$$v = \omega \frac{D}{3} \tag{23}$$

where D (ft) depends on the engine geometry (i.e., rotor eccentricity, crank shaft angle, and generating radius of the epitrochoid).

The gas velocity required to evaluate the Reynolds number should consist of the relative velocity of the rotor with respect to the housing, the velocity induced by the expanding combustion gases, and the squish velocity caused by the combustion chamber geometry.

Benson and Sierens [46], Sierens [47], and Blair and Fleck [48] used Wilmers' correlation, which gives the following film heat transfer coefficient:

$$h = \frac{590}{861}\left(\frac{\text{rpm}}{1000}p\right)^{0.786} T^{-0.525} \tag{24}$$

in kJ/m²/s/K, where p and T are to be evaluated in bars and degrees kelvin, respectively.

Sierens et al. [49, 50] modified Wilmers' correlation in order to account for the high relative velocities against the combustion chamber walls as

$$h = 0.685\left(\frac{\omega}{1000} + 0.0178v_g\right)^{0.786} p^{0.786}\, T^{-0.525} \tag{25}$$

where v_g is the gas velocity.

They also used Annand's correlation [51] where

$$h = 0.14\lambda\,\frac{\text{Re}^{0.7}}{\text{L}} \qquad \text{Re} = \rho\,\frac{vL}{\mu} \tag{26}$$

where $\lambda = \mu C_p/\text{Pr}$, with Pr being the Prandtl number (Pr $= 0.7$) and C_p the specific heat at constant pressure; $v = \omega R_R/3$; and L is a characteristic length.

2.3.1.2 Chemical heat release. Equations (3), (9), (10) or (11), (13), and (17) can be substituted into Eq. (7) or Eq. (14) in order to derive an equation for the pressure as a function of the chemical heat release [Eq. (13)] and $dV/d\theta$.

The density $\rho = m/V$ can be calculated from the equation of state:

$$p = \rho RT \tag{27}$$

Note that Eq. (27) can be differentiated with respect to θ to yield

$$\frac{d\rho}{d\theta} = \rho\left(\frac{1}{p}\frac{dp}{d\theta} - \frac{1}{T}\frac{dT}{d\theta}\right) \tag{28}$$

where $dR/d\theta = 0$ because the mixture is assumed to be homogeneous throughout the combustion chamber and the dependence of R on the combustion chamber pressure and temperature has been neglected.

Once the equation for the chemical heat release rate [cf. Eq. (7)] is expressed as a function of $dV/d\theta$, $dp/d\theta$, and $dm_{\text{inj}}/d\theta$, the resulting equation [Eq. (14)] can be used for heat release analyses by specifying experimentally determined pressure diagrams [i.e., $p(\theta)$]. Alternatively, the resulting equation can be used to predict the pressure diagram if the chemical heat release is specified as a function of the crank shaft angle. Blair and Fleck [48] used a triangular heat release rate given by the following expressions:

$$\frac{dQ}{d\theta} = \left(\frac{dQ}{d\theta}\right)_{\max} \frac{\theta - \theta_{\text{delay}}}{\theta_{\max} - \theta_{\text{delay}}} \qquad \theta_{\text{delay}} \le \theta \le \theta_{\max} \tag{29}$$

$$\frac{dQ}{d\theta} = \left(\frac{dQ}{d\theta}\right)_{\max} \frac{\theta_b - \theta}{\theta_b - \theta_{\max}} \qquad \theta_{\max} \le \theta \le \theta_b \tag{30}$$

where $(dQ/d\theta)_{\max}$ is the maximum heat release, $\theta_{\max}$ is the crank shaft angle at which the maximum heat release occurs, $\theta_b - \theta_{\text{delay}}$ is the combustion interval, and θ_{delay} is the ignition delay angle. Blair and Fleck [48] used $t_{\text{delay}} = 12$ rpm/3000.

2.3.2 Multizone Models of Spark-Ignition Engines

Single-zone models of rotary engines are clearly unrealistic since they assume homogeneous composition and temperature throughout the combustion chamber. Furthermore, single-zone models do not account for the combustion chamber geometry. These assumptions can be eliminated by considering two zones: the unburnt and burnt gases separated by an infinitesimally thin flame that propagates into the unburnt gases (Fig. 2.4). Further assumptions include a uniform pressure throughout the combustion chamber, uniform temperature in the burnt and unburnt gases, no heat transfer between the unburnt and burnt gases at the flame front, unburnt gases frozen, and burnt gases in chemical equilibrium. Under these assumptions, the following equations are obtained

$$E = m_u e_u + m_b e_b \qquad e = e^0 + \int_{T_0}^{T} C_v \, dT \tag{31}$$

$$m_u + m_b = m \qquad x_u + x_b = 1 \qquad V_u + V_b = V = \frac{m}{\rho} \tag{32}$$

$$\frac{d(m_u e_u)}{d\theta} = -p \frac{dV_u}{d\theta} - \frac{dQ_u}{d\theta} - h_u \frac{dm^*_{u,CR}}{d\theta} + h_u \frac{dm_{u,R}}{d\theta} \tag{33}$$

$$\frac{d(m_b e_b)}{d\theta} = -p \frac{dV_b}{d\theta} - \frac{dQ_b}{d\theta} - h_b \frac{dm^*_{b,CR}}{d\theta} + h_u \frac{dm_{b,R}}{d\theta} \tag{34}$$

$$m_u = \rho_u V_u = m x_u \qquad m_b = \rho_b V_b = m x_b \tag{35}$$

$$\frac{dm_b}{d\theta} = m \frac{dx_b}{d\theta} + x_b \frac{dm}{d\theta} = \frac{dm_c}{d\theta} - \frac{dm^*_{b,CR}}{d\theta} \tag{36}$$

where the subscripts b, u, and c denote burnt gases, unburnt gases, and combustion; x denotes mass fraction; the subscript R denotes reaction; $dm^*_{b,CR}/d\theta$

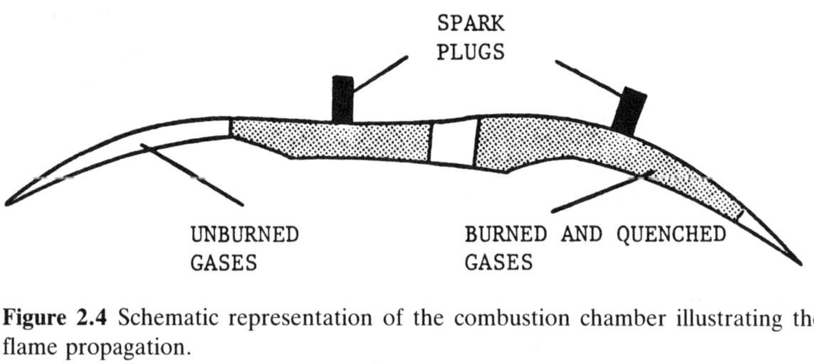

Figure 2.4 Schematic representation of the combustion chamber illustrating the spark plugs and flame propagation.

denotes the mass flow rates of the burnt gases due to leakage and flows into crevices; and $dm_c/d\theta$ denotes the combustion rate. Note that

$$\frac{dm_c}{d\theta} = \frac{dm^*_{b,CR}}{d\theta} + m\frac{dx_b}{d\theta} + x_b\frac{dm}{d\theta} = \frac{dm^*_{b,CR}}{d\theta} + \frac{dm_b}{d\theta} \tag{37}$$

Equations (33) and (34) can be written using Eq. (31) as

$$m_u C_{vu}\frac{dT_u}{d\theta} = -p\frac{dV_u}{d\theta} - \frac{dQ_u}{d\theta} + (h_u - e_u)\frac{dm_{u,R}}{d\theta} + (e_u - h_u)\frac{dm^*_{u,CR}}{d\theta} \tag{38}$$

$$m_b C_{vb}\frac{dT_b}{d\theta} = -p\frac{dV_b}{d\theta} - \frac{dQ_b}{d\theta} + (h_u - e_b)\frac{dm_{b,R}}{d\theta} + (e_b - h_b)\frac{dm^*_{b,CR}}{d\theta} \tag{39}$$

Equations (9) and (10) or (9) and (11) can be used to calculate $m^*_{u,CR}$ and $m^*_{b,CR}$, which can then be introduced into Eqs. (36), (37), (4), and (5) to obtain equations for $dm_u/d\theta$ and $dm_b/d\theta$. These equations can then be substituted into Eqs. (38) and (39), where

$$dV = dV_u + dV_b \tag{40a}$$

$$V_u = \frac{m_u R_u T_u}{p} \tag{40b}$$

$$V_b = \frac{m_b R_b T_b}{p} \tag{40c}$$

$$\frac{dV_u}{d\theta} = R_u T_u\frac{dm_u}{d\theta} + m_u R_u\frac{dT_u}{d\theta} - \frac{m_u R_u T_u}{p^2}\frac{dp}{d\theta} \qquad \frac{dV_b}{d\theta} = \frac{dV}{d\theta} - \frac{dV_u}{d\theta} \tag{41}$$

In Eq. (41) the variations of R_u with the temperature, pressure, and composition have been neglected. Equation (41) can be substituted into Eqs. (38) and (39) to obtain first-order ordinary differential equations for T_u and T_b. The pressure can be calculated from the following equation:

$$p\,(V_u + V_b) = m_u R_u T_u + m_b R_b T_b \tag{42}$$

which can be obtained by adding Eqs. (40*b*) and (40*c*).

The heat transfer losses in Eqs. (33) and (34) or Eqs. (38) and (39) can be calculated from the correlations presented in Section 2.3.1 by considering the areas of the solid walls in contact with the burnt and unburnt gases. These areas cannot be determined unless the flame location and geometry are known as a function of time.

In multizone models different film transfer coefficients may be used for the burnt and unburnt gases. For example, Eq. (18) can be used for the burnt gases since that equation accounts for the velocity induced by combustion. Equation (18) can also be used to predict heat transfer in the unburnt gases by setting $v_c = 0$.

Equations (31)–(42) do not form a closed system because they contain more unknowns than equations. Closure can be achieved by specifying the mass burning rate and the flame front geometry. Frequently, the flame is assumed to propagate spherically from the spark plugs, and the mass burning rate is either specified or calculated.

In Sections 2.3.2.1–2.3.2.3 we describe combustion models in which the mass burning rate is specified, the turbulent flame speed is specified, and the mass burning rate is calculated, respectively.

2.3.2.1 Specification of the mass burning rate. Roberts et al. [39] and Norman [40] used a combustion rate model given by a Wiebe function

$$x_b = \frac{m_b}{m} = 1 - \exp\left[-a\left(\frac{\theta - \theta_0}{\Delta\theta_b}\right)^{m+1}\right] \tag{43}$$

where θ_0 is the crank shaft angle at the start of combustion, $\Delta\theta_b$ is the combustion duration, $3 < a < 10$, $1 < m < 3$, and the combustion rate can be obtained from Eq. (37), once the leakage and flow rates into crevices are calculated from Eqs. (9) and (10) or Eqs. (9) and (11).

The geometry of the flame front needs to be specified. Frequently, it is assumed that the flame propagates spherically from the spark plug(s). However, due to the large velocities induced by the rotor motion, the flame front is distorted, particularly near the rotor trailing apex; therefore, the use of a spherical flame front is questionable. Note that the specification of the flame front geometry implies that the engine geometry is accounted for in multizone models of rotary engine combustion.

2.3.2.2 Specification of the turbulent flame speed. Sierens et al. [49, 50] used ionization probes to determine the flame front location and a turbulent flame speed that can be written as

$$S_T = fS_L \qquad S_L = 7.7808 \times 10^{-3}\left(\frac{10^4}{T_{ad}} + \frac{900}{T_u}\right)^{-4.938} p^{-0.09876} \tag{44}$$

where S_L is the laminar flame speed, T_{ad} is the adiabatic flame temperature, and the flame factor f accounts for turbulence in the mixture. The flame front was assumed to be spherical after ignition, and it separates the burnt and unburnt gases. Once the flame front reaches the side seals, three zones are created: two zones of unburnt gases separated by a zone of products. Sierens et al. [49, 50] considered both equilibrium and finite rate chemistry in the burnt-gas zone to predict the concentrations of H_2, H_2O, OH, H, N_2, NO, N, CO_2, CO, O_2, and O; the equilibrium reactions are

$$N + NO \rightleftarrows N_2 + O \tag{45a}$$

$$N + O_2 \rightleftarrows NO + O \tag{45b}$$

$$N + OH \rightleftarrows NO + H \tag{46a}$$

$$H + N_2O \rightleftarrows N_2 + OH \tag{46b}$$

$$O + N_2O \rightleftarrows N_2 + O_2 \tag{47a}$$

$$O + N_2O \rightleftarrows 2NO \tag{47b}$$

$$N_2O + X \rightleftarrows N_2 + O + X \tag{48}$$

where X is a third body.

Flame enhancement occurred at the time that the trailing flame left the rotor pocket. This phenomenon may be associated with squish and may increase the turbulence in the flame front and the flame factor f.

The fuel conversion rate is given by

$$\frac{dm_{u,R}}{d\theta} = -\rho_u A_f S_T \tag{49}$$

where A_f is the surface area of the flame front.

Sierens et al. [49, 50] calculated the state of the gases during the compression stroke, including the initiation of combustion, with a model similar to that proposed by Benson et al. [51]. Sierens et al. [49, 50] also accounted for quenching as follows. As the flame propagates through the combustion chamber, the quench layer left behind is considered as mass and enthalpy that left the system modeled by Eq. (34) or Eq. (39). The mass quenched is given by the following expression:

$$\frac{dm_q}{d\theta} = \frac{pAq}{R_u T_u} \frac{1}{V} \frac{dV_b}{d\theta} \tag{50}$$

where q is the thickness of the quench layer.

With the quenched mass m_q so computed, its enthalpy and volume are removed from the burnt gases at every time step until a converged value of x_u is obtained. The quench thickness q is calculated from

$$\frac{q}{q_0} = 0.4 \left(\frac{p_0}{p}\right)^{\alpha} \left(\frac{T_0}{T_u}\right)^{\beta} \tag{51}$$

where q_0 is the quench distance at reference pressure p_0 and temperature T_0 conditions and α and β depend on the mixture equivalence ratio [52–54].

Equation (50) represents a constraint on the burnt mass fraction. Once a certain amount of mass has been quenched, its composition is assumed to be frozen.

Danieli et al. [55] and Danieli [56] accounted for flame quenching, evaluated the burnt mass fraction using measured pressure diagrams, considered the extended Zel'dovich mechanisms given by Eqs. (45*a*), (45*b*), and (46*a*), and showed that quenching is the dominant source of unburnt hydrocarbon emissions at high speeds whereas leakage dominates at low speeds. The crevice volume was found to be a secondary source of HC emissions independent of speed and load. The model overpredicted the NO emissions due to inadequate modeling of heat transfer on the burnt-gas region, and the predicted heat transfer rates at top dead center (TDC) were much higher than the experimental values, indicating that the boundary layer structure and bulk fluid mechanics play an important role in Wankel engines.

The turbulent flame speed required to evaluate the fuel conversion rate depends on the turbulence length scales and turbulent intensity and may also depend on the mixture pressure, temperature, and equivalence ratio. Ballal and Lefebvre [57] showed that the turbulent flame speed increases with the length scale for weak turbulence and decreases with the length scale for strong turbulence. Weak turbulence was defined as $u' < 2S_L$, where u' is the root-mean-square (rms) of the velocity fluctuations and S_L is the laminar flame speed. Ballal and Lefebvre [57] also showed that the turbulent flame speed always increases with the flame turbulence. Ballal [58] studied enclosed premixed propane-air flames and showed that turbulent flame propagation is related to flame-generated turbulence.

Equation (44) does not account for the effects of the turbulent intensity and turbulent length scales on the turbulent flame speed and is not adequate for predicting rotary engine combustion unless the value of f is tuned for each engine. The importance of flame-generated turbulence in premixed flames has been demonstrated by Gokalp [59].

2.3.2.3 Calculation of the mass burning rate. Danieli et al. [60] used a turbulence flame propagation model to analyze the performance of a Wankel engine. The model was based on that of Blizard and Keck [61], which assumes that (1) there is a characteristic eddy scale below which combustion is laminar, (2) this scale is much larger than the laminar flame thickness, (3) combustion over scales larger than this scale is due to the turbulent transport of thermal energy and radicals, and (4) the volume and mass of the gases contained in the regions where fuel oxidation occurs are smaller than the mass contained in the combustion chamber.

The burning rate can be calculated from the following equation:

$$\frac{dm_b}{dt} = \frac{m_e - m_b}{\tau} \tag{52}$$

where

$$\tau = \frac{l_e}{S_L} \tag{53}$$

is the characteristic eddy burning time, l_e is a characteristic burning length scale that can be determined from experimental data, m_e is the mass entrained by the flame front, that is,

$$m_e = \int_0^t \rho_u A_f u_e \, dt \tag{54}$$

and u_e is an effective entrainment velocity. Danieli et al. [60] selected the values of the burning time τ and entrainment velocity u_e and assumed that the initial flame propagation is hemispherical. This assumption is questionable because the large velocity induced by the rotor motion may distort and advect the flame front. Danieli et al. [60] calculated the average gas-phase velocity at a given cross section by applying mass conservation to the volume enclosed by the rotor trailing apex and the section considered; this yields

$$\frac{dm}{dt} = \rho u A + \frac{dm_{l,t}}{dt} \tag{55}$$

where u is the fluid velocity; $dm_{l,t}/dt$ is the leakage l through the rotor trailing apex t; and $m = \rho V_l$, with V_l being the volume between the rotor trailing apex and the section considered. The density can be calculated from the equation of state [Eq. (27)]. A fluctuating velocity can be calculated by considering the flux of kinetic energy into the chamber and its dissipation rate, which is assumed to be isotropic. The average turbulent kinetic energy per unit mass k in the combustion chamber can be calculated as

$$mk = \int_0^t \left[\frac{1}{2} \frac{dm}{dt} \langle u \rangle^2 - \frac{m}{l} \left(\frac{2}{3} k \right)^{3/2} \right] dt \tag{56}$$

where the first and second terms within the brackets denote the flux of kinetic energy into the chamber and the dissipation rate of turbulence kinetic energy, $\langle u \rangle$ is the mean velocity of the incoming flow, and l is the length scale of the energy-containing eddies. If the turbulence is isotropic, the turbulent intensity is given by

$$u' = \left(\frac{2k}{3} \right)^{1/2} \tag{57}$$

which can be added orthogonally to the mean gas speed to obtain a local velocity vector. This velocity was used to calculate the heat transfer rates through the rotor and housing, using relative velocities and the hydraulic diameter in the Reynolds number that appears in Eq. (19).

Sierens et al. [49] also used Eq. (55) to calculate the gas average velocity v_g that appears in Eq. (25).

Danieli et al. [60] showed that the effects of leakage and heat transfer on engine performance are comparable. The turbulent entrainment model described by Eqs. (52)–(54) has been improved in recent years in order to account for the temporal evolution of the turbulence kinetic energy and its rate of dissipation. These improvements are discussed in Chapter 4.

Some sample results are shown in Fig. 2.5, which presents a comparison between the predicted and measured pressures in a homogeneous-charge rotary engine.

2.3.3 Single-Zone Models of Direct-Injection Stratified-Charge (DISC) Engines

Roberts et al. [39], Nguyen et al. [62], and Roberts [63] developed and applied a single-zone model to analyze the performance and efficiency of DISC rotary engines. The model uses a heat release rate defined by the following expressions:

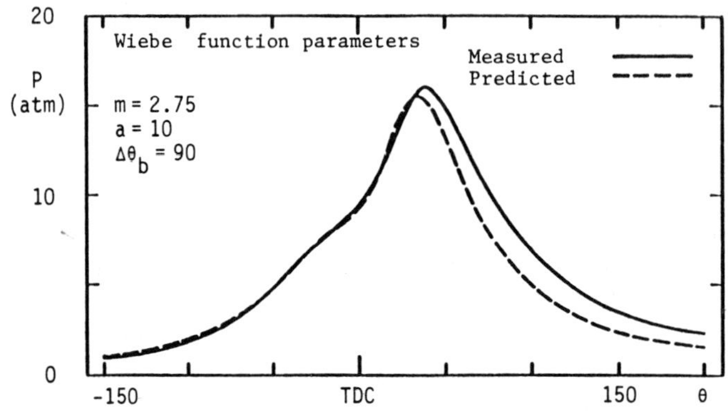

Figure 2.5 Comparison between predicted and measured pressures in a premixed rotary engine [40].

$$\frac{dQ}{d\theta} = \left(\frac{dQ}{d\theta}\right)_{\max} \frac{\theta - \theta_s}{\theta_m - \theta_s} \qquad \theta_s \leq \theta \leq \theta_m \tag{58}$$

$$\frac{dQ}{d\theta} = \left(\frac{dQ}{d\theta}\right)_{\max} \exp\left(-\frac{\theta - \theta_m}{\tau}\right) \qquad \theta_m \leq \theta \tag{59}$$

where θ_s is the crank shaft angle at the start of combustion, θ_m is the crank shaft angle corresponding to the maximum heat release, and τ is a constant in crank shaft angle degrees.

The thermodynamic DISC engine model described by Eqs. (1), (58), and (59) assumes that the mixture is uniform throughout the combustion chamber and does not account for the presence of liquid fuel droplets. However, it can account for the variation of the overall fuel equivalence ratio ϕ. The equivalence ratio increases from zero at the start of injection to the final overall equivalence ratio at which the engine operates. The equivalence ratio is defined as

$$\phi = \frac{F/A}{(F/A)_{st}} \tag{60}$$

where F/A is the fuel-air ratio and the subscript *st* denotes stoichiometric conditions.

Equation (60) can be written as

$$\phi = \frac{Y_f}{(F/A)_{st}(1 - Y_f)} \tag{61}$$

where Y_f is the fuel mass fraction (i.e., the mass of fuel divided by the total mass in the combustion chamber) and

$$\frac{d\phi}{d\theta} = \frac{dY_f/d\theta}{(F/A)_{st}\,(1 - Y_f)^2} \tag{62}$$

$$\frac{d(mY_f)}{d\theta} = \Sigma \frac{dm_j}{d\theta}(Y_f)_j \qquad \frac{dY_f}{d\theta} = \frac{1}{m}\left[-Y_f\frac{dm}{d\theta} + \Sigma \frac{dm_j}{d\theta}(Y_f)_j\right] \tag{63}$$

$$\frac{dm}{d\theta} = \Sigma \frac{dm_j}{d\theta} \tag{64}$$

where $dm_j/d\theta$ and $(Y_f)_j$ denote the mass fluxes and fuel mass fractions across the combustion chamber boundaries.

Equation (1) can be written as

$$m\frac{dh}{d\theta} = -\frac{dQ}{d\theta} + V\frac{dp}{d\theta} - h\frac{dm}{d\theta} - h_{CR}\frac{dm_{CR}}{d\theta} - h_l\frac{dm_l}{d\theta} + h_{\text{inj}}\frac{dm_{\text{inj}}}{d\theta} \tag{65}$$

and the equation of state $pV = mRT$ can be used to calculate

$$\frac{dp}{d\theta} = p\left(\frac{1}{R}\frac{dR}{d\theta} + \frac{1}{m}\frac{dm}{d\theta} + \frac{1}{T}\frac{dT}{d\theta} - \frac{1}{V}\frac{dV}{d\theta}\right) \tag{66}$$

$$\frac{dR}{d\theta} = \frac{1}{\rho T}\frac{dp}{d\theta} - \frac{p}{\rho^2 T}\frac{d\rho}{d\theta} - \frac{p}{\rho T^2}\frac{dT}{d\theta} \tag{67}$$

$$\frac{d\rho}{d\theta} = \frac{\partial\rho}{\partial T}\frac{dT}{d\theta} + \frac{\partial\rho}{\partial p}\frac{dp}{\partial\theta} + \frac{\partial\rho}{\partial\phi}\frac{d\phi}{d\theta} \tag{68}$$

$$\frac{dh}{d\theta} = C_p\frac{dT}{d\theta} + C_T\frac{dp}{d\theta} + C_\phi\frac{d\phi}{d\theta} \tag{69}$$

where

$$C_p = \left(\frac{\partial h}{\partial T}\right)_{p,\phi} \qquad C_T = \left(\frac{\partial h}{\partial p}\right)_{T,\phi} \qquad C_\phi = \left(\frac{\partial h}{\partial \phi}\right)_{p,T} \tag{70}$$

Equations (66)–(70) can be substituted into Eq. (65) to yield an equation for the combustion chamber temperature; the pressure can then be calculated from the equation of state, that is, $pV = mRT$.

In Figures 2.6 and 2.7 the predicted fuel mass burning rate, overall equivalence ratio, pressure, and temperature in a DISC rotary engine are presented [63].

2.4 DIMENSIONAL MODELS

The single- and multizone thermodynamic models presented in Section 2.3 do not account for spatial flow variations and cannot provide information about, for example, the local velocity and temperature fields. Dimensional models consider the spatial and temporal variations of the velocity, temperature, and pressure fields in one, two, or three dimensions. In this section the equations governing the flow field in Cartesian coordinates are presented. These equations

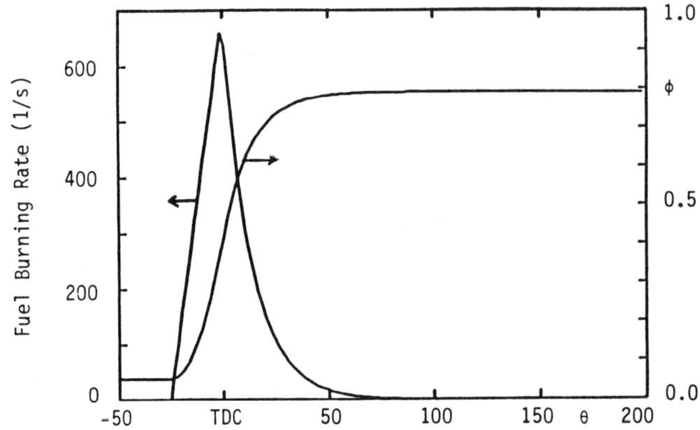

Figure 2.6 Fuel mass burning rate and overall equivalence ratio in a DISC rotary engine [63].

are then time averaged, and turbulence modeling is introduced. The time-averaged equations are then applied to study the flow field and combustion in one-, two-, and three-dimensional rotary engine models.

2.4.1 Governing Equations

The flow field in a rotary engine model is governed by the instantaneous time-dependent three-dimensional conservation equations of mass, momentum, species, and energy. These equations can be simplified by neglecting body forces, radiation, bulk viscosity, Soret and Dufour effects, and pressure-gradient diffusion and by assuming that the gas mixture is ideal and that the species diffuse according to Fick's law with equal mass diffusivities. Under these assumptions, the governing equations for a Newtonian fluid can be written in Cartesian co-

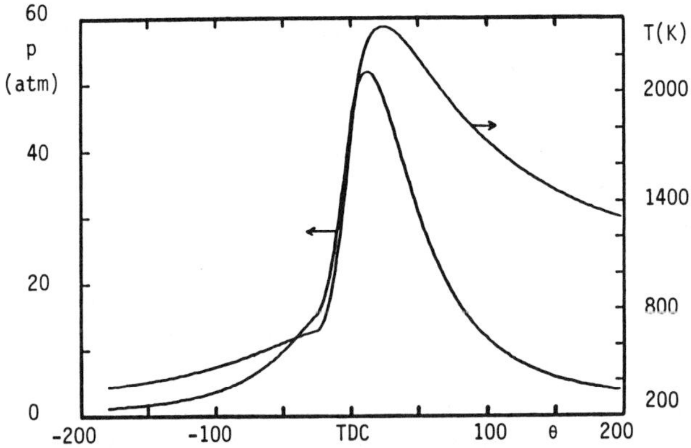

Figure 2.7 Pressure and temperature in a DISC rotary engine [63].

ordinates as [64]

$$\rho_{,t} + (\rho u_i)_{,i} = 0 \tag{71}$$

$$(\rho u_i)_{,t} + (\rho u_i u_j)_{,j} = -p_{,i} + \tau_{ij,j} \tag{72}$$

$$(\rho e)_{,t} + (\rho u_i e)_{,i} = \left[\lambda T_{,i} + \rho D \sum_{j=1}^{N} h_j (Y_j)_{,i}\right]_{,i} - p u_{,i} + \tau_{ij} u_{i,j} \tag{73}$$

$$(\rho Y_j)_{,t} + (\rho u_i Y_j)_{,i} = [\rho D (Y_i)_{,j}]_{,j} + \dot{\omega}_j \qquad j = 1, 2, \ldots, N \tag{74}$$

$$p = \rho \tilde{R} T \sum_{j=1}^{N} \left(\frac{Y_j}{M_j}\right) \tag{75}$$

$$e = \sum_{j=1}^{N} h_j Y_j - \frac{p}{\rho} \tag{76}$$

$$h_j = h_j^0 + \int_{T_0}^{T} C_{p_j}\, dT \qquad j = 1, 2, \ldots, N \tag{77}$$

where ρ, p, and T are the density, pressure, and temperature, respectively; u_i is the velocity component in the i direction; τ_{ij} is the stress tensor; e is the internal energy; λ is the thermal conductivity; D is the species mass diffusivity; h_i and Y_i are the enthalpy and mass fraction of species i, respectively; $\tilde{R}$ is the universal gas constant; N is the number of species; M_i is the molecular mass of species i; $\dot{\omega}_i$ is the reaction rate of species i; C_{p_i} is the specific heat at constant pressure of species i; T_0 is a reference temperature; h_i^0 is the enthalpy of formation of species i at temperature T_0; t is time; and the commas denote differentiation.

Equation (73) can also be written, using Eqs. (76) and (77), as

$$\rho C_p (T_{,t} + u_i T_{,i}) = p_{,t} + u_i p_{,i} + \tau_{ij} u_{i,j} - \sum_{j=1}^{N} h_j \dot{\omega}_j + (\lambda T_{,i})_{,i} + \rho D T_{,i} \sum_{j=1}^{N} C_{pj} (Y_j)_{,i} \tag{78}$$

where

$$C_p = \sum_{j=1}^{N} Y_j C_{p_j} \tag{79}$$

Equations (72) and (73) can also be used to obtain the following equation for the stagnation energy E:

$$(\rho E)_{,t} + (\rho u_i E)_{,i} = -(p u_i)_{,i} + \left[\lambda T_{,i} + \rho D \sum_{j=1}^{N} h_j (Y_j)_{,i}\right]_{,i} + (u_i \tau_{ij})_{,j} \tag{80}$$

where

$$E = e + \frac{1}{2} u_i u_i \tag{81}$$

Equation (73) can also be written as

$$(\rho h)_{,t} + (\rho u_i h)_{,i} = p_{,t} + u_i p_{,i} + \tau_{ij}\, u_{i,j} + \left[\lambda T_{,i} + \rho D \sum_{j=1}^{N} h_j (Y_j)_{,i}\right]_{,i} \tag{82}$$

where

$$h = \sum_{j=1}^{N} \left(h_j^0 + \int_{T_0}^{T} C_{p_j}\, dT\right) Y_j \tag{83}$$

Equations (72) and (82) can also be used to obtain the following equation for the stagnation enthalpy H:

$$(\rho H)_{,t} + (\rho u_i H)_{,i} = p_{,t} + \left[\lambda T_{,i} + \rho D \sum_{j=1}^{N} h_j (Y_j)_{,i}\right]_{,i} + (u_i \tau_{ij})_{,j} \tag{84}$$

where

$$H = E + \frac{p}{\rho} \tag{85}$$

For a Newtonian fluid the stress tensor is given by the following expression:

$$\tau_{ij} = \mu[u_{i,j} + u_{j,i}] - \frac{2}{3}\mu \delta_{ij} u_{k,k} \tag{86}$$

where δ_{ij} is the Kronecker delta and μ is the dynamic viscosity of the fluid.

2.4.2 Turbulence Modeling

The flow field in an internal combustion engine is turbulent and comprises many time and length scales. It is impossible with present computers to obtain a numerical solution of Eqs. (71)–(77) that accounts for all the turbulence time and length scales, and models need to be introduced. [The ratio between the time (length) scales of the Kolmogorov and energy-containing eddies is of the order of $\mathrm{Re}^{-1/2}$ ($\mathrm{Re}^{-3/4}$), where Re is the Reynolds number [65]. For a Reynolds number of 10^4 these ratios are on the order of 10^{-2} and 10^{-3}. Therefore, resolution of the small-scale eddies requires much smaller time steps and grid spacings than that of the energy-containing eddies.] We consider here time average and decompose the flow variables into the sum of mean and fluctuating values as

$$\phi = \langle \phi \rangle + \phi' \tag{87}$$

where

$$\langle \phi' \rangle = 0 \qquad \langle \phi \rangle = \lim_{\tau \to \infty} \frac{1}{\tau} \int_t^{t+\tau} \phi\, dt \tag{88}$$

and ϕ is any of the flow variables that appear in Eqs. (71)–(77).

Time average is strictly applicable to statistically stationary turbulence. The time τ must be larger than the characteristic turbulence time but smaller than the characteristic time of the mean flow [66]. These conditions may not be satisfied in internal combustion engines where the time scales of the energy-containing eddies are of the same order of magnitude as the time scales of the mean flow, and cycle times are short compared to relaxation times for important flow-adjustment phenomena [67]. In spite of these limitations, time average has been used to account for the turbulent flow field in reciprocating engines. Favre, or density-weighted, average has also been considered by some authors [68, 69]. In Favre's average the density-weighted velocity $\tilde{U}_i$ is defined by

$$\langle\rho\rangle\, \tilde{U}_i = \langle\rho u_i\rangle \tag{89}$$

where

$$\langle\rho\rangle = \rho - \rho' \qquad \langle\rho'\rangle = 0 \tag{90}$$

The main advantage of density-weighted average is simpler equations. Also, some of the turbulence measurements correspond to density-weighted (rather than time) averages. However, the modeling of the correlations that appear in the density-averaged conservation equations is similar to those for the time-averaged equations, which in turn are based on those for statistically stationary incompressible flows.

We will consider here time average and substitute Eq. (87) into Eqs. (71)–(77). If the fluctuations in density, thermal conductivity, viscosity, and mass diffusivity are neglected, if Eqs. (71)–(77) are time averaged, and if the brackets on the mean flow variables are dropped, then the following system of equations is obtained:

$$\rho_{,t} + (\rho u_i)_{,i} = 0 \tag{91}$$

$$(\rho u_i)_{,t} + (\rho u_i u_j)_{,j} = -p_{,i} + \tau_{ij,j} - (\rho\langle u_i'\, u_j'\rangle)_{,j} \tag{92}$$

$$\begin{aligned}(\rho e)_{,t} + (\rho u_i e)_{,i} &= \left[\lambda T_{,i} + \rho D\left(\sum_{j=1}^{N} h_j (Y_j)_{,i} + \sum_{j=1}^{N}\langle h_j'(Y_j')_{,i}\rangle\right)\right]_{,i} \\ &\quad - p u_{,i} - \langle p' u_{,i}'\rangle + \tau_{ij} u_{i,j} + \langle \tau_{ij}' u_{i,j}'\rangle - (\rho\langle u_i' e'\rangle)_{,i}\end{aligned} \tag{93}$$

$$(\rho Y_j)_{,t} + (\rho u_i Y_j)_{,i} = [(\rho D (Y_j)_{,i}]_{,i} + \langle\dot{\omega}_i\rangle - (\rho\langle u_i'\, Y_j'\rangle)_{,i} \tag{94}$$

Equations (75)–(77) correspond to the mean flow if the temperature-species mass fraction correlations are smaller than the product of the mean temperature and mean mass fractions.

2.4.2.1 Eddy viscosity models. Equations (92)–(94) show that the time-averaged equations contain unknown double correlations. Following Boussinesq [70], the double correlations can be modeled by means of an eddy viscosity μ_t where, for example,

$$\rho\langle u_i'\, \phi'\rangle = -\mu_t \frac{\phi_{,i}}{\sigma_\phi} \tag{95}$$

and the value of μ_t and the turbulent Prandtl/Schmidt number σ_ϕ need to be specified. Ramos et al. [71] assumed that

$$\frac{\mu_t}{\rho} = \alpha_3 \text{ rpm} \tag{96}$$

where α_3 is a constant. This model is clearly unrealistic, particularly near the solid walls, where the left-hand side of Eq. (95) is identically equal to zero, and the model assumes that the effects of turbulence are felt up to the wall. It is known that in turbulent boundary layers, there is a viscous sublayer where $\mu >> \mu_t$ and where the effects of turbulence are felt through the wall region [72]. Furthermore, a constant value of μ_t/ρ implies that the product of typical turbulent velocity and length scales is constant. This is, in general, incorrect.

2.4.2.2 k–ϵ models. Two-equation models have been introduced to account for the spatial variations of the turbulent velocity and length scales and for the history of the flow. [Equations (95) and (96) do not consider the history of the flow, since the turbulent correlations are assumed to be proportional to the local mean flow gradients with a constant proportionality factor.]

An equation for the turbulence kinetic energy can be obtained by subtracting Eq. (92) from Eq. (72), multiplying the result by u_i' and time averaging the resulting equation. This yields

$$\begin{aligned}(\rho k)_{,t} + (\rho u_i k)_{,i} &= -\rho\langle u_i' u_j'\rangle\, u_{i,j} - \rho\left\langle\frac{u_i'\, u_i'\, u_j'}{2}\right\rangle_{,j} \\ &\quad - \langle p'_{,i}\, u_i'\rangle + \langle\tau_{ij}'\, u_i'\rangle_{,j} - \langle\tau_{ij}'\, u_{i,j}'\rangle\end{aligned} \tag{97}$$

where k is the turbulence kinetic energy

$$k = \frac{1}{2}\langle u_i'\, u_i'\rangle \tag{98}$$

At high Reynolds numbers the fourth term on the right-hand side of Eq. (97) is smaller than the first and can be neglected. If the pressure-velocity correlations are neglected and a Boussinesq approximation is used, the correlations can be modeled as

$$-\rho\,\langle u_i'\, u_j'\rangle = \mu_t(u_{i,j} + u_{j,i}) - \frac{2}{3}\,\delta_{ij}(\mu_t u_{k,k} + \rho k) \tag{99}$$

$$-\rho\left\langle\frac{u_i'\, u_i'\, u_j'}{2}\right\rangle = \frac{\mu_t}{\sigma_k}\, k_{,j} \tag{100}$$

where σ_k is a constant.

The dissipation rate of turbulence kinetic energy is defined as

$$\epsilon = \frac{\langle\tau_{ij}'\, u_{i,j}'\rangle}{\rho} \tag{101}$$

Equation (97) can be closed as long as the values of μ_t and σ_k are specified and an equation for ϵ is obtained. This equation can be written by analogy with Eq. (97) as

$$(\rho\epsilon)_{,t} + (\rho u_i \epsilon)_{,i} = \left(\frac{\mu_t}{\sigma_\epsilon}\epsilon_{,i}\right)_{,i} + \frac{\epsilon}{k}(C_1 P - C_2 \rho\epsilon) + S_\epsilon \tag{102}$$

where σ_ϵ, C_1, and C_2 are constants, S_ϵ is a term that accounts for the effects of compressibility on the dissipation rate of turbulence kinetic energy, and $P = -\rho \langle u_i' u_j' \rangle u_{i,j}$.

Equation (99) can be contracted, and this results in an identity. In contrast, when Eq. (95) is used for $\phi = u_i$,

$$-\rho \langle u_i' u_j' \rangle = \mu_t (u_{i,j} + u_{j,i}) - \frac{2}{3}\mu_t \delta_{ij} u_{k,k} \tag{103}$$

If Eq. (103) is contracted, one obtains the following inconsistent result:

$$-\rho k = 0 \tag{104}$$

Substitution of Eqs. (99) and (100) into Eq. (97) yields

$$(\rho k)_{,t} + (\rho u_i k)_{,i} = \left(\frac{\mu_t}{\sigma_k} k_{,i}\right)_{,i} + P - \rho\epsilon \tag{105}$$

Equations (102) and (105) are valid for large-Reynolds-number flows away from the walls. Small-Reynolds-number effects can be introduced into Eqs. (102) and (105) and the resulting $k-\epsilon$ model can be applied up to the wall assuming that enough grid points are used to properly resolve the boundary layers [70, 73]. The effects of compressibility in the equations for k and ϵ appear through the production term P, which can be written as

$$P = \mu_t(u_{i,j} + u_{j,i})u_{i,j} - \frac{2}{3}(\mu_t u_{k,k}^2 + \rho k u_{k,k}) \tag{106}$$

and represents the effects of shear (first term) and dilatation (second term) on the turbulence kinetic energy. The dilatation effects and the term S_ϵ are identically equal to zero in incompressible flows. In compressible flows the dissipation rate of turbulence kinetic energy is affected by dilatation even in isotropic turbulence. Reynolds [67] accounted for these effects by considering rapid spherical expansion of homogeneous isotropic turbulence and obtained the following values for the constants appearing in Eq. (102):

$$C_1 = 1.44 \qquad C_2 = 11/6 \qquad S_\epsilon = C_3 \rho \epsilon u_{i,i} \tag{107}$$

where $C_3 = -2/3$.

The model used by Gosman et al. [74] employs $C_1 = 1.44$, $C_2 = 1.92$, and $C_3 = 1$. Ramos et al. [75] set $C_3 = 0$. Morel and Mansour [76] extended Reynolds' analysis [67] to more general types of compression and derived the following ϵ equation:

$$(\rho\epsilon)_{,t} + (\rho\, u_i\, \epsilon)_{,i} = \frac{\epsilon}{k}\left[\frac{1}{2} C_1\, \mu_t\, (u_{i,j} + u_{j,i})^2 - \frac{2}{3} C_1'\, \mu_t u_{k,k}^2 - \frac{2}{3} C_1''\, \rho\, k\, u_{k,k}\right]$$

$$+ \rho\, \epsilon\, u_{k,k} - C_2\, \rho \frac{\epsilon^2}{k} + \left(\frac{\mu_t}{\sigma_\epsilon} \epsilon_{,i}\right)_{,i} \qquad (108)$$

where

$$C_1' = C_1 + a(C_1 - 1.5) \qquad C_1'' = 3 + \frac{3}{2n} \qquad (109)$$

$$a = 3\frac{s_{11}^2 + s_{22}^2 + s_{33}^2}{(|s_{11}| + |s_{22}| + |s_{33}|)^2} - 1 \qquad (110)$$

$$n = 3 - (2a)^{1/2} \qquad s_{ij} = \frac{1}{2}(u_{i,j} + u_{j,i}) \qquad (111)$$

$$C_1 = 1.44 \qquad C_2 = 1.92 \qquad \mu_t = C_\mu\, \rho \frac{k^2}{\epsilon} \qquad C_\mu = 0.09 \qquad (112)$$

Grasso et al. [77] used the following values for the constants appearing in Eqs. (102) and (107): $C_1 = 1.44$, $C_2 = 1.92$, and $C_3 = -1$.

When the flow is compressed $u_{i,i} < 0$, Eq. (107) indicates that the dissipation rate of turbulence kinetic energy increases if $C_3 < 0$ and decreases if $C_3 > 0$. Morel and Mansour [76] showed that their modifications to the k–ϵ model [Eqs. (108)–(112)] yield smaller length scales during compression and larger length scales during expansion, as compared with the models proposed by Ramos et al. [75] ($C_3 = 0$) and Gosman and Watkins [78] ($C_3 = 1$). These latter models can result in length scales larger than the distance between the piston and the cylinder head (or between the housing and the rotor) during the compression stroke, a clearly unrealistic situation. Ahmadi-Befrui et al. [79] also showed that Eqs. (102) and (107) with $C_3 = 1$ result in longer length and dissipation time scales than $C_3 = 0$ and $C_3 = -0.373$ in reciprocating internal combustion engines operating under motored and firing conditions. Ramos et al. [75] used a value of $C_3 = 0$; in their discussion they noted that a positive value of C_3 is unrealistic because during the compression stroke, both the vorticity and turbulence kinetic energy increase. In order to dissipate the turbulence generated by compression, the turbulence length scales must decrease. However, a value of $C_3 > 0$ in Eq. (107) indicates that the dilatation term in the ϵ equation is negative and tends to decrease the dissipation rate of turbulence kinetic energy during compression.

For an expanding flow, that is, $u_{i,i} > 0$, a positive value of C_3 yields a greater dissipation rate [Eq. (107)], a lower turbulence kinetic energy [Eq. (105)], and smaller length scales $l = C_\mu^{3/4}\, k^{3/2}/\epsilon$ than a value of $C_3 = 0$. These results contradict the assumption $u_{i,i} > 0$, which states that the flow is expanding and that the length scales should, therefore, increase. On the other hand, a negative value of C_3 yields physically plausible results for expanding or compressing flows.

Ahmadi-Befrui et al. [79] incorrectly claim that they solve ensemble-averaged equations. In order to define ensemble averages in numerical computations, it is necessary to solve the instantaneous three-dimensional time-dependent conservation equations subject to the same boundary conditions in sufficiently small grids and with sufficiently small time steps to resolve the large and small features of the flow (a task that is impossible to achieve with present computers). If the three-dimensional time-dependent conservation equations of mass, momentum, and energy are solved a sufficient number of times to obtain meaningful statistical results and the results are averaged over a large number of realizations, ensemble-averaged quantities can be obtained.

Experimental ensemble-averaged data over many engine cycles include cycle-to-cycle variations in both the mean flow and the turbulence. These experimental ensemble-averaged data could be compared with those obtained from the solution of the instantaneous three-dimensional time-dependent conservation equations of mass, momentum, and energy.

The solution of the time-averaged or density-weighted conservation equations cannot easily be compared with ensemble-averaged or cycle-resolved measurements because these solutions are neither ensemble averaged nor cycle resolved. In addition the applicability of time averaging to engine flows is questionable [67]. Care must be therefore exercised when comparing numerical predictions and experimental data; the numerical predictions may not correspond to what is measured in the engine cylinder.

It should be noted that the $k-\epsilon$ model was developed for steady-state, incompressible, thin shear flows that are characterized by one characteristic time and one characteristic turbulent length scale. Although the $k-\epsilon$ model has been used to calculate recirculating flows with some success, it has serious limitations. In particular the model assumes that the turbulent transport is in the same direction as the mean flow gradients and does not consider the effects of the density fluctuations or pressure-velocity correlations on the k and ϵ equations. Furthermore, the ϵ equation does not have a strong physical foundation, and some authors have claimed that this equation is not valid at high Reynolds numbers [80].

There are flows where countergradient diffusion has been observed experimentally and where the pressure-velocity correlations are the main contributors to the turbulence kinetic energy balance. In premixed flames the pressure-velocity correlation is responsible for countergradient diffusion and is a source of normal stresses throughout the flame [81, 82]. Chandran et al. [83] and Chandran and Strahle [84] showed that in premixed flames the pressure-velocity correlation is negative but comparable to the values of other correlations appearing in the turbulence kinetic energy equation. Furthermore, experiments suggest that the velocity-temperature correlation in premixed flows cannot be modeled by means of the Boussinesq approximation. This is because countergradient diffusion exists at the flame front, whereas away from the flame front, the turbulence fluxes are from high to low temperatures, but these fluxes are not always perpendicular to the isotherms [85].

2.4.2.3 k–w models. Shah and Markatos [86–88] applied the k–w model developed by Ilegbusi and Spalding [89] to analyze turbulence in two- and three-dimensional flow models and showed that the standard k–ϵ model [$C_3 = 0$ in Eq. (107)] may lead to poor length-scale predictions. They also showed that the k–ϵ model proposed by Morel and Mansour [76] leads to more reasonable length-scale distributions and results in more satisfactory agreement with experimental data. The k–w model also provided better turbulence predictions than the standard k–ϵ model ($C_3 = 0$). The k–w model proposed by Ilegbusi and Spalding [89] can be written as

$$(\rho k)_{,t} + (\rho u_i k)_{,i} = \left(\frac{\mu_t}{\sigma_k} k_{,i}\right)_{,i} + P - \rho\epsilon \tag{113}$$

$$(\rho w)_{,t} + (\rho u_i w)_{,i} = \left(\frac{\mu_t}{\sigma_w} w_{,i}\right)_{,i} + C_1\mu_t |\Omega_i|^2 + C_3 \frac{wP}{k} - \rho C_2 w^{3/2} \tag{114}$$

where

$$\epsilon = 0.09\, kw^{1/2} \qquad \mu_t = \rho k w^{-1/2} \qquad C_2 = 0.17 f_1 \tag{115}$$

$$f_1 = 1 + 17.471\, l_{,i}\, l_{,i} \qquad l = \left(\frac{k}{w}\right)^{1/2} \tag{116}$$

$$\sigma_k = \sigma_w = 1 \qquad C_1 = 3.5 \qquad C_3 = 1.04 \tag{117}$$

and Ω is the local time-averaged vorticity.

Equations (113)–(117) are valid at high Reynolds numbers and near solid walls.

El Tahry [90] has derived the k–ϵ model equations for compressible flows. In the absence of combustion the k equation reduces to that for incompressible flows, and if the turbulence is assumed isotropic and some terms in the ϵ equation are neglected, a value of $C_3 = -1/3$ [Eq. (107)] is obtained. El Tahry's model predicts smaller length scales and lower turbulent intensities than the model proposed by Gosman and Watkins [78] due to the increase in the turbulence dissipation rate in the compression stroke. The value $C_3 = -1/3$ is in good agreement with that ($C_3 = -0.373$) predicted by rapid-distortion theory, where the dissipation of turbulent kinetic energy is assumed negligible and the turbulence is distorted more rapidly by changes in the mean flow than by interaction with itself [91]. Rapid-distortion theory uses local linearization, requires as inputs the strain rates and the initial state of turbulence, neglects the interactions between eddies, and assumes that angular momentum is conserved.

2.4.2.4 Subgrid-scale models. Amsden et al. [92] developed a subgrid-scale model, where the subgrid-scale turbulence is given by the following equation:

$$(\rho k)_{,t} + (\rho u_i k)_{,i} = -\frac{2}{3}\rho k\, u_{i,i} + \tau_{ij}\, u_{i,j} + (\mu_e\, k_{,i})_{,i} - \rho\frac{k^{3/2}}{L} + \dot{W}_d \tag{118}$$

where

$$\mu_e = \mu_b + \mu_a + \mu_t \tag{119a}$$

$$\mu_t = 0.05\rho L k^{1/2} \tag{119b}$$

μ_b is a background viscosity, and μ_a is the viscosity of air given by Sutherland's law, that is,

$$\mu_a = \frac{A_1 T^{3/2}}{T + A_2} \tag{120}$$

A_1 and A_2 are constants, L is a characteristic length scale approximately equal to twice the length of a typical finite difference cell size, μ_t is the subgrid-scale turbulent viscosity [93], and $\dot{W}_d$ represents the work done on the turbulent eddies by the spray droplets. Wang and Ferguson [94] used the KIVA code [95] and employed a subgrid-scale turbulent viscosity 1.6 times greater than the value given by Eq. (119*b*) in order to accurately predict the decay of angular momentum in a constant-volume cylindrical vessel.

Equation (118) does not correspond to a large-scale eddy simulation since the instantaneous time-dependent governing equations have not been averaged over a small spatial domain to remove the small eddies from the flow field and obtain an equation for the large eddies. Furthermore, the effects of the small eddies on the large ones have been incorporated in the fourth term of the right-hand side of Eq. (118). In large-scale eddy simulations, the subgrid-scale Reynolds stresses that appear in the equations for the large eddies need to be modeled; the largest eddies dominate the physics of the problem and strongly interact with the mean flow, whereas the small eddies are created mainly by nonlinear interactions among the large eddies. Most of the mass, momentum, and energy transport is due to the large eddies, which depend strongly on the geometry and nature of the flow and are highly anisotropic. On the other hand, the small eddies dissipate the mass, momentum, and energy fluctuations; affect the mean flow only slightly; are nearly isotropic; and are created and destroyed more quickly than the large eddies, which are characterized by time scales of the same order of magnitude as the mean flow [96]. Subgrid-scale models yield subgrid Reynolds stresses, which realistically simulate the transfer of large-scale variance to subgrid scales provided that the truncation errors due to the advection terms are not too large [97].

2.4.2.5 Modeling of swirl. The $k-\epsilon$ models represented by Eqs. (102), (105), (107), and (108) do not consider the effects of streamline curvature [98] and swirl on the turbulence. Complex turbulent flows such as those in internal combustion engines cannot be predicted by methods developed for thin shear layers because the longitudinal curvature and lateral divergence of the mean flow results in extra rates of strain [99].

The effects of swirl on the dissipation rate of turbulence kinetic energy have been analyzed by means of the curvature Richardson number, which affects the mixing length and eddy viscosity [100] or the dissipation rate [101]. For example, the dissipation term in the ϵ equation has been modified to account for streamline curvature as follows:

$$-C_2\,\rho\,\frac{\epsilon^2}{k}\,[1 - C_c\,\mathrm{Ri}_t] \qquad \mathrm{Ri}_t = k^2\,\frac{w(rw)_{,r}}{\epsilon^2\,r^2} \tag{121}$$

where C_c is a constant, Ri_t is the curvature Richardson number, w is the azimuthal velocity component, and r is the radial direction. Equation (121) has resulted in good agreement with experimental data for boundary layers on spinning and curved surfaces [101].

The effects of swirl on the dissipation rate of turbulence kinetic energy can also be considered by introducing a new production term in the ϵ equation that depends on the mean radial and axial velocity components [102, 103]. Swirl increases the dispersion of stratified charges. Most of the modifications introduced in two-equation models to account for the effects of swirl have been made through the ϵ equation or the turbulent length scale. However, these modifications use isotropic turbulence models, which may not be adequate for predicting swirling flows, which are typically anisotropic.

Leschziner and Rodi [104] showed that the effects of streamline curvature and the preferential influence of normal stresses on the dissipation rate of turbulence kinetic energy are important; they proposed a local equilibrium algebraic stress model in which the production of Reynolds stresses is balanced by the viscous dissipation. This results in a production term for the turbulence kinetic energy that is proportional to the derivative of the velocity along the streamlines. Their model has been used to calculate the flow field in motored Wankel engines at high rpm [68].

Rodi and Scheuerer [105] showed that the standard $k-\epsilon$ model predicts very high skin-friction coefficients, and they modified the $k-\epsilon$ to account for the generation rate of turbulence kinetic energy due to flow deceleration. Rodi and Scheuerer [105] showed that the production term in the ϵ equation has to be increased in the logarithmic law of the wall region. Hanjalic and Launder [106] introduced a new term in the ϵ equation to account for the role of irrotational strains, which promote energy transfer across the energy spectrum and increase the importance of normal stresses.

Lam and Bremhorst [107] modified the standard $k-\epsilon$ model so that it can be applied up to the solid walls. Improved versions of the $k-\epsilon$ model account for low-Reynolds-number effects and the proximity of the wall, and they assume that $\epsilon = 0$ at the wall [108]. Further references on turbulence modeling can be found in references 70, 96, and 109–127.

In Table 2.3 the governing equations for a two-dimensional flow are presented [68, 128, 129]. The $k-\epsilon$ model presented in Table 2.3 accounts for streamline curvature, preferential dissipation, low-Reynolds-number effects, and compressibility. In Table 2.4 the boundary conditions for a two-dimensional rotary engine model provided with peripheral intake and exhaust ports (Fig. 2.1) are presented [68, 129].

The $k-\epsilon$ models represented by Eqs. (102), (105), and (108) cannot be used to predict the flow field up to the solid walls. In these models the normal

Table 2.3 Two-dimensional governing equations

U	F	G	V	W	S
ρ	ρu	ρv	0	0	0
ρu	ρuu	ρuv	$\mu_e\sigma_{xx}$[a] $- \left(p + \frac{2}{3}\rho k\right)$	$\mu_e\sigma_{xy}$[b]	0
ρv	ρuv	ρvv	$\mu_e\sigma_{xy}$	$\mu_e\sigma_{yy}$[c] $- \left(p + \frac{2}{3}\rho k\right)$	0
e[d]	$\left(e + p + \frac{2}{3}\rho k\right)u$	$\left(e + p + \frac{2}{3}\rho k\right)$	$\mu_e(\sigma_{xx}u + \sigma_{xy}v) + \lambda_e T_x + \rho D_e \sum_{i=1}^{N} h_i(Y_i)_x$	$\mu_e(\sigma_{xy}u + \sigma_{yy}v) + \lambda_e T_y + \rho D_e \sum_{i=1}^{N} h_i(Y_i)_y$	0
ρY_i[e]	$\rho Y_i u$	$\rho Y_i v$	$\rho D_e(Y_i)_x$	$\rho D_e(Y_i)_y$	$\dot{\omega}_i$
ρk	ρku	ρkv	$\left(\mu + \frac{\mu_t}{\sigma_k}\right) k_x$	$\left(\mu + \frac{\mu_t}{\sigma_k}\right) k_y$	S_k[f]
$\rho\epsilon$	$\rho\epsilon u$	$\rho\epsilon v$	$\left(\mu + \frac{\mu_t}{\sigma_\epsilon}\right) \epsilon_x$	$\left(\mu + \frac{\mu_t}{\sigma_\epsilon}\right) \epsilon_y$	S_ϵ[g]

[a] $\sigma_{xx} = 2\,u_x - 2\nabla\cdot\mathbf{v}/3 \qquad \nabla\cdot\mathbf{v} = u_x + v_y.$

[b] $\sigma_{xy} = u_y + v_x.$

[c] $\sigma_{yy} = 2v_y - 2\nabla\cdot\mathbf{v}/3.$

[d] $e = \sum_{i=1}^{N} \rho h_i Y_i - p + 1/2\,\rho\mathbf{v}\cdot\mathbf{v} + \rho k.$

[e] $i = 1, 2, \ldots, N.$

[f] $S_k = G_k - \rho\epsilon + D_k \qquad G_k = \mu_e\,[2u_x^2 + 2v_y^2 + (u_y + v_x)^2] - 2\nabla\cdot\mathbf{v} \qquad (\mu_e\nabla\cdot\mathbf{v} + \rho k)/3 \qquad D_k = f_{BL}\,2\mu\,(\partial k^{1/2}/\partial n)^2 \qquad f_{BL} = 1$ in the boundary layer $f_{BL} = 0$ otherwise $\qquad f_\mu = \exp\,[-3.4\,f_{BL}/(1+R_T/50)^2]/[1 + 0.57(k^2/\epsilon^2)(u_s/r^2)(\partial/\partial r)(ru_s)] \qquad u_s =$ velocity along the streamlines $\qquad \sigma_k = 1 \qquad \sigma_\epsilon = 1.3$ $R_T = \rho k^2/\mu_t \qquad \mathrm{Pr} = \mathrm{Sc} = \mathrm{Pr}_t = \mathrm{Sc}_t = 1.$

[g] $S_\epsilon = C_1 f_1(\epsilon/k)\,(C_1'P_n + P_s) - C_2 f_2\,\rho(\epsilon^2/k) + f_3 \qquad C_1 = 1.44 \qquad C_2 = 1.92 \qquad f_1 = 1 \qquad f_2 = f_{BL}\,[1 - 0.3\exp(-R_T^2)] \qquad P_n = \mu_t\,(2u_x^2 + 2v_y^2) - 2/3\,\mu_t\,C_1''\,(\nabla\cdot\mathbf{v})^2 - 2/3\,C_1'''\,\rho\,k\nabla\cdot\mathbf{v} \qquad P_s = \mu_t(v_x + u_y)^2 \qquad f_3 = \rho\epsilon\nabla\cdot\mathbf{v} + f_{BL}\,[(2\mu\,\mu_t/\rho)\cdot(\partial^2 u/\partial n^2)]^2 \qquad C_1'' = [C_1 + a(C_1 - 1.5)]/C_1 \qquad C_1' = 2.24/C_1$ $C_1''' = (3+3/2n)/C_1 \qquad a = 3[u_x^2 + v_y^2]/[|u_x| + |v_y|]^2 - 1 \qquad n = 3 - (2a)^{1/2}.$

Table 2.4 Boundary conditions for a two-dimensional rotary engine model

Intake port	Solid walls	Fuel injection	Exhaust port
$u = 0$	$u = 0$	$u = v_0{}^a \cos \alpha$	$u = 0$
$v = -c_{\text{int}}{}^b \left[\frac{2}{\rho}(p_0 - p)\right]^{1/2}$	$v = 0$	$v = v_0 \sin \alpha$	$v = c_{\text{exh}} \left[\frac{2}{\rho}(p - p_e)\right]^{1/2}$
$T = 300$ K	$T_w = 300$ K	$T_{\text{inj}} = 300$ K	$\frac{\partial T}{\partial y} = 0$
$p_0 = 1.2 \times 10^5$ N/m^2	$\frac{\partial p}{\partial n} = 0$	$\frac{\partial p}{\partial n} = 0$	$\frac{\partial p}{\partial y} = 0$ ($p_e = 0.85 \times 10^5$ N/m^2)^c
$k = (0.03v)^2$	$k = 0$	$k = (0.03v_0)^{2\ a}$	$\frac{\partial k}{\partial y} = 0$
$\epsilon = \frac{k^{3/2}}{0.0075\ d_{\text{int}}{}^d}$	$\epsilon = 0$	$\epsilon = \frac{k^{3/2}}{0.0075\ d_{\text{inj}}}{}^d$	$\frac{\partial \epsilon}{\partial y} = 0$
Y_i specified	$\frac{\partial Y_i}{\partial n} = 0$	$\frac{\partial p}{\partial n} = 0$	$\frac{\partial Y_i}{\partial y} = 0$

[a] $v_0 = V_{\text{inj}}(\theta - \theta_1)/3°$ $\quad \theta_1 \le \theta \le \theta_1 + 3°$
$v_0 = V_{\text{inj}}$ $\quad \theta_1 + 3° \le \theta \le \theta_2 - 3°$
$v_0 = V_{\text{inj}}(\theta_2 - \theta)/3°$ $\quad \theta_2 - 3° \le \theta \le \theta_2$
where θ_1 and θ_2 are the shaft angles at the beginning and end of injection.

[b] The discharge coefficients at the intake and exhaust ports are denoted by c_{int} and c_{exh}.

[c] The exhaust back pressure is denoted by p_e.

[d] The widths of the intake port and injector are denoted by d_{int} and d_{inj}.

component of the gradient of the turbulence kinetic energy is set equal to zero at the wall, and the dissipation rate of turbulence kinetic energy is calculated by assuming an equilibrium boundary layer where production of turbulence kinetic energy is equal to its dissipation rate. Furthermore, the tangential velocity and temperature near the wall are calculated by means of linear or logarithmic laws, depending on the location of the grid points closest to the solid walls and the thickness of the viscous sublayer [72]. In these k–ϵ models the Reynolds analogy is used to calculate the heat transfer rates (see Section 2.6.2). However, the Reynolds analogy is strictly applicable to stationary, incompressible flows with similar mean velocity and temperature profiles. These conditions may not be satisfied in internal combustion engine flows. Furthermore, the ϵ equation does not have a strong physical foundation and does not account for the "spotty" character of the dissipation rate of the turbulence kinetic energy.

2.4.3 Combustion Modeling

The reaction rate term, $\dot{\omega}_i$, which appears in Eq. (74), can be written as

$$\dot{\omega}_i = M_i \sum_{k=1}^{M} (\nu''_{i,k} - \nu'_{i,k}) \left[K_{f,k}\, T^{\alpha_{f,k}} \exp\left(-\frac{E_{f,k}}{\tilde{R}T}\right) \prod_{j=1}^{N} c_j^{\nu'_{j,k}} - K_{b,k}\, T^{\alpha_{b,k}} \exp\left(-\frac{E_{b,k}}{\tilde{R}T}\right) \prod_{j=1}^{N} c_j^{\nu''_{j,k}} \right] \tag{122}$$

for M simultaneous reactions of the form

$$\sum_{j=1}^{N} \nu'_{j,k} \, \mathrm{Sp}_j \rightleftarrows \sum_{j=1}^{N} \nu''_{j,k} \, \mathrm{Sp}_j \qquad k = 1, 2, \ldots, M \tag{123}$$

where Sp denotes species; ν' and ν'' are the stoichiometric coefficients for species appearing as reactants and as products, respectively; K_f and K_b are the preexponential factors of the forward and backward reactions, respectively; α_f and α_b are the temperature exponents of the forward and backward reactions, respectively; N is the number of species; E_f and E_b are the activation energies of the forward and backward reactions, respectively; c_i denotes the concentration of species i that is related to the mass fraction by the following expression:

$$c_i = \rho \frac{Y_i}{M_i} \tag{124}$$

and Y_i and M_i denote the mass fraction and molecular mass of species i.

2.4.3.1 Mean-reaction-rate models. If Eqs. (122) and (124) are substituted into Eq. (94), the long time average of the mean reaction rates that appear in Eq. (94) yields many temperature-mass fraction-density correlations. The exponential terms [Eq. (122)] can be expanded in Taylor's series around the mean temperature. However, the expansion may not converge unless the temperature fluctuations are smaller than the mean temperature and unless the activation energy defined as $E/\tilde{R}$ is smaller than the mean temperature. These conditions are rarely satisfied in premixed turbulent chemically reacting flows; for typical hydrocarbons, the ratio of the activation temperature to the mean temperature is between 10 and 20. In addition the reaction rates given by Eq. (122) are not known except for the simplest hydrocarbons [130], and in general they involve many species and chemical reactions. These chemical reaction rates are in general stiff, that is, they are characterized by time scales that differ by many orders of magnitude from each other [131, 132]. The multidimensionality of engine flow models does not allow inclusion of complex chemical kinetics mechanisms at the present time. For these reasons, simplified reactions mechanisms have been developed to calculate the oxidation of the fuel. Most of these mechanisms involve an irreversible overall chemical reaction of the form

$$\mathrm{F\ (fuel)} + \mathrm{O\ (oxidizer)} \rightarrow \mathrm{P\ (products)} \tag{125}$$

and the reaction rates have been tuned to predict the lean and rich flammability limits, the maximum flame speed, and the equivalence ratio at which that maximum occurs [130]. For n-octane the overall reaction rate can be written as

$$\langle \dot{\omega}_f \rangle = -4.6 \times 10^{11} \exp\left(- \frac{30{,}000}{\tilde{R}T} \right) c_f^{0.25} \, c_o^{1.5} \tag{126}$$

where c denotes concentration and the subscripts f and o refer to the fuel and oxidizer, respectively. Note that the concentration exponents that appear in Eq. (126) are not related to the stoichiometric coefficients in the global reaction;

these coefficients are empirically determined from either experimental data or detailed chemical kinetics calculations. Furthermore, these exponents, the preexponential factor, the temperature exponent, and the activation energy depend on the fuel type, pressure, and equivalence ratio. Dryer and Brezinsky [133] showed that the conversion of the reaction intermediaries to carbon monoxide and carbon dioxide is much slower during the oxidation of isooctane than of *n*-octane in a flow reactor. The nature and the relative rates of oxidation of intermediaries are related to the number of primary, secondary, and tertiary hydrogen atoms present in the initial fuel.

Westbrook and Dryer [134] showed that global reaction rates with concentration exponents equal to 1 cannot yield the correct dependence of the flame speed on the equivalence ratio and pressure and cannot correctly predict the rich flammability limit. Two-step mechanisms yield information on the flame structure and burnt-gas composition, but neither global nor two-step mechanisms can accurately describe the chemical structure of the flame. Coffee [135] indicated that a single overall chemical reaction cannot do a good job of representing flames over a range of stoichiometries and that it has not been firmly established that a single reaction can even reproduce burning velocities accurately over a range of stoichiometries. Coffee [135] also indicated that even if overall reaction mechanisms are able to predict the flame speed, the flame structure cannot be accurately predicted. Asymptotic analyses based on large activation energies have been used to reduce the large number of elementary reactions that govern the oxidation of hydrocarbons to a few steps [136]. However, this reduction has been carried out for simple fuels such as ozone and methane. Future work may include the reduction of complex reaction mechanisms for *n*-octane and isooctane [137].

Westbrook and Dryer [138] modeled combustion processes by means of detailed chemical mechanisms that involve many chemical species and chemical reactions. They also developed global and quasiglobal mechanisms. In global mechanisms a single reaction is used to account for the oxidation of the fuel to carbon dioxide and water vapor. In quasiglobal mechanisms the oxidation of fuel is modeled by means of two or three chemical reactions that account for the formation of carbon monoxide and water vapor due to oxidation of the fuel. The carbon monoxide is also assumed to react with molecular oxygen, and the product of this reaction is carbon dioxide.

Duterque et al. [139] studied combustion in well-stirred reactors and used experimental data to develop quasiglobal reaction schemes for hydrocarbon combustion. Quasiglobal mechanisms can account for the formation of carbon monoxide and result in substantial computational savings as compared with detailed chemical kinetics mechanisms [130, 133–135, 138, 140].

Varma et al. [141] studied premixed laminar hydrogen-air flames using elementary and global kinetics; they showed that the flame structure depends on the kinetics scheme even for the same transport coefficients. They also claim that global models can accurately predict the burning velocity and flame thickness based on the maximum temperature slope. Reitz and Bracco [142] showed that global reactions to carbon dioxide and water vapor are adequate for equivalence ratios smaller than 0.9. These schemes yield very high product concentrations for equivalence ratios greater than 0.9 because carbon monoxide is not considered in these global schemes; for rich mixtures the hydrogen molecule must also be included. Reitz and Bracco [142] stated that the final composition depends on the rates of the following reactions:

$$C_nH_{2m} + \left(n + \frac{m}{2}\right) O_2 \rightarrow nCO_2 + mH_2O \tag{127}$$

$$CO + \frac{1}{2} O_2 \rightarrow CO_2 \tag{128}$$

$$H_2 + \frac{1}{2} O_2 \rightarrow H_2O \tag{129}$$

Reitz and Bracco [143] showed that H_2 and CO need to be included in global chemical kinetics calculations in order to accurately predict the energy release for equivalence ratios $0.5 \leq \phi \leq 3$. They also showed that the dependence of the laminar flame speed S_L on ϕ can be modeled satisfactorily for methane-air and ethane-air mixtures. Metgalchi and Keck [144] determined the laminar flame speed of propane-air mixtures for $0.4 \text{ atm} \leq p \leq 40 \text{ atm}$, $298 \text{ K} \leq T \leq 750 \text{ K}$, and $0.8 \leq \phi \leq 1.5$ in a constant-volume system. They also determined the flame speed of air-isooctane mixtures in a constant-volume system for $0.8 \leq \phi \leq 1.5$, $0.4 \text{ atm} \leq p \leq 50 \text{ atm}$, and $298 \text{ K} \leq T \leq 700 \text{ K}$ and obtained the following expression [145]:

$$S_L = S_L^0 \left(\frac{T_u}{T_0}\right)^{\alpha} \left(\frac{p}{p_0}\right)^{\beta} (1 - 2.1f) \tag{130}$$

where T_0 and p_0 are the reference temperature and pressure, respectively; f is the diluent mass fraction; T_u is the unburnt-gas temperature; and α and β depend on the equivalence ratio. The flame speed at T_0 and p_0, S_L^0, depends on the fuel and equivalence ratio. The laminar flame speeds obtained by Metgalchi and Keck [144, 145] are useful in multizone thermodynamic models of internal combustion engines.

Several approaches have been developed to model the mean-reaction-rate term, which appears in Eq. (94). In mean-reaction-rate models the density, temperature, and species fluctuations are neglected in $\langle \dot{\omega}_i \rangle$. This approximation may underestimate the mean reaction rate, particularly in premixed turbulent chemically reacting flows, where the temperature fluctuations may be of the same order of magnitude as the mean temperature. However, if the reaction rate and turbulent diffusivities are properly tuned, good agreement with experimental data can be obtained for a variety of engine loads and speeds [146, 147].

Diwakar [148] used a mean-reaction-rate model to predict combustion in a homogeneous-charge engine and showed that good agreement with experimental data required case-by-case tuning of the reaction rate when the volumetric efficiency or the equivalence ratio were varied. He showed that the preexponential factor correlates with the equivalence ratio and the residual mass fraction, and he performed the calculations with the CONCHAS code [149].

2.4.3.2 Eddy-breakup models. In mean-reaction-rate models it is implicitly assumed that the chemical kinetics time t_c is much larger than the mixing time t_m (i.e., $t_c >> t_m$) and that the reaction is controlled by chemical kinetics. Eddy-breakup models have been developed to account for mixing-controlled reactions where the characteristic time scale of the large eddies or mixing time is larger than the chemical conversion time [150, 151]. These models account for the intermittent appearance of reactants at a given point in the flow and assume that the large-scale eddies are responsible for stretching and bringing the fuel and oxidizer into contact with each other; mixing ocurs in the small scales. It must be pointed out that molecular mixing is necessary for chemical reactions to occur.

According to Magnussen and Hjertager [151], in mixing-controlled combustion the fuel conversion rate can be written as

$$\langle \dot{\omega}_f \rangle = -A \frac{\epsilon}{k} \min \left(c_f, \frac{c_0}{s}, \frac{Bc_p}{1 + s} \right) \tag{131}$$

where A and B are constants, c_p is the product concentration, and s represents the number of moles of oxidizer required to burn one mole of fuel. However, calculations in chemically reacting turbulent mixing layers have shown that A and B are not really constants [152, 153].

Grasso and Bracco [154] used the following mixing-controlled model of engine combustion:

$$\frac{Y_p}{1 + s} > Y_f \text{ and } \frac{Y_o}{s} \qquad \phi < 1 \qquad \langle \dot{\omega}_f \rangle = -C\rho\, Y_f \frac{\epsilon}{k} \tag{132}$$

$$\frac{Y_p}{1 + s} > Y_f \text{ and } \frac{Y_o}{s} \qquad \phi > 1 \qquad \langle \dot{\omega}_f \rangle = -C\rho \frac{Y_o}{s} \frac{\epsilon}{k} \tag{133}$$

$$Y_f \text{ and } \frac{Y_o}{s} > \frac{Y_p}{1 + s} \qquad \langle \dot{\omega}_f \rangle = -C\rho \frac{Y_p}{1 + s} \frac{\epsilon}{k} \tag{134}$$

where C is a constant and Y_f, Y_o, and Y_p are the mass fractions of the fuel, oxidizer, and products, respectively. Mixing-controlled models are inadequate to predict combustion near walls, self-ignition, pollutant formation, and lean and rich flammability limits; these phenomena are controlled by chemical kinetics. Ahmadi-Befrui et al. [79] used the eddy-breakup model given by Eq. (131) with $A = 4$ and $B = 0.5$ to simulate combustion in homogeneous-charge engines.

2.4.3.3 Hybrid models I. Hybrid combustion models that account for kinetics- and mixing-controlled processes have also been developed to analyze combustion in reciprocating engines. Przekwas [155] considered a second-order global reaction where

$$R_A = -A_A \rho^2 Y_f Y_o \exp\left(-\frac{E}{\tilde{R}T}\right) \tag{135}$$

and an eddy-breakup model where

$$R_E = -A_E \rho \frac{\epsilon}{k} \min\left(Y_f, \frac{Y_o}{s}, \frac{BY_p}{1+s}\right) \tag{136}$$

and set the mean reaction rate equal to the following values

$$\langle \dot{\omega}_f \rangle = R_A \qquad \text{if } t_c \geq t_m \tag{137}$$

$$\langle \dot{\omega}_f \rangle = R_E \qquad \text{if } t_c < t_m \tag{138}$$

where

$$t_c \sim \left[A_A \rho \exp\left(-\frac{E}{\tilde{R}T}\right)\right]^{-1} \qquad t_m \sim \frac{k}{\epsilon} \tag{139}$$

and A_A and A_E are constants.

Taking the harmonic average of Eqs. (137) and (138), the following mean reaction rate results:

$$\langle \dot{\omega}_f \rangle = \frac{R_A R_E}{R_A + R_E} \tag{140}$$

Equation (140) is valid for both mixing- and kinetics-controlled chemical reactions.

A continuous variation of the reaction rate can also be introduced by considering the reactedness τ as follows

$$\langle \dot{\omega}_f \rangle^{-1} = (1 - \tau) R_A^{-1} + \tau R_E^{-1} \tag{141}$$

For fast chemical kinetics, the associated two-delta function probability distribution for the reactedness is related to the variance of the scalar fluctuations g by means of the following expression [156–159]:

$$g = (1 - \tau)\tau \tag{142}$$

The value of g is related to the scalar dissipation rate through the equation

$$\epsilon_s = g_{,t} = 2D \langle \xi'_{,j} \xi'_{,j} \rangle \tag{143}$$

where

$$g = \langle \xi'^2 \rangle \tag{144}$$

ξ is the scalar, D is the molecular diffusivity, and ϵ_s is the scalar dissipation rate. A transport equation can be obtained for g [160]; Eq. (142) can then be used to calculate the reactedness required to evaluate the mean reaction rate in Eq. (141).

Combustion in premixed and nonpremixed flows can also be modeled by considering the mixture fraction and reactedness [Eq. (142)]. The mixture fraction is a conserved scalar, and the reactedness parameter can be related to the mass fractions of the burnt and unburnt gases [157].

Givi et al. [152] and Mudford and Bilger [161] studied non-premixed reacting flows seeded with nitric oxide and ozone. They showed that mean-reaction-rate and eddy-breakup models overpredict the reaction rate, whereas in second-order closure models, the triple correlations are of the order of or larger than the second-order correlations. Mudford and Bilger [161] also developed a perturbation analysis that yields reaction-rate estimates that approach zero as the flow chemistry approaches equilibrium, whereas mean-reaction-rate models may not yield zero reaction rates at equilibrium.

2.4.3.4 Chemical equilibrium. Mean-reaction-rate and eddy-breakup models do not predict chemical equilibrium if they consider only a global reaction such as the one shown in Eq. (125) and if the combustion products are assumed to be carbon dioxide and water vapor only. This deficiency can be corrected as follows. Consider the oxidation of a hydrocarbon in air, for example,

$$C_nH_{2n+2} + \frac{3n+1}{2}(O_2 + 3.76N_2) \rightarrow aCO + bCO_2 + cH_2 + dN_2 + eOH$$
$$+ fH_2O + gO + lH + mNO + nO_2 + sC(s) + tN \tag{145}$$

and assume that the products are in chemical equilibrium. A more complete list of reaction products might include CH_4, NH_3, NO^+, e^-, etc., where e^- and (s) denote electrons and solid materials, respectively. The composition of the equilibrium products depends on the temperature, pressure, and equivalence ratio.

In Eq. (145) there are 12 unknowns, that is, a, b, c, d, e, f, g, l, m, n, s, and t, and four atoms, C, H, O, and N. Conservation of these atoms requires

$$n = a + b + s \tag{146}$$

$$2n + 2 = 2c + e + 2f + l \tag{147}$$

$$3n + 1 = a + 2b + e + f + g + m + 2n \tag{148}$$

$$3.76(3n + 1) = 2d + t \tag{149}$$

Eight equilibrium equations are needed to calculate the 12 unknowns. For example, the following equilibrium equations can be considered:

$$C(s) + O_2 \rightleftarrows CO_2 \tag{150}$$

$$H_2 + \frac{1}{2}O_2 \rightleftarrows H_2O \tag{151}$$

$$C(s) + \frac{1}{2}O_2 \rightleftarrows CO \tag{152}$$

$$\frac{1}{2} N_2 + \frac{1}{2} O_2 \rightleftarrows NO \tag{153}$$

$$\frac{1}{2} O_2 \rightleftarrows O \tag{154}$$

$$\frac{1}{2} H_2 \rightleftarrows H \tag{155}$$

$$\frac{1}{2} H_2 + \frac{1}{2} O_2 \rightleftarrows OH \tag{156}$$

$$\frac{1}{2} N_2 \rightleftarrows N \tag{157}$$

Any other set of eight independent equilibrium equations can be used.

The equilibrium constants for each of these reactions can be calculated by means of the law of mass action. For example, the equilibrium constant of Eq. (151) can be written as a function of the partial pressures of oxygen, hydrogen, and water vapor as

$$K_{P,f(H_2O)} = \frac{p_{H_2O}}{p_{H_2}\, p_{O_2}^{1/2}} = \exp\left(-\frac{\Delta F^\circ}{\tilde{R}T}\right) \tag{158}$$

where ΔF° is the standard-state Gibbs free energy change, $\tilde{R}$ is the universal gas constant, and the partial pressures can be written as a function of the species concentrations or mass fractions [162] as

$$p_i = c_i\, \tilde{R}T = \rho\, Y_i \frac{\tilde{R}T}{M_i} \tag{159}$$

The equations for the conservation of atoms [Eqs. (146)–(149)] and the equilibrium constants are algebraic and can be solved to obtain the equilibrium composition of the combustion products. Numerical methods based on the minimization of the Gibbs and Helmholtz free energies can be used to calculate the composition of the combustion products [163]. Meintjes and Morgan [164] developed a procedure that algebraically combines the equilibrium equations, eliminates some unknowns, and scales the simplified equations to obtain a set of simultaneous nonlinear (polynomial) equations that are relatively easy to solve algebraically. The main advantage of the procedure developed by Meintjes and Morgan [164] is its computational speed compared with, for example, the original technique used in the KIVA code [92, 95, 165].

Meintjes and Morgan [166] combined the Newton and Gauss-Seidel methods to calculate the equilibrium compositions required in multidimensional codes [149]. The Gauss-Seidel method is fast but may be unreliable, particularly when the matrices are not diagonally dominant. On the other hand, Newton's method is reliable but slow; Newton's method has a quadratic rate of convergence when the initial guess is close enough to the solution.

Erickson and Prabhu [167] have developed a method to rapidly compute the chemical equilibrium composition of hydrocarbon combustion. They considered CO, CO_2, H_2O, O_2, H_2, N_2, H, O, OH, and NO and obtained an algebraic equation for the equilibrium composition, which results in big improvements in computer time. Their formulation does not require minimization of the free energy as in the NASA Chemical Equilibrium Program [163].

Amelio et al. [168] used a multidimensional model to evaluate the performance of a spark-ignition engine model characterized by one-step reaction, and they used an equilibrium assumption to calculate the thermodynamic properties of the mixture. Their model results in an automatic determination of the initial conditions.

2.4.3.5 Hybrid models II. Equations (146)–(157) are valid if the combustion products are in chemical equilibrium (i.e., the chemical relaxation times of the species are zero). In this section, we consider models in which the chemical relaxation times are finite.

The conversion rate of mean-reaction-rate models, that is, Eq. (137), can be expressed for species i as follows:

$$\langle\dot{\omega}_i\rangle_{\mathrm{ARR}} = \langle\dot{\omega}_i\rangle^e_{\mathrm{ARR}} + \left(\frac{\partial\langle\dot{\omega}_i\rangle}{\partial c_j}\right)^e (c_j - c_j^e) = \left(\frac{\partial\langle\dot{\omega}_i\rangle}{\partial c_j}\right)^e (c_j - c_j^e) \tag{160}$$

where the subscript ARR denotes Arrhenius kinetics and the superscript e denotes chemical equilibrium. Note that in chemical equilibrium $\langle\dot{\omega}_i\rangle^e_{\mathrm{ARR}} = 0$.

Equation (160) can also be written as

$$\langle\dot{\omega}_i\rangle_{\mathrm{ARR}} = \frac{c_j - c_j^e}{\tau_{i,j}^{\mathrm{ARR}}} \tag{161}$$

where

$$\tau_{i,j}^{\mathrm{ARR}} = \left[\left(\frac{\partial\langle\dot{\omega}_i\rangle}{\partial c_j}\right)^e\right]^{-1} \tag{162}$$

and c_j is the concentration of species j. In Eqs. (160) and (161) repeated subscripts indicate summation (i.e., Einstein's convention has been used).

The times $\tau_{i,j}^{\mathrm{ARR}}$ denote the chemical relaxation times (i.e., the times required to achieve chemical equilibrium), which are functions of the temperature, pressure, and composition. These times, in general, are of different orders of magnitude due to the stiffness of the reaction rates.

The reaction rates given by eddy-breakup models can also be expressed in a manner similar to Eq. (161) and written as

$$\langle\dot{\omega}_i\rangle_{\mathrm{EBU}} = \frac{c_j - c_j^e}{\tau_{i,j}^{\mathrm{EBU}}} \tag{163}$$

where EBU stands for eddy breakup and $\tau_{i,j}^{\mathrm{EBU}}$ can be related to k and ϵ as, for example,

$$\tau_{i,j}^{\text{EBU}} = \frac{A_{i,j}}{\epsilon} \tag{164}$$

Equations (161) and (163) can be used to calculate the mean reaction rate when $\tau_{i,j}^{\text{ARR}} > \tau_{i,j}^{\text{EBU}}$ or $\tau_{i,j}^{\text{ARR}} < \tau_{i,j}^{\text{EBU}}$ (i.e., for kinetics- or mixing-controlled combustion), by taking a harmonic average of both equations, that is,

$$\langle \dot{\omega}_i \rangle^{-1} = \langle \dot{\omega}_i \rangle_{\text{ARR}}^{-1} + \langle \dot{\omega}_i \rangle_{\text{EBU}}^{-1} \tag{165}$$

which yields

$$\langle \dot{\omega}_i \rangle = \frac{c_j - c_j^e}{\tau_{i,j}^{\text{EBU}} + \tau_{i,j}^{\text{ARR}}} \tag{166}$$

For mixing-controlled combustion $\tau_{i,j}^{\text{EBU}} > \tau_{i,j}^{\text{ARR}}$ and for kinetics-controlled combustion $\tau_{i,j}^{\text{EBU}} < \tau_{i,j}^{\text{ARR}}$, Eq. (166) reduces to Eq. (163) [Eq. (161)].

The values of $\tau_{i,j}^{\text{ARR}}$ depend on the chemical reaction and species. However, if only one global chemical reaction is considered and the relaxation times of different species are assumed to be equal, substantial simplifications result.

Reitz and Bracco [142, 143] assumed that the relaxation times for C_3H_8, O_2, H_2O, H_2, CO, and CO_2 are identical and represented the mean reaction rate by means of the following equation:

$$\langle \dot{\omega}_i \rangle = \frac{c_i - c_i^*}{\tau_R} \tag{167}$$

where c_i^* denotes the concentration of species i under local and instantaneous chemical equilibrium and the relaxation time τ_R can be written as

$$\tau_R = t_m + t_l \tag{168}$$

where t_l can be determined from chemical kinetics calculations and depends on the pressure, equivalence ratio, laminar flame speed, and temperature and where t_m is given by an eddy-breakup model as

$$t_m = Cm_2 \frac{k}{\epsilon} \left[1 - \exp\left(-\frac{t - t_s}{t_d} \right) \right] \quad \text{for } \beta \geq 1 \tag{169}$$

$$t_m = Cm_3 \frac{k}{\epsilon} (1 + s) \frac{Y_f - Y_f^*}{Y_p - Y_{p_s}} \left[1 - \exp\left(-\frac{t - t_s}{t_d} \right) \right] \quad \text{for } \beta < 1 \tag{170}$$

where

$$\beta = Cm_3 (Y_p - Y_{p_s})[Cm_2 (1 + s) (Y_f - Y_f^*)]^{-1} \tag{171}$$

$$s = \frac{Y_0 - Y_0^*}{Y_f - Y_f^*} \tag{172}$$

$$l = C_\mu^{3/4} \frac{k^{3/2}}{\epsilon} \qquad t_d = Cm_1 \frac{l}{S_L} \tag{173}$$

$$t_l = 4.59 \times 10^{-10} \left(\frac{T}{T_0}\right) \left(\frac{p}{p_0}\right)^{0.75} \exp\left[\frac{15{,}100}{T}(1 + 0.08|\phi - 1.15|)\right] \tag{174}$$

$$S_L = S_L^0 \left(\frac{T}{T_0}\right)^{1.75} \left(\frac{p}{p_0}\right)^{-1.25} \tag{175}$$

$Cm_1 = 2.75$, $Cm_2 = 0.055$, $Cm_3 = 0.092$, and $C_\mu = 0.09$; t_s is the spark timing; Y_{p_s} denotes the mass fraction of products under stoichiometric conditions; and the starred values denote quantities evaluated at the local and instantaneous chemical equilibrium conditions.

The turbulence conversion time t_m is that proposed by Magnussen and Hjertager [151] and Hjertager [169] except for the exponential term, which has been included to account for the initial laminar flame propagation from the spark plug. Note that Eq. (167) is similar to Eq. (166). However, the turbulent mixing is delayed until the laminar flame traverses a distance of the order of the length scale of the energy-containing eddies [170].

2.4.3.6 Probability density function formulations. Brandstatter and Johns [171] used a probability density function (pdf) of a progress variable that contains the effects of chemical kinetics and turbulence; they solved the pdf equation by means of an operator-splitting technique in which the diffusion terms were treated explicitly. The major problems associated with the use of a single-point pdf are the dimensionality of the pdf and the need to specify a mixing time and a mixing model [172]. The solution of the pdf equation is very sensitive to the mixing time and requires large amounts of computer storage and time; however, it may result in more accurate predictions than those of mean-reaction-rate and eddy-breakup models [173].

Some of the deficiencies of single-point pdf formulations can be eliminated by developing two-point pdf formulations. These formulations do not require estimates for the mixing length because the mixing length is already contained in the two-point pdf formulation [174]; however, they do require the specification of a mixing time.

Two-point two-time pdf formulations include length and time scales. However, their solution requires large amounts of computer storage and time.

2.4.3.7 ESCIMO. Visualization studies in mixing layers have shown the formation of large-scale structures that engulf the fluids from the low- and high-speed streams, roll up, and pair in an almost two-dimensional manner [175–177]. These structures play a central role in the development of mixing layers and the early regions of jets and wakes, and they initially roll up with a large excess of high-speed fluid in the cores [178].

Experiments with dilute reactants have shown that the combustion products exist in concentrated regions associated with the large spanwise-coherent structures [179] and that mixing and entrainment are dominated by these structures [180].

Experimental observations in reacting and nonreacting turbulent shear flows suggest that molecular transport plays an important role even in fully turbulent flows, that the dynamics of the large-scale structures dictate entrainment into the turbulent region, which in turn provides the environment for subsequent turbulent cascade and mixing processes [181], and that the fluid elements from the two streams are distributed unmixed throughout the mixing layer by large-scale inviscid motions.

Turbulence models that use a Boussinesq approximation for the modeling of the correlations require that the turbulent motions have a scale that is small compared with that over which the velocity, concentration, or temperature change significantly [114]. This condition is violated in mixing layers, and other models of turbulence have to be developed [182].

These experimental observations in mixing layers at moderate and high Reynolds numbers have been considered in the ESCIMO theory of turbulent combustion developed by Spalding [183–185]. The ESCIMO theory accounts for (1) the *E*ngulfment of the fluid, (2) the *S*tretching and distortion of the interleaved elements of fluid, (3) the *C*oherence of the large-scale structures, (4) the *I*nterdiffusion or molecular processes that occur in the interleaved elements, and (5) the Lagrangian (or *M*oving *O*bserver) characteristics of the interdiffusion processes.

The engulfment process is assumed to be controlled by velocity differences, and the folds have an initial thickness proportional to the flame thickness. The folds diminish in thickness at a stretching rate controlled by the local velocity field and flame thickness. Combustion in the folds is assumed to be laminar in premixed flows [186, 187] whereas in nonpremixed flows, the reaction rate is controlled by the stretching rate, mixture fraction, and the mass fractions of the local, completely burnt, and completely unburnt mixtures [188].

The ESCIMO theory has been applied with some success to a premixed fuel-air stream in a steady, confined baffle-stabilized flame [187], hydrogen-air diffusion flames [189], and well-stirred reactors [186].

The major criticisms of the ESCIMO theory concern the initial composition and size of the folds and the effects of the chemical reaction on the folds. In turbulent premixed flames several combustion regimes are observed, depending on $|\mathbf{u}'|/S_L$, η/δ_L, $\mathrm{Re} = \rho|\mathbf{u}'|l/\mu$, l/δ_L, and $l/|\mathbf{u}'|\tau_c$, where $|\mathbf{u}'|$ is the magnitude of the velocity fluctuations, S_L is the laminar flame speed, Re is the Reynolds number, δ_L is the laminar flame thickness, η is the length scale of the Kolmogorov eddies, ρ is the density, μ is the dynamic viscosity, l is the integral length scale, and τ_c is a representative chemical conversion time in the hot part of the laminar flame [64]. Two important combustion regimes are those corresponding to stirred reactors and reaction sheets. In the stirred-reactor regime, turbulent mixing is rapid compared with the chemistry, and combustion occurs in a distributed reaction zone. In the reaction-sheet regime, the chemistry in hot regions is rapid compared with the turbulent processes, and combustion tends to occur in thin sheets wrinkled by the turbulence. This general classification is applicable to both nonpremixed and premixed combustion [64].

Combustion in the stirred-reactor regime is a difficult problem. A sufficient criterion for the existence of the distributed-combustion regime is $\delta_L >> l$, whereas a sufficient criterion for the occurrence of the reaction-sheet regime is $\eta >> \delta_L$.

In the reaction-sheet regime, highly convoluted, folded, and multiply connected reaction sheets may occur at high turbulence intensities. The eddy-breakup [150, 190] and ESCIMO [183–189] models are applicable to flames in turbulence of high intensity or small scale.

Numerical estimates of the parameters that affect turbulent combustion in premixed-charge spark-ignition engines indicate that the combustion regime in these engines is not clear, but up to and including idle speed, combustion occurs in the reaction-sheet regime. At high engine speeds the parameters fail to satisfy the conditions known to be sufficient for occurrence of the reaction-sheet regime or the distributed-reaction regime, and the structure of the flame is not clear at this time [191]. If the reaction-sheet combustion regime prevails, combustion models of premixed-charge combustion need to be revised to accurately predict the flame structure.

Turbulent-combustion regimes are shown in Fig. 2.8, where u' is the turbulence intensity, S_L is the laminar flame speed, η is the Kolmogorov eddy size, l is the integral length scale of turbulence, Re ($= u'l/\nu$) is the Reynolds number, and Da [$= (l/u')/(\delta_L/S_L)$] is the Damköhler number. In the nonturbulent regime Re < 1, whereas the wrinkled-laminar-flames regime corresponds to $u' < S_L$.

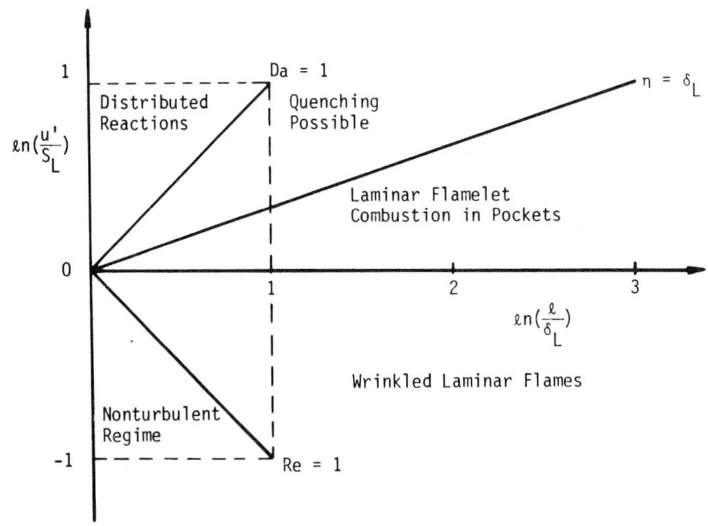

Figure 2.8 Turbulent combustion regimes.

The distributed-reaction regime corresponds to Re $>$ 1 and Da $<$ 1 [or mixing times ($t_m = l/u'$) smaller than reaction times ($t_c = \delta_L/S_L$)]. Conditions sufficient for the existence of distributed reactions are $\eta << l << \delta_L$ and $t_c > t_\eta = (\nu/\epsilon)^{1/2}$, where ϵ is the dissipation rate of turbulence kinetic energy, ν is the kinematic viscosity, and t_η is the characteristic time of the dissipation eddies.

Conditions sufficient for the existence of reaction sheets are $l >> \eta >> \delta_L$ and $t_m = l/u' >> t_\eta >> t_c$, where t_m is the characteristic time of the large eddies. Note that Da $= t_m/t_c$.

In the laminar flamelet combustion regime Re $>> 1$, $u' > S_L$, $t_\eta > t_c$, and Da $>> 1$. Flame quenching is possible if Re $>> 1$ and $t_m > t_c > t_\eta$, and very little is known about the intermediate regime $l >> \delta_L >> \eta$ [192].

2.4.3.8 Two-fluid models. Recently Spalding [193–197] developed a theory of turbulent combustion based on a two-fluid model. This theory divides the mixture into hot and cold fragments, which intermingle and exchange mass momentum and energy. Each fluid has its own velocity, composition, temperature, and density at each point. The differences in density result in different responses to pressure gradients and may be able to explain the countergradient diffusion observed in turbulent chemically reacting flows [81, 85]. Preliminary one- and two-dimensional calculations [197], with interfluid friction proportional to the square of the velocity differences of the two fluids and interfluid mass transfer proportional to the velocity difference and volume fractions of the two fluids, are encouraging. The model can be extended to n fluids in order to account for different fluid fragments.

The two-fluid model of turbulent combustion is based on two-phase-flow theory, where the major difficulties are determination of the constitutive equations and exchange laws of mass, momentum, and energy if the phases are treated as continuum media [198–200]. However, the two-fluid model has been applied with great success to analyze sedimentation problems and the Kelvin-Helmholtz and Rayleigh-Taylor instabilities [201].

2.4.3.9 Bray-Libby-Moss (BLM) model. Premixed turbulent flames typically consist of thin, moving interfaces that separate the unburnt from the fully burnt mixture, and combustion occurs only within these interfaces. At low turbulence intensities, the interface forms a single, continuous, wrinkled laminar flame but, with increasing turbulence, it can rupture to create pockets of unburnt and/or fully burnt mixture. In either case the overall rate of combustion is determined by the surface area and structure of the burning interfaces.

If turbulent mixing is sufficiently intense, the laminar flames that form these interfaces are so stretched and curved by the turbulence that the reaction rates within them are reduced and may even be quenched. The turbulent mixing rate may be characterized by u'/l, where u' is the rms turbulence velocity and l is the integral length scale. If u'/l is increased, the mean heat release rate will at first increase because the flame area increases, but flame stretch and quenching will eventually lead to a reduction in heat release.

Laminar flamelet models, in which combustion within a turbulent flame is assumed to be confined to thin, moving laminar flamelets embedded in the turbulent flow, provide a means of accounting for flame stretching and quenching. In these models the structure and properties of laminar flames under appropriate conditions of pressure, temperature, composition, strain rate, etc., are first analyzed in order to build a "library" of laminar flamelet solutions. Chemical reaction effects are then represented by an appropriate ensemble of laminar flamelet solutions from the library [202].

The Bray-Libby-Moss (BLM) model assumes that the turbulence Reynolds (Re) and Damköhler (Da) numbers are very high so that a thin-flame (sheet-combustion) regime prevails

$$\mathrm{Re} = \frac{u'l}{\nu} >> 1 \qquad \mathrm{Da} = \frac{S_L/\delta_L}{u'/l} >> 1 \tag{176}$$

where ν is the kinematic viscosity, and S_L and δ_L are the speed and thickness of an unstrained laminar flame.

Several regimes of premixed turbulent combustion can be defined, depending on u'/S_L, l/η, and l/δ_L, where η is the Kolmogorov length scale [203]:

1. Wrinkled laminar flame: $u' < S_L$.
2. Distributed reaction: $\delta_L/S_L > l/u'$.
3. Packet burning with laminar flamelets: $u' > S_L$, $u'/l < (\nu/\epsilon)^{1/2}$ (or $l < \eta$).
4. Packet burning without laminar flamelets: $l > \delta_L > \eta$.

Bray and Moss [204, 205] used thin shear flow and fast chemistry approximations, a single-step global chemical reaction, and Favre's averaging to calculate the average reaction rate. A progress variable c can be defined that obeys an equation of the form

$$\frac{\partial(\rho c)}{\partial t} + \frac{\partial}{\partial x_j}(\rho u_j c) = \frac{\partial}{\partial x_j}\left(\rho D \frac{\partial c}{\partial x_j}\right) + \dot{\omega} \tag{177}$$

where D is the diffusion coefficient.

If the mean flow is stationary and the Reynolds number is large, averaging Eq. (177) yields

$$\frac{\partial}{\partial x_j}(\langle \rho\, u_j\, c\rangle) = \langle\dot{\omega}\rangle \tag{178}$$

where the angular brackets denote time-averaged quantities.

In the laminar flamelet combustion regime, thin, moving laminar flames separate the unburnt and burnt mixtures, and the pdf of c at location x may be defined as

$$p(c; x) = \alpha(x)\,\delta(c) + \gamma(x)\,f(c; x) + \beta(x)\,\delta(1 - c) \tag{179}$$

where $\delta(c)$ and $\delta(1 - c)$ are Dirac delta functions representing the unburnt ($c = 0$) and burnt ($c = 1$) gases, respectively; α, β, and γ are the probabilities of unburnt, burnt, and intermediate burning states occurring at x, respectively; and c is a normalized progress variable.

In the laminar flamelet combustion regime $\gamma(x) << 1$, and if turbulent mixing is much slower than the reaction, $\langle\dot{\omega}\rangle$ is proportional to the rate at which turbulent motions bring unburnt gas into contact with flamelets. By combining balance equations for $\tilde{c}$ and $\widetilde{c''^2}$ and assuming that $\gamma << 1$, Bray [156] found that $\langle\dot{\omega}\rangle$ is proportional to the mean scalar dissipation function $\tilde{\epsilon}_c$, that is,

$$\langle\dot{\omega}\rangle = \langle\rho\rangle \frac{\tilde{\epsilon}_c}{2c_m - 1} \tag{180}$$

where

$$c_m = \frac{\langle c\dot{\omega}\rangle}{\langle\dot{\omega}\rangle} \tag{181}$$

$$\tilde{\epsilon}_c = 2\frac{\langle\rho\, D\, (\partial c''/\partial x_j)^2\rangle}{\langle\rho\rangle} \tag{182}$$

The main difficulty with this approach is the representation of $\tilde{\epsilon}_c$. According to classical turbulence theory,

$$\tilde{\epsilon}_c = \widetilde{c''^2}\,\frac{\tilde{\epsilon}}{\tilde{k}} \tag{183}$$

where the tildes denote Favre-averaged quantities, that is,

$$c = \tilde{c} + c'' \qquad \langle\rho c\rangle = \langle\rho\rangle\,\tilde{c} + \langle\rho c''\rangle \qquad \langle c''\rangle \neq 0 \qquad \langle\rho c''\rangle = 0 \tag{184}$$

Equation (183) ignores the fact that in laminar flamelet combustion, the composition gradient [$\partial c''/\partial x_j$ in Eq. (182)] occurs only in coherent structures (laminar flames) rather than quasirandomly, as in classical turbulence. Furthermore, Eqs. (180) and (183) suggest that $\langle\dot{\omega}\rangle$ will increase without limit if the turbulence time scale $\tilde{k}/\tilde{\epsilon}$ is reduced, that is, no allowance is made for flamelet quenching due to finite reaction rates [206]. Equation (183) corresponds to an eddy-breakup model [150].

Borghi and Dutoya [174] have developed a transport equation for $\tilde{\epsilon}_c$. However, their approach also ignores the fact that in laminar flamelet combustion, the composition gradient occurs only in laminar flames.

The BLM model uses a two-point two-time covariance correlation that accounts for single, thin, wrinkled flames or packets of reactants and products separated from each other by laminar flamelets [207, 208]. This model shows that the scalar field is anisotropic and can be extended to predict nonisenthalpic flows [209].

The laser Doppler velocimetry (LDV) and Mie scattering measurements of Shepherd et al. [210] indicate that data conditioned by the local mixture fraction permit identification of the burnt and unburnt gases. Their measurements also show that turbulence is generated by the flame, the pressure gradient is very important, and the pdf of the scalars is bimodal, indicating the presence of thin flames. Shepherd and Moss [211] showed substantial differences between most probable and average measurements and between open and confined flames.

Stretching of laminar flames can be characterized by the Karlovitz number

$$\kappa = \frac{\delta_L}{S_L}\frac{1}{A}\frac{DA}{Dt} \tag{185}$$

where A represents the flame surface area and D/Dt indicates material derivative.

In turbulent flows

$$\frac{1}{A}\frac{DA}{Dt} \approx \left(\frac{\epsilon}{\nu}\right)^{1/2} \tag{186}$$

In a turbulent flame, combustion can be represented by a finite number of laminar flames having different Karlovitz numbers; these laminar flames are assumed to occur with different probabilities and provide the temperature, composition, density, and reaction rate profiles as a function of the physical distance through the flamelet. The mean reaction rate can then be calculated from Eq. (179) as

$$\langle\dot{\omega}(x)\rangle = \gamma(x)\int_0^1 \dot{\omega}(c)\, f(c; x)\, dc \tag{187}$$

In the flamelet combustion regime, burning occurs only in the laminar flamelets, and the shape of the pdf is determined by the flamelet profiles. For a flamelet with Karlovitz number κ_s,

$$f_s\,(c) = \left[\frac{dc}{d(z/\delta_L)}\right]_s^{-1} \tag{188}$$

where z is the physical distance through the flamelet and the probabilities of laminar flamelet occurrence $p_s(x)$ of the flamelets with Karlovitz numbers κ_s are such that

$$\sum_{s=0}^{n} p_s(x) = 1 \tag{189}$$

where n is the number of strained laminar flames.

Realistic chemical reaction mechanisms can be used to determine the mean reaction rate from Eq. (187) since the reaction rates $\dot{\omega}(c)$ can be determined from laminar flame calculations [212].

Substitution of Eqs. (188) and (189) into Eq. (187) yields

$$\langle\dot{\omega}(x)\rangle = \gamma(x)\sum_{s=0}^{n} p_s(x)\int_0^1 \dot{\omega}_s(c)\, f_s(c)\, dc \tag{190}$$

and the mean fraction of species i is

$$\langle Y_i(x)\rangle = \alpha(x)\, Y_i(0) + \beta(x)\, Y_i(1) + \gamma(x)\sum_{s=0}^{n} p_s(x)\int_0^1 Y_{is}(c)\, f_s(c)\, dc \tag{191}$$

where $Y_i(0)$ and $Y_i(1)$ are the mass fraction of the ith species in the burnt and unburnt mixtures, respectively.

If γ is independent of flame quenching, Eqs. (180)–(183) and (189) yield

$$\gamma(x) = \langle\rho\rangle\, \tilde{\epsilon}_c/[(2c_m - 1)\int_0^1 \dot{\omega}_0(c)\, f_0(c)\, dc]^{-1}$$

$$p_0(x) = 1 \qquad p_s(x) = 0 \qquad \text{for } s \geq 1 \tag{192}$$

which can then be substituted into Eq. (190) to obtain the mean reaction rate, that is,

$$\langle\dot{\omega}(x)\rangle = \langle\rho\rangle\, \tilde{\epsilon}_c \sum_{s=0}^{n} p_s(x) \int_0^1 \dot{\omega}_s(c)\, f_s(c) \frac{dc}{(2c_m - 1)\int_0^1 \dot{\omega}_0(c)\, f_0(c)\, dc} \tag{193}$$

Equations (180)–(183) describe a fast chemistry limit in which burning is controlled by the rate at which turbulent motion brings unburnt gas into contact with the laminar flamelets. In Eq. (193) this eddy-breakup rate is modified by a term that describes the reduction in the reaction rate due to flamelet stretch.

The calculation of the mean reaction rate from Eq. (193) makes use of Eq. (182) to evaluate the scalar dissipation function and requires the specification of the probability density functions $p_s(x)$. Burning models that do not use Eq. (182) can also be developed [202]. However, these models require specification of the probabilities of laminar flamelet occurrence.

Laminar flamelet models raise serious doubts as to the applicability to turbulent flames of the notions developed for constant density and for nonreactive turbulence [213]; these models can be extended to include finite reaction rate (finite Damköhler number) and molecular (finite Reynolds number) transport effects by means of perturbation analyses [214].

The BLM model may be applicable to the reaction-sheet regime observed in reciprocating internal combustion engines and may avoid countergradient diffusion. However, the model has not been applied as yet to predict two-dimensional turbulent chemically reacting flows. It has been presented here in order to illustrate more physically founded approaches to the complex processes that occur in turbulent combustion.

The combustion models presented in this section are applied in the next sections to analyze the flow fields in one-, two-, and three-dimensional rotary engine models. They are also applied in Chapters 3 and 4 to predict the flow field and combustion in diesel and spark-ignition engines, respectively.

2.4.4 Two-Phase Flow Modeling

The equations presented in previous sections are valid for single-phase flows. In order to analyze the flow field in DISC rotary engines and account for the presence of vaporizing fuel droplets, it is necessary to develop mean equations for the gas and liquid phases. In this section we present a brief introduction to the modeling of two-phase flows. A more detailed presentation can be found in Chapter 3.

Ishii [215] derived mass, momentum, and energy equations for each phase by averaging the compressible flow equations over some global domain. In this approach, the mass, momentum, and energy exchanges between the gas and liquid phases have to be postulated on the basis of plausibility arguments. Information about the gas-liquid interface can be directly incorporated into the spray equation [216], which can be written as

$$\frac{\partial f}{\partial t} + \nabla_{\mathbf{x}} \cdot (f\mathbf{v}_d) + \nabla_{\mathbf{v}_d} \cdot (f\mathbf{F}_d) + \frac{\partial}{\partial r_d}(fR_d) + \frac{\partial}{\partial T_d}(f\dot{T}_d) = \dot{Q} \tag{194}$$

where $f\, dr_d\, d\mathbf{x}\, d\mathbf{v}_d\, dT_d$ is the probable number of droplets in the radius range dr_d around r_d located in the spatial range $d\mathbf{x}$ around $\mathbf{x}$ with velocities in the range $d\mathbf{v}_d$ around $\mathbf{v}_d$, and with temperatures in the range dT_d around T_d at time t; $R_d = dr_d/dt$; $\mathbf{F}_d = d\mathbf{v}_d/dt$; $\dot{T}_d = dT_d/dt$; and $\dot{Q}$ represents the rate of increase of f with time due to droplet collisions, coalescence, and breakup. Equations for determining the exchange rates can be written as [216, 217]

$$-\rho_l R_d = \frac{\lambda}{C_p} \frac{\text{Nu}}{2r_d} \frac{Y_{f,s} - Y_f}{1 - Y_{f,s}} = -\rho_l \frac{dr_d}{dt} \tag{195}$$

$$\frac{dT_d}{dt} = \frac{3}{2} \frac{\lambda_l}{\rho_l c_l r_d^2} \text{Nu}_l (T_{d,s} - T_d) + \frac{3}{r_d} R_d (T_{d,s} - T_d) \tag{196}$$

$$\dot{\mathbf{v}}_d = \dot{\mathbf{v}}_{\text{drag}} - \frac{1}{\rho_l} \nabla p - \mathbf{g} \tag{197}$$

$$\dot{\mathbf{v}}_{\text{drag}} = \frac{3}{8} \frac{\rho}{\rho_l r_d} c_d (\mathbf{u} + \mathbf{u}' - \mathbf{v}_d)|\mathbf{u} + \mathbf{u}' - \mathbf{v}_d| \tag{198}$$

$$c_d = 24 \frac{\psi^{-2.65} + \text{Re}_d^{2/3} \psi^{-1.78}/6}{\text{Re}_d} \tag{199}$$

$$\text{Re}_d = 2\rho|\mathbf{u} + \mathbf{u}' - \mathbf{v}_d| \frac{r_d}{\mu} \tag{200}$$

$$\frac{p_{v,s}}{p} = \frac{Y_{f,s}/M_f}{Y_{f,s}/M_f + Y_I/M_I} \qquad p_{v,s} = p_{\text{ref}} \exp\left[\frac{L}{R}\left(\frac{1}{T_{\text{ref}}} - \frac{1}{T_{d,s}}\right)\right] \tag{201}$$

$$M_I^{-1} = \frac{\sum_{i \neq f} Y_i/M_i}{\sum_{i \neq f} Y_i} \qquad Y_I = 1 - Y_{f,s} \tag{202}$$

$$\beta = \frac{Y_{f,s} - Y_f}{1 - Y_{f,s}} = \frac{C_p}{L(T_{d,s})}\left[T - T_{d,s} - \frac{\text{Nu}_l \lambda_l}{\text{Nu}\, \lambda}(T_{d,s} - T_d)\right] \tag{203}$$

$$\text{Nu}_l = 1 \qquad \text{Nu} = \left(2\psi^{-1.75} + 0.6 \frac{\text{Re}_d}{\psi} \text{Pr}^{1/3}\right) \frac{\ln(1 + \beta)}{\beta} \tag{204}$$

$$\psi = 1 - \iiint \frac{4}{3} \pi\, r_d^3 f\, dr_d\, d\mathbf{v}_d\, dT_d \tag{205}$$

where the subscript l denotes the liquid phase; quantities without subscript refer to the gas phase; Nu and Re are the Nusselt and Reynolds numbers, respectively; $T_{d,s}$ and $Y_{f,s}$ are the temperature of the droplet and fuel vapor mass fraction at the droplet surface, respectively; Y_f is the mass fraction of the fuel vapor; $\mathbf{g}$ is the gravity acceleration; c_d is the drag coefficient; $\mathbf{u}$ and $\mathbf{u}'$ are the mean and fluctuating velocities of the gas phase; ψ is the void fraction; M_f is the molecular mass of the fuel; β is the transfer number [218]; and L is the latent heat of vaporization of the fuel.

The effects of turbulence on the droplets can be calculated by adding a fluctuating component $\mathbf{u}'$ to the mean gas velocity $\mathbf{u}$ when computing the aerodynamic drag force and the Nusselt and Reynolds numbers. The fluctuating velocity can be chosen randomly from an isotropic Gaussian distribution of variance $(2k/3)^{1/2}$. For each droplet a new value of $\mathbf{v}'$ is chosen after a turbulent correlation time τ as follows:

$$l = \int_t^{t+\tau} |\mathbf{v}_d - \mathbf{u}|\, dt \qquad \text{or } \tau = \min\left(l\mathbf{u}'^{-1},\ l\,|\mathbf{v}_d - \mathbf{u}|^{-1}\right) \tag{206}$$

where l is proportional to $k^{3/2}/\epsilon$ ($l = C_\mu^{3/4}\, k^{3/2}/\epsilon, C_\mu = 0.09$), and $|\mathbf{u}'| = (2k/3)^{1/2}$.

Droplet collisions, coalescence, and breakup can be introduced in Eq. (194) [216, 217]. These terms as well as other spray models are discussed in Chapter 3 and are not repeated here.

The time-averaged gas-phase conservation equations can be written as

$$(\rho\psi)_{,t} + (\rho u_i\, \psi)_{,i} = -\iiint 4\pi r_d^2 R_d\, \rho_l\, f\, dr_d\, d\mathbf{v}_d\, dT_d \tag{207}$$

$$(\rho\psi Y_f)_{,t} + (\rho\psi u_i Y_f)_{,i} = (\rho D_e \psi Y_{f,i})_{,i} + \psi\langle\dot{\omega}_f\rangle - \iiint 4\pi r_d^2 R_d\, \rho_l\, f\, dr_d\, d\mathbf{v}_d\, dT_d \tag{208}$$

$$(\rho\psi u_i)_{,t} + (\rho\psi u_i u_j)_{,j} = -p_{,i} + \tau_{ij,j} - \iiint \left(4\pi r_d^2 R_d \mathbf{v}_d + \frac{4}{3}\pi r_d^3 \mathbf{F}_d\right) \rho_l f\, dr_d\, d\mathbf{v}_d\, dT_d \tag{209}$$

$$(\rho\psi h)_{,t} + (\rho\psi u_i h)_{,i} = \psi(p_{,t} + u_i p_{,i}) + (\lambda_e \psi T_{,i})_{,i} - \iiint \rho_l f \left\{4\pi r_d^2 R_d \left[h_l\,(T_d) + \frac{1}{2}(\mathbf{v}_d - \mathbf{u})^2\right] + \frac{4}{3}\pi r_d^3 [c_l \dot{T}_d + \dot{\mathbf{v}}_{\text{drag}} \cdot (\mathbf{v}_d - \mathbf{u} - \mathbf{u}')]\right\} dr_d\, d\mathbf{v}_d\, dT_d \tag{210}$$

$$\tau_{ij} = \psi \left[\mu_e(u_{i,j} + u_{j,i}) - \frac{2}{3}\,\mu_e \delta_{ij} u_{k,k} - \frac{2}{3}\,\delta_{ij}\, \rho k\right] \tag{211}$$

$$\lambda_e = \mu \frac{C_p}{\Pr} + \mu_t \frac{C_p}{\Pr_t} \tag{212}$$

$$\rho D_e = \frac{\mu}{\mathrm{Sc}} + \frac{\mu_t}{\mathrm{Sc}_t} \tag{213}$$

$$(\rho\psi k)_{,t} + (\rho u_i \psi k)_{,i} = \left[\psi\left(\frac{\mu_t}{\sigma_k} + \mu\right)k_{,i}\right]_{,i} + \tau_{ij}\, u_{i,j} - \rho\epsilon\psi \tag{214}$$

$$(\rho\psi\epsilon)_{,k} + (\rho u_i \psi\epsilon)_{,i} = \left[\psi\left(\frac{\mu_t}{\sigma_\epsilon} + \mu\right)\epsilon_{,i}\right]_{,i}$$

$$+ \frac{\epsilon}{k}\left(C_1\, \tau_{ij}\, u_{i,j} - C_2\, \psi\rho\epsilon\right) - \rho\epsilon u_{k,k}\psi \tag{215}$$

where $C_1 = 1.44$, $C_2 = 1.92$, $\sigma_k = 1.0$, and $\sigma_\epsilon = 1.3$.

The k–ϵ equations [(214) and (215)] do not account for the effects of the droplets on the turbulence kinetic energy and its dissipation rate. These effects are discussed in Chapter 3.

Equation (195) corresponds to a d^2 law modified by forced-convection effects through Eq. (204). However, Eq. (195) accounts for volumetric displacement effects through its dependence on the void fraction. These effects as well as approximations of thin and thick sprays are discussed in Chapter 3.

Volumetric displacement effects (i.e., $\psi \neq 1$) are not considered in the KIVA code developed at Los Alamos National Laboratory [95, 165]. (The void fraction generally exceeds 90%, except near the injector nozzle, and can be set equal to 1 [219].)

2.5 ONE-DIMENSIONAL MODELS

One-dimensional models of rotary engine combustion were developed at Princeton University in the early 1970s [220–222], before the thermodynamic models presented in previous sections. One-dimensional models account for the temporal and spatial variations of the velocity and temperature fields along the combustion chamber and are based on the space-averaged time-dependent mean conservation equations of mass, momentum, energy, and species presented in Section 2.4.

These equations can be rewritten in a reference frame moving with the rotor. For convenience, we introduce the coordinate x, or arc length along the rotor with origin at the trailing apex, denote the cross-sectional area between the housing and rotor by $A(x,t)$, and analyze spark- and compression-ignition rotary engines separately in the next sections.

2.5.1 One-Dimensional Models of Spark-Ignition Engines

If a second-order mean-reaction-rate model and a constant eddy diffusivity are used, if the specific heats at constant pressure of all the species are assumed equal and constant, if the pressure is assumed spatially uniform and the three-dimensional time-averaged equations presented in Section 2.4 are averaged (integrated) with respect to the cross-sectional area between the housing and rotor, then the following system of equations for $\psi = 1$ results:

$$(\rho A)_{,t} + (\rho A u)_{,x} = 0 \tag{216}$$

$$(\rho A C_p T)_{,t} + (\rho A C_p u T)_{,x} = A p_{,t} + (A \lambda_e T_{,x})_{,x} + A(\dot{q}_R - \dot{q}_L) \tag{217}$$

$$(A \rho Y_i)_{,t} + (A \rho u Y_i)_{,x} = [A \rho D_e (Y_i)_{,x}]_{,x} + A \langle \dot{\omega}_i \rangle \qquad i = 1,2, \ldots, N \tag{218}$$

$$p = \rho \tilde{R} T \sum_{i=1}^{N} \left(\frac{Y_i}{M_i} \right) \tag{219}$$

where

$$\lambda_e = \mu \frac{C_p}{\mathrm{Pr}} + \mu_t \frac{C_p}{\mathrm{Pr}_t} \tag{220}$$

$$\rho D_e = \frac{\mu}{\mathrm{Sc}} + \frac{\mu_t}{\mathrm{Sc}_t} \tag{221}$$

$$\langle \dot{\omega}_f \rangle = -K \rho^2 Y_f Y_0 \frac{\exp[-E/(\tilde{R}T)]}{M_o} \tag{222}$$

$$\dot{q}_R = -Q \langle \dot{\omega}_f \rangle \tag{223}$$

Pr and Pr_t are the laminar and turbulent Prandtl numbers, respectively; Sc and Sc_t are the laminar and turbulent Schmidt numbers, respectively; N is the number of species; M_o is the molecular mass of the oxidizer; $\dot{q}_L$ denotes the heat losses through the rotor, housing, and side walls; and $\dot{q}_R$ is the heat of combustion. The heat transfer losses can be calculated, assuming one-dimensional heat conduction through the solid surfaces, as a function of the film heat transfer coefficient of the gas mixture, thermal conductivity of the solid walls, and film coefficient of the coolant and the coolant temperature [220]. However, if the wall temperature on the combustion chamber side T_w is known and equal for all the solid surfaces, and a turbulent heat transfer correlation is used, for example, the Colburn equation [223],

$$\mathrm{Nu} = \frac{hP}{\lambda} = 0.023\ \mathrm{Re}^{0.8}\ \mathrm{Pr}^{1/3} \tag{224}$$

where

$$\lambda = \mu \frac{C_p}{\mathrm{Pr}} \qquad \mathrm{Re} = \rho P \frac{U + |u|}{\mu} \tag{225}$$

P is the perimeter of the combustion chamber and U the relative velocity between the housing and rotor. The heat transfer losses can be calculated as

$$\dot{q}_L = h(T - T_w) \tag{226}$$

If there is no leakage through the crevices, rotor leading and trailing apexes, and housing seals, the mass m of the mixture contained in the combustion chamber is constant, and the mixed hyperbolic-parabolic problem represented by Eqs. (216)–(218) can be transformed into a system of integrodifferential reaction-diffusion equations by means of the mappings

$$(t, x) \rightarrow (\tau, \eta) \qquad \phi = Tp^{(1-\gamma)/\gamma} \tag{227}$$

where

$$m\eta_{,x} = A\rho \tag{228a}$$

$$m\eta_{,t} = -A\rho u \tag{228b}$$

$$\tau = t \tag{228c}$$

and γ is the ratio of specific heats at constant pressure (i.e., $\gamma = C_p/C_v$).

Note that the physical domain $0 \leq x \leq L$ has been mapped by Eq. (228) into the domain $0 \leq \eta \leq 1$.

Substitution of Eqs. (227) and (228) into Eqs. (216)–(219) yields

$$\frac{\partial Y_i}{\partial \tau} = \frac{1}{m^2}\frac{\partial}{\partial \eta}\left(\rho^2 D_e A^2 \frac{\partial Y_i}{\partial \eta}\right) + \frac{\langle \dot{\omega}_i \rangle}{\rho} \qquad i = 1,2, \ldots, N \tag{229}$$

$$\frac{\partial \phi}{\partial \tau} = \frac{1}{m^2}\frac{\partial}{\partial \eta}\left(\rho^2 \lambda_e A^2 \frac{\partial \phi}{\partial \eta}\right) + p^{(1-\gamma)/\gamma}\frac{1}{\rho C_p}(\dot{q}_R - \dot{q}_L) \tag{230}$$

$$\rho = \frac{p^{1/\gamma}}{\phi \tilde{R} \sum_{i=1}^{N} (Y_i/M_i)} \tag{231}$$

The pressure can be calculated by integrating Eq. (231) as

$$p^{1/\gamma}\int_0^L A(x,t)\, dx = \int_0^L \rho\phi\tilde{R}\sum_{i=1}^{N}\left(\frac{Y_i}{M_i}\right) A\, dx = \int_0^1 m\tilde{R}\phi \sum_{i=1}^{N}\left(\frac{Y_i}{M_i}\right) d\eta \tag{232}$$

where L is the rotor arc length. The arc length along the rotor can be calculated from Eq. (228a) as follows:

$$x = \int_0^{\eta} \frac{m}{A}\, d\eta \tag{233}$$

Equation (228) can be substituted into Eq. (216) to yield

$$\frac{\partial u}{\partial \eta} = -\frac{m}{\rho^2 A^2}\frac{\partial}{\partial \tau}(\rho A) \qquad u = u(\eta = 0,\tau) + m\int_0^{\eta}\frac{\partial}{\partial \tau}\left(\frac{1}{\rho A}\right) d\eta \tag{234}$$

where $u(0,\tau) - 0$ denotes the velocity of the trailing apex relative to the rotor.

Bracco and Sirignano [220] assumed that $\mathrm{Pr} = \mathrm{Pr}_t = \mathrm{Sc} = \mathrm{Sc}_t = 1$ and $\mu_t >> \mu$. The eddy viscosity was considered to be a function of the local velocity and of the intake-generated turbulence as

$$\frac{\mu_t}{\rho} = \alpha_1 s\left[\frac{U}{2} + |u| + u_{\mathrm{int}} \exp\left(-\alpha_2 u_{\mathrm{int}} \frac{t - t_{\mathrm{int}}}{s_{\mathrm{int}}}\right)\right] \tag{235}$$

where α_1 and α_2 are constants, U is the tangential velocity of the rotor, s is the local chamber height, u_{int} is the intake velocity relative to the housing, s_{int} is a characteristic dimension of the intake port, and t_{int} denotes the time at which the intake port closes [224].

Equations (229)–(234) can be solved by means of a variety of finite difference and finite element methods to evaluate ρ, p, u, T, and Y_i [225, 226].

Bracco and Sirignano [220] showed that if the heat transfer losses are halved, the temperature between the two flame fronts is higher and more uniform, and faster flames result. Faster flames were also obtained by increasing the compression ratio and the heat of combustion. A reduction in the spark advance was found to result in very little effect on the flame speeds. However, the speed of the backward-moving flame was substantially reduced by reducing the magnitude of the turbulent diffusivity by 25%.

2.5.2 One-Dimensional Models of Direct-Injection Stratified-Charge (DISC) Engines

One-dimensional combustion models of rotary engines can be obtained by integrating Eq. (194) and the mean conservation equations of the gas phase between the housing and rotor (i.e., by taking cross-sectional averages from the rotor to the housing). If the coordinate x is measured along the rotor, if $A(x,t)$ denotes the cross-sectional area between the housing and rotor, if the species-specific heats at constant pressure are assumed to be identical and constant, if the temperatures of the droplets are assumed to be equal and constant, if droplet collisions, coalescence, and breakup are neglected, and if Eq. (194) is integrated with respect to r_d, $\mathbf{v}_d$, and T_d, then the following system of equations results:

$$\frac{\partial r_d}{\partial t} + v_d \frac{\partial r_d}{\partial x} = R_d \tag{236}$$

$$\frac{\partial v_d}{\partial t} + v_d \frac{\partial v_d}{\partial x} = F_d \tag{237}$$

$$\frac{\partial}{\partial t}(An_d) + \frac{\partial}{\partial x}(An_d v_d) = 0 \tag{238}$$

$$(A\rho Y_i)_{,t} + (A\rho u Y_i)_{,x} = [A\rho D_i (Y_i)_{,x}]_{,x} + A\langle\dot{\omega}_i\rangle + A\,\dot{\omega}_{\text{vap}}\,\delta_{if} \tag{239}$$

$$p = \rho \tilde{R} T \sum_{i=1}^{N} \left(\frac{Y_i}{M_i}\right) \tag{240}$$

$$\begin{aligned}(\rho A H)_{,t} + (\rho A u H)_{,x} &= (A\lambda_e T_{,x})_{,x} + \left(A\rho \sum_{i=1}^{N} D_i h_i (Y_i)_{,x}\right)_{,x} - A\dot{q}_L + Ap_{,t} \\ &- \frac{\partial}{\partial t}\iint \frac{4}{3}\pi r_d^3\, \rho_l\, Afh\, dr_d\, dv_d - \frac{\partial}{\partial x}\iint \frac{4}{3}\pi r_d^3\, \rho_l\, Afv_d h\, dr_d\, dv_d\end{aligned} \tag{241}$$

$$(\rho A u)_{,t} + (A\rho u u)_{,x} = -A p_{,x} - \frac{\partial}{\partial t}\iint \frac{4}{3}\pi r_d^3 \rho_l A f v_d \, dr_d \, dv_d - \frac{\partial}{\partial x}\iint \frac{4}{3}\pi r_d^3 \rho_l A f v_d^2 \, dr_d \, dv_d \tag{242}$$

$$H = \sum_{j=1}^{N} Y_j \left(\int_{T_0}^{T} C_{p_j} \, dT + h_j^0 \right) + \frac{u^2}{2} \tag{243}$$

$$h = L + h_f^0 + \frac{v_d^2}{2} \tag{244}$$

where n_d is the droplet number density (i.e., the number of droplets per unit volume); δ_{if} is the Kronecker delta, which is equal to zero except if $i = f$ (i.e., the fuel); h is the stagnation enthalpy of the liquid fuel; $\dot{q}_L$ represents the heat transfer losses; and $\dot{\omega}_{\text{vap}}$ is the fuel vaporization rate, which is given by the third term on the right-hand side of Eq. (208).

If the pressure is assumed to be uniform, Eq. (242) must be dropped. The pressure can then be calculated by integrating Eq. (240) throughout the combustion chamber.

Bracco [221] and Reitz and Bracco [222] used a constant eddy diffusivity; neglected droplet collisions, coalescence, and breakup; assumed constant pressure throughout the combustion chamber; and employed a second-order chemical reaction and the following correlations:

$$\dot{v}_d = \frac{3}{8}\frac{c_D}{r_d}\frac{\rho}{\rho_l}|u - v_d|(u - v_d) \qquad c_D = \frac{24}{\text{Re}_d} \tag{245}$$

$$\dot{r}_d = -\frac{\lambda^*}{8\rho r_d}(1 + 0.3\,\text{Pr}^{1/3}\,\text{Re}_d^{1/2}) \qquad \text{Re}_d = 2\rho r_d \frac{|u - v_d|}{\mu} \tag{246}$$

$$\langle \dot{\omega}_f \rangle = -\frac{K}{M_o}\rho^2 Y_f Y_o \exp\left(-\frac{E}{\tilde{R}T}\right) \tag{247}$$

$$\frac{\mu_t}{\rho} = \alpha_1 s \left[\frac{1}{2}(|U + u| + |u|) + u_{\text{int}} \exp\left(-\alpha_2 u_{\text{int}} \frac{t - t_{\text{int}}}{s_{\text{int}}}\right)\right] \tag{248}$$

$$r_d(t = 0) = r_{\text{inj}}\,\alpha_3\,(\text{Re}_{\text{inj}}\,\text{We}_{\text{inj}})^{-1/4} \tag{249}$$

$$\text{Re}_{\text{inj}} = 2 r_{\text{inj}} \rho_l \frac{v_e}{\mu_l} \qquad \text{We}_{\text{int}} = 2 r_{\text{inj}} \rho_l \frac{v_e^2}{\sigma_l} \tag{250}$$

where λ^* is a modified vaporization rate constant; α_1, α_2, and α_3 are constants; r_{inj} is the injector nozzle radius; Re and We are the Reynolds and Weber numbers, respectively; σ_l is the surface tension of the liquid fuel; and v_e is the liquid jet velocity at the injector location. Reitz and Bracco [222] assumed that the velocity of the droplets at the point of maximum penetration is given by the spray penetration rate and the angle between the injector and housing. They also assumed that the number density of the droplets n_d corresponds to a Gaussian distribution such that

$$\dot{m}_l \, \Delta t = \frac{4}{3} \pi \, r_d^3(0) \, \rho_l \int_{x_b}^{x_i} n_d \, dx \qquad v_e = \left(\frac{2 \, \Delta p}{\rho_l \, C_D} \right)^{1/2} \tag{251}$$

where $\dot{m}_l$ is the fuel injection rate, Δt is the time step employed in the calculations, $x_i - x_b$ is the injector width, x_i and x_b denote the injector location, Δp is the pressure drop along the nozzle, and C_D is a discharge coefficient.

Equations (236)–(238) are hyperbolic; their numerical solution requires artificial diffusion to obtain nonoscillatory solutions and may result in multiple solutions. Furthermore, Eq. (245) corresponds to Stokes' law, which is valid only for rigid spheres. The droplet surface motion decreases the shear stresses at the droplet surface and results in drag coefficients smaller than that given by Eq. (245). These limitations are discussed in Chapter 3.

Equation (246) corresponds to a modified d^2 law, which accounts for forced convection effects on the droplet vaporization rate.

Reitz and Bracco [222] used a one-dimensional model to select the injection parameters in a DISC Wankel engine so that an optimal spatial distribution of the fuel vapor is achieved in the combustion chamber at the time of ignition. This optimal distribution was maintained over the entire load-speed operating range. Reitz and Bracco [222] showed that a single shower-head nozzle does not allow sufficient design freedom for achieving optimal charge stratification over a broad speed-load range. A variable-orifice-size spray nozzle that produces a spray that broadens with increasing flow rate is most attractive for use in stratified-charge rotary engines. Reitz and Bracco [222] also showed that achievement of satisfactory open-chamber direct-injection stratification relies on accurate control of the liquid fuel jet breakup and on penetration and vaporization of the spray. The importance of atomization, spray penetration, and droplet sizes and velocities in DISC rotary engines was demonstrated in the calculations performed by Reitz and Bracco [222].

The Eulerian-Eulerian formulation represented by Eqs. (236)–(242) cannot resolve phenomena that occur on scales on the order of the droplet spacing because the droplet number density, radius, and velocity are functions of space and time, and they represent spatial averages over volumes that contain a large number of droplets. Furthermore, Eqs. (236)–(238) are hyperbolic and nonlinear and may yield multivalued solutions if the characteristics cross each other. The numerical solution of Eqs. (236)–(242) produces artificial diffusion, which smears the results. The Eulerian-Eulerian formulation represented by Eqs. (236)–(242) cannot account for the discrete character of the droplets observed in sprays. Eulerian-Lagrangian formulations eliminate the numerical diffusion introduced by Eulerian treatments of the liquid-phase equations. These formulations are discussed in Chapter 3.

The thermodynamic (single- and multizone) and one-dimensional models presented in previous sections cannot account for the multidimensionality of the flow field, fuel-air mixing, spray penetration, vaporization, and combustion. Although there is numerical evidence that the flow field in the compression and expansion strokes is almost unidirectional in rotary engines, two- and three-dimensional calculations in rotary engine models have shown that complex multi-dimensional fields are present in the combustion chamber during the intake stroke and injection interval. In Sections 2.6 and 2.7, two- and three-dimensional rotary engine models are presented.

2.6 TWO–DIMENSIONAL MODELS

Two-dimensional models have been developed by Shih and co-workers [68, 71, 128, 129, 227] to analyze the flow field and homogeneous-charge combustion in rotary engines. These models used the engine geometry, equations, and boundary conditions presented in Tables 2.2–2.4; a time-linearized approximate-factorization numerical technique [228, 229]; and a two-equation model of turbulence that accounts for streamline curvature, compressibility, preferential dissipation, and low-Reynolds-number effects [68, 128, 129, 227] or a constant eddy diffusivity proportional to the crank shaft rpm [71].

The equations presented in Table 2.3 can be written in vector form as

$$\mathbf{U}_t + \mathbf{F}_x + \mathbf{G}_y = \mathbf{V}_x + \mathbf{W}_y + \mathbf{S} \tag{252}$$

where **U**, **F**, **G**, **V**, **W**, and **S** are vectors and the subscripts t, x, and y denote differentiation with respect to time and with respect to the x and y coordinates, respectively.

The combustion chambers of a rotary engine vary with time and have complex geometries. A grid in (t,x,y) coordinates would have to move with the combustion chamber in such a manner that the grid points are concentrated near the walls in order to accurately resolve the boundary layers and steep flow gradients that occur near the rotor and housing. Furthermore, in order to accurately implement the Neumann boundary conditions for the species at the solid walls, the grid lines should intersect the rotor and housing perpendicularly.

In order to account for the moving combustion chambers and nonuniformly distributed grid points, the grid in (t,x,y) coordinates can be mapped onto a stationary and uniformly distributed mesh in (τ,ξ,η) coordinates by means of an algebraic grid-generation technique with transfinite interpolation in such a manner that the rotor and housing correspond to $\eta = 0$ and $\eta = 1$, respectively, and the rotor leading and trailing apexes correspond to $\xi = 0$ and $\xi = 1$, respectively. This mapping can be defined as

$$(t,x,y) \rightleftarrows (\tau,\xi,\eta) \tag{253}$$

$$t = \tau \qquad x = x(\tau,\xi,\eta) \qquad y = y(\tau,\xi,\eta) \tag{254}$$

$$x_R = x(\tau,\xi,\eta = 0) \qquad y_R = y(\tau,\xi,\eta = 0) \tag{255}$$

$$x_H = x(\tau,\xi,\eta = 1) \qquad y_H = y(\tau,\xi,\eta = 1) \tag{256}$$

where the coordinates of the rotor (x_R, y_R) and the housing (x_H, y_H) are given in Table 2.2 and the values of x and y defined in Eq. (254) can be written as

[129, 230, 231]

$$x = x_H h_1 + x_R h_2 - k_2 \frac{\partial y_H}{\partial \xi} h_3 - k_1 \frac{\partial y_R}{\partial \xi} h_4 \tag{257}$$

$$y = y_H h_1 + y_R h_2 + k_2 \frac{\partial x_H}{\partial \xi} h_3 + k_1 \frac{\partial x_R}{\partial \xi} h_4 \tag{258}$$

where the values of h_1, h_2, h_3, h_4, k_1, and k_2 are presented in references 128, 129, and 230.

Note that leakage through the rotor apexes and flows into crevices have been neglected and that there is a finite distance between the rotor and housing at the rotor leading and trailing apexes. This finite distance is required in the numerical computations in order to map the combustion chamber geometry onto a square.

Substitution of Eq. (254) into Eq. (252) yields the following vector equation:

$$\mathbf{U}_\tau = -(\xi_t \mathbf{U}_\xi + \xi_x \mathbf{P} + \xi_y \mathbf{Q} + \eta_t \mathbf{U}_\eta + \eta_x \mathbf{P}^* + \eta_y \mathbf{Q}^*) + \mathbf{S} \tag{259}$$

where $\mathbf{P}$, $\mathbf{Q}$, $\mathbf{P}^*$, and $\mathbf{Q}^*$ can be obtained from Eq. (252) and Table 2.2 and where ξ_t, ξ_x, ξ_y, η_t, η_x, and η_y are related to the Jacobian of the mapping [Eq. (254)].

Equation (259) can be linearized and discretized with respect to time, using a Crank-Nicolson approximation for the time derivatives, and written in approximate factorization form as

$$L^*_{i,j}\, \Delta\mathbf{U}^*_{i,j-1} + M^*_{i,j}\, \Delta\mathbf{U}^*_{i,j} + N^*_{i,j}\, \Delta\mathbf{U}^*_{i,j+1} = (\mathbf{RHS})^n_{i,j} \tag{260}$$

$$L_{i,j}\, \Delta\mathbf{U}^{n+1}_{i-1,j} + M_{i,j}\, \Delta\mathbf{U}^{n+1}_{i,j} + N_{i,j}\, \Delta\mathbf{U}^{n+1}_{i+1,j} = \Delta\mathbf{U}^*_{i,j} \tag{261}$$

where

$$\Delta\mathbf{U}^{n+1}_{i,j} = \mathbf{U}^{n+1}_{i,j} - \mathbf{U}^n_{i,j} \tag{262}$$

L, L^*, M, M^*, N, and N^* are block tridiagonal matrices, and **RHS** denotes the terms evaluated explicitly in Eq. (259).

Note that the approximate factorization method represented by Eqs. (260) and (261) reduces the solution of the two-dimensional equations [Eq. (252)] to a sequence of two one-dimensional linear equations, which can be easily solved by the method of Thomas [71, 129].

Equations (260) and (261) are based on central difference approximations for the convection terms and include second- and fourth-order dissipation terms to avoid nonlinear numerical instabilities. Further details on the numerical method, dissipation terms, boundary conditions, and time steps used in the calculations are given in references 71, 129, and 232. Some sample results illustrating the flow field in the combustion chamber are presented in Figs. 2.9–2.11. Figure 2.9 shows the mean velocity profiles in the intake, compression, expansion, and exhaust strokes [68]. Figure 2.10 shows the velocity field and fuel mass fraction distribution in the combustion chamber due to mixture stratification at the intake

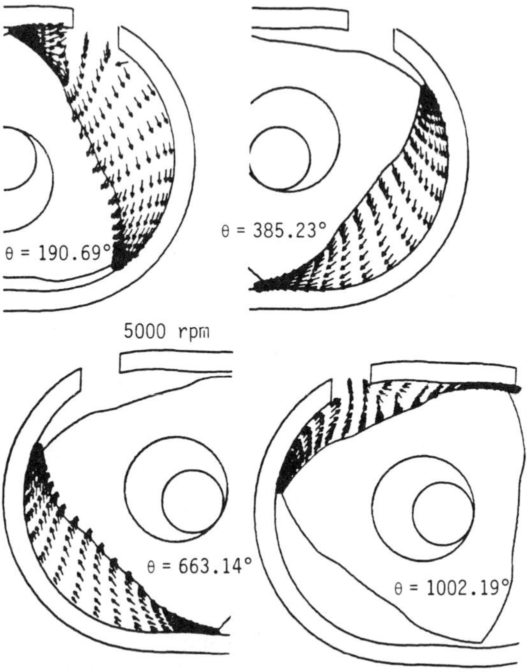

Figure 2.9 Two-dimensional velocity fields throughout the rotary engine cycle [68].

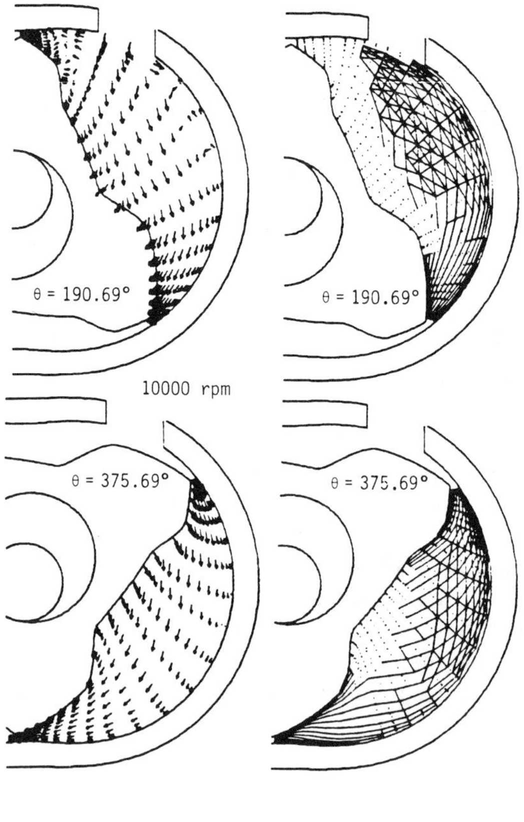

Figure 2.10 Velocity field and fuel mass fraction distribution caused by mixture stratification at the intake port in a two-dimensional rotary engine model [129].

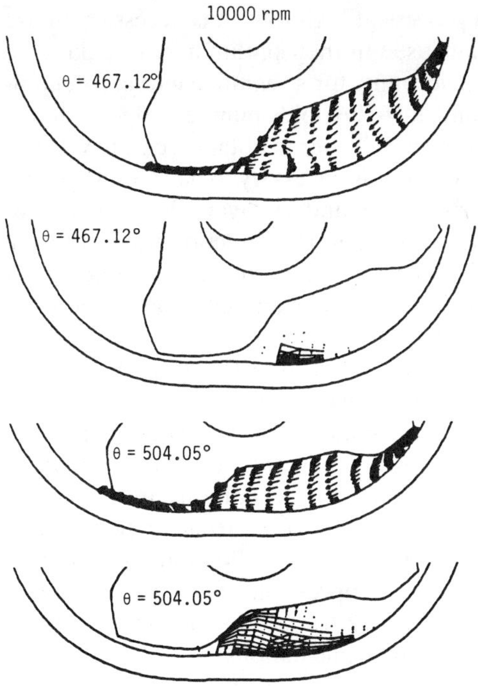

Figure 2.11 Velocity field and fuel mass fraction distribution during fuel injection in a two-dimensional rotary engine model [129].

port [129], and Fig. 2.11 shows the velocity field and the distribution of fuel mass fraction during fuel injection [129].

The two-dimensional calculations presented in references 68, 71, 128, 129, and 227 indicate that the engine speed and injection velocity have important effects on the flow field in rotary engine models; the mixture stratification at the intake port affects the fuel distribution in the intake and compression strokes; the fuel is concentrated near the housing and rotor trailing apex in the compression stroke; the fuel concentration near the rotor pocket is small; and, the fuel stratification increases as the frequency of the fluctuating fuel mass fraction at the intake port is decreased. The calculations also indicate that at 5000 and 10,000 rpm, the recirculation zones created during the intake stroke quickly disappear in the compression stroke. Recirculation zones are created during fuel injection; the size and velocity of these zones increase with the injection velocity. However, the largest fuel concentrations are located between the injector and the rotor trailing apex, and between the rotor and housing, and there is poor fuel-air mixing throughout the combustion chamber.

Shih et al. [129] extended the time-linearization technique presented in this section and attempted to calculate the flow field and flame propagation in a homogeneous-charge two-dimensional rotary engine model operating at 10,000 rpm, where the spark plug was simulated as a hot spot located in the housing. These calculations showed that the flame front is quenched by the high velocity field induced by the rotor and the large amounts of artificial dissipation required to avoid nonlinear numerical instabilities. Shih et al. [129] believe that in order to accurately simulate the combustion process at high rpm, it is necessary to use much finer grids than the 21×15 points used in their preliminary calculations.

The k–ϵ model shown in Table 2.3 accounts for streamline curvature, compressibility, preferential dissipation, and low-Reynolds-number effects, and it can be applied up to the solid walls. However, in order to obtain accurate velocity and temperature profiles near the solid walls, it is necessary to have a sufficiently large number of grid points in the turbulent boundary layer, the thickness of which is not known. If few grid points are used in the direction normal to the solid walls, between the housing and rotor, it is doubtful that accurate heat transfer calculations can be performed with the model and boundary conditions presented in Tables 2.3 and 2.4, respectively.

Most of the turbulence models that have been used to calculate the flow field in reciprocating and rotary engines do not account for low-Reynolds-number effects, preferential dissipation, and streamline curvature. Therefore, these models cannot be applied up to the solid walls, and the boundary conditions are applied near but not at the wall. These boundary conditions are analogous to those used in statistically stationary, incompressible, turbulent flows along flat plates in the absence of pressure gradients [233]. The flows in reciprocating and rotary engines are neither steady nor incompressible, but involve recirculation zones and pressure gradients. In Section 2.6.1 we describe the boundary conditions frequently used in multidimensional calculations.

2.6.1 Boundary Conditions

For a statistically stationary, incompressible, turbulent boundary layer along a smooth flat plate in the absence of axial pressure gradients, the velocity profile near the wall can be expressed as [72, 233]

$$u^+ = y^+ \qquad 0 \leq y^+ \leq 5 \tag{263}$$

$$u^+ = 2.5 \ln y^+ + 5.5 \qquad 30 \leq y^+ \leq 400 \tag{264}$$

where

$$u^+ = \frac{u}{u_f} \qquad y^+ = y\,\frac{u_f}{\nu} \tag{265}$$

ν is the dynamic viscosity, u_f is the friction velocity defined as $u_f = (\tau_w/\rho)^{1/2}$, τ_w is the shear stress at the wall, and u is the mean flow velocity parallel to the wall.

Equation (263) corresponds to the viscous sublayer, where the fluid kinematic viscosity is much larger than the eddy diffusivity, whereas Eq. (264) corresponds to the turbulent boundary layer, where the eddy diffusivity is much larger than the fluid kinematic viscosity. The regions $5 < y^+ < 30$ and $y^+ > 400$ are called the buffer layer and the wake region, respectively. In the buffer layer the eddy diffusivity is comparable to the value of the fluid kinematic viscosity.

In the buffer layer the velocity profile can be approximated by the following equation:

$$u^+ = 5 \ln y^+ - 3.05 \qquad 5 < y^+ < 30 \tag{266}$$

whereas in the wake region the velocity defect profile, or law of the wake, can be expressed as

$$\frac{u_\infty - u}{u_f} = -2.5 \ln \frac{y}{\delta} + f\left(\frac{u_f}{u_\infty}\right) \qquad 400 < y^+ \tag{267}$$

where u_∞ is the free stream velocity, δ is the turbulent boundary layer thickness, and f depends on u_f/u_∞.

In multidimensional calculations, the buffer layer and wake region are frequently lumped together into the following expressions:

$$u^+ = y^+ \qquad 0 \le y^+ \le 11.63 \tag{268}$$

$$u^+ = 2.5 \ln y^+ + 5.5 \qquad 11.63 < y^+ \tag{269}$$

Gosman et al. [74] showed that Eqs. (268) and (269) cannot predict the experimental velocity profiles observed in an oscillatory flow.

The boundary conditions near a stationary solid wall can be written as follows:

$$v_n = \mathbf{v} \bullet \mathbf{n} = 0 \tag{270}$$

$$v_t = \mathbf{v} \bullet \mathbf{t} = v_f \left(2.5 \ln \frac{v_f d}{\nu} + 5.5\right) \qquad \frac{v_f d}{\nu} > 11.63 \tag{271}$$

$$v_t = \frac{d v_f^2}{\nu} \qquad 0 \le \frac{v_f d}{\nu} \le 11.63 \qquad v_t^+ = \frac{v_t}{v_f} \qquad y^+ = \frac{d v_f}{\nu} \tag{272}$$

where v_n and v_t are the mean velocity components perpendicular and parallel to the wall, respectively; **n** and **t** are the unit vectors normal and tangential to the solid wall, respectively; v_f is the friction velocity; d is the distance from the solid wall to the grid point closest to the wall; and **v** is the gas phase velocity vector.

Equations (271) and (272) are applied near but not at the wall, and the friction velocity can be estimated from the k–ϵ model, assuming that near the wall, the convection and diffusion of turbulence kinetic energy are negligible. Thus, near the wall

$$v_f = C_\mu^{1/4} k^{1/2} \tag{273}$$

The velocity component v_t for a three-dimensional flow can be expressed as

$$v_t = (v_{t_1} + v_{t_2})^{1/2} \tag{274}$$

where v_{t_1} and v_{t_2} are two perpendicular velocity components in a plane parallel to the wall, which can be calculated by assuming that the shear stress parallel to the wall τ_t is itself parallel to the tangential velocity v_t near the wall, that is,

$$\frac{v_{t_1}}{v_{t_2}} = \frac{\tau_{t_1}}{\tau_{t_2}} \qquad \tau_t = (\tau_{t_1}^2 + \tau_{t_2}^2)^{1/2} \tag{275}$$

where $\tau_t = C_\mu^{1/2} \rho k$, and τ_{t_1} and τ_{t_2} depend on the coordinates used in the calculations. For example, in Cartesian coordinates for a wall corresponding to z = constant

$$\tau_{t_1} = \tau_{xz} \qquad \tau_{t_2} = \tau_{yz} \qquad v_{t_1} = v_x \qquad v_{t_2} = v_y \tag{276}$$

$$v_t = (v_x^2 + v_y^2)^{1/2} \qquad \tau_t = (\tau_{xz}^2 + \tau_{yz}^2)^{1/2} \tag{277}$$

$$\sin\alpha = \frac{v_x}{v_t} = \frac{\tau_{xz}}{\tau_t} \qquad \cos\alpha = \frac{v_y}{v_t} = \frac{\tau_{yz}}{\tau_t} \qquad \tan\alpha = \frac{v_x}{v_y} \tag{278}$$

$$\tau_{xz} = (\mu + \mu_t)\left(\frac{\partial u}{\partial z} + \frac{\partial w}{\partial x}\right) = (\mu + \mu_t)\frac{\partial u}{\partial z} \tag{279}$$

$$\tau_{yz} = (\mu + \mu_t)\frac{\partial v}{\partial z} \tag{280}$$

where u, v, and w are the velocity components in the x, y, and z directions, respectively; μ_t is the turbulent viscosity; and α is the angle between v_t and the y axis.

The boundary conditions for k and ϵ can be written as

$$\frac{\partial k}{\partial n} = 0 \qquad \epsilon = C_\mu^{3/4}\frac{k^{3/2}}{\chi d} \tag{281}$$

where χ is the von Karman constant ($\chi = 0.40$); χd is the mixing length; the boundary layer has been assumed to be in equilibrium (i.e., the production of turbulence kinetic energy is equal to its dissipation rate); and n denotes the direction perpendicular to the wall (i.e., $\partial k/\partial n = \partial k/\partial z$ for a solid characterized by z = constant).

2.6.2 The Reynolds Analogy

The boundary conditions for the species mass fractions can be written as

$$\frac{\partial Y_i}{\partial n} = 0 \tag{282}$$

whereas the temperature profile near the wall is frequently calculated by means of the Reynolds analogy between linear momentum and energy [234, 235]. The Reynolds analogy is strictly applicable when the Reynolds fluxes of linear momentum and energy are equal, and the profiles of the mean velocity and enthalpy are similar, that is, the laminar and turbulent Prandtl numbers are equal to 1 [234].

For constant specific heats at constant pressure, the energy and linear momentum fluxes can be written as follows:

$$\dot{q}_h = -\lambda \frac{\partial T}{\partial n} + \rho C_p \langle v'T' \rangle = -\rho C_p (\alpha_t + \alpha) \frac{\partial T}{\partial n} = G_h \, \Delta h \tag{283}$$

$$\tau = \mu \frac{\partial u}{\partial n} - \rho \langle u'v' \rangle = \rho(v_t + v) \frac{\partial u}{\partial n} = G_m \, u_f \tag{284}$$

where G_h and G_m are the Reynolds fluxes of enthalpy and momentum, respectively; α is the thermal diffusivity (i.e., $\alpha = \lambda/\rho C_p$); λ is the thermal conductivity; and $\langle v'T' \rangle$ and $\langle u'v' \rangle$ are turbulent correlations that have been modeled according to Boussinesq as

$$-\rho \langle u'v' \rangle = \mu_t \frac{\partial u}{\partial n} \tag{285}$$

$$-\rho \langle v'T' \rangle = \rho v_t \frac{\partial T}{\partial n} \tag{286}$$

where n denotes the direction of the fluctuating velocity component v'.

Dividing Eq. (284) by Eq. (283) yields the following expression:

$$\frac{G_m}{G_h} = \frac{\tau \, \Delta h}{\dot{q}_h u_f} = -\frac{v_t + v}{\alpha_t + \alpha} \frac{\Delta h}{u_f} \frac{\partial u/\partial n}{\partial h/\partial n} \tag{287}$$

where h is the enthalpy and Δh represents the enthalpy difference between the free stream and the wall. If the mean temperature and velocity profiles are similar, $G_m = G_h$.

Equation (287) represents the Reynolds analogy and can be integrated, assuming that $(v_t + v)/(\alpha_t + \alpha)$ is constant, to yield

$$T = T_w - \frac{\dot{q}_w}{C_p} \frac{v_t + v}{\alpha_t + \alpha} \frac{u}{\tau_w} = T_w - \frac{\dot{q}_w}{\rho C_p u_f^2} \frac{v_t + v}{\alpha_t + \alpha} u \tag{288}$$

where τ_w is the shear stress at the wall, T_w is the wall temperature, and $\dot{q}_w$ is the heat transfer to the wall.

In the viscous sublayer $v > v_t$, $\alpha > \alpha_t$, and Eq. (288) can be written as

$$T = T_w - \frac{\Pr \dot{q}_w}{\rho C_p u_f} u^+ = T_w - \frac{\Pr \dot{q}_w}{\rho C_p u_f} y^+ \qquad 0 \le y^+ \le 11.63 \tag{289}$$

whereas in the turbulent boundary layer $v_t > v$, $\alpha_t > \alpha$, and the temperature profile is given by

$$T = T_w - \frac{\Pr_t \dot{q}_w}{\rho C_p u_f} (2.5 \ln y^+ + 5.5) \qquad y^+ > 11.63 \tag{290}$$

where $\Pr_t = v_t/\alpha_t$ is the turbulent Prandtl number, which is a property of the flow.

Equation (290) can be modified to account for the effects of the fluid Prandtl number on heat transfer and written as [236]

$$T = T_w - \frac{\dot{q}_w \, \mathrm{Pr}_t}{\rho C_p \, u_f} (2.5 \ln y^+ + 5.5 + H) \tag{291}$$

where

$$\mathrm{H} = \frac{\pi/4}{\sin(\pi/4)} \left(\frac{A}{\chi}\right)^{1/2} \left(\frac{\mathrm{Pr}}{\mathrm{Pr}_t} - 1\right) \left(\frac{\mathrm{Pr}_t}{\mathrm{Pr}}\right)^{1/4} \tag{292}$$

A is Van Driest's constant ($A = 26$), and the turbulent Prandtl number is taken from available experimental data. Frequently it is assumed that $\mathrm{Pr}_t = 1$.

Equations (289) and (291) can also be used to calculate three-dimensional flows. However, it should be kept in mind that the analogy between linear momentum and energy may not be applicable to reciprocating and rotary engine flows. Furthermore, if the grid points closest to the solid walls are located in the turbulent boundary layer, Eqs. (290) and (291) may overestimate the heat transfer to the solid wall.

Accurate heat transfer results can be obtained only if the hydrodynamic and thermal boundary layers are resolved and the boundary conditions are applied at the walls. These accurate calculations can be performed only if enough grid points are placed in the boundary layer to resolve its structure and if low-Reynolds-number effects are introduced in the turbulence model. These calculations will be very demanding in terms of computer storage and time. Accurate calculations of the heat transfer to solid walls are essential for determining the engine efficiency.

References 233, 234, 236, and 237 provide heat transfer correlations and calculation methods for incompressible and compressible boundary layers with adverse and favorable pressure gradients. These calculation methods are based on approximate solutions of the hydrodynamic and turbulent boundary layers and have not yet been used in reciprocating and rotary engines.

2.7 THREE–DIMENSIONAL MODELS

The two-dimensional model presented in Section 2.6 has been extended to analyze the flow field in a three-dimensional rotary engine model provided with gaseous fuel injection and a rotor pocket geometry, which is independent of the axial coordinate between the front and back side walls [71, 238] (i.e., the pocket extends from the front to the back side seals). These three-dimensional calculations indicate that the fuel penetration is smaller than in the two-dimensional engine model [129]; they also indicate that gaseous fuel injection results in poor fuel-air mixing and that most of the injected fuel is entrained by the velocity field induced by the rotor and is concentrated between the injector and the rotor

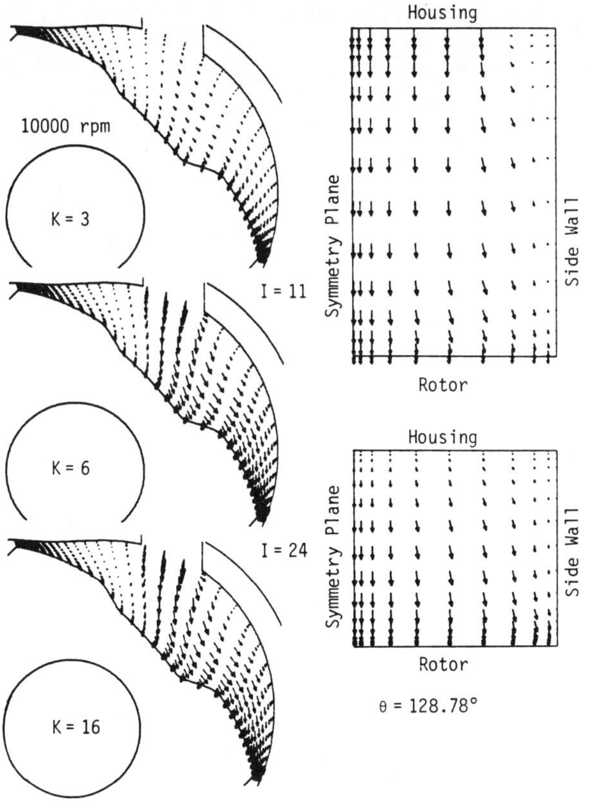

Figure 2.12 Velocity fields in a three-dimensional rotary engine model.

leading apex. Some three-dimensional velocity fields are presented in Figs. 2.12–2.15 [71]. In these figures $I = 1$ and $I = 31$ correspond to the rotor trailing and leading apexes, respectively; $J = 1$ and $J = 15$ correspond to the rotor and housing, respectively; and $K = 1$ and $K = 11$ correspond to the side seal and symmetry plane, respectively.

Recently, Grasso et al. [77] presented three-dimensional rotary engine computations of the exhaust, intake, and compression strokes, homogeneous-charge combustion, dual liquid fuel injection, and dual liquid fuel injection and combustion. These calculations used a spray model that accounts for droplet collisions and coalescence; a hybrid kinetics-mixing-controlled combustion submodel; domain decomposition to account for the wide disparity in the aspect ratios of the computational cells; grid rezoning; an analytically body-fitted grid-generation technique; a coordinate system fixed to the rotor, which requires a reevaluation of the grid nodes at each time step to account for the volume changes due to the rotor motion; and a modified version of the KIVA code [92, 95, 239].

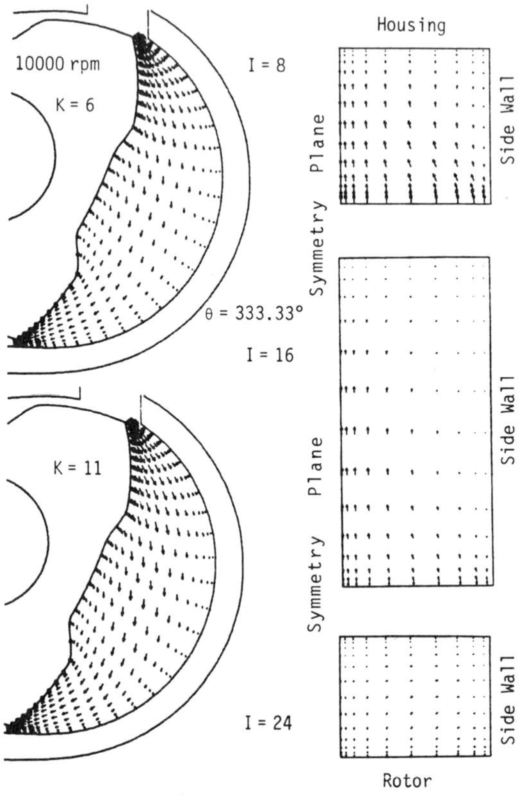

Figure 2.13 Velocity fields in a three-dimensional rotary engine model.

Grasso et al. [77] accounted for the viscous sublayer and log-wall by means of linear and logarithmic velocity and temperature profiles; they assumed that the production and dissipation rates of turbulence kinetic energy are equal near the wall and that $k = 0$ at the wall. The latter condition may not be a good approximation unless enough grid points and a low-Reynolds-number turbulence model are used in the calculations.

Grasso et al. [77] used a grid having $25 \times 12 \times 13$ points at 2000 rpm, whereas Ramos et al. [71] employed $31 \times 15 \times 11$ points at 10,000 rpm. These grids are too coarse to capture the details of the flow field, fuel-air mixing, and flame propagation. The results reported by Grasso et al. [77] are much finer than those obtained in a $25 \times 12 \times 13$ grid and seem to indicate that only one grid point was located at the injector and that interpolation was used to illustrate the flow field. Furthermore, the grid system used by Grasso et al. [77] shows a slope discontinuity at the interface between the rotor-pocket mesh and the rotor-housing mesh. A slope discontinuity is numerically undesirable, although the control-volume finite difference approach of the KIVA code may deal with such a discontinuity in a satisfactory manner. The grid lines used by Grasso et al. [77] do not intersect the boundaries perpendicularly, and the Neumann boundary

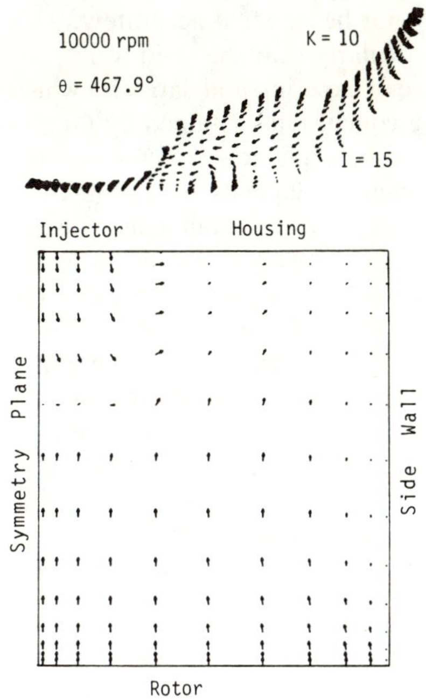

Figure 2.14 Velocity fields in a three-dimensional rotary engine model.

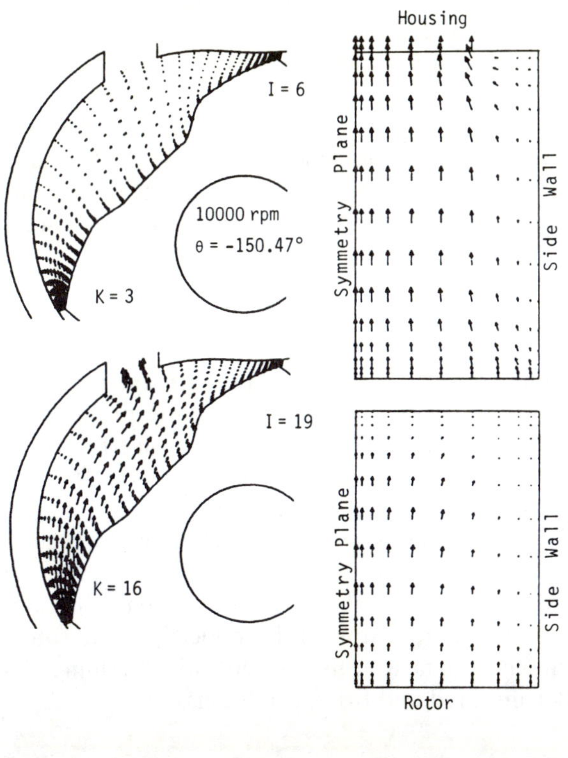

Figure 2.15 Velocity fields in a three-dimensional rotary engine model.

conditions for the species mass fractions may not be satisfied accurately. This may also result in some numerical errors, particularly near the solid walls.

In spite of all these remarks and the absence of experimental data with which the numerical predictions can be verified, the conclusions obtained by Grasso et al. [77] seem plausible. Their conclusions can be summarized as follows: (1) the burning rate in premixed-charge rotary engines depends on the mixture strength, compression ratio, combustion chamber geometry, wall heat transfer, spark plug position, and turbulent diffusivity; (2) the wall heat transfer losses increase with the burning rate and the indicated work is insensitive to the burning rate as long as combustion is completed within 45 crank angle degrees in premixed-charge rotary engines; (3) rich and lean mixtures are observed on either side of the initial premixed flames in DISC rotary engines; (4) the turbulence at TDC is insensitive to the details of the intake-generated turbulence but may depend on the initial turbulence diffusivity; (5) the distribution of turbulence at TDC is nonuniform due to squish; and (6) to obtain faster and more complete combustion, recirculation zones should be produced in DISC engines. The latter conclusion was also reached in the two- and three-dimensional calculations reported in references 71 and 129.

2.8 SUMMARY

The predictions of the thermodynamic models discussed in this chapter have been compared with global experimental data such as cylinder pressure, exhaust gas temperature, and HC and NO_x emissions. In most cases the leakage area had to be varied in order to achieve good agreement between the numerical predictions and experimental data. However, no detailed experimental data are available to assess the accuracy of multidimensional models. These models have qualitatively predicted the flame enhancement near the rotor leading apex, poor fuel-air mixing, and high HC emissions that have been observed in DISC rotary engines.

Future work on rotary engines must include detailed measurements of the velocity, temperature, and concentration fields within the combustion chamber and of the velocity, temperature, and pressure near the injector and near the exhaust and intake ports. Modeling efforts must focus on the treatment of turbulence in unsteady compressible flows with streamline curvature, preferential dissipation, compressibility, low-Reynolds-number and strong squish effects near the rotor trailing apex, leakage, flame propagation, numerical methods, grid-generation techniques, and turbulence-chemistry interactions.

The multizone models developed for spark-ignition and DISC reciprocating engines can be applied to predict combustion in rotary engines. However, these models require experimentally verified heat transfer correlations and flame propagation models that account for the flame distortion introduced by the rotor-induced velocity field. Experimental data are also needed to determine the leakage and crevice areas under motored and firing conditions.

The multizone models developed for diesel engine combustion can also be applied to DISC rotary engines provided that accurate heat transfer correlations and spray-penetration data are obtained in rotary engines. The penetration data for steady sprays or gaseous jets that have been used in multizone diesel engine models are not applicable to rotary engines due to the high velocities induced by the rotor.

Future work on single-zone models should include the development of appropriate Wiebe functions based on accurate experimental data, whereas future work on multizone models of spark-ignited rotary engines should include equations for the turbulence length and time scales and the Tabaczynski-Blizard-Keck model of turbulent combustion.

REFERENCES

1. Yamamoto, K., *Rotary Engine*, Sankaido, Tokyo, 1981.
2. Yamamoto, K., and Kuroda, T., Toyo Kogyo's Research and Development on Major Rotary Engine Problems, *SAE Transactions*, sec. 1, vol. 79, pp. 216–228, 1971.
3. Jones, C., A Progress Report on Curtiss-Wright's Rotary Stratified Charge Engine Development, SAE International Congress and Exposition, Detroit, Mich., SAE Paper No. 741206, 1974.
4. Jones, C., The Curtiss-Wright Rotating Combustion Engines Today, *SAE Transactions*, vol. 73, pp. 127–147, 1965.
5. Fujimoto, Y., Tatsutomi, Y., Ozeki, H., and Tadokoro, A., Present and Prospective Technologies of Rotary Engine, SAE International Congress and Exposition, Detroit, Mich., SAE Paper No. 870446, 1987.
6. Kohno, T., Ito, R., Morita, M., and Mizuno, N., Analysis of Light-Load Performance in Rotary Engines, *SAE Transactions*, sec. 2, vol. 88, pp. 1637–1652, 1980.
7. Kamo, R., Yamada, T. Y., and Hamada, Y., Starting Low Compression Ratio Rotary Wankel Diesel Engine, SAE International Congress and Exposition, Detroit, Mich., SAE Paper No. 870449, 1987.
8. Mount, R. E., and Greiner, W. L., High Performance, Stratified Charge Rotary Engines for General Aviation, AIAA/ASME/SAE/ASEE 22nd Joint Propulsion Conference, Huntsville, Ala., Paper No. AIAA-86-1553, 1986.
9. Ward, H. M., Griffith, M. J., Miller, G. E., and Stephenson, D. K., Outboard Marine Corp.'s Production Rotary Combustion Snowmobile Engine, *SAE Transactions*, vol. 82, pp. 461–472, 1974.
10. Willis, E. A., and Olcott, J. B., Development Potential of Intermittent Combustion (I.C.) Aircraft Engines for Commuter Transport Applications, SAE Commuter Aircraft and Airline Operations Meeting, Savannah, Ga., SAE Paper No. 820718, 1982.
11. Willis, E. A., and Wintucky, W. T., An Overview of NASA Intermittent Combustion Engine Research, AIAA/ASME/SAE 20th Joint Propulsion Conference, Cincinnati, Ohio, Paper No. AIAA-84-1393, 1984.
12. Schock, H. J., Willis, E. A., and Lee, C.-M., Stratified-Charge Rotary Engine (SCRE) Program at NASA Lewis Research Center, SAE International Congress and Exposition, Detroit, Mich., SAE Paper No. 870407, 1987.
13. Schock, H. J., A Review of Internal Combustion Engine Combustion Chamber Process Studies at NASA Lewis Research Center, AIAA/ASME/SAE 20th Joint Propulsion Conference, Cincinnati, Ohio, Paper No. AIAA-84-1316, 1984.
14. Jones, C., Ellis, D. R., and Meng, P. R., Multi-Fuel Rotary Engine for General Aviation

Aircraft, AIAA/SAE/ASME 19th Joint Propulsion Conference, Seattle, Wash., Paper No. AIAA-83-1340, 1983.

15. Jones, C., and Berkowitz, M., Multifuel Rotary Aircraft Engine, AIAA/SAE/ASME 16th Joint Propulsion Conference, Hartford, Conn., Paper No. AIAA-80-1237, 1980.
16. Jones, C., Lamping, H. D., Myers, D. M., and Loyd, R. W., An Update of the Direct Injected Stratified Charge Rotary Combustion Engine Developments at Curtiss-Wright, SAE International Congress and Exposition, Detroit, Mich., SAE Paper No. 770044, 1977.
17. Blair, G. P., Personal communication, Dubrovnik, Yugoslavia, August 27, 1987.
18. Loyd, R., Curtiss-Wright Stratified Charge Rotary Engine Developments, *Combustion Science and Technology*, vol. 12, pp. 47–61, 1976.
19. Cichanowicz, J. E., and Sawyer, R. F., Rotary Engine Combustion with Hydrogen Addition, SAE International Congress and Exposition, Detroit, Mich., SAE Paper No. 760611, 1976.
20. Shimamura, K., and Tadokoro, T., Fuel Economy and Improvement of Rotary Engine by Using Catalyst System, SAE International Congress and Exposition, Detroit, Mich., SAE Paper No. 810277, 1981.
21. King, S. R., Durability of Natural Gas Fueled Rotary Engines, SAE International Congress and Exposition, Detroit, Mich., SAE Paper No. 870448, 1987.
22. Sierens, R., and Verdonck, W., Development of a Natural Gas Stratified Charge Rotary Engine, SAE International Congress and Exposition, Detroit, Mich., SAE Paper No. 850034, 1985.
23. Eberle, M. K., and Klomp, E. D., An Evaluation of the Potential Performance Gain from Leakage Reduction in Rotary Engines, *SAE Transactions*, vol. 82, pp. 454–460, 1974.
24. Knoll, J., Vilmann, C. R., Schock, H. J., and Stumpf, R. P., A Dynamic Analysis of Rotary Combustion Engine Seals, SAE International Congress and Exposition, Detroit, Mich., SAE Paper No. 840035, 1984.
25. Orlandea, N. V., Weinert, M. S., and Keleher, D. B., Computer Simulation of the Rotary Engine Apex Seal System, SAE International Congress and Exposition, Detroit, Mich., SAE Paper No. 870410, 1987.
26. Hicks, Y. R., Schock, H. J., Craig, J. E., Umstatter, H. L., and Lee, D. Y., Visualization of Flows in a Motored Rotary Combustion Engine Using Holographic Interferometry, AIAA/ASME/SAE/ASEE 22nd Joint Propulsion Conference, Huntsville, Ala., Paper No. AIAA-86-1557, 1986.
27. Schock, H. J., Rice, W. J., and Meng, P. R., Experimental Analysis of IMEP in a Rotary Combustion Engine, SAE International Congress and Exposition, Detroit, Mich., SAE Paper No. 810150, 1981.
28. Meng, P. R., Rice, W. J., Schock, H. J., and Pringle, D. P., Preliminary Results on Performance Testing of a Turbocharged Rotary Combustion Engine, SAE International Congress and Exposition, Detroit, Mich., SAE Paper No. 820354, 1982.
29. Chun, K. S., Synchronization Trigger Control System for Flow Visualization, SAE International Congress and Exposition, Detroit, Mich., SAE Paper No. 870450, 1987.
30. Lee, C.-M., MacDonald, P. S., and Purnell, D. J., Thermal and Structural Analysis of a Low Heat Rejection Rotary Engine Combustion Chamber, ADAPCO, Melville, N.Y., 1986.
31. Bayer, R. J., DeNagel, S. F., and Steiner, J. C., Rotary Combustion Engine Hydrocarbon Source Studies, *SAE Transactions*, sec. 4, vol. 87, pp. 3619–3636, 1979.
32. Lamping, R. K., Manning, I., Myers, D., and Tjoa, B., Performance, Emissions, and Physical Characteristics of a Rotating Combustion Aircraft Engine, NASA CR-135119, suppl. A, 1980.
33. Burley, H. A., Meloeny, M. R., and Stark, T. L., Sources of Hydrocarbon Emissions in Rotary Engines, SAE International Congress and Exposition, Detroit, Mich., SAE Paper No. 780419, 1978.
34. Ferguson, C. R., Danieli, G. A., Heywood, J. B., and Keck, J. C., Time Resolved Measurements of Exhaust Composition and Flow Rate in a Wankel Engine, *SAE Transactions*, sec. 1, vol. 84, pp. 141–154, 1976.
35. Ito, R., Morita, M., Mizuno, N., and Hukuda, T., A Study of Combustion in Rotary Engines—Observation of Combustion Flames, *Toyota Technology*, vol. 27, no. 1, special issue, pp. 1–15, 1977.

36. Wankel, F., *Rotary Piston Machines*, Iliffe, London, 1965.
37. Ansdale, R. F., *The Wankel RC Engine*, Iliffe, London, 1968.
38. Norbye, J. P., *The Wankel Engine*, Chilton, Philadelphia, 1971.
39. Roberts, J. A., Norman, T. J., Ekchian, J. A., and Heywood, J. B., Computer Models for Evaluating Premixed and DISC Wankel Engine Performance, SAE International Congress and Exposition, Detroit, Mich., SAE Paper No. 860613, 1986.
40. Norman, T. J., A Performance Model of a Spark Ignition Wankel Engine: Including the Effects of Crevice Volumes, Gas Leakage, and Heat Transfer, M. S. thesis, Massachusetts Institute of Technology, 1983.
41. Rohsenow, W. M., and Choi, H. Y., *Heat, Mass and Momentum Transfer*, Prentice-Hall, Englewood Cliffs, N.J., 1961.
42. Woschni, G., A Universally Applicable Equation for the Instantaneous Heat Transfer Coefficient in the Internal Combustion Engine, SAE International Congress and Exposition, Detroit, Mich., SAE Paper No. 670931, 1967.
43. Atesmen, K. M., Heat Transfer in Rotary Combustion Engines, *ASME Journal of Heat Transfer*, vol. 97, ser. C, pp. 288–293, 1975.
44. Annand, W. J. D., Heat Transfer in the Cylinders of Reciprocating Engines, *Proceedings of the Institution of Mechanical Engineers*, vol. 177, no. 36, pp. 963–990, 1967.
45. Lee, C. M., Addy, H. E., Bond, T. H., Chun, K. S., and Lu, L. Y., Estimation of Instantaneous Heat Transfer Coefficients for a Direct-Injection Stratified-Charge Rotary Engine, SAE International Congress and Exposition, Detroit, Mich., SAE Paper No. 870444, 1987.
46. Benson, R. S., and Sierens, R., Simulation of the Wave Action in the Inlet and Exhaust System and Calculation of the Burnt Mass Fraction in a Rotary Wankel Engine, *Israel Journal of Technology*, vol. 15, pp. 209–221, 1977.
47. Sierens, R., Numerische Simulatie van de Verbranding in Een Wankelmotor, Ph.D. thesis, vols. 1 and 2, University of Delft, Delft, The Netherlands, 1981.
48. Blair, G. P., and Fleck, P., The Unsteady Gas Flow Behaviour in a Charge Cooled Rotary Piston Engine, SAE Off-Highway Vehicle Meeting and Exhibition, Milwaukee, Wis., SAE Paper No. 770763, 1977.
49. Sierens, R., Baert, R., Winterbone, D. E., and Baruah, P. C., A Comprehensive Study of Wankel Engine Performance, SAE International Congress and Exposition, Detroit, Mich., SAE Paper No. 830332, 1983.
50. Sierens, R., Baert, R., Winterbone, D. E., and Baruah, P. C., Combustion Studies in a Rotary Engine, *International Conference on Combustion Engineering*, vol. 2, pp. 31–42, I. Mech. E. Conference Publications 1983-3, London, 1983.
51. Benson, R. S., Annand, W. J. D., and Baruah, P. C., A Simulation Model Including Intake and Exhaust Systems for a Single Cylinder Four-Stroke Cycle Spark Ignition Engine, *International Journal of Mechanical Sciences*, vol. 17, pp. 97–124, 1975.
52. Friedman, R., and Johnston, W. C., Pressure Dependence of Quenching Distance of Normal Heptane, Iso-Octane, Benzene, and Ethyl Ether Flames, *Journal of Chemical Physics*, vol. 20, pp. 919–920, 1952.
53. Agnew, J. T., and Green, K. A., Quenching Distances for Propane-Air Flames in a Constant-Volume Bomb, *Combustion and Flame*, vol. 15, pp. 189–191, 1970.
54. Friedman, R., and Johnston, W. C., The Wall-Quenching of Laminar Propane Flames as a Function of Pressure, Temperature, and Air-Fuel Ratio, *Journal of Applied Physics*, vol. 21, pp. 791–795, 1950.
55. Danieli, G. A., Ferguson, C. R., Heywood, J. B., and Keck, J. C., Predicting the Emissions and Performance Characteristics of a Wankel Engine, *SAE Transactions*, vol. 83, pp. 805–820, 1975.
56. Danieli, G. A., A Performance Model of a Wankel Engine, Including the Effects of Burning Rates, Heat Transfer, Leakage and Quenching Compared with Measured Pressure Time Histories, Ph.D. thesis, Massachusetts Institute of Technology, Cambridge, 1972.
57. Ballal, D. R., and Lefebvre, A. H., Turbulence Effects on Enclosed Flames, *Acta Astronautica*, vol. 1, pp. 471–483, 1974.

58. Ballal, D. R., An Experimental Study of Flame Turbulence, *Acta Astronautica*, vol. 5, pp. 1095–1112, 1978.
59. Gokalp, I., On the Turbulence in Premixed Flames, *Acta Astronautica*, vol. 6, nos. 7/8, pp. 847–860, 1979.
60. Danieli, G. A., Keck, J. C., and Heywood, J. B., Experimental and Theoretical Analysis of Wankel Engine Performance, SAE International Congress and Exposition, Detroit, Mich., SAE Paper No. 780416, 1978.
61. Blizard, N. C., and Keck, J. C., Experimental and Theoretical Investigation of Turbulent Burning Model for Internal Combustion Engines, SAE International Congress and Exposition, Detroit, Mich., SAE Paper No. 740191, 1974.
62. Nguyen, H. L., Addy, H. E., Bond, T. H., Lee, C. M., and Chun, K. S., Performance and Efficiency Evaluation and Heat Release Study of a Direct-Injection Stratified-Charge Rotary Engine, SAE International Congress and Exposition, Detroit, Mich., SAE Paper No. 870445, 1987.
63. Roberts, J. M., Heat Release Estimation and Prediction of Wankel Stratified-Charge Combustion Engine, M.S. thesis, Massachusetts Institute of Technology, Cambridge, 1985.
64. Williams, F. A., *Combustion Theory*, 2nd ed., pp. 1–9, Benjamin-Cummings, Menlo Park, Calif., 1985.
65. Tennekes, H., and Lumley, J. L., *A First Course in Turbulence*, p. 21, MIT Press, Cambridge, Mass., 1972.
66. Hinze, J. O., *Turbulence*, 2nd ed., p. 6, McGraw-Hill, New York, 1975.
67. Reynolds, W. C., Modeling of Fluid Motions in Engines—An Introductory Overview, in *Combustion Modeling in Reciprocating Engines*, eds. J. N. Mattavi and C. A. Amann, pp. 41–68, Plenum Press, New York, 1980.
68. Shih, T. I-P., Yang, S. L., and Schock, H. J., A Two-Dimensional Numerical Study of the Flow inside the Combustion Chamber of a Motored Rotary Engine, SAE International Congress and Exposition, Detroit, Mich., SAE Paper No. 860615, 1986.
69. Favre, A., Statistical Equations of Turbulent Gases, in *Problems of Hydrodynamics and Continuum Mechanics*, pp. 231–266, SIAM, Philadelphia, 1969.
70. Launder, B. E., and Spalding, D. B., *Lectures on Mathematical Models of Turbulence*, Academic Press, New York, 1972.
71. Ramos, J. I., Shih, T. I-P., and Schock, H. J., Wankel Engine Modelling, in *Heat and Mass Transfer in Gasoline and Diesel Engines*, ed. D. B. Spalding, Hemisphere, Washington, D.C., 1988.
72. Schlichting, H., *Boundary-Layer Theory*, 6th ed., p. 565, McGraw-Hill, New York, 1968.
73. Jones, W. P., and Launder, B. E., The Prediction of Laminarization with a Two-Equation Model of Turbulence, *International Journal of Heat and Mass Transfer*, vol. 15, pp. 301–314, 1972.
74. Gosman, A. D., Johns, R. J. R., and Watkins, A. P., Development of Prediction Methods for In-Cylinder Processes in Reciprocating Engines, in *Combustion Modeling in Reciprocating Engines*, eds. J. N. Mattavi and C. A. Amann, pp. 62–129, Plenum Press, New York, 1980.
75. Ramos, J. I., Humphrey, J. A. C., and Sirignano, W. A., Numerical Prediction of Axisymmetric Laminar and Turbulent Flows in Motored, Reciprocating Internal Combustion Engines, SAE International Congress and Exposition, Detroit, Mich., SAE Paper No. 790356, 1979.
76. Morel, T., and Mansour, N. N., Modeling of Turbulence in Internal Combustion Engines, SAE International Congress and Exposition, Detroit, Mich., SAE Paper No. 820040, 1982.
77. Grasso, F., Wey, M.-J., Bracco, F. V., and Abraham, J., Three-dimensional Computations of Flows in a Stratified-Charge Rotary Engine, SAE International Congress and Exposition, Detroit, Mich., SAE Paper No. 870409, 1987.
78. Gosman, A. D., and Watkins, A. P., A Computer Prediction Method for Turbulent Flow and Heat Transfer in Piston/Cylinder Assemblies, *Proceedings of the First Symposium on Turbulent Shear Flows*, pp. 5.23–5.30, Pennsylvania State University, State College, 1977.
79. Ahmadi-Befrui, B., Gosman, A. D., Lockwood, F. C., and Watkins, A. P., Multidimensional Calculation of Combustion in an Idealised Homogeneous Charge Engine: A Progress Report, SAE International Congress and Exposition, Detroit, Mich., SAE Paper No. 810151, 1981.

80. Ramos, J. I., The $k-\epsilon$ Model of Turbulence in the Limiting Case Re $\rightarrow \infty$, Report CO/80/3, Department of Mechanical Engineering, Carnegie-Mellon University, Pittsburgh, Pa., 1980.
81. Bray, K. N. C., Libby, P. A., Masuya, G., and Moss, J. B., Turbulence Production in Premixed Turbulent Flames, *Combustion Science and Technology*, vol. 25, pp. 127–140, 1981.
82. Strahle, W. C., Estimation of Some Correlations in a Premixed Reactive Turbulent Flow, *Combustion Science Technology*, vol. 29, pp. 243–260, 1982.
83. Chandran, S. B. S., Komerath, N. M., and Strahle, W. C., Scalar-Velocity Correlations in a Turbulent Premixed Flame, *Twentieth Symposium (International) on Combustion*, pp. 429–435, The Combustion Institute, Pittsburgh, Pa., 1985.
84. Chandran, S. B. S., and Strahle, W. C., The Effect of Pressure-Velocity Correlation in a Premixed, Planar, Turbulent Flame, *Combustion and Flame*, vol. 51, pp. 313–324, 1983.
85. Tanaka, H., and Yanagi, T., Cross-Correlation Velocity and Temperature in a Premixed Turbulent Flame, *Combustion and Flame*, vol. 51, pp. 183–191, 1983.
86. Shah, P., and Markatos, N. C., On the 2D and 3D Computer Analyses of Turbulence in Internal Combustion Engines, *International Journal of Numerical Methods in Fluids*, in press.
87. Shah, P., and Markatos, N. C., On the 2D and 3D Analyses of Turbulent Flow in Internal Combustion Engines, *Proceedings of the First International PHOENICS Users Conference on Numerical Simulation of Fluid Flow and Heat and Mass Transfer Processes*, Dartford, U.K., 1985.
88. Markatos, N. C., and Shah, P., Turbulence Modelling in Internal Combustion Engines, in *Numerical Methods in Laminar and Turbulent Flow*, eds. C. Taylor, M. D. Olson, P. M. Gresho, and W. G. Habashi, pt. 2, pp. 1439–1954, Pineridge Press, Swansea, U.K., 1985.
89. Ilegbusi, J. O., and Spalding, D. B., An Improved Version of the $k-w$ Model of Turbulence, *ASME Journal of Heat Transfer*, vol. 107, no. 1, pp. 63–69, 1985.
90. El Tahry, S. H., $k-\epsilon$ Equation for Compressible Reciprocating Engine Flows, *Journal of Energy*, vol. 7, no. 4, pp. 345–353, 1983.
91. Wong, V. W., and Hoult, D. P., Rapid Distortion Theory Applied to Turbulent Combustion, SAE International Congress and Exposition, Detroit, Mich., SAE Paper No. 790357, 1979.
92. Amsden, A. A., Ramshaw, J. D., O'Rourke, P. J., and Dukowicz, J. K., KIVA: A Computer Program for Two- and Three-dimensional Fluid Flows with Chemical Reactions and Fuel Sprays, Report No. LA-10245-MS, Los Alamos National Laboratory, Los Alamos, N.M., 1985.
93. Deardorff, J. W., On the Magnitude of the Subgrid Scale Eddy Coefficient, *Journal of Computational Physics*, vol. 7, pp. 120–133, 1971.
94. Wang, Q. C., and Ferguson, C. R., Multidimensional Modeling of the Decay of Angular Momentum and Internal Energy in a Constant Volume Cylindrical Vessel, *ASME Journal of Heat Transfer*, vol. 108, no. 3, pp. 700–703, 1986.
95. Amsden, A. A., Butler, T. D., O'Rourke, P. J., and Ramshaw, J. D., KIVA-A Comprehensive Model for 2-D and 3-D Engine Simulations, SAE International Congress and Exposition, Detroit, Mich., SAE Paper No. 850554, 1985.
96. Ferziger, J. H., Higher-Level Simulations of Turbulent Flows, in *Computational Methods for Turbulent, Transonic, and Viscous Flows*, ed. J. A. Essers, pp. 93–182, Hemisphere, Washington, D.C., 1983.
97. Deardorff, J. W., The Use of Subgrid Transport Equations in a Three-dimensional Model of Atmospheric Turbulence, *ASME Journal of Fluids Engineering*, vol. 95, pp. 429–438, 1973.
98. Bradshaw, P., Effects of Streamline Curvature on Turbulent Flows, AGARDograph No. 169, 1973.
99. Bradshaw, P., Review—Complex Turbulent Flows, *ASME Journal of Fluids Engineering*, vol. 97, pp. 146–154, 1975.
100. Sharma, B. I., Launder, B. E., and Scott, C. J., Computation of Annular, Turbulent Flow with Rotating Core Tube, *ASME Journal of Fluids Engineering*, vol. 98, pp. 753–758, 1976.
101. Launder, B. E., Priddin, C. H., and Sharma, B. I., The Calculation of Turbulent Boundary Layers on Spinning and Curved Surfaces, *ASME Journal of Fluids Engineering*, vol. 99, pp. 231–239, 1977.
102. Mansour, N. N., Dispersion of Stratified Charges in the Engine Cylinder during the Compres-

sion Stroke, in *Flows in Internal Combustion Engines*, ed. T. Uzkan, pp. 35–43, ASME, New York, 1982.

103. Elkotb, M. M., Abou-Ellail, M. M. M., and Salem, I. S., Rotating Flow and Nonisotropic Turbulence in Reciprocating Engines, in *Flows in Internal Combustion Engines*, ed. T. Uzkan, pp. 81–91, ASME, New York, 1982.
104. Leschziner, M. A., and Rodi, W., Calculation of Annular and Twin Parallel Jets Using Various Discretization Schemes and Turbulence-Model Variations, *ASME Journal of Fluids Engineering*, vol. 103, pp. 352–360, 1981.
105. Rodi, W., and Scheuerer, G., Scrutinizing the $k-\epsilon$ Model under Adverse Pressure Gradient Conditions, *ASME Journal of Fluids Engineering*, vol. 108, pp. 174–179, 1986.
106. Hanjalic, K., and Launder, B. E., Sensitizing the Dissipation Equation to Irrotational Strains, *ASME Journal of Fluids Engineering*, vol. 102, pp. 34–40, 1980.
107. Lam, C. K. G., and Bremhorst, K., A Modified Equation of the $k-\epsilon$ Model for Predicting Wall Turbulence, *ASME Journal of Fluids Engineering*, vol. 103, pp. 456–460, 1981.
108. Nagano, Y., and Hishida, M., Improved Form of the $k-\epsilon$ Model for Wall Turbulent Shear Flows, *ASME Journal of Fluids Engineering*, vol. 109, pp. 156–160, 1987.
109. Harlow, F. H., and Nakayama, P. I., Turbulent Transport Equations, *Physics of Fluids*, vol. 10, pp. 2323–2332, 1967.
110. Reynolds, W. C., Computation of Turbulent Flows, *Annual Review of Fluid Mechanics*, vol. 8, pp. 183–208, 1976.
111. Reynolds, W. C., Recent Advances in the Computation of Turbulent Flows, *Advances in Chemical Engineering*, vol. 9, pp. 193–246, 1974.
112. Rodi, W., Progress in Turbulence Modeling for Incompressible Flows, AIAA 19th Aerospace Sciences Meeting, St. Louis, Mo., Paper No. AIAA-81-0045, 1981.
113. Harlow, F. H., *Turbulence Transport Modeling*, AIAA Selected Reprint Series, vol. 14, AIAA, New York, 1973.
114. Corrsin, S., Limitations of Gradient Transport Models in Random Walks and in Turbulence, in *Turbulent Diffusion in Environmental Pollution*, eds. F. N. Frenkiel and R. E. Munn, *Advances in Geophysics*, vol. 18A, pp. 25–60, Academic Press, New York, 1974.
115. Bradshaw, P., Compressible Turbulent Shear Layers, *Annual Review of Fluid Mechanics*, vol. 9, pp. 33–54, 1977.
116. Lumley, J. L., and Khajeh-Nouri, B., Computational Modeling of Turbulent Transport, in *Turbulent Diffusion in Environmental Pollution*, eds. F. N. Frenkiel and R. E. Munn, *Advances in Geophysics*, vol. 18A, pp. 169–192, Academic Press, New York, 1974.
117. Mellor, G. L., and Herring, H. J., A Survey of the Mean Turbulent Field Closure Models, *AIAA Journal*, vol. 11, no. 5, pp. 590–599, 1973.
118. Dussauge, J. P., Gaviglio, J., and Favre, A., Density Changes and Turbulence Production in the Expansion or Compression of a Turbulent Flow at Supersonic Speed, in *Structure and Mechanisms of Turbulence II*, ed. H. Fiedler, *Lecture Notes in Physics*, vol. 76, pp. 385–395, Springer-Verlag, New York, 1977.
119. Launder, B. E., and Spalding, D. B., The Numerical Computation of Turbulent Flows, *Computer Methods in Applied Mechanics and Engineering*, vol. 3, pp. 269–289, 1974.
120. Bradshaw, P., ed., *Turbulence*, 2nd ed., *Topics in Applied Physics*, vol. 12, Springer-Verlag, New York, 1978.
121. Markatos, N. C., The Mathematical Modeling of Turbulent Flows, *Applied Mathematical Modelling*, vol. 10, no. 3, pp. 190–220, 1986.
122. Kumar, S., Mathematical Modelling of Natural Convection in Fires—A State of the Art Review of the Field Modelling of Variable Density Flows, *Fire and Materials*, vol. 7, no. 1, pp. 1–24, 1983.
123. Lumley, J. L., Turbulence Modeling, *ASME Journal of Applied Mechanics*, vol. 50, pp. 1097–1183, 1983.
124. Lumley, J. L., Computational Modeling of Turbulent Flows, in *Advances in Applied Mechanics*, ed. C. S. Yih, vol. 18, pp. 124–176, Academic Press, New York, 1978.
125. Kollmann, W., ed., *Prediction Methods for Turbulent Flows*, Hemisphere, Washington, D.C., 1980.

126. Launder, B. E., Reece, G., and Rodi, W., Progress in the Development of a Reynolds Stress Turbulence Closure, *Journal of Fluid Mechanics*, vol. 68, pp. 537–566, 1975.
127. Frost, W., and Moulden, T. H., eds., *Handbook of Turbulence*: vol. 1. *Fundamentals and Applications*, Plenum Press, New York, 1977.
128. Shih, T. I-P., Nguyen, H. L. and Stegeman, J., Fluid Flow and Fuel-Air Mixing in a Motored Two-Dimensional Wankel Rotary Engine, AIAA/ASME/SAE/ASEE 22nd Joint Propulsion Conference, Huntsville, Ala., Paper No. AIAA-86-1556, 1986.
129. Shih, T. I-P., Schock, H. J., and Ramos, J. I., Fuel-Air Mixing and Combustion in a Two-Dimensional Wankel Engine, SAE International Congress and Exposition, Detroit, Mich., SAE Paper No. 870408, 1987.
130. Westbrook, C. K., and Dryer, F. L., Chemical Kinetic Modeling of Hydrocarbon Combustion, *Progress in Energy and Combustion Science*, vol. 10, no. 1, pp. 1–57, 1984.
131. Byrne, G. D., and Hindmarsh, A. C., Stiff ODE Solvers: A Review of Current and Coming Attractions, *Journal of Computational Physics*, vol. 70, no. 1, pp. 1–62, 1987.
132. Ramos, J. I., Numerical Methods for One-dimensional Reaction-Diffusion Equations Arising in Combustion Theory, *Annual Review of Numerical Fluid Mechanics and Heat Transfer*, vol. 1, pp. 150–261, 1987.
133. Dryer, F. L., and Brezinsky, K., A Flow Reactor Study of the Oxidation of *n*-Octane and Iso-Octane, *Combustion Science and Technology*, vol. 45, pp. 199–212, 1986.
134. Westbrook, C. K., and Dryer, F. L., Simplified Reaction Mechanisms for the Oxidation of Hydrocarbon Fuels in Flames, *Combustion Science and Technology*, vol. 27, pp. 31–43, 1981.
135. Coffee, T. P., On "Simplified Reaction Mechanisms for the Oxidation of Hydrocarbon Fuels in Flames" by C. K. Westbrook and F. L. Dryer, *Combustion Science and Technology*, vol. 43, pp. 333–339, 1985.
136. Peters, N., Numerical and Asymptotic Analysis of Systematically Reduced Reaction Schemes for Hydrocarbon Fuels, in *Numerical Simulation of Combustion Phenomena*, eds. R. Glowinski, B. Larrouturou, and R. Temam, *Lecture Notes in Physics*, vol. 241, pp. 90–109, Springer-Verlag, New York, 1985.
137. Peters, N., and Williams, F. A., The Asymptotic Structure of Stoichiometric Methane/Air Flames, *Combustion and Flame*, vol. 68, pp. 185–207, 1987.
138. Westbrook, C. K. and Dryer, F. L., Chemical Kinetics and Modeling of Combustion Processes, *Eighteenth Symposium (International) on Combustion*, pp. 749–767, The Combustion Institute, Pittsburgh, Pa., 1981.
139. Duterque, J., Borghi, R., and Tichtinsky, H., Study of Quasi-global Schemes for Hydrocarbon Combustion, *Combustion Science and Technology*, vol. 26, pp. 1–15, 1981.
140. Warnatz, J., The Mechanism of High Temperature Combustion of Propane and Butane, *Combustion Science and Technology*, vol. 34, pp. 177–200, 1983.
141. Varma, A., Chatwani, A. U., and Bracco, F. V., Studies of Premixed Laminar Hydrogen-Air Flames Using Elementary and Global Kinetics, *Combustion and Flame*, vol. 64, pp. 233–236, 1986.
142. Reitz, R. D., and Bracco, F. V., Global Kinetics Models and Lack of Thermodynamic Equilibrium, *Combustion and Flame*, vol. 53, pp. 141–143, 1983.
143. Reitz, R. D., and Bracco, F. V., Toward the Formulation of a Global Local Equilibrium Kinetics Model for Laminar Hydrocarbon Flames, in *Numerical Methods in Laminar Flame Propagation*, eds. N. Peters and J. Warnatz, pp. 130–151, Friedr. Vieweg & Sohn, Braunschweig/Wiesbaden, F.R.G., 1982.
144. Metgalchi, M., and Keck, J. C., Laminar Burning Velocity of Propane-Air Mixtures at High Temperature and Pressure, *Combustion and Flame*, vol. 38, pp. 143–154, 1980.
145. Metgalchi, M., and Keck, J. C., Burning Velocities of Mixtures of Air with Methanol, Iso-octane, and Indolene at High Pressure and Temperature, *Combustion and Flame*, vol. 48, pp. 191–210, 1982.
146. Gupta, H. C., Steinberger, R. L., and Bracco, F. V., Combustion in a Divided Chamber, Stratified Charge, Reciprocating Engine: Initial Comparisons of Calculated and Measured Flame Propagation, *Combustion Science and Technology*, vol. 22, pp. 27–61, 1980.
147. Syed, S. A., and Bracco, F. V., Further Comparisons of Computed and Measured Divided-

Chamber Engine Combustion, SAE International Congress and Exposition, Detroit, Mich., SAE Paper No. 790247, 1979.

148. Diwakar, R., Assessment of the Ability of a Multidimensional Computer Code to Model Combustion in a Homogeneous-Charge Engine, SAE International Congress and Exposition, Detroit, Mich., SAE Paper No. 840230, 1984.
149. Butler, T. D., Cloutman, L. D., Dukowicz, J. K., and Ramshaw, J. D., CONCHAS: An Arbitrary Lagrangian-Eulerian Computer Code for Multicomponent Chemically Reactive Fluid Flow at All Speeds, Report No. LA-8129-MS, Los Alamos National Laboratory, Los Alamos, N.M., 1979.
150. Spalding, D. B., Development of the Eddy-Break-Up Model of Turbulent Combustion, *Sixteenth Symposium (International) on Combustion*, pp. 1657–1663, The Combustion Institute, Pittsburgh, Pa., 1977.
151. Magnussen, B. F., and Hjertager, B. H., On Mathematical Modeling of Turbulent Combustion with Special Emphasis on Soot Formation and Combustion, *Sixteenth Symposium (International) on Combustion*, pp. 719–729, The Combustion Institute, Pittsburgh, Pa., 1977.
152. Givi, P., Ramos, J. I., and Sirignano, W. A., Turbulent Reacting Concentric Jets: Comparison between PDF and Moment Calculations, in *Dynamics of Flames and Reactive Systems*, eds. J. R. Bowen, N. Manson, A. K. Oppenheim, and R. I. Soloukhin, *Progress in Astronautics and Aeronautics*, vol. 95, pp. 384–418, AIAA, New York, 1984.
153. Ramos, J. I., Two-dimensional Modelling of Spark-Ignition Engines, in *Modelado en Motores de Combustion Interna*, eds., F. Payri, J. M. Desantes, and F. Boada, pp. 321–332, Universidad Politecnica de Valencia, Valencia, Spain, 1987.
154. Grasso, F., and Bracco, F. V., Evaluation of a Mixing-Controlled Model for Engine Combustion, *Combustion and Science and Technology*, vol. 28, pp. 185–210, 1982.
155. Przekwas, A., Modeling of Fluid Flow, Heat Transfer and Flame Propagation in Reciprocating Engines, in *Flows in Internal Combustion Engines*, ed. T. Uzkan, pp. 73–79, ASME, New York, 1982.
156. Bray, K. N. C., The Interaction between Turbulence and Combustion, *Seventeenth Symposium (International) on Combustion*, pp. 223–233, The Combustion Institute, Pittsburgh, Pa., 1979.
157. Lockwood, F. C., The Modelling of Turbulent Premixed and Diffusion Combustion in the Computation of Engineering Flows, *Combustion and Flame*, vol. 29, pp 111–122, 1977.
158. Jones, W. P., and Whitelaw, J. H., Calculation Methods for Reacting Turbulent Flows: A Review, *Combustion and Flame*, vol. 48, pp. 1–26, 1982.
159. Jones, W. P., and Whitelaw, J. H., Modelling and Measurements in Turbulent Combustion, *Twentieth Symposium (International) on Combustion*, pp. 233–249, The Combustion Institute, Pittsburgh, Pa., 1985.
160. Spalding, D. B., Mixing and Chemical Reaction in Steady Confined Turbulent Flames, *Thirteenth Symposium (International) on Combustion*, pp. 649–657, The Combustion Institute, Pittsburgh, Pa., 1971.
161. Mudford, N. R., and Bilger, R. W., Examination of Closure Models for Mean Chemical Reaction Rate Using Experimental Results from an Isothermal Turbulent Reacting Flow, *Twentieth Symposium (International) on Combustion*, pp. 387–394, The Combustion Institute, Pittsburgh, Pa., 1985.
162. Strehlow, R. A., *Combustion Fundamentals*, pp. 28–46, McGraw-Hill, New York, 1984.
163. Gordon, S., and McBride, B. J., Computer Program for Calculation of Complex Chemical Equilibrium Compositions, Rocket Performance, Incident and Reflected Shocks, and Chapman-Jouguet Detonations, NASA SP-273, National Aeronautics and Space Administration, Washington, D.C., 1971.
164. Meintjes, K., and Morgan, A. P., Equilibrium Equations for a Model of Combustion, Report No. GMR-4361, General Motors Research Laboratories, Warren, Mich., 1983.
165. O'Rourke, P. J., The KIVA Computer Program for Multidimensional Chemically Reactive Fluid Flows with Fuel Sprays, in *Numerical Simulation of Combustion Phenomena*, eds., R. Glowinski, B. Larrouturou and R. Temam, *Lecture Notes in Physics*, vol. 241, pp. 74–89, Springer-Verlag, New York, 1985.

166. Meintjes, K., and Morgan, A. P., Performance of Algorithms for Calculating the Equilibrium Composition of a Mixture of Gases, *Journal of Computational Physics*, vol. 60, pp. 219–234 1985.
167. Erickson, W. D., and Prabhu, R. K., Rapid Computation of Chemical Equilibrium Composition: An Application to Hydrocarbon Combustion, *AIChE Journal*, vol. 32, no. 7, pp. 1079–1087, 1986.
168. Amelio, M., Belli, M., and Danieli, G. A., A Multidimensional Performance Model of a Spark Ignition Engine Characterized by a Single Step Reaction to Equilibrium, *Combustion Science and Technology*, vol. 45, pp. 115–128, 1986.
169. Hjertager, B. H., Simulation of Transient Compressible Turbulent Reactive Flows, *Combustion Science and Technology*, vol. 27, pp. 159–170, 1982.
170. Abraham, J., Bracco, F. V., and Reitz, R. D., Comparisons of Computed and Measured Premixed Charge Engine Combustion, *Combustion and Flame*, vol. 60, pp. 309–322, 1985.
171. Brandstatter, W., and Johns, R. J. R., The Application of a Probability Method to Engine Combustion Modelling, *International Conference on Combustion Engineering*, vol. I, pp. 113–124, I. Mech. E. Conference Publications 1983-3, London, 1983.
172. Borghi, R., and Moreau, P., Turbulent Combustion in a Premixed Flow, *Acta Astronautica*, vol. 4, pp. 321–341, 1977.
173. Givi, P., Ramos, J. I., and Sirignano, W. A., Probability Density Function Calculations in Turbulent Chemically Reacting Round Jets, Mixing Layers and One-Dimensional Reactors, *Journal of Non-Equilibrium Thermodynamics*, vol. 10, pp. 75–104, 1985.
174. Borghi, R., and Dutoya, D., On the Scales of the Fluctuations in Turbulent Combustion, *Seventeenth Symposium (International) on Combustion*, pp. 235–244, The Combustion Institute, Pittsburgh, Pa., 1979.
175. Brown, G. L., and Roshko, A., On Density Effects and Large Structure in Turbulent Mixing Layers, *Journal of Fluid Mechanics*, vol. 64, pp. 775–816, 1974.
176. Roshko, A., Structure of Turbulent Shear Flows: A New Look, *AIAA Journal*, vol. 14, pp. 1349–1357, 1976.
177. Dimotakis, P. E., and Brown, G. L., The Mixing Layer at High Reynolds Number: Large-Structure Dynamics and Entrainment, *Journal of Fluid Mechanics*, vol. 78, pp. 535–560, 1976.
178. Koochesfahani, M. M., and Dimotakis, P. E., Laser Induced Fluorescence Measurements of Concentration in a Plane Mixing Layer, AIAA 22nd Aerospace Sciences Meeting, Reno, Nev., Paper No. AIAA-84-0198, 1984.
179. Breidenthal, R. E., Structure in Turbulent Mixing Layers and Wakes Using a Chemical Reaction, *Journal of Fluid Mechanics*, vol. 109, pp. 1–24, 1981.
180. Mungal, M. G., Dimotakis, P. E., and Hermanson, J. C., Reynolds Number Effects on Mixing and Combustion in a Reacting Mixing Layer, AIAA 22nd Aerospace Sciences Meeting, Reno, Nev., Paper No. AIAA-84-0371, 1984.
181. Broadwell, J. E., and Dimotakis, P. E., Implications of Recent Experimental Results for Modeling Reactions in Turbulent Flows, AIAA 22nd Acrospace Sciences Meeting, Reno, Nev., Paper No. AIAA-84-0545, 1984.
182. Broadwell, J. E., and Breidenthal, R. E., A Simple Model of Mixing and Chemical Reaction in a Turbulent Shear Layer, *Journal of Fluid Mechanics*, vol. 125, pp. 397–410, 1982.
183. Spalding, D. B., Chemical Reactions in Turbulent Fluids, in *Physicochemical Hydrodynamics*, ed. D. B. Spalding, pp. 321–338, Advanced Publications, London, 1977.
184. Spalding, D. B., A General Theory of Turbulent Combustion: The Lagrangian Aspects, AIAA 15th Aerospace Sciences Meeting, Los Angeles, Calif., Paper No. 77-141, 1977.
185. Spalding, D. B., A General Theory of Turbulent Combustion, *Journal of Energy*, vol. 2, pp. 16–23, 1978.
186. Spalding, D. B., The Theory of Turbulent Reacting Flows—A Review, AIAA 17th Aerospace Sciences Meeting, New Orleans, La., Paper No. 79-0213, 1979.
187. Ma, A. S. C., Noseir, M. A., and Spalding, D. B., An Application of the ESCIMO Theory of Turbulent Combustion, AIAA 18th Aerospace Sciences Meeting, Pasadena, Calif., Paper No. AIAA-80-0014, 1980.

188. Spalding, D. B., A Simple Model for the Rate of Turbulent Combustion, in *Turbulent Combustion*, ed. L. A. Kennedy, *Progress in Astronautics and Aeronautics*, vol. 58, pp. 105–116, AIAA, New York, 1977.
189. Ma, A.S.C., Spalding, D. B., and Sun, R. L. T., Application of "ESCIMO" to the Turbulent Hydrogen-Air Diffusion Flame, *Nineteenth Symposium (International) on Combustion*, pp. 393–402, The Combustion Institute, Pittsburgh, Pa., 1982.
190. Spalding, D. B., Mixing and Chemical Reaction in Steady Confined Turbulent Flames, *Thirteenth Symposium (International) on Combustion*, pp. 649–657, The Combustion Institute, Pittsburgh, Pa., 1971.
191. Abraham, J., Williams, F. A., and Bracco, F. V., A Discussion of Turbulent Flame Structure in Premixed Charges, SAE International Congress and Exposition, Detroit, Mich., SAE Paper No. 850345, 1985.
192. Libby, P. A., and Williams, F. A., Eds., *Turbulent Reacting Flows*, *Topics in Applied Physics*, vol. 44, Springer-Verlag, New York, 1980.
193. Spalding, D. B., Turbulence Modelling: A State-of-the-Art Review, Report CFD/83/3, Computational Fluid Dynamics Unit, Imperial College of Science and Technology, London, 1983.
194. Spalding, D. B, The Two-Fluid Model of Turbulence Applied to Combustion Phenomena, *AIAA Journal*, vol. 24, pp. 876–884, 1986.
195. Spalding, D. B., A Turbulence Model for Buoyant and Combusting Flows, *International Journal for Numerical Methods in Engineering*, vol. 24, pp. 1–23, 1987.
196. Spalding, D. B., Towards a Two-Fluid Model of Turbulent Combustion in Gases, with Special References to the Spark Ignition Engine, *International Conference on Combustion Engineering*, vol. I, pp. 135–142, I. Mech. E. Conference Publications 1983-3, London, 1983.
197. Spalding, D. B., A 2-Fluid Model of Turbulent Combustion in the Reciprocating Engine, Advanced Course on Computer Simulation of Fluid Flow, Heat and Mass Transfer and Combustion in Reciprocating Engines, International Centre for Heat and Mass Transfer, Dubrovnik, Yugoslavia, 1987.
198. Dobran, F., Theory of Multiphase Mixtures: A Thermomechanical Formulation, *International Journal of Multiphase Flow*, vol. 11, pp. 1–30, 1985.
199. Drew, D. A., Mathematical Modelling of Two-Phase Flow, *Annual Review of Fluid Mechanics*, vol. 15, pp. 261–291, 1983.
200. Dobran, F., Constitutive Equations for Multiphase Mixtures of Fluids, *International Journal of Multiphase Flow*, vol. 10, pp. 273–305, 1984.
201. Spalding, D. B., Numerical Simulation of Transient Two-Phase Flow, Advanced Course on Computer Simulation of Fluid Flow, Heat and Mass Transfer and Combustion in Reciprocating Engines, International Centre for Heat and Mass Transfer, Dubrovnik, Yugoslavia, 1987.
202. Bray, K. N. C., Scales and Burning Rates in Premixed Turbulent Flames, 9th Australasian Fluid Mechanics Conference, Auckland, New Zealand, 1986.
203. Borghi, R., Mise au Point sur le Structure des Flames Turbulentes, *Journal de Chimie Physique et de Physico-Chimie Biologique*, vol. 81, pp. 361–370, 1984.
204. Bray, K. N. C., and Moss, J. B., A Unified Statistical Model of the Premixed Turbulent Flame, *Acta Astronautica*, vol. 4, pp. 291–319, 1977.
205. Bray, K. N. C., and Moss, J. B., A Closure Model for the Turbulent Premixed Flame with Sequential Chemistry, *Combustion and Flame*, vol. 30, pp. 125–131, 1977.
206. Bray, K. N. C., Laminar Flamelet Modelling of Turbulent Combustion, in *Numerical Simulation of Combustion Phenomena*, eds. R. Glowinski, B. Larrouturou, and R. Temam, *Lecture Notes in Physics*, vol. 241, pp. 3–19, Springer-Verlag, New York, 1985.
207. Bray, K. N. C., Libby, P. A., and Moss, J. B., Scalar Length Scale Variations in Premixed Turbulent Flames, *Twentieth Symposium (International) on Combustion*, pp. 421–427, The Combustion Institute, Pittsburgh, Pa., 1985.
208. Bray, K. N. C., Libby, P. A., and Moss, J. B., Flame Crossing Frequencies and Mean Reaction Rates in Premixed Turbulent Combustion, *Combustion Science and Technology*, vol. 41, pp. 143–172, 1984.
209. Bray, K. N. C., Champion, M., Dave, N., and Libby, P. A., On the Thermochemistry of

Premixed Turbulent Combustion in Variable Enthalpy Systems, *Combustion Science and Technology*, vol. 46, nos. 1 & 2, pp. 31–43, 1986.

210. Shepherd, I. G., Moss, J. B., and Bray, K. N. C., Turbulent Transport in a Confined Premixed Flame, *Nineteenth Symposium (International) on Combustion*, pp. 423–431, The Combustion Institute, Pittsburgh, Pennsylvania, 1982.
211. Shepherd, I. G., and Moss, J. B., Characteristic Scales for Density Fluctuations in a Turbulent Premixed Flame, *Combustion Science and Technology*, vol. 33, pp. 231–243, 1983.
212. Liew, S. K., Moss, J. B., and Bray, K. N. C., Predicted Structure of Stretched and Unstretched Methane-Air Diffusion Flames, in *Dynamics of Flames and Reactive Systems*, eds. J. R. Bowen, N. Manson, A. K. Oppenheim, and R. I. Soloukhin, *Progress in Astronautics and Aeronautics*, vol. 95, pp. 305–319, AIAA, New York, 1984.
213. Libby, P. A., and Bray, K. N. C., Implications of the Laminar Flamelet Model in Premixed Turbulent Combustion, *Combustion and Flame*, vol. 39, pp. 33–41, 1980.
214. Libby, P. A., Bray, K. N. C., and Moss, J. B., Effects of Finite Reaction Rate and Molecular Transport in Premixed Turbulent Combustion, *Combustion and Flame*, vol. 34, pp. 285–301, 1979.
215. Ishii, M., *Thermo-Fluid Dynamic Theory of Two-Phase Flows*, Eyrolles, Paris, 1975.
216. O'Rourke, P. J., Collective Drop Effects on Vaporizing Liquid Sprays, Ph.D. thesis, Princeton University, Princeton, N.J., 1981.
217. O'Rourke, P. J., and Bracco, F. V., Modelling of Drop Interactions in Thick Sprays and a Comparison with Experiments, *Stratified Charge Automotive Engines*, pp. 101–116, I. Mech. E. Conference Publications 1980-9, London, 1980.
218. Glassman, I., *Combustion*, pp. 181–185, Academic Press, New York, 1977.
219. O'Rourke, P. J., Personal communication, Dubrovnik, Yugoslavia, September 1, 1987.
220. Bracco, F. V., and Sirignano, W. A., Theoretical Analysis of Wankel Engine Combustion, *Combustion Science and Technology*, vol. 7, pp. 109–123, 1973.
221. Bracco, F. V., Theoretical Analysis of Stratified, Two-Phase Wankel Engine Combustion, *Combustion Science and Technology*, vol. 8, pp. 69–84, 1973.
222. Reitz, R. D., and Bracco, F. V., Studies toward Optimal Charge Stratification in a Rotary Engine, *Combustion Science and Technology*, vol. 12, pp. 63–74, 1976.
223. Incropera, F. P., and DeWitt, D. P., *Fundamentals of Heat Transfer*, p. 406, Wiley, New York, 1981.
224. Sirignano, W. A, One-dimensional Analysis of Combustion in a Spark-Ignition Engine, *Combustion Science and Technology*, vol. 7, pp. 99–108, 1973.
225. Ramos, J. I., Numerical Studies of Laminar Flame Propagation in Spherical Bombs, *AIAA Journal*, vol. 21, no. 3, pp. 415–422, 1983.
226. Lee, D. N., and Ramos, J. I., Application of the Finite Element Method to One-Dimensional Flame Propagation Problems, *AIAA Journal*, vol. 21, no. 2, pp. 262–269, 1983.
227. Shih, T. I-P., Schock, H. J., Nguyen, H. L., and Stegeman, J. D., Numerical Simulation of the Flowfield in a Motored Two-Dimensional Wankel Engine, *AIAA Journal of Propulsion and Power*, vol. 3, pp. 269–276, 1987.
228. Beam, R. M., and Warming, R. F., An Implicit Factored Scheme for the Compressible Navier-Stokes Equations, *AIAA Journal*, vol. 16, pp. 393–402, 1978.
229. Beam, R. M., and Warming, R. F., An Implicit Finite-Difference Algorithm for Hyperbolic Equations in Conservation-Law Form, *Journal of Computational Physics*, vol. 22, pp. 87–100, 1976.
230. Yang, S. L., and Shih, T. I-P., An Algebraic Grid Generation Technique for Time-Varying Two-Dimensional Spatial Domains, *International Journal for Numerical Methods in Fluids*, vol. 6, pp. 291–304, 1986.
231. Smith, R. E., Two-Boundary Grid Generation for the Solution of the Three-Dimensional Compressible Navier-Stokes Equations, NASA Technical Memorandum 83123, 1981.
232. Shih, T. I-P., Smith, G. E., Springer, G. S., and Rimon, Y., Boundary Conditions for the Solution of the Compressible Navier-Stokes Equations by an Implicit Factored Method, *Journal of Computational Physics*, vol. 52, pp. 54–79, 1983.

233. Rotta, J. C., Turbulent Boundary Layers in Incompressible Flow, in *Boundary Layer Problems*, eds. A. Ferri, D. Kuchemann, and L. H. G. Sterne, *Progress in Aeronautical Sciences*, vol. 2, pp. 1–219, Pergamon Press, New York, 1962.
234. Reynolds, A. J., *Turbulent Flows in Engineering*, Wiley, New York, 1974.
235. Davies, J. T., *Turbulence Phenomena*, Academic Press, New York, 1972.
236. Cebeci, T., and Smith, A. M. O., *Analysis of Turbulent Boundary Layers*, Academic Press, New York, 1974.
237. Kutateladze, S. S., and Leont'ev, A. I., *Turbulent Boundary Layers in Compressible Gases*, Academic Press, New York, 1964.
238. Steinthorsson, E., Shih, T. I-P., Schock, H. J., and Stegeman, J., Calculations of the Unsteady, Three-Dimensional Flow Field inside a Motored Wankel Engine, SAE International Congress and Exposition, Detroit, Mich., SAE Paper No. 880625, 1988.
239. Amsden, A. A., Ramshaw, J. D., Cloutman, L. D., and O'Rourke, P. J., Improvements and Extensions to the KIVA Computer Program, Report No. LA-10534-MS, Los Alamos National Laboratory, Los Alamos, N.M., 1985.

CHAPTER

THREE

MATHEMATICAL MODELS OF DIESEL ENGINES

3.1 INTRODUCTION

The objective of this chapter is to present and review some of the mathematical models that have been used to analyze diesel engine combustion. The emphasis here is on the major assumptions of the models and their limitations.

Models of diesel engine combustion can be classified into two groups: thermodynamic and multidimensional. Thermodynamic models can in turn be classified into two subgroups: single- and multizone. In single-zone models the cylinder charge (i.e., the mixture in the cylinder) is assumed to be uniform in both composition and temperature, and the first law of thermodynamics is used to calculate the mixture energy accounting for the enthalpy flux due to fuel injection. The fuel injected into the cylinder is assumed to mix instantaneously with the cylinder charge, which is assumed to behave as an ideal gas.

Single-zone models can be used to analyze the heat release rate if experimentally determined pressure diagrams are specified in the first law of thermodynamics. Alternatively, single-zone models can be used as predictive tools if the heat release rate or the fuel mass burning rate is specified. The heat release rate may account for both premixed and diffusive burning by means of, for example, a Wiebe function. Premixed burning occurs in the first stages of combustion, where the fuel is vaporized and mixed with the fresh mixture. Once the premixed fuel-air mixture is consumed, diffusive burning takes place and governs most of the combustion duration.

Single-zone models yield a system of ordinary differential equations for the mixture pressure, temperature, and mass. However, they do not account for the presence of vaporizing liquid droplets, air entrainment, combustion chamber geometry, and spatial variations of the mixture composition and temperature.

Multizone models account for the temporal and spatial distributions of temperature and concentration by dividing the injected liquid fuel into parcels assumed to have uniform composition and temperature. Spray tip and spray width correlations are used to calculate the location of each parcel. These correlations are frequently based on experimental data for steady-state gaseous fuel jets and may account for deflection of the fuel jet by swirl and fuel impingement on solid walls.

Most of the multizone models discussed in this chapter consider a gaseous fuel jet (i.e., the liquid fuel jet is injected into the cylinder under critical or supercritical conditions). Other multizone models divide the liquid fuel jet into droplets, which are assigned to parcels. Air entrainment and droplet vaporization and combustion in each parcel is accounted for by means of droplet vaporization models that consider forced-convection effects.

In multidimensional models the time-dependent, instantaneous conservation equations of mass, momentum, energy, and species are time averaged, and the turbulent correlations are considered to be proportional to the gradients of the mean flow. The details of the atomization process (i.e., liquid jet breakup into ligaments and droplets) are neglected, and the mass injected into the cylinder at each time step is assigned to a droplet distribution function, which in turn is discretized into a finite number of droplet packets. All the droplets contained in a packet have the same diameter, velocity, and temperature, and a Lagrangian formulation is employed to account for the mass, momentum, and energy exchanges between the gas phase and the droplets.

Stochastic models are introduced in multidimensional models in order to consider the effects of the gas-phase turbulence on the liquid droplets; however, the effects of the droplets on turbulence have, for the most part, been ignored. These effects depend on the droplet diameter, droplet residence time in a typical eddy, eddy turnover time, length scale of the energy-containing eddies, and number of droplets in a turbulent eddy.

In the next sections, single-zone, multizone, and multidimensional models of diesel engine combustion are discussed, and comparisons between numerical predictions and experimental data are presented. Multidimensional models of direct-injection stratified-charge (DISC) engines that consider a gaseous charge are also discussed; DISC engines offer great potential in terms of both efficiency and pollutant emissions.

3.2 SINGLE-ZONE MODELS

In single-zone models of diesel engine combustion the cylinder charge is assumed to be a homogeneous mixture of ideal gases at all times. The instantaneous state of the mixture can be described by the mixture pressure p, temperature T, and equivalence ratio ϕ. Fuel is added to the cylinder throughout the combustion period to increase the energy and fuel-air ratio of the charge; however, the simulated fuel addition has no physical relationship with the actual fuel-injection process except that its total mass is equal to the actual total fuel mass. The fuel burns instantaneously as it is added to the cylinder so that the effect of the unburnt-fuel vapor on ignition delay or combustion is ignored.

In this section a model that accounts for the flow through the intake and exhaust valves is described. Quasisteady one-dimensional equations are used to calculate the flow rates through the intake and exhaust valves.

If leakage through the piston rings and the mass flows into crevices are neglected, conservation of mass applied to the cylinder yields

$$\frac{dm}{dt} = \dot{m}_{\text{int}} + \dot{m}_f - \dot{m}_{\text{exh}} \qquad m = \rho V \qquad p = \rho RT \tag{1}$$

where m is the mass contained in the cylinder; $\dot{m}_{\text{int}}$ and $\dot{m}_{\text{exh}}$ are the instantaneous mass flow rates through the intake and exhaust valves, respectively; $\dot{m}_f$ is the rate of the "imaginary" fuel addition; t is time; V is the cylinder volume; p, ρ, and T are the mixture pressure, density, and temperature, respectively; and R is the mixture gas constant.

The first law of thermodynamics applied to the mixture contained in the cylinder can be written as

$$\frac{d}{dt}(mu) = -p\frac{dV}{dt} + \dot{q}_w + \Sigma\, h_j\, \dot{m}_j \tag{2}$$

where u is the specific internal energy of the mixture; $\dot{q}_w$ is the instantaneous heat transfer rate into the cylinder; and $\Sigma\, h_j\, \dot{m}_j$ represents the enthalpy fluxes through the intake and exhaust valves as well as the enthalpy flux associated with fuel injection.

The internal energy and the gas constant are functions of T, p, and ϕ [1]. The mixture equivalence ratio can be calculated by considering that the total mass in the cylinder can be expressed as the sum of the mass of air m_a and the mass of fuel m_f, that is,

$$m = m_a + m_f \tag{3}$$

The equivalence ratio is defined as

$$\phi = \frac{m_f/m_a}{(m_f/m_a)_{st}} = \frac{f}{f_{st}} \tag{4}$$

where the subscript st denotes stoichiometric conditions and f denotes the fuel-air ratio.

Equation (4) can be written as

$$\phi = m_f\frac{1 + \phi f_{st}}{m f_{st}} \tag{5}$$

which can be differentiated with respect to time to yield, after some algebraic manipulations,

$$\frac{d\phi}{dt} = \frac{1 + \phi f_{st}}{m f_{st}} \left[(1 + \phi f_{st}) \frac{dm_f}{dt} - \phi f_{st} \frac{dm}{dt} \right] \tag{6}$$

Equation (2) can also be written as [2]

$$\frac{dT}{dt} = \left\{ \left[\frac{p}{m} \frac{dV}{dt} + \frac{1}{M} \left(\dot{q}_w + \Sigma\, h_j\, \dot{m}_j - u\, \dot{m} - m \frac{\partial u}{\partial \phi} \frac{d\phi}{dt} \right) \right] - A \right\} \times \left(\frac{\partial u}{\partial T} + \frac{B}{C} \frac{p}{T} \frac{\partial u}{\partial p} \right)^{-1} \tag{7}$$

where

$$A = \frac{p}{C} \frac{\partial u}{\partial p} \left(\frac{\dot{m}}{m} - \frac{\dot{V}}{V} + \frac{1}{R} \frac{\partial R}{\partial \phi} \dot{\phi} \right) \tag{8}$$

$$B = 1 + \frac{T}{R} \frac{\partial R}{\partial T} \tag{9}$$

$$C = 1 - \frac{p}{R} \frac{\partial R}{\partial T} \tag{10}$$

$$pV = m\, RT \tag{11}$$

and the dots denote differentiation with respect to time.

3.2.1 Premixed-Diffusive Models

According to the correlations obtained by Watson et al. [3], combustion in a direct-injection (DI) diesel engine is considered to start at the dynamic injection point, defined as the crank shaft angle at which the injector needle starts to lift, and to consist of two distinct phases: the ignition delay and the heat release period (Fig. 3.1). The ignition delay is considered to be the time interval between the actual dynamic injection point and ignition. The ignition delay time t_d can be related to the cylinder pressure and temperature through the following semi-empirical equation [4]:

$$t_d = a_1\, p_m^{a_3} \exp\left(\frac{a_2}{T_m} \right) \tag{12}$$

where a_1, a_2, and a_3 are empirical constants and p_m and T_m are the arithmetic mean pressure and temperature in the cylinder during ignition delay, which can be calculated from the following expressions [5]:

$$T_m = \sum_{j=1}^{J} \frac{T_j}{J} \qquad p_m = \sum_{j=1}^{J} \frac{p_j}{J} \tag{13}$$

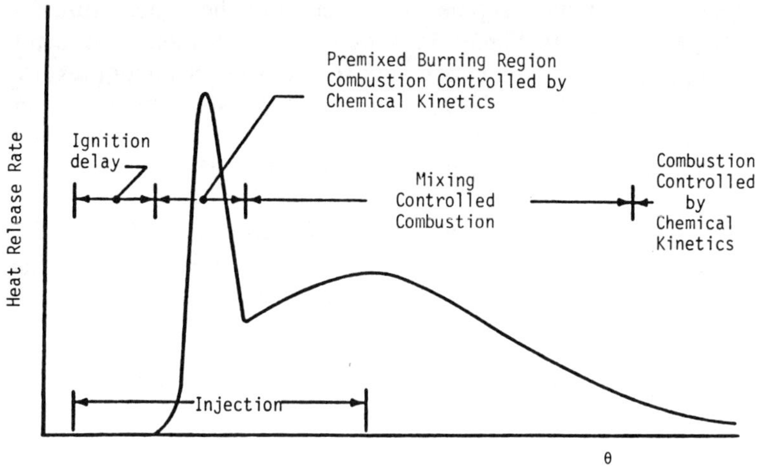

Figure 3.1 Schematic of the heat release rate in a direct-injection diesel engine.

where J is the number of time steps during the ignition delay, p_j and T_j are the cylinder pressure and temperature at time $j\ \Delta t$ after injection, and Δt is the time step.

Expressions for the ignition delay time [Eq. (12)] have been determined in constant-volume bombs by Hurn and Hughes [6] and by Baev et al. [7]. More recently, Ikegami et al. [8] and Tsao et al. [9] measured the induction period of diesel sprays in a rapid-compression machine and in a reciprocating engine, respectively. Gerrish and Ayer [10], Spadaccini [11], Walsh and Cheng [12], and Scharnweber and Hoppie [13] showed that preheating the fuel before injection decreases the induction (or ignition delay) time. Parker et al. [14] examined the effects of preheating both the fuel and compressed air in a constant-volume bomb to supercritical temperatures of pure n-dodecane and showed that the induction time can be expressed as

$$t_d = A \exp\left(\frac{T_a}{T}\right) \tag{14a}$$

$$T_a = 9032 \text{ K} \tag{14b}$$

$$A = 2.96 \times 10^{-8} \tag{14c}$$

for 4.4 atm $\leq p \leq$ 28 atm and 600 K $\leq T \leq$ 875 K, where A is the preexponential factor and T_a is the activation temperature.

Parker et al. [14] showed that the induction period is insensitive to the combustion chamber pressure and that preheating the fuel before injection has a significant influence in shortening the induction period by diminishing the vaporization and mixing times.

Figure 3.2 shows the induction time as a function of the temperature for various chamber pressures [14]. Figure 3.3 presents the preexponential factor and activation temperature T_a as a function of the combustion chamber pressure [14]. These figures indicate that the induction time decreases as the chamber temperature is increased; they also indicate that the preexponential factor and the activation temperature [Eq. (14*a*)] are functions of the combustion chamber pressure.

Shipinski et al. [15] used the method proposed by Krieger and Borman [1] to analyze the heat release in a 1.2-liter diesel engine of compression ratio equal to 16 at 2000 rpm and obtained the following crank angle for the beginning of combustion:

$$\theta = \theta_{\text{inj}} + \frac{6\ \text{rpm}}{1000}\,\frac{0.8216}{p_m^{0.386}}\exp\left(\frac{4644}{T_m}\right)\left(\frac{40}{\text{CN}}\right)^{0.69} \tag{15}$$

where θ_{inj} is the crank angle at which injection starts, CN is the fuel cetane number, rpm denotes the engine rpm, and p_m and T_m are the average pressure and temperature from the beginning of injection until the beginning of combustion.

Sitkei [16] developed an expression for the ignition delay time that accounts for cold and blue flames as

$$t_d = 0.5 + 0.133\,p^{-0.7}\exp\left(\frac{3930}{T}\right) + 0.00463\,p^{-1.8}\exp\left(\frac{3930}{T}\right) \tag{16}$$

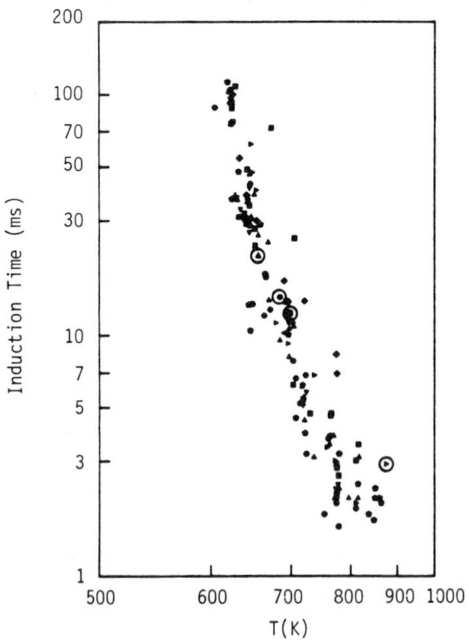

Figure 3.2 Induction time as a function of the combustion chamber temperature [14].

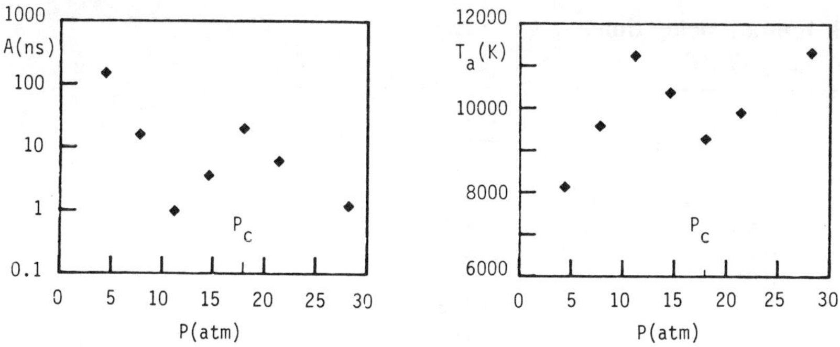

Figure 3.3 Preexponential factor and activation temperature of the induction time [Eq. (14*a*)] as a function of the combustion chamber pressure [14].

and a Wiebe function that depends on the combustion duration, rpm, and cylinder pressure and on the temperature at intake valve closing (IVC).

Equations (12), (14), (15), and (16) have been derived from experimental tests in uniform air environments where the pressure and temperature changed only due to the cooling effect of the fuel-vaporization and fuel-heating process. However, in a diesel engine, the pressure and temperature change considerably during the ignition delay time due to the compression resulting from the piston motion. Furthermore, the form of the correlations for the ignition delay time does not allow all the influencing fuel and engine parameters to be included in the calculation of t_d.

Hardenberg and Hase [17] developed an empirical formula for predicting the ignition delay in DI engines as a function of the fuel characteristics, engine parameters, and ambient conditions. Dent and Mehta [18] showed that this formula can give good agreement with experimental data over a wide range of engine conditions. However, the pressure and temperature used in their correlation are identified as the corresponding conditions at top dead center (TDC) and estimated using a polytropic model for the compression stroke.

The ignition delay time can be calculated as the difference between the time at which combustion starts t_{ig} and the time at which injection starts t_{inj}. The time t_{ig} can be obtained from the following expression:

$$\int_{t_{\text{inj}}}^{t_{\text{ig}}} \frac{dt}{t_d} = 1 \tag{17}$$

which accounts for the pressure and temperature variations resulting from compression. In Table 3.1, the ignition delay time is presented for a variety of fuels [19].

Baev et al. [7] studied the induction time of gasoline and diesel fuel sprays injected into a reservoir containing air at high pressure and temperature, and they correlated their results using Arrhenius's expressions similar to those shown in Table 3.1. Their results reveal the existence of a critical temperature at which

Table 3.1 Ignition delay time, $t_d = A\,p^{-n} \exp(T_a/T)$

Fuel	A	n	T_a (K)
Diesel	3.45	1.02	2100
Diesel	53.5	1.23	676.5
n-Cetane	0.872	1.24	4050
n-Heptane	0.748	1.44	5270
Diesel	0.0405	0.757	5470
Diesel	2.43×10^{-6}	2	20,915.3
Kerosene	1.68×10^{-5}	2	19,008.7
Cetane	4.04×10^{-10}	2	25,383

the ignition mechanism undergoes a drastic change. At the critical temperature, the induction time is governed not so much by the reaction of the fuel-air mixture as by the development of the concentration field and the initial temperature in the fuel-air mixing zone. As a consequence, the activation energy determined from the measurements of the ignition time cannot be considered as an expression for the ignition delay time in the usual sense (i.e., the ignition time depends on the fluid mechanics of the fuel-air mixing).

Gulder and Glavincevski [20] studied the relations between ignition quality and hydrogen distribution in the fuel and between ignition delay time and cetane number. Their studies indicate that the ignition delay time is not constant but depends on the pressure, temperature, and fuel chemistry. Kirsch et al. [21] showed that at low temperatures the cetane number is insufficient to characterize the ignition quality of fuels with significant aromatic content, and they observed a two-stage process leading to ignition. In the first stage the liquid jet breaks up and there is mixing between the fuel and air, whereas chemical reactions characterize the second stage.

Belardini et al. [22] showed that the ignition delay time is a function of the compression ratio, swirl, and rpm in DI diesel engines and that no existing ignition delay time correlation fits the available experimental data. They also showed that the illumination delay is not a good indicator of the ignition delay time.

The heat release period consists of two different modes of burning: premixed and diffusive burning (Fig. 3.1). These two modes are assumed to proceed concurrently from the ignition point during a small part of the burning period. Initially, most of the fuel burns in the premixed mode until the fuel-air mixture that was prepared and ready to burn prior to ignition is exhausted. After this point burning is assumed to proceed only in the diffusive mode until the end of combustion.

The apparent fuel-burning rate can be expressed in the term of the premixed $\dot{m}_p$ and diffusive $\dot{m}_d$ burning rates as

$$\dot{m}_f = \dot{m}_p + \dot{m}_d \tag{18}$$

In order to indicate the burning mode, a factor β defined by

$$\beta = \frac{m_p}{m_f} = 1 - 0.926\, \phi_{ig}^{0.37}\, t_d^{-0.26} \tag{19}$$

and the following expressions

$$\frac{dm_k}{dt} = \frac{dM_k}{d\tau}\frac{m_k}{\Delta\theta_b}\frac{360\text{ rpm}}{60} \qquad k = f,p,d \tag{20}$$

are introduced into Eq. (18). The nondimensional apparent fuel-burning rate can then be written as

$$\frac{dM}{d\tau} = \beta\frac{dM_p}{d\tau} + (1-\beta)\frac{dM_d}{d\tau} \tag{21}$$

where

$$\frac{dM_p}{d\tau} = c_1\, c_2\, \tau^{c_1 - 1}\,(1 - \tau^{c_1})^{c_2 - 1} \tag{22}$$

$$\frac{dM_d}{d\tau} = c_1'\, c_2'\, \tau^{c_2' - 1} \exp\,(-c_1'\, \tau^{c_2'}) \tag{23}$$

$$\tau = \frac{\theta - \theta_{ig}}{\Delta\theta_b} \qquad \Delta\theta_b = \theta_{end} - \theta_{ig} \tag{24}$$

$$c_1 = 2 + 1.25 \times 10^{-8}\,(\text{rpm} \cdot t_d)^{2.4} \qquad c_2 = 5000 \tag{25}$$

$$c_1' = 14.2\, \phi_{ig}^{-0.644} \qquad c_2' = 0.79(c_1')^{0.25} \tag{26}$$

$$\beta = 1 - 0.926\, \phi_{ig}^{0.37}\, t_d^{-0.26} \tag{27}$$

θ_{ig} is the ignition angle; θ_{end} is the crank shaft angle at the end of combustion; θ is the crank shaft angle ($\theta = \text{rpm} \cdot t$); ϕ_{ig} is the equivalence ratio at the time of ignition; $\Delta\theta_b$ is the combustion duration, which is a function of the engine load and speed; and the values of c_1, c_2, c_1', and c_2' depend on the diesel engine considered.

Equations (1), (6), (7), (17), and (21) can be solved for m, ϕ, T, t_{ig}, and $dm_f/d\theta$ once the enthalpy of the fuel and the heat transfer losses are specified. The enthalpy of the fuel can be calculated by specifying the low heating value (LHV) of the fuel at 25°C and 1 atm, as [1]

$$h_f = 2.326[\text{LHV} - 19{,}183 + 0.5(T_f - 537)] \tag{28}$$

where T_f and h_f are the temperature and enthalpy of the fuel.

Wu [23] used the measured cylinder pressure to determine the fuel-burning rate in diesel engines and showed that the peak specific fuel burning rate in the premixed phase of combustion decreases as the engine speed is increased, and decreases as the overall fuel-air ratio is increased. The ignition delay time increased with the engine speed and air-fuel ratio. However, no correlation was found between the premixed burning period and the ignition delay time.

3.2.2 Heat Transfer Correlations

The instantaneous wall heat transfer can be calculated by means of a correlation such as the one developed by Woschni [24]:

$$\dot{q}_w = hA(T_w - T) \tag{29}$$

$$h = 0.00326\, p^{0.8} \frac{(v_{\text{mot}} + v_{\text{comb}})^{0.8}}{B^{0.2}T^{0.53}} \tag{30}$$

$$v_{\text{mot}} = c_1'' v_{\text{pis}} \qquad v_{\text{comb}} = c_2'' \frac{V_d T_1}{p_1 V_1} (p - p_{\text{mot}}) \tag{31}$$

$$v_{\text{pis}} = \frac{2\, S \cdot \text{rpm}}{60} \tag{32}$$

where h is the film heat transfer coefficient; T_w is the wall temperature; B is the cylinder bore; v_{pis} is the average piston speed; V_d is the displacement volume $(= \pi B^2 S/4)$; p_1, V_1, and T_1 are the pressure, volume, and temperature at ignition; p_{mot} is the value of the pressure under motored conditions; S is the stroke; and the subscripts comb and mot denote combustion and motored conditions, respectively. The values of c_1'' and c_2'' are shown in Table 3.2; $v_{\text{comb}} = 0$ for motored conditions.

Many other convective heat transfer correlations have been proposed in the literature and can be summarized as [25, 26]

$$\text{Nu} = \frac{hL}{\lambda} = a\, \text{Re}^d\, \text{Pr}^e \tag{33}$$

where Nu, Re, and Pr are the Nusselt, Reynolds $(= \rho v L/\mu)$, and Prandtl $(= \mu C_p/\lambda)$ numbers, respectively; λ is the mixture thermal conductivity; L is a characteristic length; v is a characteristic velocity; ρ is the mixture density; μ is the mixture dynamic viscosity; C_p is the mixture specific heat at constant pressure; and a, d, and e are constants adjusted to fit experimental data.

Trifunovic [27] generalized Eq. (33) and accounted for the liquid fuel film along the cylinder liner, swirl, and squish. Local heat transfer coefficients can also be calculated by means of the von Karman-Pohlhausen method [28]. These

Table 3.2 Parameters in Woschni's correlation

	c_1''	c_2''
Compression	2.28	0
Combustion and expansion	2.28	0.00324
Gas exchange process	6.18	0

calculations are generally based on steady-state, incompressible flows with zero pressure gradients, and they provide both the values of the hydrodynamic and thermal boundary layer thicknesses and the local film heat transfer coefficient h along the solid walls. It is questionable whether integral methods for analyzing boundary layers are accurate enough to predict wall heat transfer because the characteristics of engine flows may not be similar to those along flat plates.

Morel et al. [29, 30] used a thermodynamic engine cycle model, a steady-state heat conduction model of the engine structure, and a one-dimensional time-dependent model along the combustion chamber surfaces to show that heat is absorbed by the wall during the power stroke and returned to the gas during the rest of the cycle in low-heat-rejection or insulated engines.

The variation of the Prandtl number of the air and combustion products is small, and Eq. (33) can be written as

$$\mathrm{Nu} = a\,\mathrm{Re}^{d} \tag{34}$$

In order to calculate the film heat transfer coefficient from Eq. (34), the instantaneous values of L, λ, ρ, and μ are needed. It is currently impossible to accurately predict the spatial and temporal variations of λ, ρ, and μ in internal combustion engines. This difficulty can be overcome by using representative values of the characteristic length scale, velocity, and mixture pressure, temperature, and equivalence ratio at which the gas properties are to be evaluated. For example, Eq. (30) is based on the cylinder bore and the velocities due to the piston motion and combustion.

Many attempts have been reported to determine a and d through curve-fitting experimental data [25, 26]. The suggested values are $0.035 \leq a \leq 0.13$ and $0.7 \leq d \leq 0.8$, depending on the intensity of the charge motion. The density and transport properties that appear in Eq. (34) are evaluated at the mean gas pressure, temperature, and equivalence ratio of the cylinder mixture. For example, Mansouri and Heywood [31] used the NASA equilibrium program [32] to calculate the viscosity of hydrocarbon-air combustion products as a function of T, p, and ϕ and found that

$$\mu_{\mathrm{prod}} = \frac{3.3 \times 10^{-7}\, T^{0.7}}{1 + 0.027\,\phi} \qquad (\mathrm{kg/m \cdot s}) \tag{35}$$

for $500\ \mathrm{K} \leq T \leq 4000\ \mathrm{K}$ and $0 \leq \phi \leq 4$.

The equilibrium Prandtl number (Pr) of hydrocarbon-air combustion products can be correlated with the ratio of specific heats γ as

$$\mathrm{Pr} = 0.05 + 4.2\,(\gamma - 1) - 6.7(\gamma - 1)^2 \tag{36}$$

for lean mixtures ($\phi \leq 1$) and $500\ \mathrm{K} \leq T \leq 4000\ \mathrm{K}$; the following correlation applies:

$$\mathrm{Pr} = \frac{0.05 + 4.2(\gamma - 1) - 6.7(\gamma - 1)^2}{1 + 0.015 \times 10^{-6}\,(\phi T)^2} \tag{37}$$

for rich mixtures ($1 < \phi \leq 4$) and $2000\ \mathrm{K} \leq T \leq 3500\ \mathrm{K}$.

In Eqs. (36) and (37)

$$\gamma = \frac{C_p}{C_v} \qquad \mathrm{Pr} = \mu \frac{C_p}{\lambda} \tag{38}$$

where C_p and C_v are the specific heats at constant pressure and at constant volume, respectively.

The characteristic length and velocity scales that appear in Eq. (34) can be estimated by means of a turbulence model [33, 34], which consists of a zero-dimensional energy cascade as follows. The mean flow kinetic energy K supplied to the cylinder through the valves is converted to turbulence kinetic energy k through a turbulent dissipation process. The turbulence kinetic energy is converted to heat through viscous dissipation. When mass flows out of the cylinder, it carries with it both mean and turbulence kinetic energy.

At any time during the cycle the mean velocity U and the turbulence intensity u' can be found from K and k respectively, as follows:

$$K = \frac{mU^2}{2} \tag{39a}$$

$$k = 3m \frac{u'^2}{2} \tag{39b}$$

where we have assumed that the turbulence is isotropic.

The time rates of change of K and k are

$$\frac{dK}{dt} = \Sigma \frac{1}{2} \dot{m}_{\mathrm{int}} U_{\mathrm{int}}^2 - P - K \frac{\dot{m}_{\mathrm{exh}}}{m} \tag{40}$$

$$\frac{dk}{dt} = P - m\epsilon - k \frac{\dot{m}_{\mathrm{exh}}}{m} + RD \tag{41}$$

where $\dot{m}_{\mathrm{int}}$ and $\dot{m}_{\mathrm{exh}}$ are the mass flow rates into and out of the cylinder, respectively; P is the production of turbulence kinetic energy; U_{int} denotes the mean flow velocity into the cylinder; RD is the turbulence amplification due to rapid distortion; and ϵ is the dissipation rate of turbulence kinetic energy given by

$$\epsilon = \frac{u'^3}{l} = \frac{(2k/3m)^{1/2}}{l} \tag{42}$$

where l is the length scale of the energy-containing eddies.

If the turbulence production within the cylinder engine is similar to that of a boundary layer over a flat rate [35]

$$P = \mu_t c_3 \left(\frac{U}{L}\right)^2 \qquad \mu_t = \frac{C_\mu k^2}{m\epsilon} \qquad C_\mu = 0.09 \tag{43}$$

where c_3 is a constant and L can be identified as the characteristic size of the large-scale eddies l, which is assumed to be given by

$$l = L = \frac{4V}{\pi B^2} \tag{44}$$

subject to the restriction that $L \leq B/2$.

Equation (43) can be written as

$$P = 0.3307\, c_3 \frac{K}{L} \left(\frac{k}{m}\right)^{1/2} \tag{45}$$

During compression and combustion, the turbulence kinetic energy decays due to viscous dissipation but is amplified due to the rapid distortion that the cylinder charge undergoes with rising cylinder pressures [36–38].

The rate of turbulence amplification due to rapid distortion can be calculated by assuming that conservation of mass and angular momentum can be applied to the large-scale eddies during the rapid-distortion period, that is,

$$\rho \text{ Vol} = \text{constant} = (\rho \text{ Vol})_0 \tag{46a}$$

$$\text{Vol} \sim L^3 \tag{46b}$$

$$u'L = (u'L)_0 \tag{46c}$$

Substitution of Eq. (46*c*) into Eq. (46*a*) and differentiation with respect to time yield

$$\frac{du'}{dt} = \frac{u'}{3\rho}\frac{d\rho}{dt} \qquad RD = 3mu'\frac{du'}{dt} = \frac{2}{3}\frac{k}{\rho}\frac{d\rho}{dt} \qquad \rho = \frac{m}{V} \tag{47}$$

where Vol is the volume of the large-scale eddies.

The rapid-distortion term RD from Eq. (47) can be substituted into Eq. (41), which can then be used to calculate u' through Eq. (39*b*). The turbulence intensity can then be employed to calculate a characteristic velocity v_c as follows:

$$v_c = \left[U^2 + u'^2 + \left(\frac{v_{\text{pis}}}{2}\right)^2\right]^{1/2} \tag{48}$$

The values of v_c and l can then be used to calculate the Reynolds number ($\text{Re} = \rho\, v_c\, l/\mu$) and the film heat transfer coefficient from Eq. (34).

Morel and Keribar [39] extended the zero-dimensional turbulence model represented by Eqs. (40) and (41) to study heat transfer in bowl-in-piston combustion chambers. They divided the combustion chamber into three zones (multizone model) and derived equations for the turbulence kinetic energy, turbulence dissipation rate, and angular momentum in each zone. Their model predicts that the heat transfer coefficients vary from surface to surface, depending on the engine geometry and flow field.

Figures 3.4–3.7 show the p–V diagram, the pumping losses, the cylinder temperature, and the heat transfer rate in a DI diesel engine [5]. The results presented in Figs. 3.4–3.7 were obtained with the model represented by Eqs. (21)–(27) and Woschni's heat transfer correlation [24].

In addition to convective heat transfer [Eqs. (33) and (34)], the primary sources of radiative heat transfer in a diesel engine are the high temperature of the burnt gases and the soot particles formed as an intermediate step in the turbulent diffusion-controlled combustion. The radiative heat transfer due to soot depends on the particle size distribution, number density, temperature, and flame geometry, and it is difficult to estimate and measure. Estimates of the relative importance of radiation in cooled engines vary between 5% and 50% of the total heat transfer [25, 40–48].

The radiant heat flux depends on the location in the combustion chamber, crank angle (time), and engine load and geometry. At high loads, measurements [40–48] indicate that the radiative heat transfer is between 25% and 45% of the total heat flux. However, due to the complexity of the problem, there are no good predictive formulas for radiative heat transfer in diesel engines. Annand [25] proposed an expression based on an adjustable empirical constant C_R to predict the radiation term $\dot{q}_R$ as

$$\dot{q}_R = C_R A(T^4 - T_w^4) \tag{49}$$

where $C_R = C_R^* \sigma$, with σ being the Stefan-Boltzmann constant and $0.6 \leq C_R^* \leq 3.1$, depending on the engine load [49]. Flynn et al. [47] employed a monochromator to measure the radiation intensity at several wavelengths and

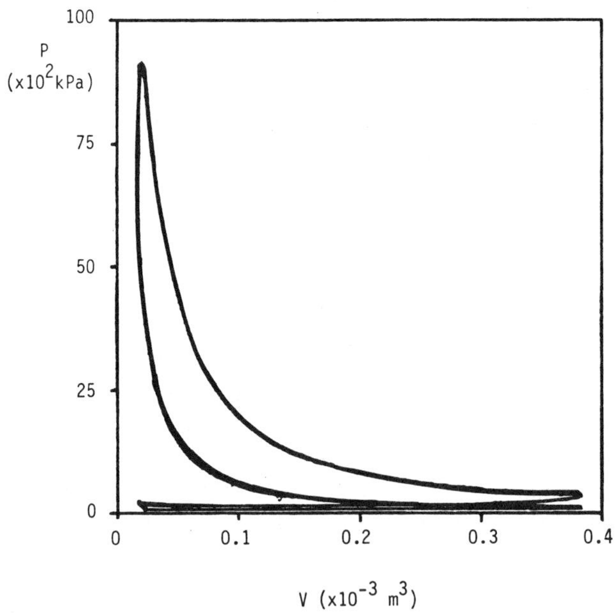

Figure 3.4 Pressure–volume (p–V) diagram in a direct-injection diesel engine [5].

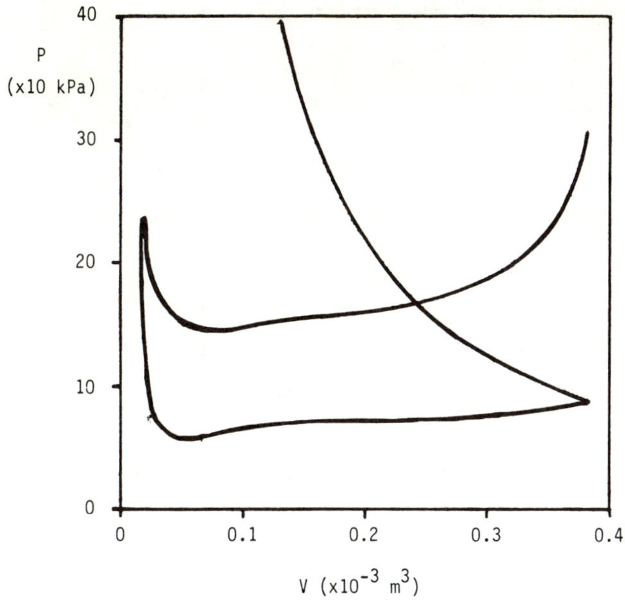

Figure 3.5 Pumping losses in a direct-injection diesel engine [5].

obtained an apparent radiant temperature, an apparent gray body emissivity, and a total radiant heat transfer rate at each crank angle. Flynn et al. [47] showed that the apparent radiant temperature T_R is much higher than the average mixture temperature. The apparent emissivity was about 0.8 to 0.9 during the period of maximum radiation and then dropped to zero by the end of combustion.

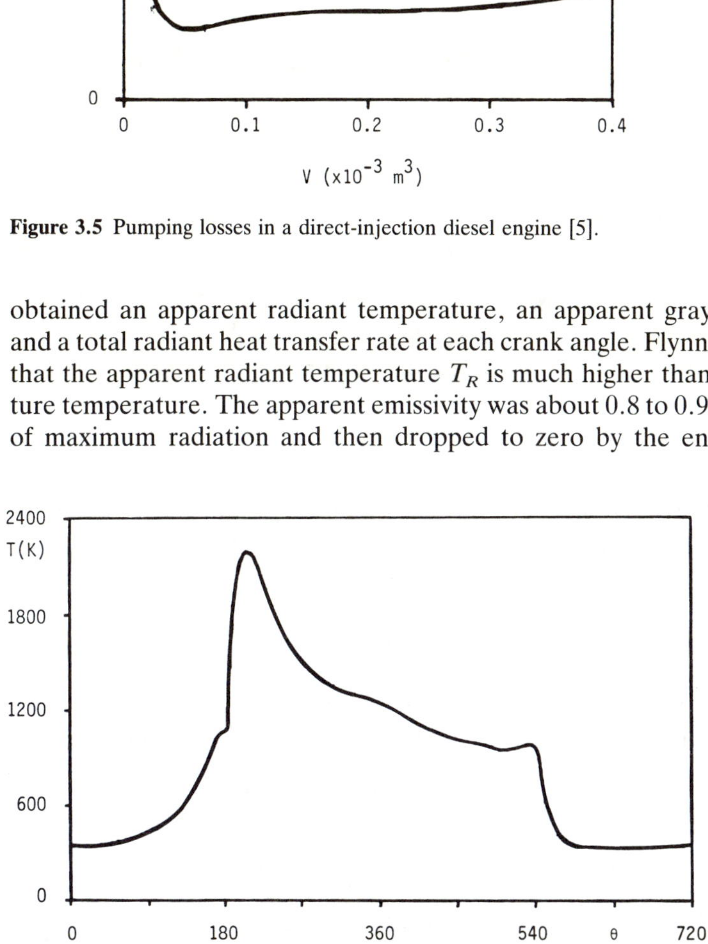

Figure 3.6 Temperature profile in a direct-injection diesel engine as a function of the crank shaft angle [5].

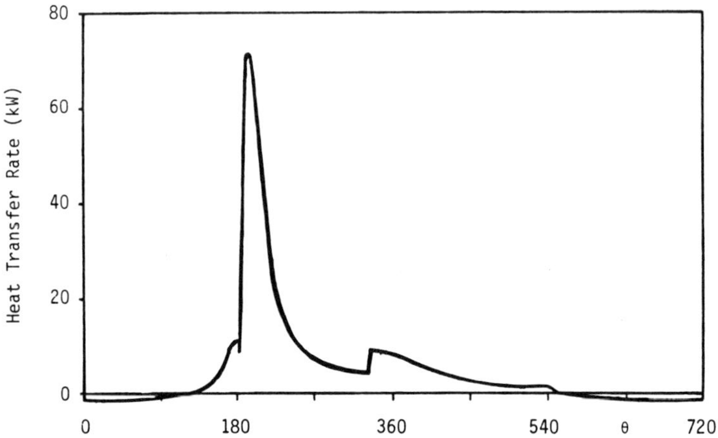

Figure 3.7 Wall heat transfer rate as a function of the crank shaft angle in a direct-injection diesel engine [5].

These results can be summarized as

$$\dot{q}_R = \epsilon_a \sigma A (T_R^4 - T_w^4) \tag{50}$$

where ϵ_a is the apparent gray body emissivity and T_R is close to the adiabatic flame temperature during the period of peak heat release.

Assanis [19] proposed the following expression for the apparent radiant temperature:

$$T_R = \frac{T + T(\phi = 1.1)}{2} \tag{51}$$

where $T(\phi = 1.1)$ is the temperature of the combustion products at an equivalence ratio of 1.1, which can be calculated as a function of the instantaneous air temperature and pressure as follows. First,

$$T_{\text{air}} = T_{\text{air,ig}} \left(\frac{p}{p_{\text{ig}}}\right)^{(\gamma-1)/\gamma} \quad \text{adiabatic compression} \tag{52}$$

The value of ϵ_a is assumed to vary linearly between its maximum value (0.9) at the time of ignition t_{ig} and 0 over the duration of the combustion process, and

$$T(\phi = 1.1) = [1 + 0.0002317\,(T_{\text{air}} - 950)]\,[2726.3 + 0.9306p - 0.003233p^2] \tag{53}$$

for $800\ \text{K} < T_{\text{air}} \leq 1200\ \text{K}$.

$$T(\phi = 1.1) = [1 + 0.000249\,(T_{\text{air}} - 650)]\,[2497.3 + 4.7521p - 0.11065p^2 + 0.000898p^3] \tag{54}$$

for $450\ \text{K} < T_{\text{air}} < 800\ \text{K}$.

Annand [25] proposed the following heat transfer equation:

$$\frac{dq_w}{d\theta} = -\frac{A}{6\text{ rpm}}\left[\frac{0.45\lambda}{B}\text{Re}^{0.7}(T - T_w) + 7.8 \times 10^{-6}(T^4 - T_w^4)\right] \quad (55)$$

where B is the cylinder bore, $\text{Re} = \rho v_{\text{pis}} B/\mu$, and A is the surface area of the combustion chamber.

Transient heat flux measurements in a divided-chamber diesel engine have shown that the spatial nonuniformities of the fluid flow and combustion cause different temperature histories at different points in the cylinder head [50]. These measurements also indicate that the time-averaged and local heat transfer rates decrease when the rpm and load are decreased and that retarding the combustion timing beyond TDC decreases the peak heat flux in the combustion chamber and increases it in the main chamber because of the relative increase in the amount of fuel that burns in the main chamber with retarded combustion timing.

Alkidas and Cole [51] showed that if the coolant temperature is increased, the unburnt hydrocarbon (HC) emissions increase. They also showed that if the prechamber surface temperature decreases, the gaseous emissions increase, whereas the emissions of soot and particulates decrease.

Multidimensional calculations of radiation heat transfer in diesel enginelike environments [52] have shown that the use of an average homogeneous absorption coefficient cannot be justified and that the total heat losses can be underpredicted by as much as 60%. Experiments in constant-volume bombs [53] have shown that NO_x (nitric oxide) and particulate formation occur at the beginning of the combustion period. The NO_x concentration varied little in the middle and final stages of combustion, but the particulate concentration decreased. The peak particle concentration increased with the air swirl and air temperature, but the combustion interval decreased.

Radiation heat transfer is treated in more detail in Section 3.7.3.

The air mass flow rates through the valves, which appear in Eq. (1), can be calculated by assuming quasisteady, one-dimensional flow and using discharge coefficients to account for the differences between the effective and actual areas, as

$$\dot{m}_j = c_d A_v \frac{p_0}{RT_0}(\gamma R T_0)^{1/2}\left\{\frac{2}{\gamma - 1}\left[\left(\frac{p_s}{p_0}\right)^{2/\gamma} - \left(\frac{p_s}{p_0}\right)^{(\gamma+1)/\gamma}\right]\right\}^{1/2} \quad (56)$$

and for choked flow

$$\frac{p_0}{p_s} > \left(\frac{2}{\gamma + 1}\right)^{\gamma/(\gamma-1)} \qquad \dot{m}_j = c_d A_v \frac{p_0}{RT_0}(\gamma R T_0)^{1/2}\left(\frac{2}{\gamma + 1}\right)^{(\gamma+1)/2(\gamma-1)} \quad (57)$$

where c_d is a discharge coefficient; A_v is the valve area; the subscripts 0 and s denote stagnation and static conditions, respectively; γ is the ratio of specific heats; R is the gas constant; and T is the temperature.

Figures 3.8–3.10 show the fuel-burning rate normalized with respect to the total fuel injected per cycle per cylinder, the pressure diagram, and the mean

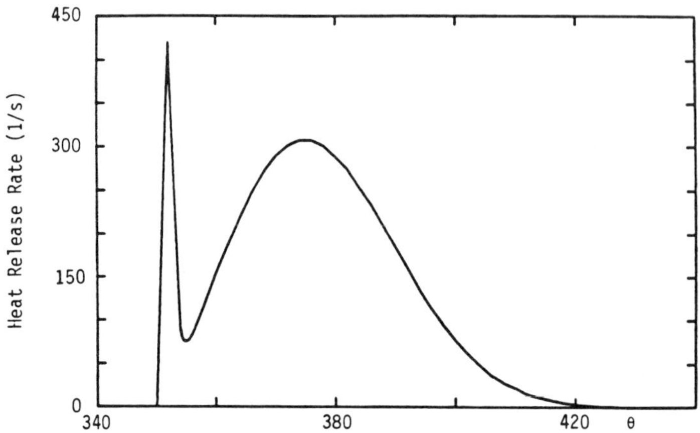

Figure 3.8 Fuel-burning rate normalized with respect to the total fuel injected per cycle per cylinder as a function of the crank shaft angle in a turbocharged turbocompounded diesel engine [19].

gas temperature and the heat transfer rate as a function of the crank shaft angle for a turbocharged turbocompounded diesel engine [19]. The results presented in Figs. 3.8–3.10 were obtained with the premixed-diffusive combustion model represented by Eqs. (21)–(27).

Figure 3.11 shows that the radiative heat transfer rate is approximately four times smaller than the convective heat transfer rate to the piston and cylinder head, whereas the radiative heat transfer rate to the cylinder liner is approximately half of the convective heat transfer rate [19]. Figure 3.12 indicates that the convective and total heat transfer rates in a conventional water-cooled diesel engine are much higher than those in a partially insulated engine. However, conventional water-cooled and partially insulated engines have approximately

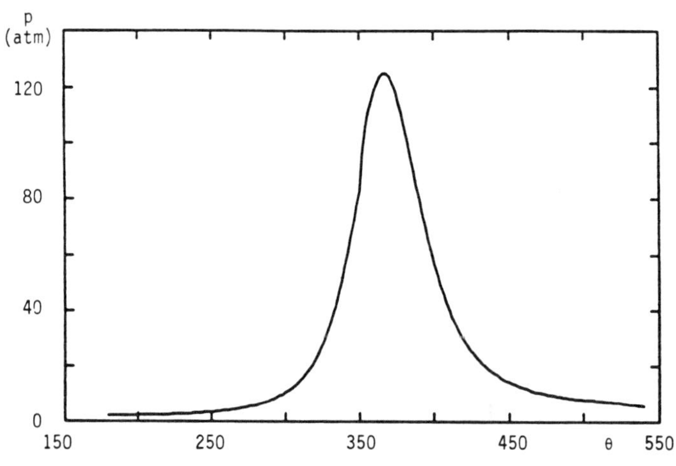

Figure 3.9 Cylinder pressure as a function of the crank shaft angle in a turbocharged turbocompounded diesel engine [19].

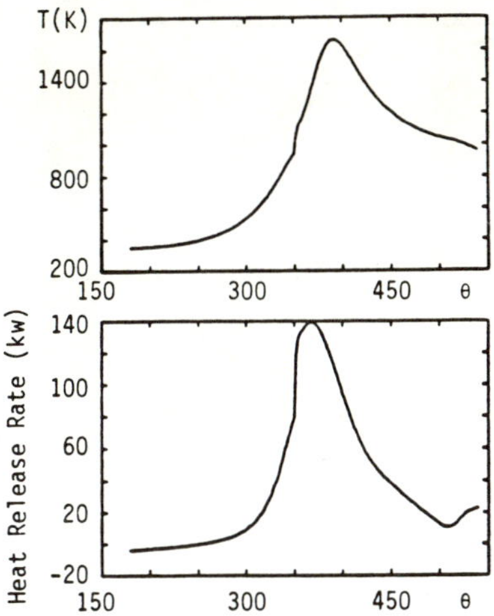

Figure 3.10 Mean gas temperature (top) and heat release rate (bottom) as a function of the crank shaft angle in a turbocharged turbocompounded diesel engine [19].

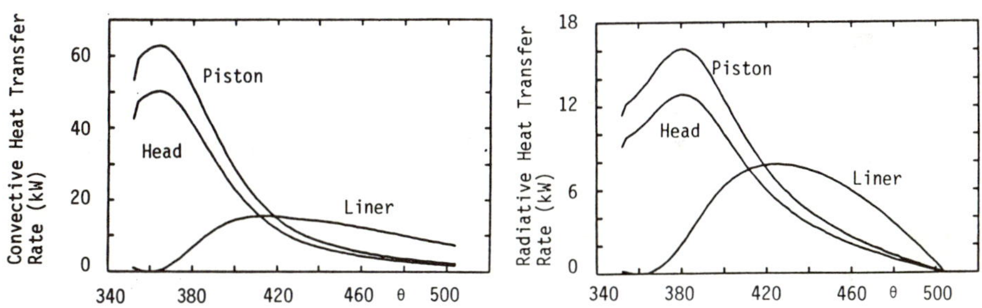

Figure 3.11 Convective (left) and radiative (right) heat transfer rates to the cylinder liner, cylinder head, and piston in a turbocharged turbocompounded diesel engine [19].

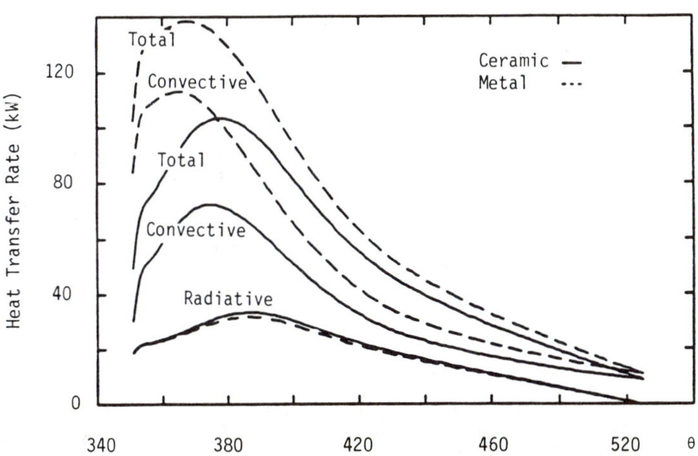

Figure 3.12 Convective, radiative, and total heat transfer rates in ceramic (solid lines) and conventional water-cooled (dashed lines) turbocharged turbocompounded diesel engines [19].

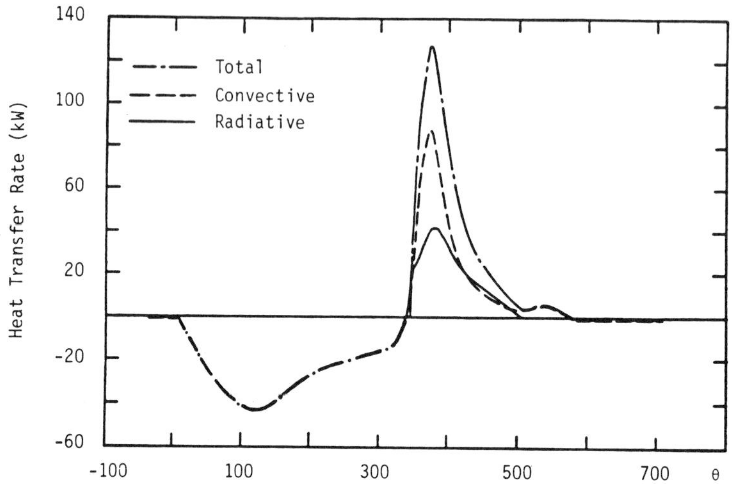

Figure 3.13 Radiative (solid lines), convective (dashed lines), and total (dashed-dotted lines) heat transfer rates in fully insulated turbocharged turbocompounded diesel engines [19].

the same radiative heat transfer rates. Figure 3.13 indicates that the convective heat transfer rate represents about two-thirds of the total heat transfer rate.

3.2.3 Wiebe Function Analysis

Equation (1) can be written as

$$\frac{dQ}{d\theta} = \frac{1}{\gamma - 1}\left(p\gamma \frac{dV}{d\theta} + V\frac{dp}{d\theta}\right) - \frac{pV}{(\gamma - 1)^2}\frac{d\gamma}{d\theta} - \frac{dq_w}{d\theta} \tag{58}$$

where the term on the left-hand side represents the heat release rate. In Eq. (58) the sensible enthalpy of the injected fuel and the variation of C_v with T have been neglected, and the last term on the right-hand side of Eq. (58) represents the heat losses. The specific heat ratio can be calculated from thermochemical data [54], assuming that the mixture composition is fixed by simple stoichiometry and is linearly related to the degree of reaction. The heat transfer losses can be calculated by using Woschni's [24] or Annand's [25] correlation.

Equation (58) can be used to perform heat release analyses if the pressure diagram is known. It can also be used to predict the cylinder pressure and temperature if the heat release rate $dQ/d\theta$ is specified. In order to describe the premixed and diffusive combustion periods observed in diesel engines, two Wiebe functions [55–57] can be used as follows:

$$\begin{aligned}\frac{dQ}{d\theta} = {} & 6.9\frac{Q_p}{\theta_p}(M_p + 1)\left(\frac{\theta}{\theta_p}\right)^{M_p} \exp\left[-6.9\left(\frac{\theta}{\theta_p}\right)^{M_p+1}\right] \\ & + 6.9\frac{Q_d}{\theta_d}(M_d + 1)\left(\frac{\theta}{\theta_d}\right)^{M_d} \exp\left[-6.9\left(\frac{\theta}{\theta_d}\right)^{M_d+1}\right]\end{aligned} \tag{59}$$

where the subscripts p and d refer to premixed and diffusive combustion, respectively; M_p and M_d are shape factors corresponding to premixed and diffusive combustion, respectively; θ_p and θ_d are the durations of the energy release in premixed and diffusive combustion, respectively; and Q_p and Q_d characterize the heat release in premixed and diffusive combustion, respectively. Typical values of the premixed and diffusive combustion parameters are shown in Fig. 3.14 [56].

Figure 3.15 shows the combustion rates computed with two Wiebe functions [Eq. (59)] as a function of the crank shaft angle for three different injection timings [56]. Figure 3.16 shows a schematic of the premixed combustion parameter Q_p and the amount of fuel injected during the ignition delay time Q_I [56]. Figure 3.16 also shows that $Q_p = 0.50\ Q_I$ for a variety of fuels in a DI diesel engine. Figure 3.17 indicates that the diffusive combustion duration parameter θ_d correlates linearly with the diffusive parameter Q_d for a variety of fuels in DI and indirect-injection (IDI) diesel engines [56]. Figure 3.18 shows that the premixed and diffusive combustion parameters are functions of the injection timing [56].

Miyamoto et al. [56] performed experiments in DI and IDI diesel engines for a range of speeds, loads, injection timings, fuels, and steady and transient conditions; they analyzed the combustion rate data by means of a double Wiebe function [Eq. (59)]. They found that M_p, M_d, and θ_p remained almost constant in both DI and IDI engines ($M_p = 3$ for DI and IDI engines; $M_d = 0.5$ for DI

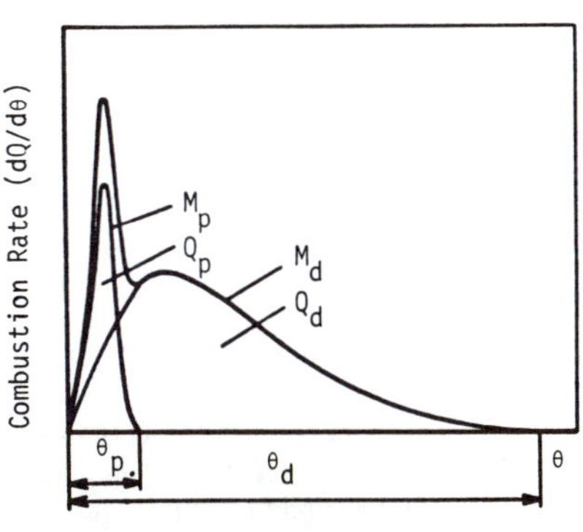

Figure 3.14 Combustion rate represented by two Wiebe functions [56].

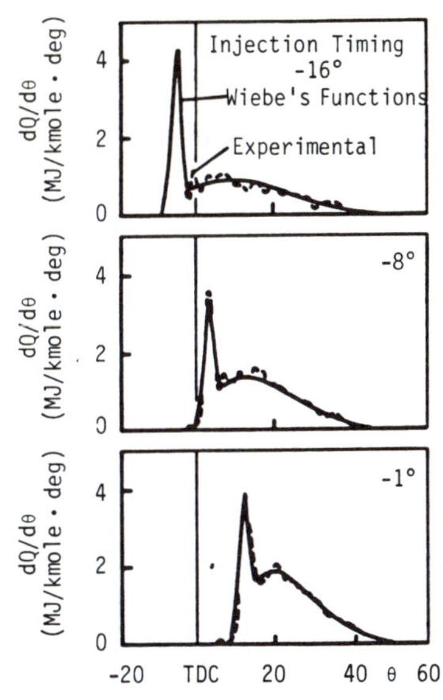

Figure 3.15 Experimental (dashed lines) and predicted (solid lines) heat release rates in a diesel engine for three different injection timings [56].

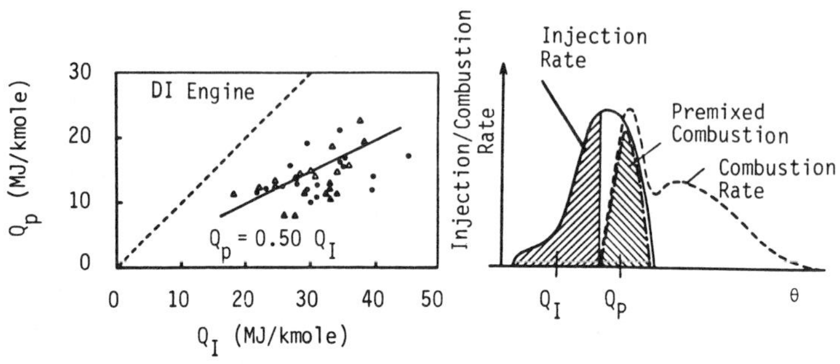

Figure 3.16 Correlation between the premixed Q_p combustion parameters and the enthalpy of the fuel injected during the ignition delay time Q_I (left) and schematic of Q_p and Q_I (right) [56].

engines; $M_d = 0.9$ for IDI engines; and $\theta_p = 7°$CA (crank angle degrees) for DI and IDI engines). However, the values of Q_p, Q_d, and θ_d were found to depend on the fuel-injection rate and ignition delay time. The premixed combustion parameter Q_p correlated linearly with the amount of fuel injected during the ignition delay time, and the duration of the diffusive burning θ_d was a linear function of the diffusive combustion parameter Q_d. The engine smoke, noise, and thermal efficiency were found to correlate with Q_p and θ_p. Both the smoke and the noise can be reduced by decreasing Q_p while maintaining a value of θ_d less than 60°CA.

Ghojel [58] used Eq. (58) and two Wiebe functions similar to Eq. (59) to predict the pressure diagram in DI diesel engines.

Shipinski et al. [15], Sitkei [16], and Tinaut [59] used a single exponential Wiebe function to predict the average pressure and temperature in DI and IDI engines. In these studies the burning rate was expressed as

$$\frac{dQ}{d\theta} = \frac{H_c m_{f,\text{inj}}}{\Delta\theta_b} a(m + 1) \left(\frac{\theta - \theta_{ig}}{\Delta\theta_b}\right)^m \exp\left[-a\left(\frac{\theta - \theta_{ig}}{\Delta\theta_b}\right)^{m+1}\right] \tag{60}$$

$$\Delta\theta_b = \theta_{\text{end}} - \theta_{ig} \tag{61}$$

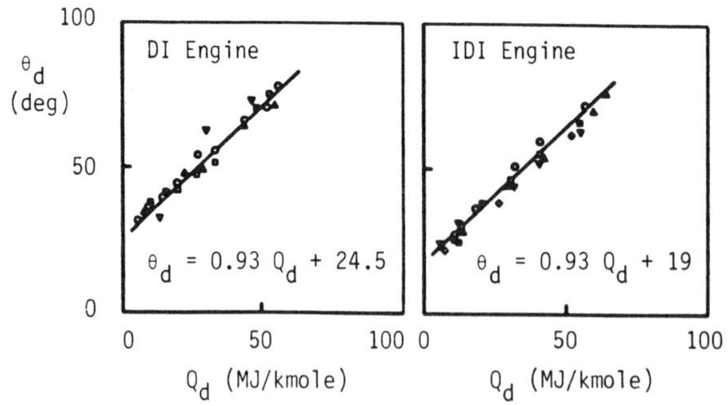

Figure 3.17 Correlation between the diffusive combustion duration parameter θ_d and the diffusive combustion parameter Q_d in direct-injection and indirect-injection diesel engines [56].

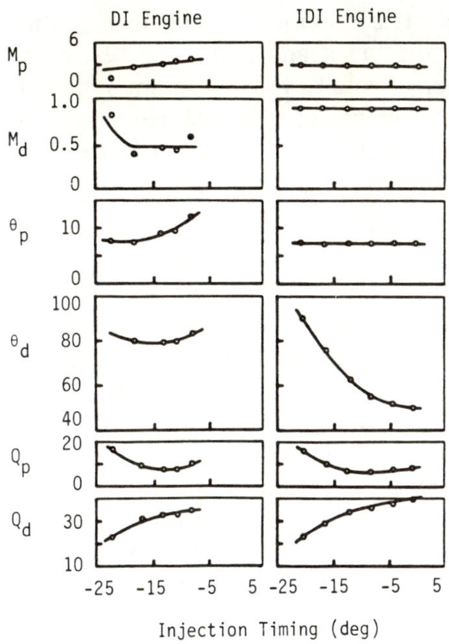

Figure 3.18 Dependence of the premixed and diffusive combustion parameters on the injection timing in direct-injection and indirect-injection diesel engines [56].

where H_c is the enthalpy of combustion, $m_{f,\text{inj}}$ is the total amount of fuel injected into the cylinder each cycle, θ_{ig} is the crank shaft at the start of combustion, θ is the crank shaft angle, and a and m are the Wiebe function parameters. The value of parameter a depends on the fuel-air ratio; $\Delta\theta_b$ depends on the equivalence ratio and rpm; and m depends on the engine rpm and the pressure and temperature at intake valve closing.

The models presented in Sections 3.2.1 and 3.2.3 have been used to analyze the transient performance of turbocharged diesel engines [60], IDI diesel engines (single-zone model with experimentally determined heat release rates for IDI diesel engines) [61], friction and combustion rates during diesel engine transients [62], two-stroke turbocharged diesel engines [63], and turbocharged turbocompounded diesel engines [19].

Watson [60] used the empirical fuel-burning rate of Watson et al. [3] and submodels for the turbocharger, charge air cooler, friction, governor, and fuel controller to analyze the transient performance of turbocharged diesel engines. However, the parameters that appear in the Wiebe functions [Eqs. (59) and (60)] are a function of the engine-operating conditions (rpm, load), combustion chamber geometry, etc. Experimental best-fit correlations can be derived for a variety of engine-operating conditions [64].

Brandstatter [65] used two Wiebe functions to analyze combustion in a prechamber stratified-charge engine with injection in the prechamber. He included the flows through the valves, intake and exhaust systems, used quasi-steady adiabatic flow expressions to calculate the mass flow rate through the throat that separates the prechamber and main chamber, simulated the entire engine cycle, assumed perfect mixing of the burnt and unburnt gases in the prechamber and main chamber (which were treated as two different systems), but did not consider wave phenomena in the exhaust and intake manifolds. He also used real gas effects and considered that the mixture specific heat at constant pressure is a function of the pressure, temperature, and equivalence ratio.

3.2.4 Whitehouse-Way Model

Winterbone and Tennant [62] and Winterbone and Loo [63] also accounted for engine friction, compressor, blower, intercooler, turbine, and turbocharger in their analyses of two-stroke turbocharged diesel engines, but used the single-zone combustion model of Whitehouse and Way [66]. In this model the atomization of fuel into fuel droplets, vaporization of the fuel, entrainment of air, and micromixing of fuel and air are collectively known as preparation of fuel. If the burning time is negligible compared with the rate of fuel preparation, the heat release rate can be considered as a measure of the fuel preparation rate for most of the burning period [67]. Chemical kinetics is a rate-controlling process at the beginning of burning when the temperature is low. As the temperature increases, the burning rate and the heat release rate increase, but they decrease as the prepared fuel is deleted. In the model proposed by Whitehouse and Way [66] combustion is assumed to occur at the droplet surface. If the droplets are of uniform diameter, the rate of fuel preparation is given by the following expression:

$$\frac{dm_{fp}}{d\theta} = C\, m_{\text{inj}}^{a_1}\, m_{up}^{a_2}\, p_{\text{O}}^{m} \tag{62}$$

where m_{inj} is the mass of the fuel injected into the cylinder, m_{up} is the mass of injected unprepared fuel, p_{O} is the oxygen partial pressure, $a_1 = 1/3$, $a_2 = 2/3$, $m = 0.4$, and the values of C and m need to be adjusted in order to achieve good agreement with experimental data for a given engine. Equation (62) indicates that the fuel preparation rate depends on the availability of oxygen in the cylinder.

During premixed combustion the burning rate is controlled by chemical kinetics, and during this period the fuel-burning rate is given by the following equation:

$$\frac{dm_{fb}}{d\theta} = \frac{C'}{\text{rpm}}\frac{p_{\text{O}}}{T^{1/2}} \exp\left(-\frac{E}{\tilde{R}T}\right) \int \left(\frac{dm_{fp}}{d\theta} - \frac{dm_{fb}}{d\theta}\right) d\theta \tag{63}$$

where C' and $E/\tilde{R}$ are constants that control the ignition delay period, and m_{fb} denotes the burnt fuel.

The fuel-burning rate is less than the preparation rate at the start of combustion. This results in accumulation of prepared fuel, which causes an increase in the burning rate.

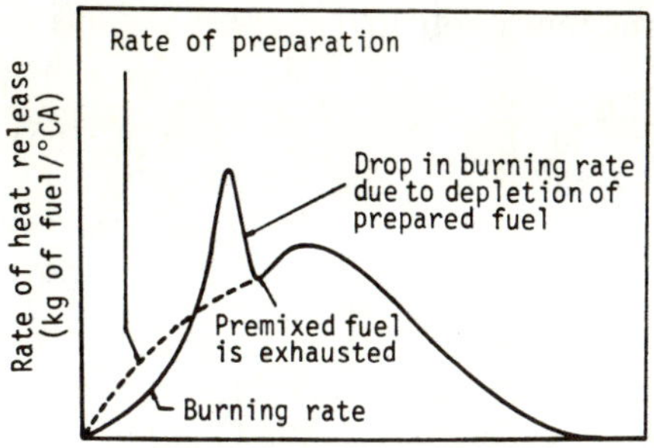

Figure 3.19 Rates of fuel preparation and heat release [72].

Figure 3.19 shows the fuel-burning rate as a function of the crank shaft angle. The first peak in the figure corresponds to the combustion of premixed fuel and air. The decrease in the burning rate is due to the depletion of prepared fuel. After the prepared fuel is exhausted, the heat release rate is equal to the fuel preparation rate.

Although no ignition delay time is considered in the model proposed by Whitehouse and Way [66], the small burning rate during the initial stages of combustion can be considered to represent the delay time. Thus, C and $E/\tilde{R}$ effectively control the delay period.

The Whitehouse-Way model [66] is consistent with experimental observations [67] showing that the major factors affecting the burning rate are the injection rate, engine speed, and combustion chamber design, all of which affect the fuel-air mixing.

The fuel-air mixing affects the burning processes in a diesel engine. These processes proceed in three phases. In the first phase the premixed part of the fuel burns, and the burning rate is highest in a typical DI engine. In the second phase the majority of the heat is released in a crank angle interval, which may be eight times greater than that of the premixed burning. The burning rate decays, and the fuel jet burns as a highly turbulent diffusion flame in the second phase. The burning rate is quite low in the third phase, which may last through the expansion stroke. The third phase accounts for about 10% to 20% of the total heat input. The nature of combustion in the third phase is not clear, however, but appears to be an inherent part of diesel engine combustion.

Recently, Carapanayotis and Salcudean [68] used a modified version of the Whitehouse-Way model that accounts for the ignition delay time, fuel vaporization, and combustion as follows. The ignition delay was experimentally determined as a function of the engine rpm and cylinder pressure at the beginning of injection, and the following expression was obtained:

$$t_d = 2.5\, p^{-0.009x} \qquad (\mu\text{s}) \qquad x = (\text{rpm})^{1/2} \tag{64}$$

The amount of fuel vapor in the cylinder at any crank angle after the start of injection was calculated, assuming that the spray was characterized by a single droplet of Sauter mean diameter equal to SMD. For a single droplet in the absence of combustion, the following expressions apply [69]:

$$d^2 = \text{SMD}^2 - \frac{8\,\rho_s\,Dt}{\rho_l} \ln(1 + B) \tag{65}$$

$$B = C_p \frac{T - T_s}{L} + 1 \tag{66}$$

where L is the latent heat of vaporization, T is the gas temperature far from the droplet, T_s is the temperature at the droplet surface, ρ_l is the liquid droplet density, ρ_s is the density at the droplet surface, and D is the mass diffusivity.

During the ignition delay period, some of the injected fuel is evaporated and mixed with the air, forming a combustible mixture. In the early stages, combustion occurs under premixed conditions at a rate equal to

$$RR_p = 10^5\, Y_{\text{O}}^{1.2}\, Y_f^{1.1} \exp\left(-\frac{20{,}000}{T}\right) \qquad (\text{kg/°CA}) \tag{67}$$

where Y_{O} and Y_f denote the mass fractions of oxygen and fuel vapor, respectively, and RR_p is the premixed reaction rate.

Equation (67) is assumed to be valid until the amount of fuel evaporated at the end of the ignition delay period has been consumed. When all the initial fuel vapor has been consumed, $RR_p = 0$ and combustion is assumed to be controlled by the fuel-air mixing rate, according to the following expression:

$$RR_m = 0.017\, m_{\text{inj}}\, p_{\text{O}} \tag{68}$$

where m_{inj} is the total mass of fuel injected at a crank angle, p_{O} is the oxygen partial pressure, and RR_m is the mixing-controlled reaction rate.

The mass burning rate during the premixed and diffusive periods can be expressed as

$$BR = RR_p + RR_m \tag{69}$$

where BR is the burning rate. Note that during premixed combustion $RR_p \gg RR_m$.

Carapanayotis and Salcudean [68] used Woschni's heat transfer correlation [24] [cf. Eq. (30)] with

$$v_{\text{mot}} = 6.18\, v_{\text{pis}} \qquad v_{\text{comb}} = 0 \qquad \text{during scavenging} \tag{70}$$

$$v_{\text{mot}} = 2.28\, v_{\text{pis}} \qquad v_{\text{comb}} = 0 \qquad \text{during compression} \tag{71}$$

$$v_{\text{mot}} = 2.28\, v_{\text{pis}} \qquad v_{\text{comb}} = 3.24 \times 10^{-5} \frac{V_d\, T_1}{p_1\, V_1} (p - p_{\text{mot}}) \qquad \text{during combustion and expansion} \tag{72}$$

3.3 MULTIZONE MODELS

The assumption of homogeneous dispersion of the injected fuel is clearly unrealistic. It is well established that the fuel is dispersed in the form of droplets and that there are fuel-rich and fuel-free zones in the combustion chambers of a diesel engine. This heterogeneity affects the temperature and composition within the combustion chamber and the fuel-burning rate.

Several multizone models have been developed to analyze combustion in DI diesel engines. Most of these models use experimental and theoretical correlations for the fuel jet penetration [70] and divide the chamber into burning and nonburning zones. The effects of swirl on the fuel deflection can be empirically introduced into the models.

In the next sections, some multizone models are described.

3.3.1 Two-Zone Models

We consider here the model proposed by Whitehouse and Sareen [71] and extended by Kumar et al. [72]. This model divides the mixture charge into burnt and unburnt zones; the burnt zone corresponds to the injected fuel, and the unburnt zone corresponds to the surrounding air (Fig. 3.20). The model is based on the following assumptions: (1) the combustion chamber is a circular cylinder with the injector located at the center of the cylinder and with the sprays from the nozzle moving horizontally; (2) the combustion products are CO_2, N_2, H_2O, and O_2, that is, dissociation is not considered; (3) air entrainment is due to the shearing action of the fuel spray; (4) combustion does not produce additional air entrainment; (5) boundary layer effects on the wall jet are neglected; (6) the fuel jets, which constitute the burning zone, are assumed to be homogeneous; (7) the heat transfer from the burning zone is by radiation only until the jet strikes on the cylinder wall; after that, the heat transfer is due to both radiation and convection; (8) the heat transfer from the nonburning zone is by convection only; (9) no heat transfer occurs between the burning and nonburning zones; and (10) the pressure is uniform throughout the combustion chamber though the temperature and composition of the two zones are different.

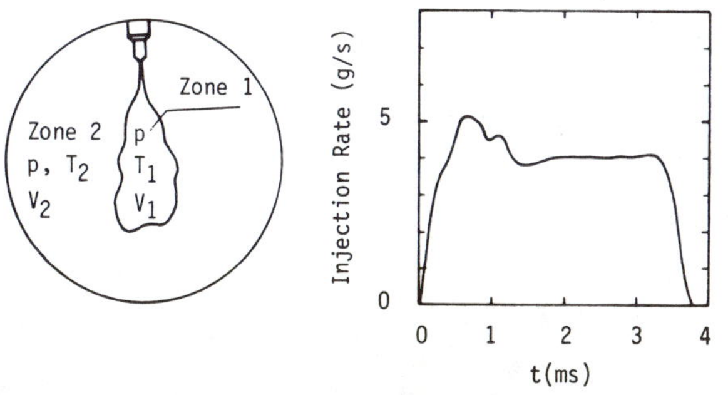

Figure 3.20 Schematic of a two-zone model and fuel injection rate.

The steady-state jet penetration x can be written as

$$x = 17.28 \left[\left(\frac{\Delta p}{\rho_b} \right)^{1/2} d_{\text{inj}}\, t \right]^{1/2} \tag{73}$$

where ρ_b is the density of the burning zone, Δp is the pressure drop in the nozzle, t is time and d_{inj} is the nozzle diameter. Equation (73) is not strictly applicable to engines because the fuel jet is confined; the jet is also subjected to additional resistance. Borman and Johnson [73] suggested that the spray tip velocity is about 0.72 times that of a steady-state jet.

The spray cone angle depends on the density of the mixture in the cylinder and, therefore, is a function of time. However, we will assume here that it is constant.

The volumetric flow rate through the jet is

$$\frac{dV_j}{dt} = 2\pi \int_0^{r_0} rv\, dr \tag{74}$$

where the velocity distribution for a two-phase jet is

$$\frac{v}{v_{\text{max}}} = (1 - y^{1.5})^2 \qquad y = \frac{r}{r_0} \tag{75}$$

with r_0 being the local jet radius; the maximum velocity v_{max} can be calculated by differentiating Eq. (73) with respect to time.

Equations (74) and (75) are valid as long as there is fuel injection. When the spray detaches from the nozzle, the air entrainment rate is the difference between Eq. (74) and the rate of decrease of volume due to the detachment of the jet from the nozzle.

The rate of change of mass in the burning zone is equal to the air entrainment rate plus the fuel burning rate, that is,

$$\frac{dm_b}{dt} = \frac{dm_e}{dt} + \frac{dm_{fb}}{dt} \tag{76}$$

where

$$\frac{dm_e}{dt} = \left(\pi x^2 \tan^2 \alpha \frac{dx}{dt} - \frac{dV_b}{dt} \right) \rho_a \tag{77}$$

α is the jet half angle, V_b is the volume of burnt gases, and ρ_a is the density of the nonburning gases.

The first law of thermodynamics applied to the burning and nonburning zones yields

$$\frac{d}{dt}(m_b\, u_b) = -\frac{dq_b}{dt} - p\frac{dV_b}{dt} + h_u \frac{dm_e}{dt} \qquad u_b = f(T_b, \phi_b) \tag{78}$$

$$\phi_b = \frac{m_{fb}/m_{ab}}{(m_f/m_a)_{st}} \qquad \frac{dm_{ab}}{dt} = \frac{dm_e}{dt} \qquad pV_b = m_b\, R_b\, T_b \tag{79}$$

$$\frac{d}{dt}(m_u u_u) = -\frac{dq_u}{dt} - p\frac{dV_u}{dt} - h_u\frac{dm_e}{dt} \qquad u_u = f(T_u, \phi_u) \tag{80}$$

$$\frac{dm_u}{dt} = -\frac{dm_e}{dt} \qquad pV_u = m_u R_u T_u \qquad R_u = f(T_u, \phi_u) \qquad R_b = f(T_b, \phi_b) \tag{81}$$

$$V = V_b + V_u \tag{82}$$

where the subscripts u and b denote the unburnt and burnt zones, respectively; dq/dt represents the heat transfer losses; and m_{ab} is the mass of air in the burning zone.

Equations (73)–(82) can be written as a system of first-order differential equations for T_b, T_u, V_b, p, and V_u.

Kuo and Bracco [74, 75] used a two-dimensional model with a $k-\epsilon$ submodel of turbulence to scale transient laminar, turbulent, and spray jets. For thick sprays, where droplet collisions and coalescence are important, they obtained the following results:

$$\frac{x}{d_{\text{eff}}} = 0.686\left(\frac{tv_{\text{inj}}}{d_{\text{eff}}}\right) \qquad \frac{x}{d_{\text{eff}}} \leq 0.6 \tag{83}$$

$$\frac{x}{d_{\text{eff}}} = 0.538\left(\frac{tv_{\text{inj}}}{d_{\text{eff}}}\right)^{0.676} \qquad \frac{x}{d_{\text{eff}}} \geq 0.6 \tag{84}$$

where v_{inj} is the injection velocity and d_{eff} is a characteristic length that accounts for droplet vaporization as

$$d_{\text{eff}} = d_{\text{inj}}(\tan\alpha)^{-0.25}\left(\frac{\rho_l}{\rho}\right)(1 + B_T)^{-0.45} \tag{85}$$

$$B_T = C_{pl}\frac{T - T_l}{L} \tag{86}$$

where α is the spray cone half angle, the subscript l refers to the liquid phase, L is the latent heat of vaporization of the fuel, C_{pl} is the specific heat at constant pressure of the inert species, and B_T is the Spalding transfer number.

Kuo and Bracco [75] showed that vaporizing sprays reach steady state around the back of their head vortices, and for many applications the stems of such transient jets can be considered to be in steady state. The correlations obtained by Kuo and Bracco [74, 75] are strictly valid for spray jets injected into quiescent, constant-pressure environments. In diesel engines there may be strong swirl and radial (squish) flows produced by bowl-in-piston combustion chambers; further, in diesel engines the presure does not remain constant, and the spray jets are confined. Therefore, Eqs. (83) and (84) may not be valid in diesel engines.

A comparison between Eqs. (73) and (84) reveals that the temporal dependences of the spray tip penetration are 0.5 and 0.676, respectively. Note that the injection velocity v_{inj} can be defined as $v_{inj} = C_D (\Delta p/\rho_l)^{1/2}$, where C_D is a discharge coefficient.

The spray cone angle in Eqs. (77) and (85) can be calculated from an aerodynamic theory that attributes jet breakup to the aerodynamically induced growth of unstable surface waves [76]. This theory can be derived by considering infinitesimal surface waves on the liquid-gas interface; it predicts

$$\tan \alpha = \frac{4\pi}{A}\left(\frac{\rho}{\rho_l}\right)^{1/2} f\left[\frac{\rho_l}{\rho}\left(\frac{\mathrm{Re}_l}{\mathrm{We}_l}\right)^2\right] \tag{87}$$

where the value of the constant A must be obtained from experimental data,

$$\mathrm{Re}_l = \rho_l\, v_{inj}\, \frac{d_{inj}}{\mu_l} \qquad \mathrm{We}_l = \rho_l\, v_{inj}^2\, \frac{d_{inj}}{\sigma} \tag{88}$$

σ is the surface tension of the liquid, $f = 3^{1/2}/6$ for $(\rho_l/\rho)\,(\mathrm{Re}_l/\mathrm{We}_l)^2 >> 1$, and Re and We are the Reynolds and Weber numbers, respectively.

The results of Reitz and Bracco [77] indicate that the variation of the spray angle with the gas density can be predicted only if A has a different value for each nozzle; the spray angle was found to increase with decreasing liquid viscosity [78, 79]; the spray angle decreases with increased nozzle passage length for nozzle length-to-diameter ratios greater than 10; for the same nozzle length, rounded inlet nozzles produce less divergent jets than do sharp-edge nozzles [80]. A summary of the available experimental data for the atomization regime can be found in references 81 and 82.

Equation (73) is not valid when the spray jet strikes on the cylinder wall and a wall jet is formed. Furthermore, the use of only two zones to represent the fuel combustion is unrealistic. These assumptions are relaxed in Section 3.3.2, on multizone models.

The total linear momentum of an undeflected jet with a velocity distribution of the form given by

$$\frac{v}{v_{max}} = 1 - \left(\frac{r}{r_0}\right)^2 \tag{89}$$

is

$$\mathrm{LM} = \int_0^{r_0} \rho v^2\, 2\pi r\, dr = \frac{1}{3}\rho\pi v_{max}^2\, x^2 \tan^2 \alpha \tag{90}$$

The drag on the head of the jet is given by [83]

$$\mathrm{DF} = \frac{1}{2}\rho C_f \frac{\pi}{3} v_{max}^2\, x^2 \tan^2 \alpha \tag{91}$$

where C_f is a drag coefficient.

Conservation of linear momentum implies that

$$\mathrm{DF} + \mathrm{LM} = \rho\pi\, d_{inj}^2\, \frac{v_{inj}^2}{4} \tag{92}$$

which yields

$$v_{max} = 3^{1/2} d_{inj} v_{inj} \frac{1 - y^2}{2(1 + 0.5 C_f)^{1/2} x \tan \alpha} \tag{93}$$

where $y = r/r_0$.

Substitution of Eq. (93) into Eq. (89) yields

$$v(x,y) = 3^{1/2} d_{inj} v_{inj} \frac{1 - (r/r_0)^2}{2(1 + 0.5C_f)^{1/2} x \tan \alpha} \tag{94}$$

Introducing the distance s defined by

$$x = s \cos \theta \qquad r = x \tan \theta \qquad r_0 = x \tan \alpha \qquad 0 \leq \theta \leq \alpha \tag{95}$$

into Eq. (94) and integrating the result ($ds/dt = v$) yields

$$s = \left\{ 3^{1/2} v_{inj} d_{inj} \left[\frac{[1 - (\tan \theta/\tan \alpha)^2]t}{(1 + 0.5 C_f)^{1/2} \tan \alpha \cos \theta} \right] \right\}^{1/2} + \frac{d_{inj}}{\cos \theta} \tag{96}$$

Equation (96) allows determination of the jet envelope for given values of d_{inj}, v_{inj}, t, α, θ, and C_f. A typical value of C_f is 2.

The mass entrainment rate can be calculated by integrating the mass flux at a distance x and subtracting the original mass flux through the nozzle, that is,

$$\frac{dm_e}{dt} = \int_0^{r_0} \rho \, 2\pi r \, v \, dr - \frac{\pi}{4} \rho \, d_{inj}^2 \, v_{inj} \tag{97}$$

The spray penetration can be calculated from Eq. (96) with $r = 0$ or $\theta = 0$; this yields

$$x = (A + Bt)^{1/2} \tag{98}$$

$$B = 3^{1/2} \frac{d_{inj} v_{inj}}{2(1 + 0.5 C_f)^{1/2} \tan \alpha} \tag{99}$$

where A is an integration constant that corresponds to the virtual origin of the jet (i.e., $A = (d_{inj} \tan \alpha)/2$).

When the free jet strikes on the cylinder wall, a wall jet results. The geometry of a wall jet is illustrated in Fig. 3.21. Figure 3.22 shows the growth of the jet envelope and the jet deflection upon impingement on the wall. Conservation of mass into and out of the deflection region yields the maximum velocity of the wall jet. Further simplifying assumptions (e.g., the maximum jet velocity decreases inversely from the wall jet pole) can be introduced in order to determine the penetration of the wall jet as a function of time [83].

Packer et al. [84] studied spray penetration in diesel engines with high swirl levels and included the cross-sectional distortion of the spray and nonsimilar concentration profiles. Mehta and Gupta [85] developed a continuum model that accounts for the spray centerline velocity to calculate spray-swirl interactions

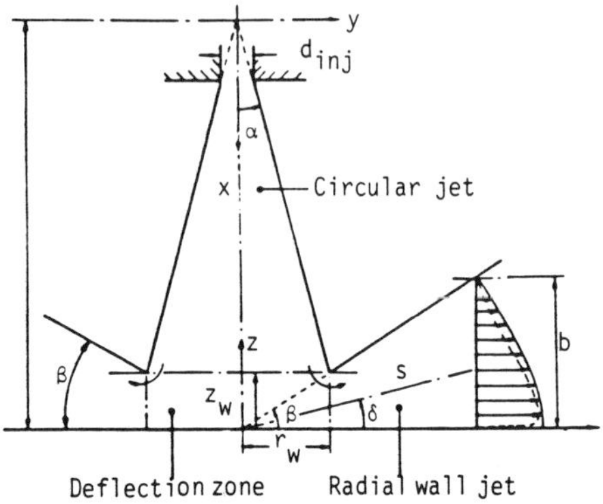

Figure 3.21 Schematic of a wall jet [83].

in DI diesel engines. Their model includes three-dimensional features of the fuel spray motion and yields good agreement with expermental data for various air swirl, fuel injection, and cylinder charge conditions.

Ha et al. [86] studied air entrainment into single-shot engine sprays, measured the gas velocity around the spray with hot wires, and used smoke tracers to determine the velocity direction. They also developed a model for the fully developed spray that assumes the spray momentum to be constant. Their experimental and theoretical results indicate that the entrainment velocity is proportional to a power of the radial distance, whereas the entrainment coefficient ranges from 0.2 to 0.3.

The effects of swirl on the fuel jet penetration can be analyzed by considering the jet deflection by pressure and viscous forces. Sinnamon et al. [87] developed a model based on the continuity and momentum equations for a steady-state gaseous jet, which contained adjustable entrainment and drag parameters and

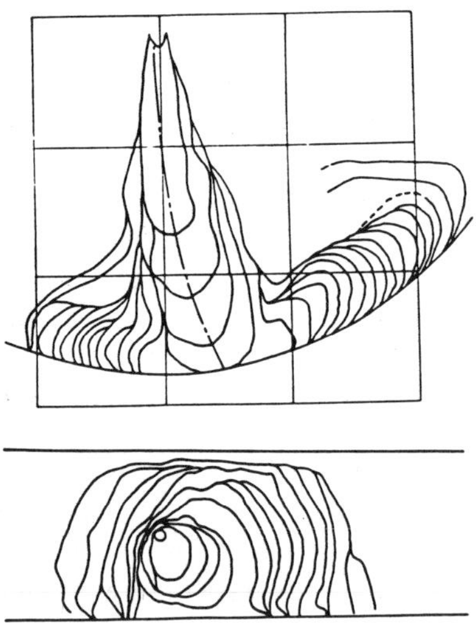

Figure 3.22 Growth of jet envelope and jet deflection on a solid wall [83].

accounted for mass entrainment by means of the expression proposed by Ricou and Spalding [88]. The model is based on the following set of equations (Fig. 3.23):

$$\frac{d}{ds}\left(\int_0^{r_0} \frac{m_f}{m} \rho\, v\, 2\pi r\, dr\right) = 0 \tag{100}$$

$$\frac{d}{ds}\left(\int_0^{r_0} \rho\, v\, 2\pi r\, dr\right) = [\rho(r=0)\,\rho_\infty]^{1/2}\, 2\pi r_0 [\alpha|v - V_t| + \lambda|V_n|] = \frac{d\dot{m}}{ds} \tag{101}$$

where v is the velocity along the jet axis; θ is the angle the spray forms with the horizontal (x) direction; V_t and V_n are the air velocity components parallel and normal to the jet axis, respectively; ρ_∞ is the air density outside the jet; $\dot{m}$ is the air entrainment rate; s is the coordinate along the jet; and α and λ are adjustable axial and normal entrainment parameters.

Equation (100) shows that the fuel flow rate at any cross section of the jet is constant, whereas Eq. (101) expresses the conservation of the total mass of air and fuel and uses the entrainment rate given by Ricou and Spalding [88].

Conservation of linear momentum in the horizontal and vertical directions can be written as

$$\frac{d}{ds}\left(\cos\theta \int_0^{r_0} \rho\, v^2\, 2\pi r\, dr\right) = V\cos\phi\,\frac{d\dot{m}}{ds} + F_d\cos\theta \tag{102}$$

$$\frac{d}{ds}\left(\sin\theta \int_0^{r_0} \rho\, v^2\, 2\pi r\, dr\right) = V\sin\phi\,\frac{d\dot{m}}{ds} + F_d\sin\theta \tag{103}$$

where

$$v - V_t = (v_{\max} - V_t)\left[1 - \left(\frac{r}{r_0}\right)^{1.5}\right]^2 \tag{104}$$

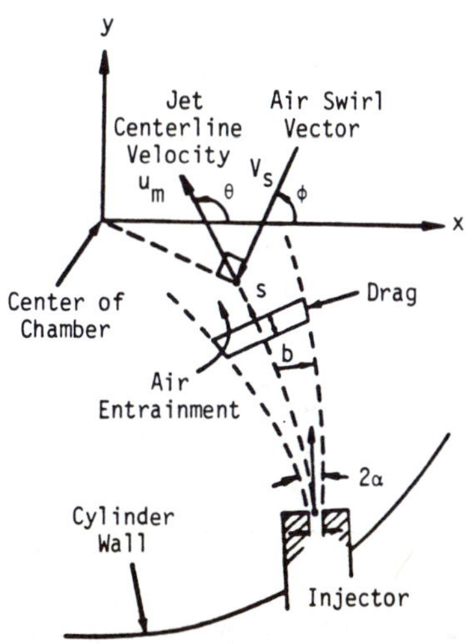

Figure 3.23 Geometry of a spray model [87].

and F_d is the drag force. The first term on the right-hand sides of Eqs. (102) and (103) accounts for the jet momentum changes due to entrainment.

The local concentration in the jet is

$$\frac{c}{c_{max}} = 1 - \left(\frac{r}{r_0}\right)^{1.5} \tag{105}$$

and the local density can be expressed as

$$\rho = \frac{\rho_\infty}{1 - (c/c_{max})(1 - \rho_\infty/\rho_f)} \tag{106}$$

where ρ_f is the density of the fuel.

Equations (100)–(106) yield ordinary differential equations for v_{max}, c_{max}, r_0, and θ, which can then be substituted into Eqs. (104) and (105) to obtain the velocity and concentration distributions in the fuel jet.

Figure 3.24 shows some sample results that illustrate the spray trajectory and penetration in a combustion chamber with and without swirl [87].

3.3.2 Multizone Models

The two-zone models presented in the preceding section divide the cylinder mixture into the burning and the nonburning zones, and they account for air entrainment. However, the assumption of uniform fuel composition in the burning zone is unrealistic. In multizone models the fuel jet is divided into many elements, and the combustion process in each element is analyzed as a process

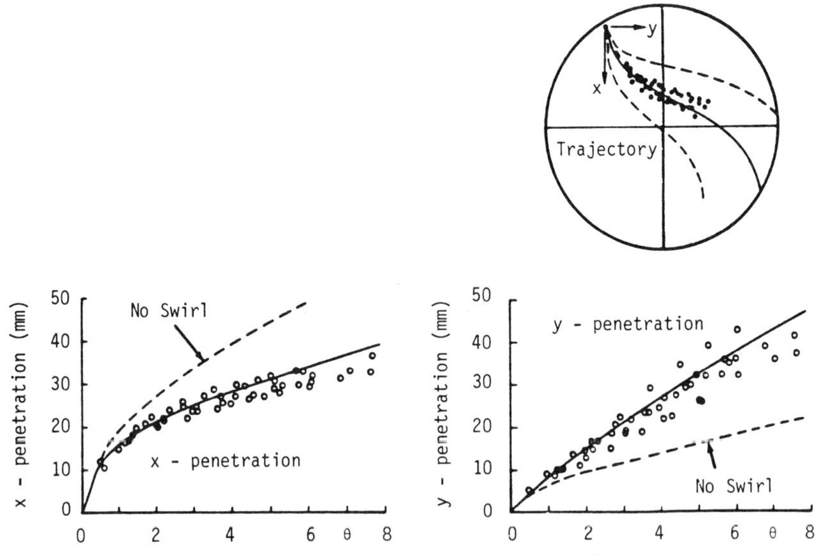

Figure 3.24 Spray trajectory and penetration with (solid lines) and without (dashed lines) swirl [87].

of mixing between the jet and surrounding air, entrainment into the flame front, and subsequent combustion. Each individual element is assumed to be homogeneous, with two temperatures corresponding to the burnt and unburnt mixture. The unburnt gas is composed of a low-temperature mixture of air, nonreacting fuel, and residual gases, whereas the burnt zone is composed of high-temperature unburnt mixture and combustion products. There may be mixing between any element and the surrounding air, but there is no mixing between the jet elements themselves.

If the air and fuel are divided into N elements, the following equations apply:

$$pV_j = m_j R_j T_j \qquad R_j = R_j(\phi_j, T_j, p) \tag{107}$$

$$\frac{dm_j}{dt} = \dot{m}_{j,\text{in}} - \dot{m}_{j,\text{out}} \tag{108}$$

$$\frac{d}{dt}(m_j u_j) = -p\frac{dV_j}{dt} - \frac{dq_{L,j}}{dt} + \dot{m}_{j,\text{in}} h_{j,\text{in}} - \dot{m}_{j,\text{out}} H_{j,\text{out}} \tag{109}$$

$$m_j u_j = m_j h_j - p V_j \qquad \frac{dh_j}{dt} = C_{p_j}\frac{dT_j}{dt} \qquad C_{p_j} = C_{p_j}(\phi_j, T_j, p) \tag{110}$$

$$\frac{dV}{dt} = \sum_{j=1}^{N} \frac{dV_j}{dt} \tag{111}$$

where the subscript j refers to the element j, the subscripts in and out refer to fluxes into and out of the element, respectively, the subscript L denotes heat losses, h denotes the specific enthalpy, and V is the volume.

Hiraki and Rife [89] used Eqs. (107)–(111) and a jet model similar to that presented in Section 3.3.1 to calculate the fuel jet trajectory and an entrainment rate of the surrounding mixture given by

$$\frac{dm_e}{dt} = (\rho_u \rho_b)^{1/2} A\, u_e \tag{112}$$

where A is the flame surface area, the subscripts u and b denote unburnt and burnt gases, and u_e is an entrainment velocity that can be taken as a fraction (22%) of the fuel jet velocity. Note that Eq. (112) is applied to each burning or jet zone.

Figures 3.25 and 3.26 show schematics of the spray and ignition models used by Hiraki and Rife [89].

Hiraki and Rife [89] also used the Blizard-Keck-Tabaczynski model [90, 91], in which the burning rate of an individual eddy is assumed to be proportional to the mass of unburnt gas in the flame

$$\frac{dm_b}{dt} = \frac{m_u}{\tau} = \frac{m_e - m_b}{\tau} \tag{113}$$

The value of m_e can be calculated from the spray model, and τ is a characteristic

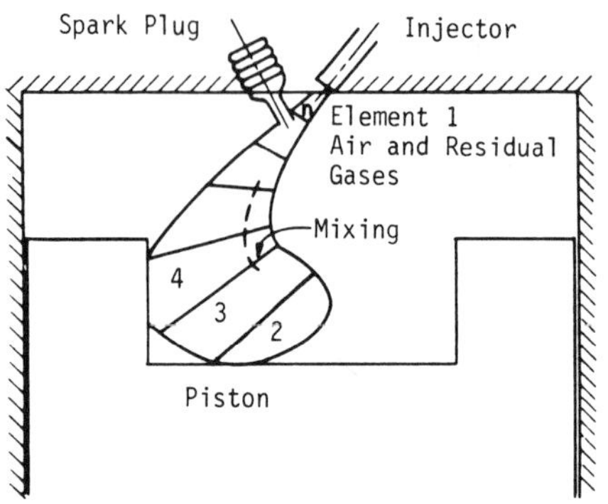

Figure 3.25 Schematic of multizone model [89].

reaction time equal to the ratio of the Taylor microscale to the laminar flame speed, $\tau = l_T/S_L$, and

$$l_T = 0.17 \frac{\text{intake valve lift}}{\text{compression ratio}} \tag{114}$$

Hiraki and Rife [89] used the extended Zel'dovich mechanism [92, 93]

$$N + NO \rightleftarrows N_2 + O \tag{115}$$

$$N + O_2 \rightleftarrows NO + O \tag{116}$$

$$N + OH \rightleftarrows NO + H \tag{117}$$

where O, O_2, H, OH, and N_2 are assumed to be in equilibrium and $d[N]/dt = 0$ (i.e., a steady-state assumption was used for [N]). The equilibrium assumption can be justified because the hydrocarbon oxidation reactions are faster than the NO formation process.

Figure 3.27 shows the mass history of an element as a function of time from injection and the exhaust concentration of nitric oxide as a function of the injection timing [89]. Figure 3.28 indicates that the NO_2 emissions increase monotonically with the equivalence ratio [89].

Other multizone models include those of Kono et al. [94]. Meguerdichian and Watson [95], Hiroyasu et al. [96], Kobayashi et al. [97], Shahed et al. [98], and Tinaut [59].

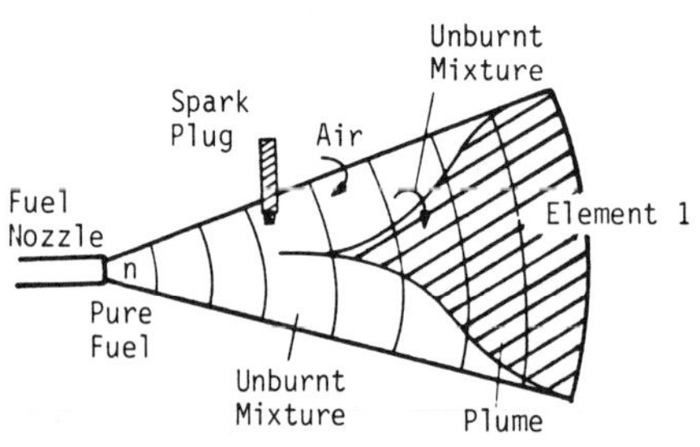

Figure 3.26 Schematic of ignition model [89].

3.3.3 Kono's Model

Kono et al. [94] accounted for the air entrainment rates, which are different between the center and outer portion of the jet, by dividing the spray into conical elements and applying the mass and momentum conservation equations to each element, so that the rate of change of fuel mass in element j can be written as

$$\frac{dm_{f,j}}{dt} = (\rho_{f,j}\,\rho_a)^{1/2}\, 2\pi R_j\, (x_{j+1} - x_j)\, C_x\, |v_j - V_{a,j}| \qquad (118)$$

where the subscripts a and f denote air and fuel, respectively, R_j and x_j denote the radius and location of each element, v is the jet velocity, V_a is the air velocity, and C_x is a drag coefficient. The mean flow and turbulence kinetic energies were calculated by a two-equation $k-\epsilon$ model of turbulence that accounts for turbulence generation by compression and shear.

In the absence of swirl, the $k-\epsilon$ model can be written as

$$\rho \frac{dk}{dt} = \frac{2}{3} k \frac{d\rho}{dt} - \rho\epsilon \qquad (119)$$

$$\rho \frac{d\epsilon}{dt} = \frac{4}{3} \epsilon \frac{d\rho}{dt} - c_2\, \rho \frac{\epsilon^2}{k} \qquad (120)$$

where $c_2 = 1.92$.

The fuel mass in element j given by Eq. (118) is different from that proposed by Kono et al. [94]; their equation is dimensionally incorrect.

The combustion model used by Kono et al. [94] accounts for the ignition delay time t_{ig} by means of the following expression:

$$\int_{t_{inj,j}}^{t_{inj,j}+t_{ig,j}} \frac{dt}{t_{d,j}} = 1 \qquad t_{d,j} = a\, p^b\, \phi_j^c \exp\left(\frac{T^*}{T_j}\right) \qquad (121)$$

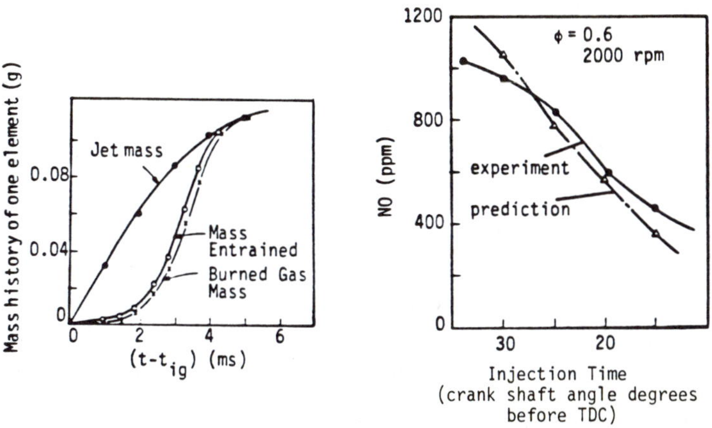

Figure 3.27 Mass history of one element and exhaust NO concentration [89].

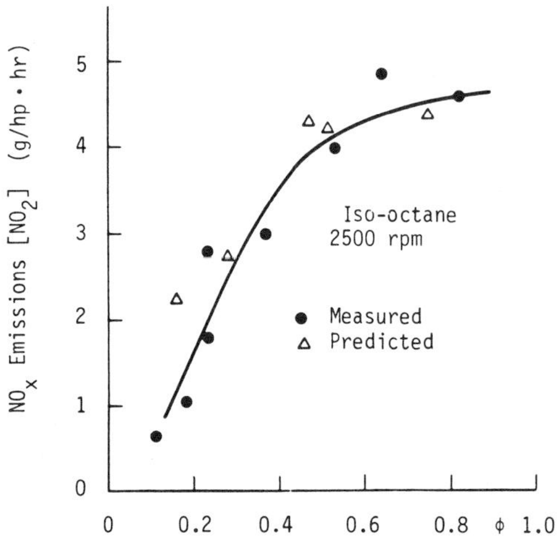

Figure 3.28 Emissions of NO_2 as a function of the equivalence ratio [89].

where T^* is an activation temperature and where a, b, and c are constants. Kono et al. [94] assumed that the air entrainment and combustion processes are independent and that the rate of combustion is proportional to the total amount of air entrained during premixed combustion. Kono et al. [94] employed the Blizard-Keck-Tabaczynski [90, 91] model of turbulent combustion and expressions similar to those of Eqs. (107)–(111) to calculate the temperatures and volumes of the burnt and unburnt gases.

During premixed combustion the entrainment and combustion rates can be expressed as

$$\frac{dm_e}{dt} = \sum_j (\rho_{f,j}\,\rho_a)^{1/2}\, 2\pi\, R_j\, (x_{j+1} - x_j)\, C_x\, |v_j - V_{a,j}| \tag{122}$$

$$\frac{dm_b}{dt} = c_1 \frac{m_e}{\tau_c} \qquad \tau_c = \frac{\lambda}{S_L} \tag{123}$$

where c_1 is a constant, λ is the Taylor microscale, and S_L is the laminar flame speed.

The entrainment and combustion rates during diffusive combustion can be expressed as

$$\frac{dm_e}{dt} = c_2 \frac{m_e}{\tau_e} \qquad \tau_e = \left(\frac{l^2}{\epsilon}\right)^{1/3} \qquad l = C_\mu^{3/4} \frac{k^{3/2}}{\epsilon} \tag{124}$$

$$\frac{dm_b}{dt} = \frac{m_e}{\tau_c} \tag{125}$$

where l is the length scale of the energy-containing eddies and $C_\mu = 0.09$.

The NO formation was calculated by means of the extended Zel'dovich mechanism [Eqs. (115)–(117)], and soot formation was calculated by means of the following expression:

$$\frac{dm_s}{dt} = A_f\, m_{fg}\, p^{0.5} \exp\left(-\frac{E_f}{\tilde{R}T}\right) - A_c\, m_s \frac{p_O}{p}\, p^{1.8} \exp\left(-\frac{E_c}{\tilde{R}T}\right) \tag{126}$$

where A_f and A_c are the preexponential factors, m_{fg} is the mass of fuel vapor, p_O is the oxygen partial pressure, E_f = 12,500 cal/mol, E_c = 14,000 cal/mol, m_s is the mass of soot, and the subscripts c and f denote combustion and formation, respectively.

Figure 3.29 shows some sample results calculated with Kono's model [94]. These results indicate that the NO emissions increase with the crank shaft angle whereas the soot emissions peak at about 40° after top dead center (ATDC). Figure 3.29 also shows that the calculated heat release rate is in good agreement with the measured one except near TDC.

3.3.4 Meguerdichian and Watson's Model

Meguerdichian and Watson [95] divided the spray into a number of burning zones at the same pressure but at different temperatures and applied expressions similar to those of Eqs. (107)–(111) where the burning rate was specified as

$$\frac{d}{dt}\left(\frac{m_{fb,j}}{m_{f,j}}\right) = A\rho_{a,j}\left(1 - \frac{m_{fb,j}}{m_{f,j}}\right)\left(1 - \phi_j\frac{m_{fb,j}}{m_{f,j}}\right) \tag{127}$$

for $\phi_{\text{lean}} < \phi_j < \phi_{\text{rich}}$, and

$$\frac{d}{dt}\left(\frac{m_{fb,j}}{m_{f,j}}\right) = 0 \tag{128}$$

outside the flammability limits, where $m_{fb,j}$ is the mass of burnt fuel in the jth zone and the subscripts lean and rich denote the flammability limits.

The total mass in the jth zone can be expressed as

$$m_j = m_{a,j} + m_{f,j} = m_{a,j} + m_{fu,j} + m_{fb,j} \tag{129}$$

where the subscripts a, fu, and fb refer to air, unburnt and burnt fuel, respectively.

Meguerdichian and Watson [95] considered the fuel-air mixing process and the free and wall jets; they accounted for swirl, which deflects the burning elements, by means of experimental correlations for the spray penetration.

Meguerdichian and Watson [95] assumed that the jet front has a bell shape identical to that of the velocity profile [cf. Eq. (104)]. This means that the fluid elements closer to the jet surface lag behind those located near the jet axis. The

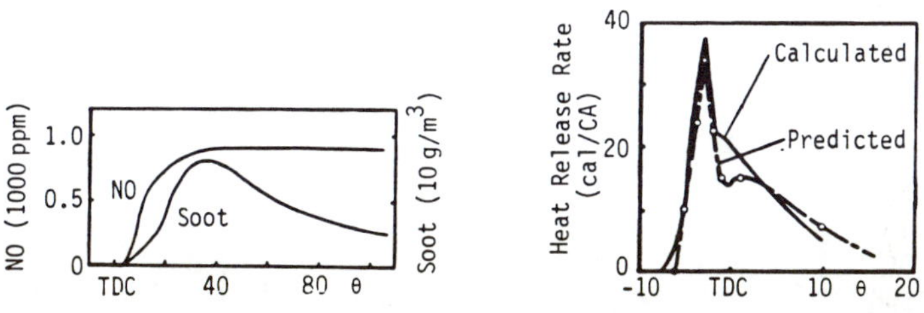

Figure 3.29 Nitric oxide and soot profiles (left) and heat release rate (right) as functions of the crank shaft angle (CA) [94].

jet was divided into zones (moving control volumes) representing motion of the same fuel element as follows. The velocity profile given by Eq. (104) was used to define the zone boundaries along the jet axis. The thickness of each zone was obtained by dividing the jet into concentric layers of unequal width. The fast-moving zones reaching the jet tip were assumed to change layers and settle as slow-moving zones. Additional zones were created at the nozzle exit at each time step while there was injection.

Figure 3.30 shows the motion of the jet zones and their deflection upon impingement on solid walls. Figure 3.31 illustrates the growth of a particular zone [95].

The model proposed by Meguerdichian and Watson [95] is unrealistic and does not truly represent the complex situation that exists behind the jet tip. Furthermore, the concentration values of a fully developed jet were used, and the zone velocities along the layers were approximated by the axial velocities of a fully developed jet at the corresponding positions. However, the model proposed by Meguerdichian and Watson [95] accounts for the jet deflection due to air velocity components normal to the fuel jet and for centerline concentration decay by means of experimental correlations.

Kobayashi et al. [97] developed a two-zone model where the fuel injected into the cylinder is assumed to vaporize instantaneously. Their two-zone model consists of the gaseous fuel jet and the surrounding atmosphere. Kobayashi et al. [97] measured the cylinder pressure and volume occupied by the fuel jet as a function of time, and they determined the temperature in the surrounding atmosphere by assuming that the heat transfer to the solid walls is related to the rate of change of the combustion chamber pressure.

3.3.5 Hiroyasu's Model

In the model proposed by Hiroyasu and co-workers [96, 99–103], the injected fuel spray is divided into several elements. These elements entrain air, vaporize,

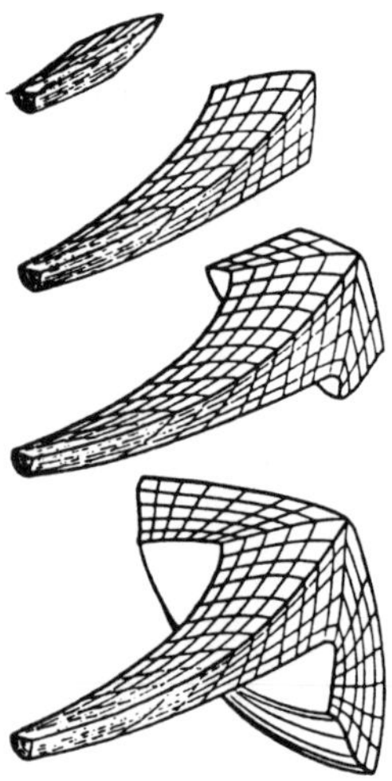

Figure 3.30 Jet zone motion, deflection, and impingement [95].

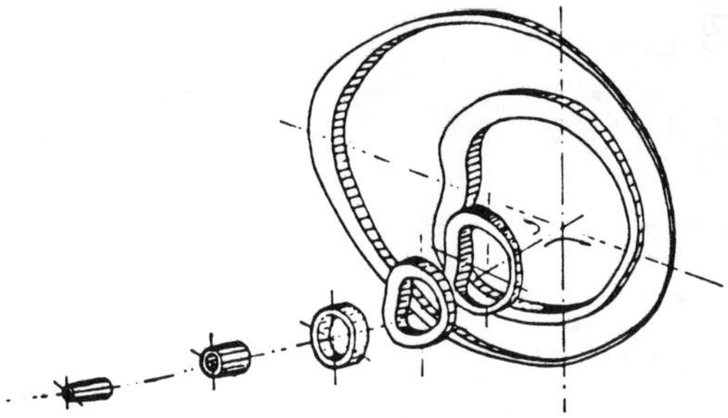

Figure 3.31 Schematic of zone growth [95].

and mix before igniting and reacting. During injection and combustion, the elements expand and entrain air. After ignition, the fuel droplets evaporate, and fuel vapor is mixed with air and reacts. Schematics of these processes are shown in Figs. 3.32 and 3.33.

Hiroyasu and Arai [102] used the breakup theory of Levich [104], which is based on the following assumptions: (1) $\rho_l >> \rho_a$; (2) the relative velocity between the liquid jet and the gas is large; (3) as the amplitude of the surface perturbation increases, the jet becomes unstable and breaks into droplets; and (4) the breakup length is given by the following expression:

$$L = v_{\text{inj}}\, t_{\text{break}} \cong c_1 \left(\frac{\rho_l}{\rho_a}\right)^{1/2} d_{\text{inj}} \tag{130}$$

where c_1 is a constant, and ρ_a is the density of the surrounding air.

If the jet velocity within the intact core length is equal to the injection velocity

$$v_{\text{inj}} = c_2 \left(\frac{2\,\Delta p}{\rho_l}\right)^{1/2} \tag{131}$$

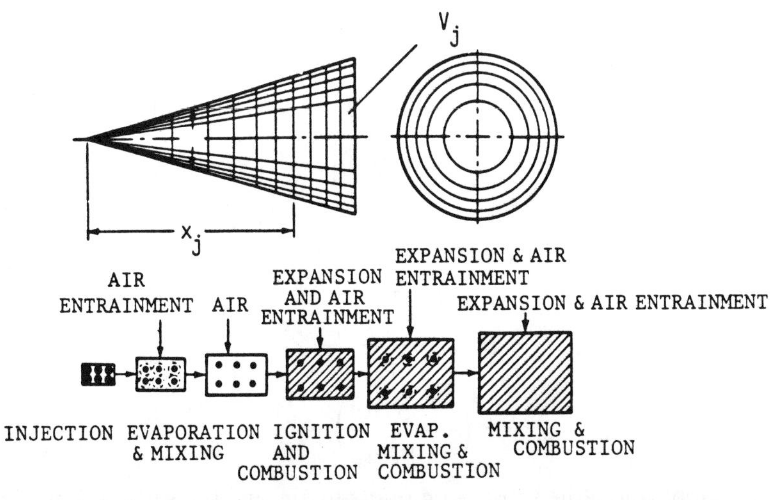

Figure 3.32 Schematic of spray combustion [96].

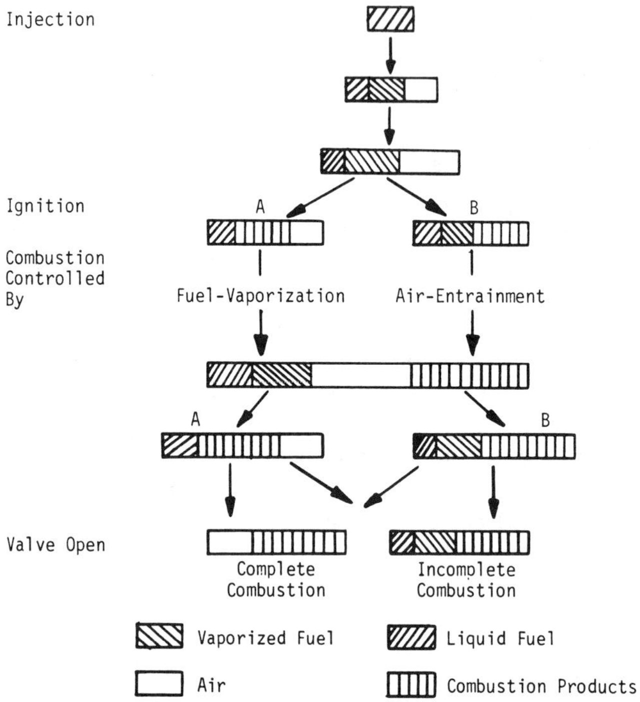

Figure 3.33 Evolution of a spray element [96].

then

$$t_{\text{break}} = c_1 \, \rho_l \frac{d_{\text{inj}}}{(2 \, c_2^2 \, \rho_a \, \Delta p)^{1/2}} \tag{132}$$

where c_1 and c_2 are constants ($c_1 = 15.8$, $c_2 = 0.39$).

The spray penetration and angle are given by the following expression [96]:

$$0 < t < t_{\text{break}} \qquad x = 0.39 \, t \left(\frac{2 \, \Delta p}{\rho_l}\right)^{1/2} \tag{133}$$

$$t \geq t_{\text{break}} \qquad x = 2.95 \left[\frac{(d_{\text{inj}} \, t)^2 \, \Delta p}{\rho_a}\right]^{1/4} \tag{134}$$

$$t_{\text{break}} = 28.65 \, \rho_l \frac{d_{\text{inj}}}{(\rho_a \, \Delta p)^{1/2}} \tag{135}$$

$$\alpha = 0.05 \left(\frac{d_{\text{inj}}^2 \, \rho_a \, \Delta p}{\mu_a^2}\right)^{1/4} \tag{136}$$

where α is the spray angle, the subscript a refers to air, and the subscript break denotes liquid jet breakup.

Equations (133)–(136) are valid for fuel elements on the spray centerline. For fuel elements in the radial direction away from the centerline

$$x_R = x \exp\left[-4.403 \times 10^{-3} \times (L - 1)^2\right] \tag{137}$$

where $L = 2, 3, \ldots$, denotes the number of fuel elements injected per injection interval.

Arai et al. [99] accounted for the jet deflection by a crossflow velocity $v_\theta(x)$ by applying a correction factor to Eqs. (133) and (134) as follows:

$$x_s = xC_x \qquad \theta_s = \theta C_\theta \tag{138}$$

where x_s and θ_s denote the spray penetration and angle due to crossflow velocities, and C_x and C_θ depend on the swirl ratio, rpm, and injection velocity. For example,

$$C_x = \left(1 + \frac{\pi\, x \text{ rpm } S}{30\, v_{\text{inj}}}\right)^{-1} \tag{139}$$

where S is the swirl ratio, v_{inj} is the initial spray-tip penetration velocity, that is,

$$v_{\text{inj}} = c_1 \left(\frac{2\,\Delta p}{\rho_l}\right)^{1/2} \tag{140}$$

and c_1 is a constant.

Hiroyasu et al. [100] recommended a value of $c_1 = 1$. However, the numerical results of Kuo [101] indicate that a value of $c_1 = 1$ results in a spray tip penetration rate that is too fast. Kuo [101] recommended $c_1 = 0.39$ based on comparisons between numerical predictions and experimental data.

Conservation of linear momentum applied to an element of the fuel jet yields

$$\dot{m}_f\, v_{\text{inj}} = (\dot{m}_f + \dot{m}_a)\,\frac{dx}{dt} \tag{141}$$

which corresponds to the air entrainment rate if the fuel jet is free.

The air entrainment into a fuel packet before the start of combustion and wall impingement can be expressed as

$$\dot{m}_{ae} = \dot{m}_a \tag{142}$$

After the start of combustion and before the spray impinges on the wall

$$\dot{m}_{ae} = c_f\, \dot{m}_a \tag{143}$$

where $c_f = 0.7$ and 0.4 for 2- and 0.52-liter-per-cylinder engines, respectively [101].

However, the jet momentum decreases upon impingement on the wall. The air entrainment for the wall jet can be calculated as

$$\frac{dm_{aw}}{dt} = c_w\,\frac{dm_a}{dt} \tag{144}$$

where $c_w = 1.5$ and 1.4 for 2- and 0.52-liter engines, respectively [101].

The tangential deflection of the spray due to swirl can be calculated by means of the following expression:

$$r_{\text{tang}} = x \frac{m_a}{m_a + m_f} \frac{2\pi\, x \text{ rpm } C_x}{30\,[v_{\text{inj}} + (dx/dt)]} \tag{145}$$

The liquid fuel in each element is postulated to be composed of many droplets with a Sauter mean diameter given by

$$\text{SMD} = A\, \Delta p^{-0.135}\, \rho_a^{0.12}\, B^{0.131} \tag{146}$$

where B is the mass of fuel injected in the cylinder per cycle and A (= 23.9) is a constant that depends on the nozzle geometry.

The droplets are assumed to evaporate according to the following expressions:

$$\frac{dm_l}{dt} = -\pi d^2 k^* \frac{Y_{fs}}{1 - (1+\xi)\, Y_{fs}} \tag{147}$$

$$\frac{d}{dt}(d) = \frac{2}{\pi \rho_l d^2}\left(\frac{dm_l}{dt} - \frac{\pi\, d^3}{6}\frac{d\rho_l}{dt}\right) \qquad \frac{d\rho_l}{dt} = \frac{d\rho_l}{dT_l}\frac{dT_l}{dt} \tag{148}$$

$$\frac{dT_l}{dt} = \frac{1}{m_l\, C_l}\left[\pi d^2 h^*(T - T_l) + L\frac{dm_l}{dt}\right] \tag{149}$$

where d is the droplet diameter; k^* and h^* are the mass and heat transfer coefficients, respectively; ξ is the air-to-fuel flow ratio at the droplet surface; L is the latent heat of vaporization; Y_{fs} is the mass fraction of the gaseous fuel at the droplet surface; and

$$k^* = -\frac{\rho_a D}{d}\frac{1 - (1+\xi)\, Y_{fs}}{(1+\xi)\, Y_{fs}} \text{Sh} \ln\,[1 - (1+\xi)\, Y_{fs}] \tag{150}$$

$$h^* = \frac{X}{\pi d^2}\left[\exp\left(\frac{X}{\text{Nu}\ \pi\ d\lambda}\right) - 1\right]^{-1} \tag{151}$$

$$\xi = -\rho_a \frac{1 - Y_{fs}}{\rho_l + (d/6)\,[d\rho_l/d(d)]} \tag{152}$$

$$X = -(C_{p_f} + C_{p_a})\frac{dm_l}{dt} \tag{153}$$

where D is the binary diffusion coefficient; C_{p_a} and C_{p_f} are the specific heats at constant pressure of the air and fuel, respectively; Nu is the Nusselt number; and Sh is the Sherwood number, which can be written using the Ranz-Marshall correlation [105] as

$$\text{Sh} = 1 + 0.276\ \text{Re}^{1/2}\ \text{Sc}^{1/3} \tag{154}$$

$$\text{Nu} = 1 + 0.276\ \text{Re}^{1/2}\ \text{Pr}^{1/3} \tag{155}$$

where the Reynolds (Re), Schmidt (Sc), and Prandtl (Pr) numbers are defined as

$$\mathrm{Re} = \rho_a \frac{du}{\mu_a} \tag{156}$$

$$\mathrm{Sc} = \frac{\mu_a}{\rho_a D} \tag{157}$$

$$\mathrm{Pr} = \mu_a \frac{C_{pa}}{\lambda_a} \tag{158}$$

ρ_a, μ_a, and λ_a denote the density, dynamic viscosity, and thermal conductivity of the surrounding air, respectively; u is the relative velocity between the droplets and surrounding air, which can be calculated from spray-penetration data.

In order to solve Eqs. (147)–(149) the droplet size at the nozzle exit must be specified. Hiroyasu et al. [96] found that the droplet distribution function depends on the nozzle employed in the experiments, and can be written as

$$\frac{d\,\mathrm{Vol}}{\mathrm{Vol}_t} = 13.5 \left(\frac{d}{\mathrm{SMD}}\right)^3 \exp\left[-3\left(\frac{d}{\mathrm{SMD}}\right)\right] d\left(\frac{d}{\mathrm{SMD}}\right) \tag{159}$$

where Vol is the volume of the droplets, Vol_t is the total droplet volume, and SMD can be determined from Eq. (146). The mass of fuel injected at each time step can be allocated to the droplet distribution function ($\mathrm{Vol} = \dot{m}_{\mathrm{inj}}\,\Delta t$, where $\dot{m}_{\mathrm{inj}}$ is the fuel injection rate). The continuous droplet distribution function can then be discretized into droplet groups characterized by the same droplet radius and temperature. These droplet groups can then be assigned to zones.

A substantial simplification results if all the droplets in each group have the same diameter, that is, if $\mathrm{Vol} = \pi(\mathrm{SMD})^3/6$, since in this case, each zone contains droplets of the same diameter.

The mass of vaporized fuel m_{fg} is

$$m_{fg} = \frac{\pi}{6}\,(\rho_{lo}\, d_0^3 - \rho_l\, d^3)\mathrm{N} \tag{160}$$

where N denotes the number of droplets in an element, and the subscript 0 refers to the previous time step.

During fuel vaporization the temperature falls. However, after a short time t_{ig} the gaseous mixture is ignited,

$$\int_0^{t_{\mathrm{ig},j}} \frac{dt}{\tau_{d,j}} = 1 \qquad \tau_d = 2 \times 10^{-2}\, p^{-2.5}\, \phi^{-1.04} \exp\left(\frac{4000}{T}\right) \tag{161}$$

where 0 refers to the beginning of injection, the subscript j denotes the jth zone, and

$$\phi = \frac{m_{fg}}{f_{st}\, m_g} \tag{162}$$

f_{st} is the stoichiometric air-fuel ratio and m_g is the mass of gas in the fuel element.

The model proposed by Hiroyasu and co-workers [99–103] assumes that there is no single droplet surrounded by a flame and that the evaporated fuel reacts with the air entrained into the element forming a diffusion flame around a group of droplets. The combustion of an element is controlled by the rate of fuel vaporization or by air entrainment. Hiroyasu and co-workers [99–103] postulated that combustion occurs at stoichiometric conditions and that some residual portion of fuel or air in the element participates in the next step of combustion.

The heat release rate for the jth element can be written as

$$\frac{dQ_{c,j}}{dt} = h_u \frac{dm_{f,j}}{dt} \tag{163}$$

where h_u is the calorific value of the fuel and

$$\frac{dm_{f,j}}{dt} = \frac{dm_{fu,j}}{dt} \ (\phi_j \leq 1) \qquad \text{or} \qquad \frac{dm_{f,j}}{dt} = \frac{dm_{fc,j}}{dt} \ (\phi_j > 1) \tag{164}$$

where $m_{fu,j}$ corresponds to the vaporized fuel, $m_{fc,j}$ is the gaseous fuel determined from air entrainment by a stoichiometric consideration, and ϕ_j denotes the equivalence ratio of the jth element.

The total heat release rate can be expressed as

$$\frac{dQ_c}{dt} = h_u \sum_{j=1}^{N} \frac{dm_{f,j}}{dt} \tag{165}$$

where N is the number of fuel elements in the cylinder. The heat transfer losses can be calculated, using Newton's cooling law, as

$$\dot{Q}_w = A\ h\ (T - T_w) \tag{166}$$

where A is the cylinder surface area, T_w is the wall temperature,

$$h = 265\ B^{-0.2}\ p^{0.8}\ V_p^{0.8}\ T^{-0.53} \tag{167}$$

and V_p is the instantaneous piston speed.

The net rate of heat release and pressure change can be written as

$$\frac{dQ}{d\theta} = \frac{dQ_c}{d\theta} - \frac{dQ_w}{d\theta} \tag{168}$$

and

$$\frac{dp}{d\theta} = \frac{1}{V}\left(\frac{\gamma - 1}{A}\frac{dQ}{d\theta} - \gamma\, p \frac{dV}{d\theta}\right) \qquad \gamma = \frac{C_p}{C_v} \tag{169}$$

Hiroyasu and co-workers [99–103] considered NO, CO, CO_2, O_2, H_2, H_2O, OH, H, O, N_2, and N, that is, 11 species. All the species except N and NO were assumed to be in chemical equilibrium, and the NO was calculated by means of the Zel'dovich mechanism [Eqs. (115)–(117)] with a steady-state approximation for N, that is, $d[\mathrm{N}]/dt = 0$, where the brackets denote concentration. Hiroyasu and co-workers [99–103] studied the formation of soot by means of Eq. (126).

Kuo [101] used Eq. (126) with $A_f = 500$ and with $A_c = 4.66$ and 2.80 for 2- and 0.52-liter-per-cylinder engines, respectively, in order to obtain good agreement betwen the predicted and measured soot emissions in the engine exhaust.

A model similar to that proposed by Hiroyasu et al. [96] has been developed by Tinaut [59], who studied the fuel jet penetration and the jet deflection by the solid walls. Tinaut [59] also used a Wiebe function to predict the heat release rate in DI diesel engines. Some sample results are shown in Fig. 3.34.

Kuo [101] showed that the model proposed by Hiroyasu et al. [100] predicts the indicated mean effective pressure and indicated thermal efficiency within 5% of the experimental data for 2- and 0.52-liter-per-cylinder engines (2-liter engines are heavy duty, and 0.52-liter engines are light duty). The pressure-time and heat release rate agreed within 5% to 15% with experimental data; however, the nitric oxide and soot predictions were in poor agreement with experimental measurements, especially for the light-duty engine. Inadequate modeling of the spray impingement upon the walls may be the reason for the observed disagreement.

3.3.6 Cummins Engine Model

In the model proposed by Shahed and co-workers [98, 106–108]—here referred to as the Cummins engine model—the spray is treated as a vapor jet in the spray-mixing calculation. The fuel vapor concentration is assumed to be continuous, and the vapor jet is divided into a series of discrete combustion zones. Energy conservation, chemical equilibrium, and nitric oxide finite rate chemistry are applied to each zone.

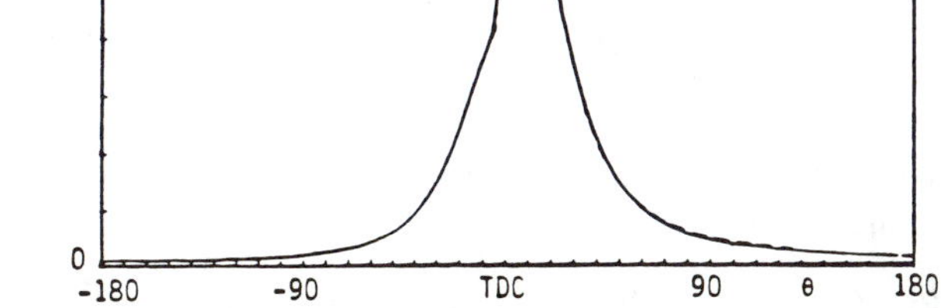

Figure 3.34 Measured and predicted pressure diagrams [59].

At the moment of ignition the premixed combustible mixture is considered to be bounded by the lean flammability limit at the outer edge of the spray and by the rich flammability limit near the core. This region is divided into several zones at appropriate fuel-air ratio values; the properties of the mixture in each zone are considered to be uniform.

After ignition more fuel from the spray becomes combustible as a result of continuous air entrainment. The rate at which the combustible fuel-air mixture is formed is called the preparation rate, and the mixture prepared in successive crank angle intervals is assigned to progressive combustion zones until all fuel is prepared. Thus, the number of combustion zones increases with time.

Equilibrium calculations are carried out for each zone by tracking its mass, composition, and energy. Identification of combustion zones requires a detailed temporal and spatial fuel-air distribution. Fuel-penetration and combustion dilution rates can also be obtained by means of a spray-mixing model. The Cummins engine model uses empirical correlations for the spray penetration, width growth rate, and trajectory to describe the complete spray geometry. The fuel vapor concentration in the spray can be specified as a normal distribution across the spray and a hyperbolic profile along the spray axis.

The spray tip penetration along the radial direction with swirl can be calculated as follows

$$\frac{x - x_\omega}{x} = 0.35\left(\frac{x_\omega\, Q_a}{d_{\text{inj}}\, Q_f}\right)^{0.44} \tag{170}$$

$$x = \frac{450\, d_{\text{inj}}^{0.5}\, (\rho_f/\rho_d)^{0.4}}{(1 + \rho_a/\rho_{\text{atm}})^{0.85}}\left(\frac{\rho_a}{\rho_{\text{atm}}}\right)^{0.5} (p_{\text{inj}} - p_a)^{0.25}\, t^{0.6} \tag{171}$$

where x and x_ω are the tip penetrations without and with swirl, respectively; d_{inj} is the injector diameter; ρ_a, ρ_f, ρ_d, and ρ_{atm} are the densities of air, reference fuel, diesel fuel, and atmosphere, respectively; p_{inj} and p_a are the injection and cylinder pressures, respectively; and, Q_a and Q_f are the momenta of the air and fuel, that is,

$$Q_a = \rho_a\, x_\omega^2\, \omega^2 \qquad Q_f = \rho_f\, v_{\text{inj}}^2 \qquad v_{\text{inj}} = c_d\left(2\frac{p_{\text{inj}} - p_a}{\rho_a}\right)^{1/2} \tag{172}$$

where ω is the air angular speed.

The tangential displacement of the spray is

$$x_t = d_{\text{inj}}\left(\frac{Q_a}{Q_f}\right)\left(\frac{r}{d_{\text{inj}}}\right)^{2.217} \tag{173}$$

where r is the radial coordinate.

The cross-sectional area of the spray can be approximated, by two ellipses or an equivalent circle of radius b, as

$$b = \left(b_3\,\frac{b_1 + b_2}{2}\right)^{1/2} \qquad \frac{db_1}{dx} = 0.12\left(1 + \frac{\rho_a}{\rho_f}\right) \tag{174}$$

$$b_3 = b_1 + 0.11x\left(\frac{\rho_a}{\rho_f}\right) \qquad b_2 = b_1\,(1+0.0016\ \mathrm{Re}^{0.66}) \tag{175}$$

$$\mathrm{Re} = 2\pi\,\rho_a \frac{r\omega d_{\mathrm{inj}}}{\mu_a} \tag{176}$$

and the fuel concentration distribution across the equivalent circular spray is

$$c = c_m\left[1-\left(\frac{y}{b}\right)^{1.5}\right] \tag{177}$$

where

$$c_m = [\alpha(t)\,x+1]^{-1} \qquad x_1 < x < x_{\mathrm{tip}} \tag{178}$$

$$c_m = 0 \qquad x_1 > x \qquad x > x_{\mathrm{tip}} \tag{179}$$

and $\alpha(t)$ can be determined from mass conservation, as follows. The total injected fuel up to time t is

$$\int_0^t \frac{dm_f}{dt}\,dt = 2\pi \int_{x_1}^{x_{\mathrm{tip}}} \int_0^b \rho Y_f\, y\, dx\, dy \tag{180}$$

$$\rho = \frac{p_a}{[(1-Y_f)R_a + Y_f R_f]\,T_a} \tag{181}$$

Kuo et al. [109] used a stochastic two-dimensional, thick spray model to analyze fuel vaporization in the quiescent chamber of DI diesel engines; they showed that an increase in the vaporization rate does not necessarily imply an increase in the fuel-air mixing rate. They also showed that combustion in such engines can be controlled by mixing rather than vaporization. Their numerical results also indicate that the concentration distribution across the spray can be expressed as [cf. Eq. (177)]

$$c = c_m \exp\left[-0.693\left(\frac{r}{R_{1/2}}\right)^{2.5}\right] \tag{182}$$

where $R_{1/2}$ is the radial location at which the concentration reaches a value equal to $c_m/2$.

The approach followed by Kuo et al. [109] is unique in that they used multidimensional thick spray models to obtain correlations that can be used in thermodynamic models. A similar approach was used by Elkotb et al. [110], who assumed that droplet breakup occurs when the Weber number is larger than the square root of the Reynolds number (see Sections 3.5.4 and 3.5.5).

At incipient ignition the combustion chamber is divided into $(n + 2)$ zones, which correspond to the medium into which the spray is injected A, the rich core of the spray C, and n zones B_i of combustible mixture (Fig. 3.35). The combustion zones B_i are bounded by the lean and rich flammability limits and

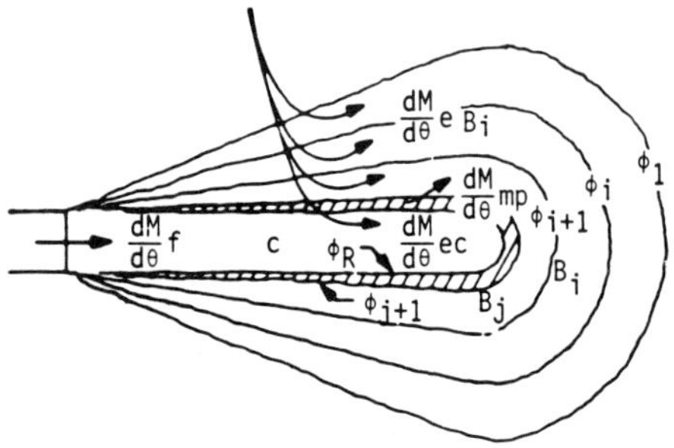

Figure 3.35 Schematic of the progressive evolution of the combustion zones and entrainment rates [108].

intermediate equivalence ratios ϕ_i as follows:

$$\phi_R \geq \phi_i \geq \phi_L \tag{183}$$

$$m_{f,B_i} = 2\pi \int_{x_1}^{x_{\text{tip}}} \int_{y(\phi_{i+1})}^{y(\phi_i)} \rho Y_f y \, dy \, dx \tag{184}$$

$$m_{A,B_i} = 2\pi \int_{x_1}^{x_{\text{tip}}} \int_{y(\phi_{i+1})}^{y(\phi_i)} \rho(1 - Y_f) y \, dy \, dx \tag{185}$$

$$m_{f,C} = 2\pi \int_{x_1}^{x_{\text{tip}}} \int_{y(\infty)}^{y(\phi_R)} \rho Y_f y \, dy \, dx \tag{186}$$

After ignition, which is determined by an ignition delay time correlation, air entrainment continues, and the fresh mixture prepared for combustion is assigned to a new combustion zone B_J. The fuel mass in each combustion zone, except in B_J, is fixed.

The air entrainment into the core zone C results in mixture preparation for combustion, and the air entrainment rate into this zone is

$$\frac{dm_{e,C}}{d\theta} = 2\pi \frac{d}{d\theta} \int_{x_1}^{x_{\text{tip}}} \int_{y=0}^{y(\phi_{J+1})} (1 - Y_f)\rho y \, dy \, dx \tag{187}$$

where ϕ_{J+1} is the boundary of the richest combustion zone.

The mixture prepared for combustion is defined as the fresh mixture crossing the rich limit of combustion ϕ_R and can be written as

$$m_{f,p} = 2\pi \int_{x_1}^{x_{\text{tip}}} \int_{y(\phi_R)}^{y(\phi_{J+1})} Y_f \rho y \, dy \, dx \tag{188}$$

$$m_{a,p} = 2\pi \int_{x_1}^{x_{\text{tip}}} \int_{y(\phi_R)}^{y(\phi_{J+1})} (1 - Y_f)\rho y \, dy \, dx \tag{189}$$

The total mixture prepared is

$$m_{m,p} = m_{f,p} + m_{a,p} \tag{190}$$

where the subscripts a, f, m, and p denote air, fuel, mixture, and prepared fuel, respectively.

Since each combustion zone contains a fixed mass of fuel, the air entrainment into zone B_i is

$$\frac{dm_{e,B_i}}{d\theta} = \frac{dm_{B_i}}{d\theta} \tag{191}$$

Conservation of energy for all the zones can be written as

$$\frac{d}{d\theta}(m_C u_C) = -p\frac{dV_C}{d\theta} + \frac{dq_C}{d\theta} + h_f\frac{dm_f}{d\theta} + h_A\frac{dm_{e,C}}{d\theta} - h_m\frac{dm_{m,p}}{d\theta} \tag{192}$$

$$\frac{d}{d\theta}(m_A u_A) = -p\frac{dV_A}{d\theta} + \frac{dq_A}{d\theta} - h_A\frac{dm_A}{d\theta} \tag{193}$$

$$\frac{d}{d\theta}(m_{B_i} u_{B_i}) = -p\frac{dV_{B_i}}{d\theta} + \frac{dq_{B_i}}{d\theta} + h_A\frac{dm_{e,B_i}}{d\theta} \qquad i = 1, 2, \ldots, J-1 \tag{194}$$

$$\frac{d}{d\theta}(m_{B_J} u_{B_J}) = -p\frac{dV_{B_J}}{d\theta} + \frac{dq_{B_J}}{d\theta} + h_m\frac{dm_{m,p}}{d\theta} - h_A\frac{dm_{e,B_J}}{d\theta} \tag{195}$$

$$V = V_A + V_C + \sum_{i=1}^{J} V_{B_i} \tag{196}$$

$$\frac{dm_C}{d\theta} = \frac{dm_f}{d\theta} + \frac{dm_{e,C}}{d\theta} - \frac{dm_{m,p}}{d\theta} \tag{197}$$

$$\frac{dm_A}{d\theta} = -\frac{dm_{e,C}}{d\theta} - \sum_{i=1}^{J}\frac{dm_{e,B_i}}{d\theta} \tag{198}$$

$$\frac{dm_{B_i}}{d\theta} = \delta_{iJ}\frac{dm_{m,p}}{d\theta} + \frac{dm_{e,B_i}}{d\theta} \tag{199}$$

$$\frac{dm_C}{d\theta} + \frac{dm_A}{d\theta} + \sum_{i=1}^{J}\frac{dm_{B_i}}{d\theta} = \frac{dm_f}{d\theta} \tag{200}$$

where $dm_f/d\theta$ is the burning rate, and δ_{iJ} is the Kronecker delta (i.e., $\delta_{iJ} = 0$ if $i \neq J$ and $\delta_{iJ} = 1$ if $i = J$).

The Cummins engine model uses Annand's correlation [Eq. (55)] to calculate convective and radiative heat transfer based on a bulk average temperature defined as

$$T_{\text{av}} = \frac{\sum_{i=1}^{J} m_{B_i} T_{B_i} + m_A T_A + m_C T_C}{\sum_{i=1}^{J} m_{B_i} + m_A + m_C} \tag{201}$$

and the total heat transfer losses are apportioned to the various zones in relation to their mass and temperature as

$$\frac{dq_{B_i}}{d\theta} = \frac{dQ}{d\theta} \frac{m_{B_i} T_{B_i}}{\sum_{i=1}^{J} m_{B_i} T_{B_i} + m_A T_A + m_C T_C} \tag{202}$$

where $dQ/d\theta$ represents the total heat transfer losses.

Figure 3.35 shows a schematic of the progressive evolution of the combustion zones and entrainment rates [108]. Figure 3.36 shows the NO concentration and pressure as a function of the crank shaft angle [98]. The NO concentration was calculated by means of the Zel'dovich mechanism [Eqs. (115) and (116)], and all the species except NO were considered to be in local chemical equilibrium. The production rate of NO in zones A and C is equal to zero, whereas in the other zones

$$\frac{d}{d\theta}(m_{B_i} Y_{NO,B_i}) = M_{NO} \frac{d([NO]V_{B_i})}{d\theta} + \frac{dm_{e,B_i}}{d\theta} Y_{NO,A} \tag{203}$$

$$\frac{d}{d\theta}(m_{B_J} Y_{NO,B_J}) = M_{NO} \frac{d([NO]V_{B_J})}{d\theta} + \frac{dm_{m,p}}{d\theta} Y_{NO,C} + \frac{dm_{e,B_J}}{d\theta} Y_{NO,A} \tag{204}$$

$$\frac{d}{dt}\{[NO]V_{B_i}\} = V_{B_i}\{k_1[N_2][O] + k_2[O_2][N] - k_3[NO][N] - k_4[NO][O]\} \tag{205}$$

where k_1, k_2, k_3, and k_4 are the reaction constants.

Figure 3.37 indicates that the Cummins engine model can predict with some success the NO emissions in DI diesel engines [107].

The assumption of a gaseous fuel jet in the Cummins engine model is valid if the combustion chamber conditions are near or above the thermodynamical critical point of the fuel; when this occurs, droplets appear only near the nozzle. However, under normal operating conditions the cylinder pressure is above criticality, whereas the cylinder temperature is usually below criticality. This means that a fuel spray, not a gaseous fuel jet, is injected into the cylinder. Moreover, for the same injection velocity, the penetration of a fuel spray is larger than that of a gaseous fuel, and case-by-case tuning may be required to

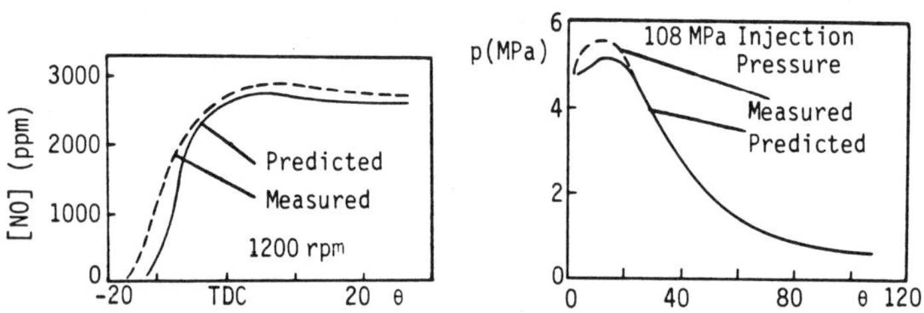

Figure 3.36 Nitric oxide concentration (left) and pressure (right) as functions of the crank shaft angle [98].

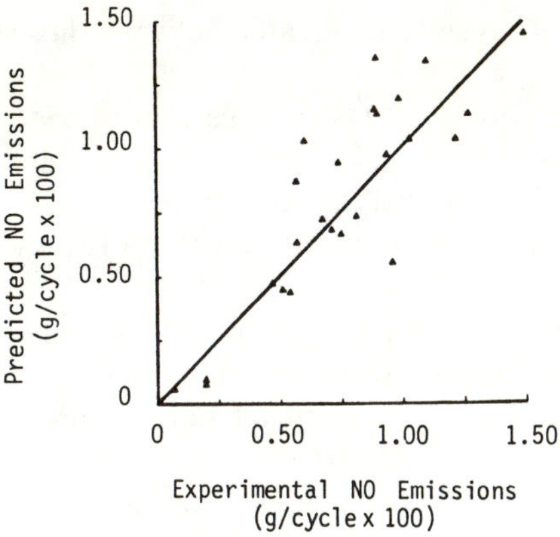

Figure 3.37 Comparison between predicted and measured nitric oxide emissions [107].

estimate the fuel penetration of gaseous fuel jets so that the numerical predictions are in good agreement with experiments.

Chapman et al. [111] used the Cummins engine model to study radiative heat transfer in diesel engines. Assuming that the gaseous medium is optically thin and that the soot behaves as a gray body, they employed a zonal method to calculate the view factors and heat transfer exchanges between different zones; they showed that the primary radiation source is from soot formed on fuel-rich regions. They also observed significant spatial and temporal variations in the incident radiation.

The gaseous jet penetration used in the Cummins engine model [Eq. (171)] is different from that obtained by Kuo and Bracco [112], Gaillard [113], Turns and Ran [114], and others.

For impulsively started incompressible turbulent round jets, Kuo and Bracco [112] derived the following penetration rates

$$x^* = 0.5t^* \qquad \text{for } x^* \leq 8 \tag{206}$$

$$x^* = 2(t^*)^{1/2} \qquad \text{for } x^* > 8 \tag{207}$$

where

$$t^* = t\,\frac{u(0,0)}{d_{\text{inj}}}\,\text{Re}^{0.053} \tag{208}$$

$$t^* = \frac{x}{d_{\text{inj}}\,\text{Re}^{0.053}} \qquad \text{Re} = \rho u(0,0)\,\frac{d_{\text{inj}}}{\mu} \tag{209}$$

and $u(0,0)$ is the jet centerline velocity at the nozzle exit.

Gaillard [113] studied impulsively started round free jets of methane in confined and unconfined geometries and obtained the following penetration data for unconfined jets:

$$x^* = 2t^{*0.5} \qquad \text{for } x^* \geq 7 \qquad \text{and } 70\%\ V_{CL}, \text{ steady} \tag{210}$$

$$x^* = 1.64t^{*0.5} \qquad \text{for } x^* \geq 6 \qquad \text{and } 99\%\ V_{CL}, \text{ steady} \tag{211}$$

where V_{CL} is the jet centerline velocity. These results are located between those proposed by Kuo and Bracco [112].

Turns and Ran [114] studied the penetration of unsteady carbon dioxide jets in confined swirling air flows and showed that the penetration times were twice those of a quasisteady model. They also showed that the tip penetration is not a function of the jet-to-air swirl momentum ratio and that there is flow asymmetry for jet-to-air swirl momentum ratios smaller than 1.

Varde et al. [115] analyzed sprays from plain orifices using a single-pulse mode; they showed that the spray angle is a function of the orifice dimensions, injection pressure, ambient density, and nozzle length-to-diameter ratio, whereas the droplet Sauter mean diameter is mainly a function of the injection pressure and orifice diameter for constant fuel pressures. Varde et al. [115] obtained the following expressions from their experimental data:

$$\tan \alpha = A\left(\frac{\rho}{\rho_l}\right)^{0.33} \tag{212}$$

$$\frac{\text{SMD}}{d_{\text{inj}}} = 8.7\ (\text{Re}_l\ \text{We}_l)^{-0.28} \tag{213}$$

where A is a constant, the subscript l denotes the liquid jet, and Re and We denote the Reynolds and Weber numbers, respectively. These expressions do not agree with those predicted by aerodynamic theories of liquid jet breakup [81, 82].

Wu et al. [116] used spark photography to study the penetration of water, n-hexane, and n-tetradecane liquid jets at injection pressures of 11.1 MPa to 107.6 MPa into gaseous nitrogen at pressures of 0.1 MPa to 4.2 MPa; they showed that the spray angle is a function of the nozzle geometry and ρ/ρ_l, but not a function of the injection velocity. Wu et al. [116] also showed that their experimental results are compatible with an aerodynamic theory of liquid jet breakup. (Atomization and liquid breakup regimes are discussed in Section 3.4.)

When liquid fuels are injected into the cylinders of diesel engines, the thermodynamic state of the mixture (pressure and temperature) and the injection characteristics determine the droplet size distribution and spray regime. In most cases atomization into fine droplets is observed and three characteristic times—the mixing, vaporization, and chemical kinetics times—must be considered. Chemical kinetics may be important only in the initial stages of combustion; thereafter, the fuel conversion rate is controlled by the fuel vaporization rate and by fuel-air mixing. These two processes are controlled by the cylinder pressure, temperature, and turbulence. Thus, the assumption of uniform composition and temperature in each "global" combustion zone is not valid.

3.3.7 Other Models

Kobayashi et al. [117] and Kamimoto et al. [118] developed a model that accounts for air entrainment, mixing, and combustion. These authors considered the increase in air entrainment due to the thermal expansion during the initial combustion period, a probability density function (pdf) for the fuel, a mixing rate that depends on the gradient of the fuel concentration, and a combustion model that takes into account the flammability limits. Kobayashi et al. [117] assumed that the fuel vaporization rate is equal to the fuel injection rate but shifted by the spray breakup and vaporization delay and that mixing takes place between the fuel vapor and the air. Kamimoto et al. [118] showed that their pdf model predicts quite satisfactorily the heat release rate observed in rapid compression machines.

The Cummins engine model is somewhat related to those proposed by Kobayashi et al. [117] and Kamimoto et al. [118]. These models consider only fuel vapor, account for the spatial and temporal variations of composition and temperature by means of combustion zones or a pdf formulation, and use experimentally determined correlations to obtain the spray penetration.

Ikegami et al. [119] developed a thermodynamic model of diesel engine combustion that uses an equation for the turbulence kinetic energy and a collision-dispersion model. Their model assumes that the spray formation is controlled by turbulent mixing, not by vaporization, and it satisfactorily predicts the heat release rate and NO emissions.

The major drawbacks of the model proposed by Ikegami et al. [119] are the specification of the turbulent length scale and mixing time in their stochastic mixing model. A turbulent length scale can be calculated from a transport equation for the dissipation rate of turbulence kinetic energy. However, the mixing time or mixing frequency can be calculated only by using a two-point two-time probability density function in their stochastic mixing model.

Plee and Ahmad [120] studied the relative roles of premixed and diffusive burning in a diesel engine. They used an equation for the turbulence kinetic energy and assumed homogeneous isotropic turbulence to calculate the Taylor microscale. Their model accounts for the ignition delay t_{ig}, combustion t_c, and fuel-air mixing t_m times. The ignition delay time was calculated from experimental correlations (Table 3.1) and

$$t_c = \frac{\delta_L}{S_L} \qquad t_c = \frac{K}{\rho_u C_p S_L^2} \tag{214}$$

$$t_m = \frac{\lambda^2}{6\nu} \qquad \frac{\lambda}{l} = \left(\frac{15}{\mathrm{Re}}\right)^{1/2} \qquad \mathrm{Re} = \frac{lu'}{\nu} \tag{215}$$

where λ_L is the laminar flame thickness, S_L is the laminar flame speed, K is the thermal conductivity, C_p is the specific heat at constant pressure, ρ_u is the density of the unburnt gases, λ is the Taylor microscale, l is the integral length scale, ν is the kinematic viscosity, and u' is the fluctuating velocity rms, which can be calculated from the turbulence kinetic energy equation.

In diesel engines

$$t_c << t_m \qquad \text{during diffusive combustion} \tag{216}$$

$$t_m << t_{ig} \qquad \text{during the ignition phase} \tag{217}$$

Multidimensional models account for the local variations of composition and temperature within the combustion chamber and for the history of each individual droplet packet injected into the cylinder.

Before describing multidimensional models, it is convenient to discuss liquid jet breakup and atomization.

3.4 LIQUID JET ATOMIZATION AND BREAKUP REGIMES

The injection of a liquid fuel into a gas eventually results in the formation of droplets, but the liquid jet breakup process depends on various conditions. Two different lengths can be identified in breakup processes: the intact core length and the breakup length. The breakup length is defined as the distance from the nozzle exit to the point where droplets are formed on the liquid jet surface, whereas the intact core length refers to the distance from the nozzle exit to the point where the liquid jet breaks up into droplets at the jet centerline. The breakup length can be observed experimentally; however, it has been difficult to obtain accurate experimental data for the intact core length due to the non-transparent characteristics and large numbers of droplets (dense sprays) that exist near the nozzle exit.

Five regimes have been observed in experiments on liquid fuel injection (Figs. 3.38 and 3.39). These regimes have been labeled drip flow, Rayleigh, first wind-induced, second wind-induced, and atomization. The drip flow regime occurs at low injection velocities, and the jet surface breakup length or, simply, the jet breakup length increases with the injection velocity.

The breakup length increases linearly with the injection velocity in the Rayleigh regime. Thereafter, it reaches a maximum and then decreases in the first wind-induced breakup regime. Both the Rayleigh and first wind-induced breakup regimes are well understood. In these two regimes, droplets are pinched off from the end of the liquid jet, and their sizes are comparable to that of the jet and to the nozzle diameter.

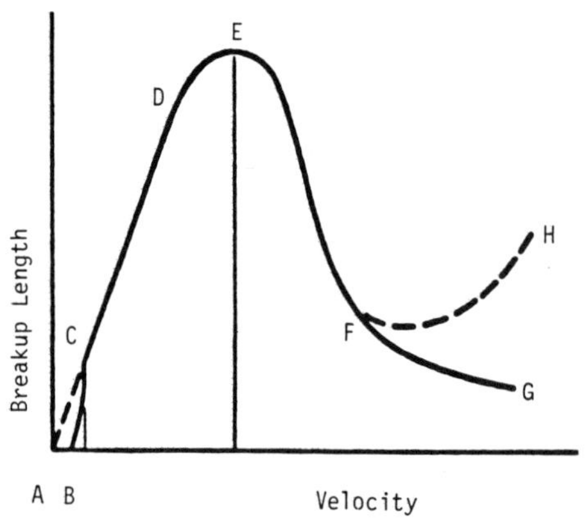

Figure 3.38 Jet surface breakup length as a function of the jet velocity: ABC, drip flow; CD, Rayleigh breakup regime; EF, first wind-induced breakup regime; FG (FH), second wind-induced breakup regime; beyond G (H), atomization regime [81].

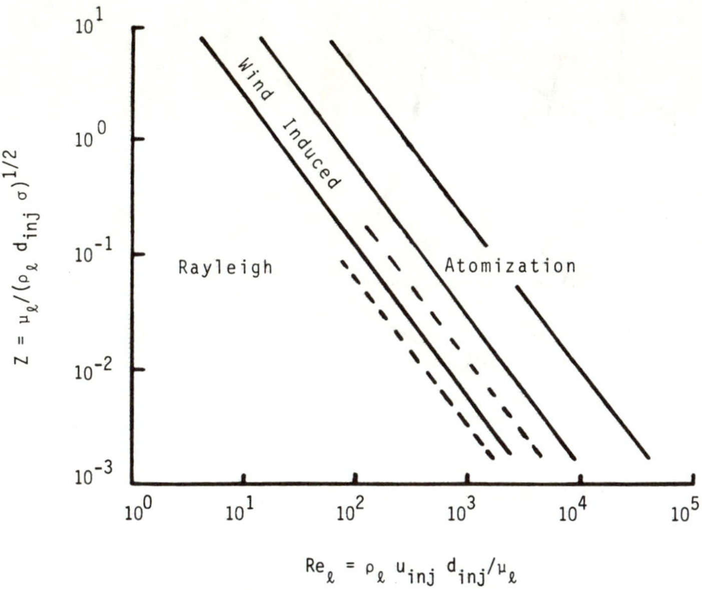

Figure 3.39 Jet breakup regime boundaries [81].

There is some controversy about the variation of the breakup length with the injection velocity beyond the first wind-induced regime. Some experiments indicate that the breakup length beyond the first wind-induced regime remains constant or decreases slightly with the injection velocity, and then it abruptly reduces to zero at high injection velocities [121]. However, other experiments show that the breakup length first increases and then decreases with the injection velocity [122].

In the Rayleigh and first wind-induced regimes the jet breaks up simultaneously over the entire cross section, and the surface breakup length coincides with the intact core length. However, in the second wind-induced regime and in the atomization regime the jet surface breaks up before the jet axis, and both the breakup length and the intact core length are different.

The atomization regime is defined as the regime where the breakup length is zero (i.e., droplets are formed on the surface of the liquid jet at the exit nozzle) and the size of the droplets is much smaller than the nozzle diameter [123, 124]. In the second wind-induced regime, the breakup length is not zero, and the size of the droplets is also much smaller than the nozzle diameter [81].

Figure 3.40 shows the influence of the gas density on breakup regime boundaries [81]. Theoretical studies of liquid jet breakup have, for the most part, assumed cylindrical jets issuing from a circular orifice into a stationary incompressible gas and studied the temporal stability of the liquid jet surface to perturbations using a first-order linear theory, which is valid only in the early stages of the instability. When the amplitude of the unstable waves reaches a critical

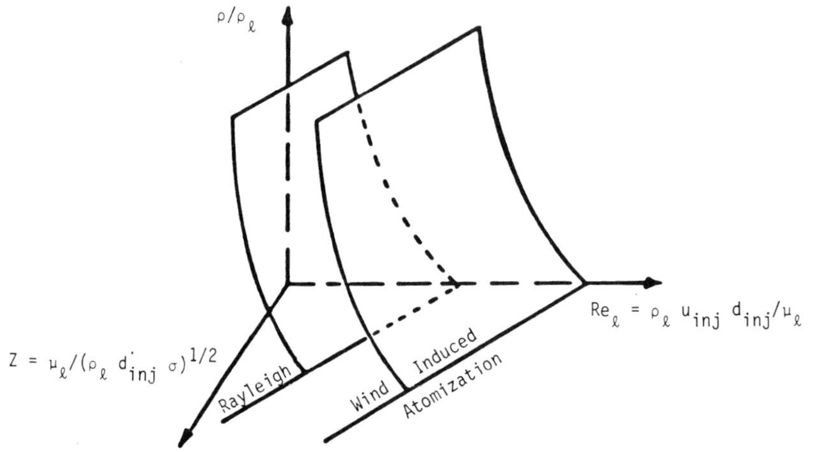

Figure 3.40 Schematic of the influence of the gas density on breakup regime boundaries [81].

value, nonlinear effects become important, and the linear theory is no longer valid. Furthermore, temporal stability theories have been developed for infinite liquid jets, and they, therefore, ignore the presence of the nozzle. In addition, instabilities are developed in both space and time, and theories which account only for temporal instabilities are only approximations to the physics of atomization.

Perhaps, the most complete temporal instability theory available is that of Reitz and Bracco [77], who verified the previous studies of Rayleigh, Taylor [76], Levich [104], and others and obtained a dispersion relation that depends on Bessel's functions of the third kind, the nozzle diameter, and

$$\beta = \omega\left(\frac{\rho_l d_{\mathrm{inj}}^3}{\sigma}\right)^{1/2} \tag{218}$$

$$Z = \frac{\mu_l}{(\rho_l\, \sigma d_{\mathrm{inj}})^{1/2}} \tag{219}$$

$$\mathrm{We} = \rho_g\, u_{\mathrm{inj}}^2\, \frac{d_{\mathrm{inj}}}{\sigma} \tag{220}$$

where ρ is the density; μ is the dynamic viscosity; d_{inj} is the nozzle diameter; σ is the surface tension; u_{inj} is the uniform velocity of the undisturbed jet; the subscripts l and g denote the liquid and gas, respectively; We is the Weber number; Z is the Ohnesorge number; and ω is the frequency of the linear perturbation.

In the second wind-induced regime, the liquid jet breakup is due to the unstable growth of short-wavelength surface waves ($kd_{\mathrm{inj}} > 1$, where k is the wave number), which are induced by the relative motion between the jet and the ambient gas [77, 81].

Taylor [76] assumed that $\rho_g << \rho_l$ and $kd_{\text{inj}} >> 1$ and found the following wave growth rate:

$$\frac{\omega}{ku_{\text{inj}}} = 2\left(\frac{\rho_l}{\rho_g}\right)^{1/2} X\,g(\Gamma,X) \tag{221}$$

where

$$X = \rho_g \frac{u_{\text{inj}}^2}{\sigma k} \qquad \Gamma = \frac{\rho_l}{\rho_g}\left(\frac{\text{Re}_l}{\text{We}_l}\right)^2 \tag{222}$$

$$\text{Re}_l = \rho_l \frac{u_{\text{inj}}\, d_{\text{inj}}}{\mu_l} \qquad \text{We}_l = \rho_l \frac{u_{\text{inj}}^2\, d_{\text{inj}}}{\sigma} \tag{223}$$

and $g(\Gamma,X)$ is a correction factor that accounts for the liquid viscosity.

Taylor [76] calculated the intact core length by computing the rate at which the droplets remove mass from the liquid core and by assuming that the droplet size is proportional to the wavelengths of the unstable surface waves. The analysis of Reitz and Bracco [77] yields

$$L_{\text{ic}} = c_1 d_{\text{inj}} \frac{(\rho_l/\rho_g)^2}{f(\Gamma)} \qquad f(\Gamma) = X\,g(\Gamma,X) \tag{224}$$

where c_1 is a constant and $f(\Gamma)$ tends asymptotically to $3^{1/2}/6$ as Γ tends to infinity.

Reitz and Bracco [81] estimated the breakup length as

$$L_{\text{bu}} = c_2 \frac{L_{\text{ic}}}{\text{We}_l} \tag{225}$$

where c_2 is another constant, and the subscripts bu and ic denote breakup and intact core, respectively.

According to Eq. [225] the breakup length decreases as the velocity increases, and surface tension opposes the breakup process.

Ranz [125] proposed that the droplet size is related to the wavelengths of the most unstable waves and argued that the spray angle θ of atomizing jets is given by

$$\tan\frac{\theta}{2} = \frac{1}{A}\, 4\pi\left(\frac{\rho_g}{\rho_l}\right)^{1/2} f(\Gamma) \tag{226}$$

where the constant A depends on the experiment.

Figure 3.41 illustrates the theoretical dependence of the spray angle as a function of the aerodynamic shear [125].

Reitz and Bracco [81] assumed that atomization occurs when the breakup

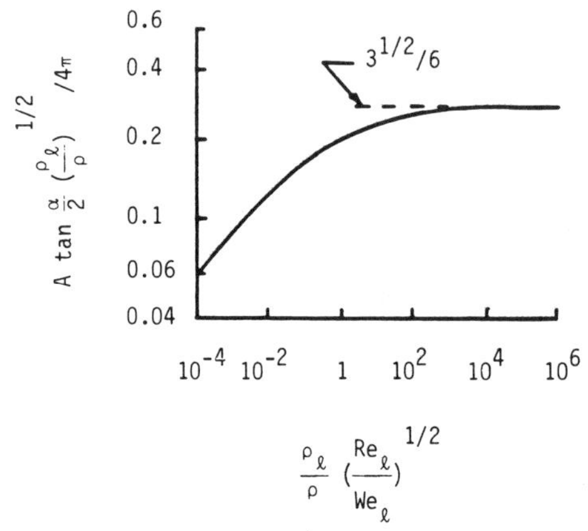

Figure 3.41 Theoretical dependence of the spray angle as a function of the aerodynamic shear [125].

length approaches zero and obtained the following atomization criteria:

$$\left(\frac{\rho_l}{\rho_g}\right)^{1/2} < K \qquad \text{for } \Gamma > 1 \tag{227}$$

$$\left(\frac{\rho_l\,\mathrm{We}_l}{\rho_g\,\mathrm{Re}_l}\right)^{1/3} < K \qquad \text{for } \Gamma < 1 \tag{228}$$

where K depends on the nozzle geometry and has to be found experimentally.

The experimental results of Reitz and Bracco [78, 79] indicate that the spray angle increases with the gas density, not with the gas pressure, and decreases with the liquid viscosity and with the nozzle length-to-diameter ratio for ratios greater than 10 or 20. Rounded inlet nozzles result in smaller spray angles than sharp-edge inlet nozzles. Reitz and Bracco [77] concluded from their experimental and theoretical results that aerodynamic effects, liquid turbulence, jet velocity profile arrangement effects, and liquid supply pressure oscillations cannot alone explain the experimental results.

The experimental results of Wu et al. [80] were found to agree with those predicted by an aerodynamic theory of surface breakup if, between the jet surface where the droplets are formed and the edge of the spray, drops undergo collisions and coalescence. However, the locations where the droplets were formed were not visible.

Clark and Dombrowski [126] used a second-order analysis to analyze the aerodynamic growth of sinuous waves on parallel inviscid sheets; they showed that the thinning of the sheets is caused by the growth of a harmonic wave. This second-order theory uses a perturbation technique where the perturbation parameter is the initial amplitude of the disturbance, represents an improvement over the strictly linear theory [127–129], and may be used to estimate the breakup length and droplet size in hollow sprays.

Biswas et al. [130] employed a linear stability theory to study the stability of sheets of finite thickness moving with a uniform velocity in still air; they showed that the sheets are stable for $\mathrm{We}_g > 0.4$. They also found that the wavelength corresponding to the maximum growth rate decreases with the reciprocal of the Weber number.

The theoretical and experimental results of Dombrowski and Hooper [131] and of Dombrowski and Johns [132] on the disintegration of sheets produced by single-hole spray nozzles show that under some circumstances the droplet size increases with the ambient density. This result is rather surprising and should be compared with that of Wu et al. [80], that is,

$$d_d = B \frac{2\pi\sigma}{\rho_g u_{\text{inj}}^2} X_{\max}(\Gamma) \tag{229}$$

where $X_{\max}(\Gamma)$ approaches 3/2 for Γ greater than 1, d_d is the droplet diameter, and B is a constant. Equation (229) indicates that the droplet diameter decreases with the ambient density.

Recently, Lin and Kang [133] studied the atomization of liquid jets by means of a linear spatial instability theory. They showed that wavelengths smaller than the capillary length $(\sigma/\rho_g u_{\text{inj}}^2)$ are stable; they also showed that both temporal and spatial instability theories predict a maximum growth rate proportional to the reciprocal of the square of the liquid velocity.

The analysis of Lin and Kang [133] shows that the maximum growth rate predicted by the spatial instability theory is proportional to the square root of the gas-to-liquid density ratio when the ratio of the surface tension force to the viscous force exceeds 100. The temporal instability theory predicts a maximum growth rate proportional to the square root of the gas-to-liquid density ratio when the ratio of the surface tension force to the viscous force exceeds 1. The spatial theory shows that the spray angle varies along the liquid jet axis (the spray angle is constant according to the temporal theory). The growth rates predicted by the temporal and spatial theories deviate considerably from each other when the ambient pressure is higher than 10 atm for liquids at room temperature and the ratio of the surface tension force to the viscous force is larger than 100.

Surface instabilities on liquid jets are developed in both space and time; therefore, temporal or spatial theories cannot capture the complete physics of the atomization caused by aerodynamic forces. In some circumstances temporal theories can be related to spatial theories through the group velocity [134].

Although aerodynamic theories of liquid jet breakup can in some cases predict the experimentally observed trends [80], it must be pointed out that other factors affect the atomization process. Some of these factors are cavitation, nozzle geometry (rounded or sharp), nozzle length-to-diameter ratio, boundary layer instabilities, liquid turbulence, turbulence in the ambient gas, nozzle roughness, and relaxation of the liquid velocity profiles from a nonslip condition at the nozzle walls to stress and kinematic conditions outside the nozzle. At present there are no theories that account for both internal and external effects. The internal effects associated with the nozzle geometry are frequently assigned a constant value, which is determined experimentally for each nozzle and used in aerodynamic theories of liquid jet breakup.

Nurick [135] studied the effects of cavitation on spray mixing and showed that cavitation in sharp-edged orifices results in a substantial reduction in mixing uniformity for circular orifices but not for rectangular orifices.

Lafrance [136] developed a model to study the breakup of liquid jets. His model is based on the idea that random turbulent fluctuations may be amplified by capillary instabilities. Ingebo [137] developed a theory of liquid jet breakup based on acceleration waves and showed that for water jets in swirling and nonswirling flows, and for We Re $> 10^6$, the ratio between the orifice diameter and the Sauter mean diameter (SMD) can be expressed as

$$\frac{d_{\text{inj}}}{\text{SMD}} = c(\text{We Re})^{0.4} \tag{230}$$

where c is a constant.

Equation (230) should be compared with the results of the aerodynamic theory of liquid jet breakup developed by Reitz and Bracco [77], which predicts

$$\text{SMD} = B\frac{4\pi\sigma}{\rho u_{\text{inj}}^2}\frac{3}{2} \tag{231}$$

Equations (230) and (231) clearly show that different models of liquid jet breakup result in different Sauter mean diameters. However, aerodynamic theories of liquid jet breakup have predicted results in good agreement with experimental data [80].

It must be emphasized that linear theories of temporal or spatial instabilities only yield a dispersion relation that relates the frequency and wavelength of the disturbance, but they cannot predict the amplitude of these instabilities. Amplitudes can be predicted only by means of nonlinear theories.

The supplemented aerodynamic theory of surface breakup developed by Reitz and Bracco [77, 81] provides estimates about the droplet size, assuming that most of the droplets are formed by the fastest growing unstable surface waves. However, it is based on the linearized Navier-Stokes equations; therefore, it cannot yield accurate droplet sizes that are governed by nonlinear processes.

The results of a linear theory of aerodynamic breakup will be used in the next section, on multidimensional models of diesel engines.

3.5 MULTIDIMENSIONAL MODELS

Multidimensional models of diesel engine combustion account for the temporal and spatial variations of the flow field, temperature, composition, pressure, and turbulence within the combustion chamber. Most of the multidimensional models that have appeared in the literature are based on the spray equation [138], which depends on time and on the radius, velocity, and temperature of the droplets and may account for thick spray effects (droplet collisions, coalescence, and volumetric displacement of the gas phase) and for droplet breakup.

Two-phase flows with nonequilibrium processes can also be analyzed by developing continuum approaches for both phases [139–141]. However, the major difficulties in continuum theories are the determination of the constitutive equations [142], pressure relations [143], interfacial momentum transfer and interfacial geometry [144], shear stress on the droplets [145], etc.

In continuum two-phase-flow models, the pressures of the gas and liquid phases are different [146]; however, if the volume fraction (i.e., the ratio of the volume occupied by the droplets to that of the gas) is small, the pressure gradient in the liquid momentum equations can be neglected [147].

In continuum two-phase-flow models, the gas phase and droplet properties are considered to be continuous functions of space and time. This implies that the local properties represent average values over some volume around a point in space and time, and the size of that volume is much larger than the average spacing between the droplets [148, 149]. This averaging procedure allows for the superposition of a continuum gas field and a continuum droplet field; however, continuum two-phase-flow models are not able to resolve phenomena on scales on the order of the droplet diameter since the averaging is performed over a larger distance [150].

Rigorous formulation of the equations governing two-phase flow has been a challenge due to the difficulty associated with the presence of a large number of interfaces and the apparently random nature of the flow field. Two-phase-flow models are usually derived from the macroscopic conservation equations for each phase and for interphase boundary conditions. The latter set of equations specifies the interphase mass, momentum, energy, and entropy transfers. Integration of the macroscopic equations over space and time yields averaged two-phase-flow conservation equations. Some examples of averaging techniques include time, volume, area, and one-dimensional averages [151]. Two key issues in continuum formulations for two-phase flows are the specification of the constitutive equations for the time and space operators and the selection of the averaging time interval and space regions. Different choices of these values will give rise to qualitatively and quantitatively different solutions [152, 153].

Continuum turbulent two-phase-flow formulations have been used to predict two-phase jets and internal flows [154, 155]. Elghobashi and Abou-Arab [154] eliminated the need to simulate in an ad hoc manner the effects of the dispersed phase on the turbulence structure by closing the time-averaged conservation equations up to third order by means of a k/ϵ model of turbulence. This model is based on those for single-phase flows (see Chapter 2) and uses a linearized Lagrangian formulation to estimate the particle response to turbulent fluctuations.

The model proposed by Gibeling et al. [155] uses an equation for the gas-phase turbulence kinetic energy, specifies the length scale algebraically, and employs interphase drag relationships derived from experiments on and theories for isolated spheres and packed beds.

Mostafa and Elghobashi [156] developed a two-equation model of turbulence for steady two-phase flows. Their model is based on a set of conservation equations for the mass and momentum of each phase and on equations for the concentration, turbulence kinetic energy, and dissipation rate of the carrier fluid. The model accounts for volumetric displacement effects (void fraction $\neq 1$) and for the effects of the droplets on the turbulence kinetic energy and on its rate of dissipation. Mostafa and Elghobashi [156] calculated the interphase transport by means of correlations similar to those presented in Sections 3.5.1–3.5.7.

Two approaches have been used to analyze the flow field in diesel engines: (1) solution of the spray equation with a continuum gas-phase formulation and (2) Lagrangian-Eulerian formulations, where Lagrangian equations are employed for groups of droplets and Eulerian equations are employed for the gas phase. Lagrangian-Eulerian formulations can also account for thick spray effects such as droplet collisions, coalescence, and breakup.

3.5.1 Models Based on the Spray Equation

The spray equation can be written as [157]

$$\frac{\partial f}{\partial t} + \nabla_{\mathbf{x}} \cdot (f\mathbf{v}_d) + \nabla_{\mathbf{v}_d} \cdot (f\mathbf{F}_d) + \frac{\partial}{\partial r_d}(fR_d) + \frac{\partial}{\partial T_d}(f\dot{T}_d) = \dot{Q} \tag{232}$$

where $\mathbf{v}_d$, $\mathbf{F}_d$, R_d, and T_d denote the velocity, force, rate of change of the radius, and temperature of the droplets, respectively; r_d is the droplet radius; $f\, dr_d\, d\mathbf{v}_d\, d\mathbf{x}\, dT_d$ is the probable number of droplets in the radius range dr_d around r_d located in the spatial range $d\mathbf{x}$ around $\mathbf{x}$ with velocities in the range $d\mathbf{v}_d$ around $\mathbf{v}_d$ and with temperatures in the range dT_d around T_d; $\dot{Q}$ represents the rate of increase of f due to collisions, coalescence, and breakup; and $\dot{T}_d$ represents the rate of change of the droplet temperature with respect to time.

Equation (232) generalizes the spray equation presented in Williams [138] and accounts for variations of the droplet temperature. The values of $\mathbf{F}_d$, R_d, $\dot{T}_d$, and $\dot{Q}$ need to be specified in order to solve Eq. (232). These values are frequently taken from experiments for isolated droplets, and they account for the effects of vaporization on $\mathbf{F}_d$, R_d, and $d\mathbf{v}_d/\mathrm{dt}$.

The void fraction can be calculated as

$$\psi = 1 - \int \frac{4}{3}\pi r_d^3 f\, dr_d\, d\mathbf{v}_d\, dT_d \tag{233}$$

where the integral is to be performed with respect to all the velocities, radii, and temperatures of the droplets.

The gas-phase conservation equations can be written, accounting for volumetric displacement effects, as

$$(\rho\psi)_{,t} + (\rho u_i \psi)_{,i} = -\int [4\pi r_d^2\, R_d\, \rho_l]\, f\, dr_d\, d\mathbf{v}_d\, dT_d \tag{234}$$

$$(\rho\psi u_i)_{,t} + (\rho u_i\, u_j \psi)_{,j} = -p_{,i} + \tau_{ij,j} - \int\left[\left(4\pi r_d^2 R_d \mathbf{v}_d + \frac{4}{3}\pi r_d^3 \mathbf{F}_d\right)\rho_l\right] f\, dr_d\, d\mathbf{v}_d\, dT_d \tag{235}$$

$$(\rho\psi h)_{,t} + (\rho\psi h u_i)_{,i} = \psi(p_{,t} + u_i\, p_{,i}) + \tau_{ij}\, u_{i,j} + (\psi\lambda T_{,i})_{,i}$$
$$- \int \rho_l \left\{ 4\pi r_d^2\, R_d \left[h_l(T_d) + \frac{1}{2}\,(\mathbf{v}_d - \mathbf{u})^2 \right] + \frac{4}{3}\,\pi r_d^3 \left[c_l \dot{T}_d + \mathbf{F}_d^d(\mathbf{v}_d - \mathbf{u}) \right] \right\} f\, dr_d\, d\mathbf{v}_d\, dT_d \quad (236)$$

$$h = \sum_{i=1}^{N} h_i\, Y_i \qquad h_i = h_i^0 + \int_{T_0}^{T} C_{p_i}\, dT$$
$$\tau_{ij,j} = \psi\left[\mu(u_{i,j} + u_{j,i}) - \frac{2}{3}\,\mu\delta_{ij}u_{k,k} \right] \quad (237)$$

$$\mathbf{F}_d = \mathbf{F}_d^d - \frac{1}{\rho_l}\,\nabla p + \mathbf{g} \quad (238)$$

$$(\rho\psi Y_j)_{,t} + (\rho\psi u_i Y_j)_{,i} = [\psi\, \rho D (Y_j)_{,i}]_{,i}$$
$$- \delta_{jf} \int [4\pi r_d^2\, R_d\, \rho_l]\, f\, dr_d\, d\mathbf{v}_d\, dT_d + \dot{\omega}_j \quad (239)$$

where ρ, p, T, and $\mathbf{u}$ denote the gas-phase density, pressure, temperature, and velocity, respectively; μ and λ are the gas-phase dynamic viscosity and thermal conductivity, respectively; h, h_i^0, and h_i denote the total enthalpy of the mixture, the enthalpy of formation of species i, and the specific enthalpy of species i, respectively; C_{p_i} and Y_i are the specific heat at constant pressure and mass fraction of species i, respectively; T_0 is a reference temperature; δ_{if} is the Kronecker delta ($\delta_{if} = 1$ if $i = f$ and $\delta_{if} = 0$ if $i \neq f$); $\dot{\omega}_j$ is the reaction rate of species j; the subscript f denotes fuel; and D is the diffusion coefficient, which has been assumed equal for all the species.

In Eqs. (233)–(239) the species have been assumed to diffuse according to Fick's law, with equal diffusion coefficients for all of them; bulk viscosity, Soret and Dufour effects, and thermal radiation have been neglected.

The terms in brackets in Eqs. (234), (235), and (239) can also be written, if the density of the liquid phase is constant, as

$$4\pi r_d^2\, R_d\, \rho_l = \frac{d}{dt}\left(\frac{4}{3}\,\pi r_d^3 \rho_l \right) \quad (240)$$

$$\left(4\pi r_d^2\, R_d \mathbf{v}_d + \frac{4}{3}\,\pi r_d^3 \mathbf{F}_d \right) \rho_l = \frac{d}{dt}\left(\frac{4}{3}\,\pi r_d^3 \rho_l \mathbf{v}_d \right) \quad (241)$$

and they correspond to the changes of mass and momentum of each droplet.

Equations (234)–(239) can be time averaged to yield a similar system of equations if the density fluctuations, pressure-velocity correlations, and temperature-species mass fraction correlations are neglected and if a Boussinesq approximation is used to model double and triple correlations (see Chapter 2). The resulting time-averaged equations are similar to Eqs. (234)–(239) if μ, λ, $\dot{\omega}_j$, and D are replaced by μ_e, λ_e, $\langle\dot{\omega}_j\rangle$, and D_e, where the subscript e denotes effective (laminar plus turbulent contributions) and the angular brackets denote time-averaged quantities. The values of these quantities can be written as

$$\mu_e = \mu + \mu_t \qquad \lambda_e = \left(\frac{\mu}{\mathrm{Pr}} + \frac{\mu_t}{\mathrm{Pr}_t}\right) C_p \qquad \rho D_e = \frac{\mu}{\mathrm{Sc}} + \frac{\mu_t}{\mathrm{Sc}_t} \tag{242}$$

where the subscript t denotes turbulent, Pr and Sc are the Prandtl and Schmidt numbers, and

$$C_p = \sum_{i=1}^{N} C_{p_i} \tag{243}$$

where N denotes the number of species.

3.5.1.1 Droplet-turbulence interactions I. The effects of the gas-phase turbulence on the droplets can be represented by considering that the gas-phase velocity is the sum of the time-averaged and fluctuating gas-phase velocities, that is,

$$\mathbf{u} = \langle \mathbf{u} \rangle + \mathbf{u}' \tag{244}$$

where $\mathbf{u}'$ can be obtained from a turbulence model (see Chapter 2), and assuming that in a turbulent flow, $f = f(t, \mathbf{x}, \mathbf{v}_d, r_d, T_d, \mathbf{u}')$.

In this case $f\, dr_d\, d\mathbf{v}_d\, dT_d\, d\mathbf{u}'$ represents the probable number of droplets per unit volume at time t and point $\mathbf{x}$ with velocities in the interval $(\mathbf{v}_d, \mathbf{v}_d + d\mathbf{v}_d)$, radii in the interval $(r_d, r_d + dr_d)$, temperature in the interval $(T_d, T_d + dT_d)$, and gas-phase turbulence velocities in the interval $(\mathbf{u}', \mathbf{u}' + d\mathbf{u}')$. If each component of $\mathbf{u}'$ is distributed according to a Gaussian distribution with a mean square deviation $|\mathbf{u}'|^2 = 2k/3$ calculated from isotropic turbulence, the distribution function f can be written as [158, 159]

$$f(t, \mathbf{x}, r_d, \mathbf{v}_d, T_d, \mathbf{u}') = \left(\frac{4}{3}\pi k\right)^{-3/2} \exp\left[\frac{-3|\mathbf{u}'|^2}{4k}\right] f^*(t, \mathbf{x}, r_d, \mathbf{v}_d, T_d) \tag{245}$$

where k is the turbulence kinetic energy of the gas phase and f^* can be obtained from f by integrating over all values of $\mathbf{u}'$.

Under these assumptions, the spray equation [Eq. (232)] can be written as

$$\frac{\partial f}{\partial t} + \nabla_{\mathbf{x}} \cdot (\mathbf{v}_d f) + \frac{\partial}{\partial r_d}(f R_d) + \nabla_{\mathbf{v}_d} \cdot (f \mathbf{F}_d)$$
$$+ \frac{\partial}{\partial T_d}(f \dot{T}_d) + \nabla_{\mathbf{u}'}\left(f \frac{\partial \mathbf{u}'}{\partial t}\right) = \dot{Q} \tag{246}$$

We will consider the k–ϵ model (see Chapter 2) where

$$(\rho\psi k)_{,t} + (\rho\psi u_i k)_{,i} = \left[\psi\left(\frac{\mu_t}{\sigma_k} + \mu\right) k_{,i}\right]_{,i} + \tau_{ij} u_{i,j} - \rho\psi\epsilon \tag{247}$$

$$(\rho\psi\epsilon)_{,t} + (\rho\psi u_i\,\epsilon)_{,i} = \left[\psi\left(\frac{\mu_t}{\sigma_\epsilon} + \mu\right)\epsilon_{,i}\right]_{,i}$$

$$+ \frac{\epsilon}{k}(c_1\,\tau_{ij}\,u_{i,j} - c_2\,\rho\psi\epsilon) - \rho\,\psi\,\epsilon\,u_{k,k} \qquad (248)$$

$$\begin{aligned} \mu_t &= c_\mu \frac{\rho k^2}{\epsilon} \qquad c_\mu = 0.09 \qquad \sigma_k = 1 \\ \sigma_\epsilon &= 1.3 \qquad c_1 = 1.44 \qquad c_2 = 1.92 \end{aligned} \qquad (249)$$

and k and ϵ denote the turbulence kinetic energy and its dissipation rate, respectively.

The effects of the droplets on the gas-phase turbulence can be accounted for by introducing the following new terms on the right-hand sides of Eqs. (247) and (248):

$$S_k = -\int \frac{4}{3}\pi r_d^3\,\rho_l\,\mathbf{F}_d \cdot \mathbf{u}'\,dr_d\,d\mathbf{v}_d\,dT_d \qquad (250)$$

$$S_\epsilon = -C'\frac{\epsilon}{k}S_k \qquad (251)$$

where C' is a constant.

Equations (250) and (251) indicate that the effect of the droplets on the gas-phase fluctuating velocity field is to decrease the turbulence kinetic energy and to increase the dissipation rate because of friction.

The term S_k can be interpreted as the work done by the droplets on the fluctuating gas-phase velocity field, whereas the term S_ϵ has been modeled in a manner analogous to that of other terms that appear in the ϵ equation (see Chapter 2).

Experimental studies of turbulent two-phase round jets have shown that the spread rate of two-phase flow jets is smaller than that of single-phase jets. These studies have also shown that the levels of turbulence in two-phase jets depend on the mass loading and are smaller than those observed in single-phase jets [160–162]. Equations (250) and (251) are consistent with these experimental observations.

Equations (247) and (248) do not account for the effects of the droplets on k and ϵ except through the void fraction. However, in addition to the void fraction, k and ϵ are affected by the loading (i.e., the number of droplets within a turbulent eddy of length $l = c_\mu^{3/4}\,k^{3/2}/\epsilon$). Droplet loading may be very important near the injector, where the void fraction is on the order of 10%. For dilute or thin sprays the droplet-gas-phase turbulence interactions can be modeled using linear approximations to account for the turbulence modulation terms in the k and ϵ equations [163]. These terms depend on the velocity difference between the gas phase and the droplets, the droplet diameter, the drag coefficient, the kinematic viscosity of the gas phase, the density ratio, k, and ϵ, and they are based on the droplet response to the gas-phase turbulence spectrum. For dense or thick sprays (e.g., near the injector), droplet-gas-phase turbulence interactions are difficult to model, and analyses developed for dilute sprays are not valid. Both k and ϵ are affected by the number and size of the droplets in an energy-containing eddy.

Amsden et al. [158] used the following $k-L$ model of turbulence:

$$(\psi \rho k)_{,t} + (\rho k u_i \psi)_{,i} = -\frac{2}{3} \psi \rho k \, u_{j,j} + \tau_{ij} u_{i,j} + (\psi \mu_e k_{,j})_{,j} - \psi c_1 \rho L^{-1} k^{3/2} + DT\psi \quad (252)$$

where

$$\mu_e = \rho \nu_0 + \mu_a + \mu_t \qquad \mu_t = c_2 \rho L k^{1/2} \quad (253)$$

$$\mu_a = c_3 \frac{T^{3/2}}{T + c_4} \quad (254)$$

c_1, c_2, c_3, and c_4 are constants; ν_0 is a constant uniform background turbulent diffusivity; μ_t is a subgrid turbulent viscosity; L is a characteristic length approximately equal to twice the length of a typical finite difference cell size; and DT represents the effects of the droplets on the gas-phase turbulence, which can be calculated as

$$DT = -\int \frac{4}{3} \pi r_d^3 \rho_l \mathbf{F}_d \cdot \mathbf{u}' \, dr_d \, d\mathbf{v}_d \, dT_d \, d\mathbf{u}' \quad (255)$$

It must be pointed out that in the KIVA program, volumetric displacement effects are neglected and $\psi = 1$ [158, 159]. Volumetric displacement effects are important near the nozzle exit. The pressure gradient and gravity terms in Eq. (238) are also neglected in the KIVA code [158, 159]. The gravity term is expected to be small in high-speed sprays.

O'Rourke [157] and O'Rourke and Bracco [164] suggested that droplet collisions are important in dense sprays; droplet collisions can be a mechanism for droplet growth, but they can also contribute to droplet breakup. Collisions have also been suggested as the mechanism responsible for rapid radial growth of the droplet-containing region observed in sprays. Wu et al. [80] showed that a supplemented aerodynamic theory of droplet breakup can accurately predict the droplet size at the spray edge if the droplets undergo collisions and coalescence. In this region, droplet coalescence leads to a rapid increase in drop size. However, in thick, high-speed sprays the droplet-gas-phase turbulence interactions (i.e., droplet dispersion) play an important role in the turbulence kinetic energy and its rate of dissipation. Droplet dispersion contributes to the radial growth of the spray and may be as important as droplet collisions and coalescence. Furthermore, high levels of turbulence may be generated at the spray edge due to the high slip velocities between the gas phase and the droplets; this injection-generated turbulence is a mechanism that also contributes to the radial growth of the spray.

The droplet response to the gas-phase velocity fluctuations depends, in a complex manner, on the droplet response time and number of droplets in a typical turbulent eddy [165]. A conservative estimate is to assume that slip is negligible and that the velocities of the droplets and gas phase are identical. If this assumption is valid, the effects of the droplets on the gas phase can be treated by means of an effective turbulent diffusivity. Choi and Chung [166] showed that small particles decrease the turbulence fluctuations and increase the dissipation rate of turbulence kinetic energy. These results indicate that friction at the particle surface plays an important role in determining k and ϵ. (Note that the droplet surface area-to-volume surface tends to infinity as the size of the droplet tends to zero.)

Hinze [165] estimated the number of collisions per unit time and per unit volume of uniform particles as

$$\frac{dn_c}{dt} \sim |\mathbf{v}_d - \mathbf{u}| \frac{(1 - \psi)^2}{r_d^4} \tag{256}$$

where n_c the number of collisions per unit volume. For polydisperse sprays the large droplets have significant relative velocities with respect to both the gas phase and small particles. Equation (256) indicates that the collision rate is proportional to the relative velocity between the droplets and the gas phase but inversely proportional to the fourth power of the droplet diameter. The strong dependence on the droplet diameter shows why collisions may be important in aerodynamic processes yielding small droplets.

Hinze [167] showed that if the particle response time t_r to turbulence fluctuations is smaller than the characteristic time of the Kolmogorov eddies $t_k = (\nu/\epsilon)^{1/2}$, the particles follow the smallest eddies. He also showed that the particle response to turbulent fluctuations depends on t_r, t_k, the Kolmogorov eddy size η, a characteristic dimension of the particles l_p, and ρ_p/ρ_f, where ρ_p and ρ_f denote the densities of the particles and fluid, respectively. Hinze's analysis contrasts with that of Tchen [168, 169], who assumed $l_p \ll \eta$ and specular collisions between the particles and solid surfaces. Tchen [169] used Oseen's approximation to study the motion of curved and elongated particles.

Ahmadi and Goldschmidt [170] studied the diffusion (dispersion) of particles in random fields simulated as the superposition of randomly distributed vortices (eddies) accounting for kinematic effects only; they showed that the relative mass diffusivity first increases and ultimately decreases to zero as the particle size is increased. The experimental data of Goldschmidt et al. [171] show that the turbulent dispersion of liquid droplets in plane two-dimensional jets increases with droplet size.

In low-frequency turbulent flows, the droplets do not follow the gas-phase turbulent fluctuations, and the droplet response depends on the Stokes number, $\mathrm{St} = \nu/\omega d_d^2$, where ν is the kinematic viscosity of the fluid, ω is the angular frequency of the turbulent fluctuations, and d_d is the droplet diameter. If $\rho_p/\rho_f >> 1$ or $\rho_p << \rho_f$, virtual mass, pressure forces, and history effects on the droplet motion are negligible [172]. Note that in engine sprays $\rho_l/\rho >> 1$ and the droplet motion is affected mainly by friction.

3.5.1.2 Droplet collisions and coalescence. O'Rourke [157] and O'Rourke and Bracco [164] developed a method for computing droplet collisions, coalescence, and shattering based on a discrete element analysis. Collisions may result in many new droplet groups that may be controlled by modifying the properties of colliding groups, while satisfying conservation principles. Droplet collisions may result in coalescence, separation, or shattering. Droplet separation occurs if the rotational energy of the coalesced droplet pair exceeds the surface energy required to re-form the original droplets from the coalesced pair. The coalescence efficiency depends on the droplet radii before collision and the Weber number based on the droplet velocity difference.

The term $\dot{Q}$ in Eq. (232) or Eq. (246) can be written as

$$\dot{Q} = \frac{1}{2}\int f_1^* f_2^* \pi(r_{d_1} + r_{d_2})^2 |\mathbf{v}_{d_1} - \mathbf{v}_{d_2}| [\sigma - \delta(r_d - r_{d_1})\,\delta(\mathbf{v}_d - \mathbf{v}_{d_1})\,\delta(T_d - T_{d_1}) - \delta(r_d - r_{d_2})\,\delta(\mathbf{v}_d - \mathbf{v}_{d_2})\,\delta(T_d - T_{d_2})]\, dr_{d_1}\, d\mathbf{v}_{d_1}\, dT_{d_1}\, dr_{d_2}\, d\mathbf{v}_{d_2}\, dT_{d_2} \tag{257}$$

where

$$f_1^* = f^*(t, \mathbf{x}, r_{d_1}, \mathbf{v}_{d_1}, T_{d_1}) \qquad f_2^* = f^*(t, \mathbf{x}, r_{d_2}, \mathbf{v}_{d_2}, T_{d_2}) \tag{258}$$

σ is a transition probability, and δ is the Dirac delta function.

Droplet collisions may result in droplet coalescence, shattering, and grazing collisions, depending on the velocity of the colliding droplets. In grazing collisions, a droplet is formed that immediately breaks up into two big droplets and many small ones.

The values of r_d, $\mathbf{v}_d$, and T_d that appear in Eq. (232) or Eq. (246) can be calculated from theoretical and experimental results that account for droplet vaporization and combustion. As the droplets vaporize, the pressure (or form) drag increases and the friction drag decreases; the droplets are also subjected to Basset forces, virtual mass effects, Magnus forces, and Saffman [173] effects. The Basset forces account for the history of the droplet motion; virtual mass effects are simply due to the fact that when the droplets accelerate, they also have to accelerate the surrounding fluid; Magnus forces are due to the droplet rotation, whereas Saffman effects are associated with the forces that a particle experiences in a shear flow. In addition, the fluid acceleration introduces pressure effects on the droplets [172]. In most sprays these effects are negligible compared with the forces due to friction, gas-phase pressure gradient, and gravity [174].

Brownian motion [175] in diesel engine sprays is expected to be small. However, Brownian motion may be important for very small droplets, whereas large droplets with large inertia may not follow the gas-phase turbulent motion

[176]. The effects of Brownian motion and particle density on two-phase laminar pipe flow were analyzed by Cox and Mason [177], who showed that a spherical particle more dense than a fluid in an upflow (or a spherical particle less dense than a fluid in a downflow) migrates to the tube axis; however, it migrates toward the wall if the sphere is less (more) dense than the fluid.

For thick sprays, O'Rourke [157], O'Rourke and Bracco [164], and Grasso et al. [178] proposed the following expressions:

$$\rho_l R_d = \rho_l \frac{dr_d}{dt} = -\frac{\lambda}{C_p} \frac{\text{Nu}}{2r_d} \beta \tag{259}$$

$$\mathbf{F}_d = \frac{d\mathbf{v}_d}{dt} = \mathbf{F}_d^d - \frac{1}{\rho_l} \nabla p - \mathbf{g} \tag{260}$$

$$\mathbf{F}_d^d = \frac{3}{8} \frac{\rho}{\rho_l r_d} |\mathbf{u} + \mathbf{u}' - \mathbf{v}_d| (\mathbf{u} + \mathbf{u}' - \mathbf{v}_d) C_d \tag{261}$$

$$\frac{dT_d}{dt} = \frac{3}{2} \frac{\lambda_l}{\rho_l c_l r_d^2} \text{Nu}_l (T_{ds} - T_d) + \frac{3}{r_d} R_d (T_{ds} - T_d) \tag{262}$$

$$C_d = 24 \frac{\psi^{-2.65} + \text{Re}_d^{2/3} \psi^{-1.78}/6}{\text{Re}_d} \tag{263}$$

$$\text{Re}_d = 2\rho|\mathbf{u} + \mathbf{u} - \mathbf{v}_d| \frac{r_d}{\mu} \tag{264}$$

$$\text{Nu}_l = 1 \qquad \text{Nu} = \left[2\psi^{-1.75} + 0.6\left(\frac{\text{Re}_d}{\psi}\right) \text{Pr}^{1/3}\right] \frac{\ln (1 + \beta)}{\beta} \tag{265}$$

$$\beta = \frac{C_p}{L(T_{ds})} \left[T - T_{ds} - \frac{\text{Nu}_l \lambda_l}{\text{Nu} \lambda} (T_{ds} - T_d)\right] = \frac{Y_{fs} - Y_f}{1 - Y_{fs}} \tag{266}$$

$$\frac{p_{vs}}{p} = \frac{Y_{fs}/M_f}{Y_{fs}/M_f + Y_I/M_I} \tag{267}$$

$$M_I^{-1} = \sum_{i \neq f} \frac{Y_i/M_i}{\sum_{i \neq f} Y_i} \qquad Y_I = 1 - Y_{fs} \tag{268}$$

where the subscripts l, f, and fs denote the liquid, fuel, and fuel vapor at the droplet surface, respectively; β is the transfer number; λ and C_p are the thermal conductivity and specific heat at constant pressure, respectively; L is the latent heat of vaporization; and p_{vs} is the equilibrium fuel vapor pressure at temperature T_{ds} and can be calculated by means of the Clausius-Clapeyron equation as

$$p_{vs} = p_{\text{ref}} \exp \left[\frac{L}{\tilde{R}} \left(\frac{1}{T_{\text{ref}}} - \frac{1}{T_{ds}}\right)\right] \tag{269}$$

where $\tilde{R}$ is the universal gas constant, L is the latent heat of vaporization (assumed constant), and the subscript ref denotes reference conditions. If the latent heat of vaporization is not constant, the Clausius-Clapeyron equation can be written as

$$\frac{p_{vs}}{p_{\text{ref}}} = \exp\left(\int_{T_{\text{ref}}}^{T_{ds}} \frac{L}{\tilde{R}T^2}\, dT\right) \tag{270}$$

In Sections 3.5.2, 3.5.3, 3.5.4, and 3.5.5, models based on the solution of the spray equation [Eq. (232)] or of Eq. (246) and thin and thick spray models are discussed.

Equations (263) and (265) account for volumetric displacement effects on the drag coefficient and Nusselt number. Theoretical and experimental data indicate that for a pair of spheres in viscous flows, the drag coefficient and average Nusselt numbers of either sphere are always less than that of a single isolated sphere [179]. Also, the vaporization of nondilute sprays depends on the distance between droplets [180]; if this distance is much larger than the droplet diameter, the mass transfer correlations derived for single isolated droplets in infinite surroundings may be applicable.

Using statistical mechanics concepts, Samson et al. [181, 182] developed simple models of fuel spray burning that include the effects of heat and mass transfer but neglect hydrodynamic effects. They calculated the dependence of the burning rate of a droplet on the burning rates of all the droplets in the spray, and they demonstrated that for linear arrays of burning droplets, the burning rates strongly depend on the initial droplet diameter and on the separation between droplets.

3.5.2 Spray Equation Models

The earliest multidimensional models of diesel engine combustion were based on the numerical solution of Eq. (232) and neglected the source term $\dot{Q}$ and the last term on the right-hand side of Eq. (232) [183–187].

Equation (232) is hyperbolic, and its solution is unique if

$$f(t, r_{d_{\max}}, \mathbf{x}, \mathbf{v}_d) = f(t, r_d, \mathbf{x}, \mathbf{v}_{d_{\min}}) = 0 \tag{271}$$

$$f(t, r_d, \mathbf{x}, \mathbf{v}_{d_{\max}}) = 0 \tag{272}$$

Bracco et al. [184] assumed that

$$f(0, r_d, \mathbf{x}, \mathbf{v}_d) = 0 \tag{273}$$

they specified the value of f at the injector, neglected the effects of turbulence ($\mathbf{u}' = 0$), employed Stokes' law to calculate the drag coefficient, that is,

$$\mathbf{F}_d = \frac{3}{8}\frac{C_D}{r_d}\frac{\rho}{\rho_l}\,|\mathbf{u}-\mathbf{v}_d|\,(\mathbf{u}-\mathbf{v}_d) \qquad C_D = \frac{24}{\text{Re}_d} \tag{274}$$

$$\text{Re}_d = 2\rho\,|\mathbf{u}-\mathbf{v}_d|\,\frac{r_d}{\mu} \qquad R_d = -\frac{\lambda}{8r_d\rho}\,(1+0.3\,\text{Re}_d^{1/2}\,\text{Pr}^{1/3}) \tag{275}$$

and solved Eq. (232) by means of a first-order accurate, explicit numerical method that introduces large amounts of artificial viscosity and smooths the droplet distribution function. Furthermore, the use of Stokes' law to calculate the friction drag is unjustified, for the slip at the droplet surface decreases the shear stress and results in lower drag. (Stokes' law is valid for rigid spheres at low Reynolds numbers [188, 189].)

Experimental studies of droplet evaporation in high-temperature air streams have shown that evaporation affects the heat transfer rates through changes in composition and temperature of the surrounding gases [190]. Experimental studies have also shown that evaporation reduces the heat transfer rates and that the Nusselt number can be expressed as

$$\mathrm{Nu}_f = (2 + 0.57\,\mathrm{Re}_d^{1/2}\,\mathrm{Pr}_f^{1/3})\,(1 + B_f)^{-0.7} \tag{276}$$

where the subscript f denotes film conditions and B is Spalding's transfer number.

Numerical studies in the range $10 \leq \mathrm{Re}_d \leq 100$ have shown that mass transfer decreases the friction drag and increases the presssure (or form) drag in vaporizing droplets [190]. These studies have also shown that the standard curve for the drag on solid spheres can be used to predict the drag on vaporizing droplets and that the mass efflux and variable properties decrease the heat transfer rates to liquid droplets in agreement with available experimental data [191].

Crowe et al. [192] showed that evaporation can reduce the drag coefficient due to the mass flux from the droplet surface, and they proposed that

$$C_D = \frac{\mathrm{Re}_d}{24}\,(1 + 0.15\,\mathrm{Re}_d^{0.687})\,(1 + B)^{-1} \tag{277}$$

where the last term in Eq. (277) accounts for the drag reduction due to evaporation.

Prakash and Sirignano [193, 194], Law et al. [195], and Sirignano [196] developed models for convective droplet vaporization that account for unsteady heat transfer in the liquid phase. Aggarwal et al. [197] evaluated a simplified version of the model proposed by Prakash and Sirignano [193, 194], a conduction limit model, where the temperature distribution within the droplet is assumed to be spherically symmetric, and an infinite conductivity model, where the droplet temperature is uniform. Aggarwal et al. [197] showed that the Ranz-Marshall correlation [Eq. (275)] [105] overpredicts the vaporization rate as compared with a vortex model [193], a higher heat loss from the gas phase, and lower gas-phase temperatures. However, since the vortex model [193] is valid only at relatively high Reynolds numbers, it predicts an unacceptably small vaporization rate at low Reynolds numbers.

Jin and Borman [198] analyzed the vaporization of multicomponent droplets at elevated pressures and temperatures by means of a model that accounts for the variable properties of the gas phase, assumes phase equilibrium at the gas-liquid interface, and uses an effective diffusivity to consider the internal motion within the droplet. Tong and Sirignano [199, 200] developed a vaporization model for multicomponent droplets that accounts for the liquid-phase internal circulation, transient droplet diffusion, and axisymmetric gas-phase convection. However, these models have not as yet been used to predict the flow field in diesel engines.

Haselman and Westbrook [186] and Westbrook [187] also solved the spray equation [Eq. (232)] and the gas-phase conservation equations of mass, linear momentum, species, and energy; they used a turbulent diffusivity that depends on the strain rate of the mean flow field (see Chapter 2). Some sample results are presented in Fig. 3.42, which shows the contours of constant equivalence ratio at different times.

The solution of the spray equation yields the fuel vaporization rate, but it does not provide resolution of scales of the order of or larger than the droplet spacing. However, the spray equation provides a basis for developing the thin and thick spray models presented in Sections 3.5.3 and 3.5.5.

A probability density function (pdf) for spray analyses was developed by Kerstein [201]. However, this pdf is based on the gas concentration and is related to the quasidimensional models described in references 117 and 118.

3.5.3 Thin Spray Models

In thin or dilute spray models volumetric displacement effects ($\psi \neq 1$) can be neglected, as can droplet collisions, coalescence, and breakup; the interactions between individual droplets can be disregarded or highly simplified [202]. These assumptions correspond to dilute sprays for which the droplet loading (i.e., the

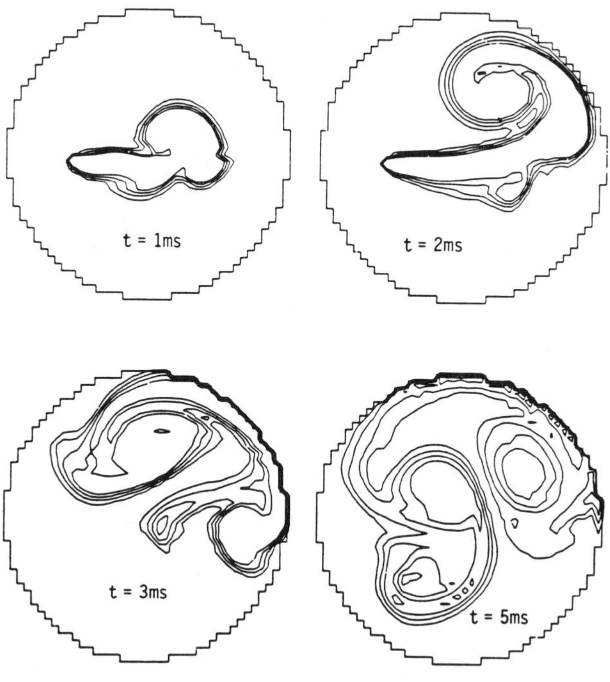

Figure 3.42 Contours of constant equivalence ratio at different times [186].

number of droplets in a turbulent eddy is small) and correlations can be derived to estimate the effects of the droplets on the gas-phase equations for the turbulence kinetic energy and its rate of dissipation [163]. However, in most cases these effects have been neglected, and only the effects of the gas-phase turbulence have been considered through the fluctuating velocity $\mathbf{u}'$ in Eqs. (261) and (264).

Dilute two-phase-flow mixtures are not very much affected by the presence of droplets except in boundary layers. If the characteristic time scale in the boundary layer is commensurable with the particle response time, the boundary layer thickens, decreasing the friction and enhancing the heat transfer coefficient. Away from the walls, the effect of the droplets is to shift the energy spectrum to high wave numbers [203].

Boysan et al. [204] used a thin spray model, an eddy breakup model of turbulent combustion (see Chapter 2), a radiation model, and the following expressions:

$$\frac{d\mathbf{v}_d}{dt} = -\frac{9\mu}{2\rho r_d}\frac{C_D\,\mathrm{Re}_d}{24}(\mathbf{v}_d - \mathbf{u}) \qquad Re_d = \frac{\rho 2r_d|\mathbf{v}_d - \mathbf{u}|}{\mu} \tag{278}$$

$$\begin{aligned} C_D &= 27\,\mathrm{Re}_d^{-0.84} \qquad 0 \le \mathrm{Re}_d \le 80 \\ C_D &= 0.271\,\mathrm{Re}_d^{0.271} \qquad 80 \le \mathrm{Re}_d \le 10^4 \end{aligned} \tag{279}$$

$$R_d = -\frac{C_b}{8r_d}(1 + 0.23\,\mathrm{Re}_d^{1/2}) \qquad C_b = \frac{8\lambda}{\rho_l\,C_p}\ln\left[1 + \frac{C_p}{L}(T - T_d)\right] \tag{280}$$

$$\frac{dT_d}{dT} = 6\lambda(2+0.6\,\mathrm{Re}_d^{1/2}\,\mathrm{Pr}^{1/3})\frac{T - T_d}{4\rho_l\,r_d^2\,c_l} \tag{281}$$

Equation (281) accounts for the droplet heat-up period; evaporation is assumed to begin when the droplet reaches the boiling point.

Boysan et al. [204] used the following eddy-breakup model:

$$\langle \dot{w}_f \rangle = -\,C_{\mathrm{EBU}}\,\rho\,k^{1/2}\,|\nabla Y_f| \tag{282}$$

where $C_{\mathrm{EBU}} = 6$, Y_f is the fuel mass fraction, and $\langle \dot{w}_f \rangle$ is the fuel mean reaction rate. Boysan et al. [204] used a radiation model that accounts for the heat fluxes in three spatial directions. This model can be written as

$$\left(\frac{1}{a + s}R_{i,j}\right)_{,j} = a(R_i - E) + \frac{s}{3}\left(2R_i - \sum_{j=1,k\neq i}^{3} R_k\right) \tag{283}$$

where R_i denotes the radiative heat flux in the ith direction, a is the absorption coefficient, s is a scattering coefficient, $E = \sigma T^4$ is the blackbody emissive power at the fluid temperature, and σ is the Stefan-Boltzmann constant.

The contribution of the radiation fluxes to the enthalpy equation of the gas phase can be considered as an additional source term, which can be written as

$$S_{h,\mathrm{rad}} = 2a\left(\sum_{j=1}^{3} R_j - 3E\right) \tag{284}$$

Equations (280) and (281) are based on the Ranz-Marshall correlation [105] and may overpredict the vaporization rate. Furthermore, the effects of the gas-phase turbulence have been neglected in Eq. (278), and droplet dispersion may be underpredicted.

For hollow cone sprays, Boysan and Binark [205] used the following drag coefficient:

$$C_D = \frac{k_1}{\mathrm{Re}_d} + \frac{k_2}{\mathrm{Re}_d^2} + k_3 \tag{285}$$

where k_1, k_2, and k_3 are constants. They also used a mixing length model and a Rosin-Rammler droplet distribution function at the nozzle exit.

Gosman and loannides [206], Gosman and Johns [207], and Gosman and Harvey [208] employed the Ranz-Marshall correlation to calculate the vaporization rate and accounted for droplet dispersion effects as follows:

$$\begin{aligned} C_D &= 0.44 \qquad \mathrm{Re}_d > 1000 \\ C_D &= \frac{1+0.15\,\mathrm{Re}_d^{0.687}}{\mathrm{Re}_d/24} \qquad \mathrm{Re}_d \leq 1000 \end{aligned} \tag{286}$$

$$\frac{d\mathbf{v}_d}{dt} = \frac{3}{8} C_D \frac{\rho}{\rho_l r_d} |\mathbf{v}_d - \mathbf{u} - \mathbf{u}'| (\mathbf{u} + \mathbf{u}' - \mathbf{v}_d) \tag{287}$$

$$m_d c_l \frac{dT_d}{dt} = 2\pi r_d \lambda (T - T_d) \,\mathrm{Nu} \frac{Z}{\exp Z - 1} + L \frac{dm_d}{dt} \tag{288}$$

$$Z = -C_p \frac{dm_d/dt}{2\pi r_d \lambda \,\mathrm{Nu}} \tag{289}$$

$$\frac{dm_d}{dt} = -4\pi r_d D \frac{p}{RT} \mathrm{Sh} \ln \left(\frac{p - p_f}{p - p_{fs}} \right) \tag{290}$$

$$\mathrm{Nu} = 2 + 0.6\,\mathrm{Re}_d^{1/2}\,\mathrm{Pr}^{1/3} \qquad \mathrm{Sh} = 2 + 0.6\,\mathrm{Re}_d^{1/2}\,\mathrm{Sc}^{1/3} \tag{291}$$

where R is the gas constant and D is the mass diffusivity.

Equation (288) accounts for both the heat-up and evaporation periods [209]. It also considers the effects of mass transfer on droplet evaporation through the factor Nu Z/(exp Z − 1).

The use of a uniform droplet temperature in Eq. (288) can be justified if the thermal capacity of the surrounding flow is much smaller than the heat capacity of the droplets. In liquid-liquid systems, the fluid-to-droplet capacitance ratio is on the order of 0.3, and the droplet temperature may not be uniform [210, 211].

3.5.4 Droplet-Turbulence Interactions II

The effects of the gas-phase turbulence on the droplets can be simulated as a stochastic process as follows. If the gas-phase turbulence is isotropic and has a

Gaussian probability distribution for the fluctuating velocity $\mathbf{u}'$ and a standard deviation SD $= (2k/3)^{1/3}$, then the probability distribution of $\mathbf{u}'$ is given by the following expression:

$$f(\mathbf{u}') = \pm \text{SD erfc } (|x|) \tag{292}$$

where erfc denotes the complementary error function and x is a random variable having uniform distribution in the range $|x| < 1$.

The velocity $\mathbf{u}'$ can be calculated by randomly sampling the distribution [Eq. (292)] at appropriate times. The time interval over which the droplet interacts with the randomly sampled gas-phase velocity fluctuation is determined by the slip velocity and the droplet residence time within a typical turbulent eddy. This residence time can be estimated by linearizing the droplet momentum equations and is given by the following expression:

$$t_R = -\tau \ln \left[1 - \frac{l}{\tau} |\mathbf{u} - \mathbf{v}_d|\right] \tag{293}$$

where τ is the droplet relaxation time defined by

$$\tau = \frac{8}{3} \frac{\rho_l r_d}{\rho C_D |\mathbf{u} - \mathbf{v}_d|} \qquad l = C_\mu^{3/4} \frac{k^{3/2}}{\epsilon} \tag{294}$$

and $C_\mu = 0.09$.

The eddy-turnover time can be estimated from the following expression:

$$t_T = \frac{l}{|\mathbf{u}'|} \tag{295}$$

and the droplet-turbulence interaction time is

$$t_{\text{INT}} = \min(t_T, t_R) \tag{296}$$

Note that if $t_T >> t_R$ (i.e., if the eddy-turnover time is much larger than the droplet residence time in a large-scale eddy), the interaction time between the gas-phase turbulence and the droplets is the residence time. If $t_T << t_R$ (i.e., if the eddy-turnover time is much smaller than the droplet residence time in a large-scale eddy), the interaction time between the gas-phase turbulence and the droplets is the eddy-turnover time.

Equation (296) represents an approximation to the complex problem of droplet-turbulence interactions. These interactions depend not only on the characteristic droplet residence time and eddy-turnover time, but also on the droplet diameter-to-eddy size ratio, the droplet-to-turbulence velocity ratio, the number of droplets per unit volume (or the number of droplets in a typical eddy), etc. Most of these interactions are not well known because they involve determination of the droplet response to fluctuating velocity fields and the effects of the droplets on the gas-phase turbulence.

Equation (296) is an approximation to droplet-turbulence interactions that does not account directly for length-scale effects and loading in the interaction time.

Gosman and Johns [207] showed that the spray induces velocities and turbulence levels in the gas phase that are comparable to those produced by swirl and squish. Gosman and Harvey [208] used a hybrid mixing-chemical kinetics-controlled model (see Chapter 2) to predict the flow field and combustion in DI diesel engines with bowl-in-piston combustion chambers.

Nguyen et al. [212, 213] and Carpenter and Ramos [214, 215] used a thin spray approximation and a mean-reaction-rate model to study the flow field, combustion, and heat transfer losses in DI diesel engines. Nguyen et al. [212, 213] and Carpenter and Ramos [214, 215] used a modified $k-\epsilon$ model of turbulence that accounts for compressibility effects and a very small time step so that the turbulent fluctuating velocity is applied at each time step to the gas-phase mean velocity in order to calculate the vaporization rate. They also used an integration scheme where the droplet mass, momentum, and energy exchanges with the gas phase were assigned to the computational cells that the droplets cross as a function of the time that the droplets spend in each cell. The numerical calculations were compared with the experimental data of Hiroyasu and Kadota [216], and good agreement was obtained. Some sample results are shown in Figs. 3.43 and 3.44. Figure 3.43 shows a comparison between the predicted and measured penetration of sprays injected into stagnant combustion chambers at different pressures [213, 216], whereas Fig. 3.44 shows the droplet locations and mean gas velocity vectors for a spray injected into a stagnant combustion chamber at 11 atm.

Nguyen et al. [212, 213] showed that the spray penetration and vaporization increase with the intake swirl number, and the spray vaporization increases with engine speed due to the higher turbulence kinetic energy levels at TDC of the compression stroke. Nguyen et al. [212, 213] also showed that the spray cone angle has a strong effect on the spray penetration and vaporization in combustion chambers provided with flat pistons. For the same amount of injected liquid fuel, early injection results in deeper spray penetration, whereas late injection results in faster vaporization and smaller spray penetration.

Figures 3.45 and 3.46 show the velocity vectors and droplet locations, and the Sauter mean diameter distributions in a four-stroke diesel engine at several crank shaft angles [215].

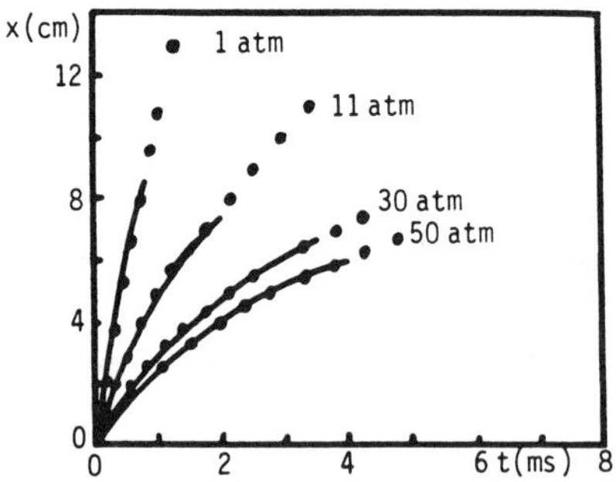

Figure 3.43 Comparison between predicted (solid line) and measured (symbols) spray tip penetration at several chamber pressures [213].

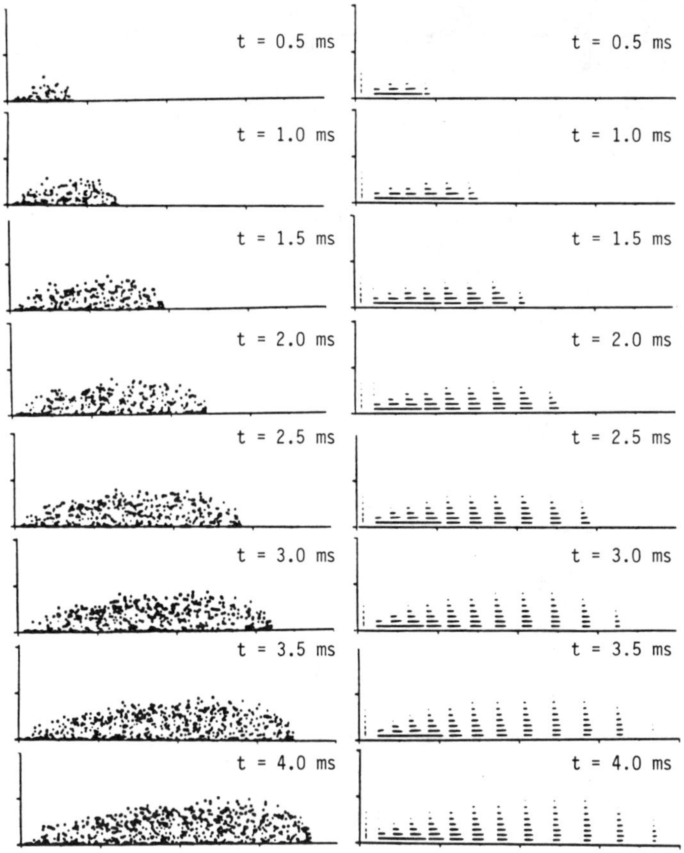

Figure 3.44 Droplet locations (left) and mean gas velocity vectors (right) for a spray injected into a stagnant combustion chamber at 11 atm.

3.5.5 Thick Spray Models

Thick or dense spray models that account for volumetric displacement effects seem to have originated with Dukowicz's work [217], where a Lagrangian-Eulerian formulation was used to compute the behavior of atomized, nonevaporating liquid sprays injected into a gaseous environment. Since a Lagrangian formulation was employed to calculate the droplet trajectories, numerical diffusion was minimized.

Dukowicz [217] used a simplified expression for the drag force

$$\mathbf{F}_d^d = \left(6\pi\ \mu\ r_d + \frac{1}{2}\ \pi r_d^2\ \rho\ C_D|\mathbf{u}-\mathbf{v}_d|\right)(\mathbf{u}-\mathbf{v}_d) \qquad C_D = 0.4 \qquad (297)$$

where the first term represents the Stokes drag, and he introduced a forcing term in the droplet momentum equations in order to simulate droplet dispersion.

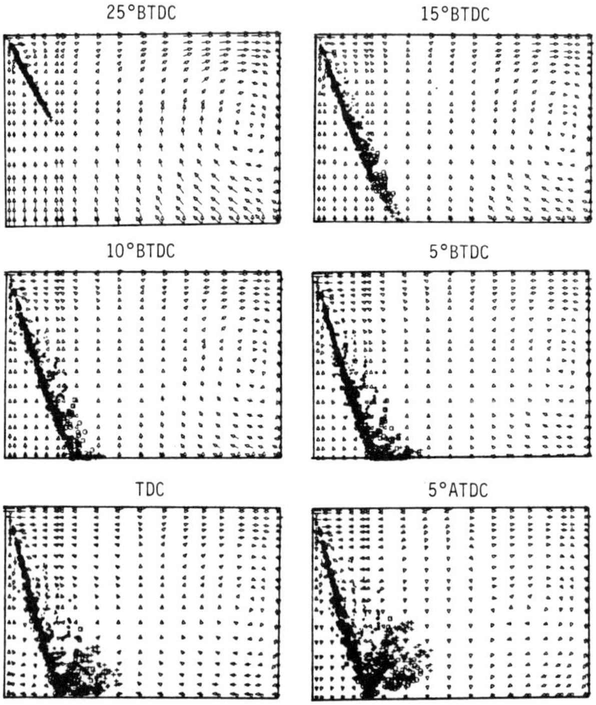

Figure 3.45 Velocity fields and droplet locations in a four-stroke diesel engine [215].

He also compared his predictions with the experimental data of Hiroyasu and Kadota [216] and obtained good agreement for a variety of chamber pressures. However, this agreement was obtained by using small droplets at the injector nozzle exit.

Cloutman et al. [218] and Butler et al. [219] used the stochastic model proposed by Dukowicz [217], laws of the wall to account for the turbulent boundary layers and heat transfer to solid walls, and a mean-reaction-rate model (see Chapter 2) to analyze the flow fields, fuel sprays, and flame propagation in DI diesel engines. The calculations were performed with the CONCHAS-SPRAY code [220], which was used by Bassoli et al. [221, 222] to analyze combustion in DI diesel engines. Bassoli et al. [221] showed that with reentrant combustion chambers, optimum fuel-air mixing can be achieved with low swirl levels at the intake port.

Figure 3.47 shows the velocity vectors and fuel sprays in open chambers and in reentrant bowls [221]. Figure 3.48 illustrates the temperature distributions in a reentrant bowl at several crank shaft angles [222].

Butler et al. [223] analyzed the compression and expansion strokes of DISC engines by means of the CONCHAS-SPRAY code [220], which uses as implicit-continuous fluid Eulerian (ICE) technique and an arbitrary Lagrangian-Eulerian (ALE) formulation [224, 225]; a global kinetics model; a discrete droplet model; and the Zel'dovich mechanism for NO formation. Their model (1) does not consider Basset, virtual mass, and droplet internal circulation effects; (2) assumes that the droplet temperature is uniform; (3) accounts for form drag and friction on the droplets and volumetric displacement effects; and (4) uses a predictor corrector method to calculate the mass, velocity, and momentum of the droplets. Butler et al. [223] increased the gas temperature in a few computational cells rapidly to simulate ignition and showed that the maximum temperatures in spark-ignition and DISC engines are approximately 2920 K and 2600 K, respectively. The DISC engines have high efficiency and low emissions for lean mixture operation.

In the ICE formulation used in the CONCHAS-SPRAY and RICE codes [226], the energy equation is solved explicitly, whereas the species conservation equations are solved in three steps (explicit convection, explicit diffusion, and implicit reaction) and a pressure correction is calculated to satisfy the continuity equation.

Kuo and Bracco [75, 227], Wu et al. [80], O'Rourke [157], and O'Rourke and Bracco [164] used the thick spray model represented by Eqs. (232)–(239) with and without a two-equation model of turbulence to calculate transient axisymmetric pulsating and evaporating sprays, assess the importance of droplet collisions and coalescence near the spray edge, and compute the spray tip penetration and droplet size distributions in sprays injected into high-pressure cham-

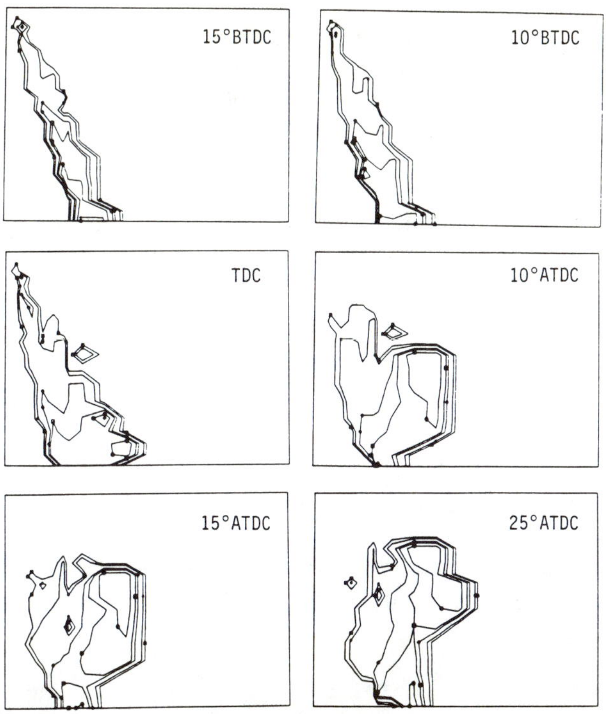

Figure 3.46 Sauter mean diameter (SMD) distributions in a four-stroke diesel engine [215].

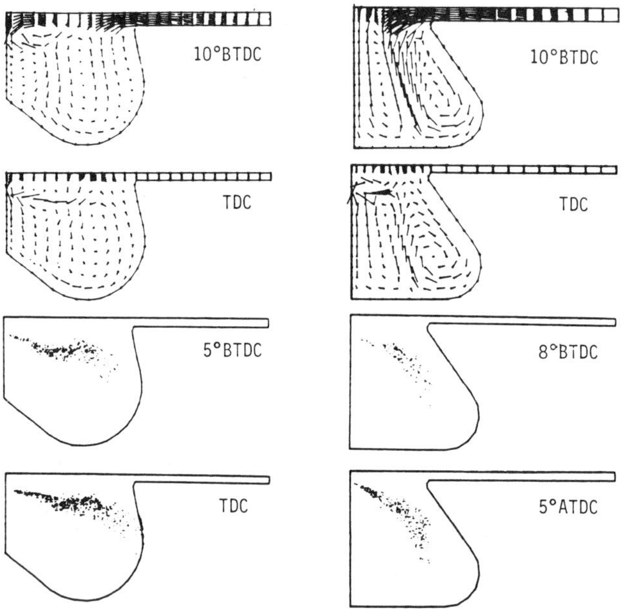

Figure 3.47 Velocity vectors and fuel sprays in open chambers (left: 2700 rpm, swirl ratio = 2.5) and in reentrant bowls (right: 2700 rpm, swirl ratio = 1.0) [221].

bers. Kuo and Bracco [227] showed that pulsating injection results in smaller droplets than continuous injection particularly at high gas densities. They attributed these differences to the inhibition of droplet collisions and coalescence due to the gas gap generated between the pulses. The length of the liquid core was found to be sensitive to the injection pressure, gas temperature, and gas density.

Kuo and Bracco [227] used the correlations proposed by Reitz and Bracco [77] for the spray cone angle and initial mean droplet size, that is,

$$\tan \alpha = \frac{1}{c_\alpha} 4\pi \left(\frac{\rho}{\rho_l}\right)^{1/2} \frac{3^{1/2}}{6} \tag{298}$$

$$\text{SMR} = B\, 4\pi \frac{\sigma_l}{\rho\, u_{\text{inj}}^2} \frac{3}{2} \tag{299}$$

where α is the spray half angle u_{inj} is the injection velocity, c_α is the constant that depends on the nozzle geometry, B is a constant independent of the nozzle geometry, and SMR is the Sauter mean radius.

Kuo and Bracco [227] chose $\mathbf{u}'$ by randomly sampling an isotropic Gaussian distribution with mean square deviation $2k/3$. After a turbulent correlation time τ, a new value of $\mathbf{u}'$ was chosen so that

$$l = \int_t^{t+\tau} |\mathbf{v}_d - \mathbf{u}|\, dt \qquad l = C_\mu^{3/4} \frac{k^{3/2}}{\epsilon} \qquad C_\mu = 0.09 \tag{300}$$

where k is the turbulent kinetic energy or a fraction of the mean flow kinetic energy

$$k = c\frac{3}{2}\langle \mathbf{u}'^2 \rangle \qquad c = 0.20 \tag{301}$$

Dukowicz [217] used a constant correlation time, whereas O'Rourke [157] assumed that τ is the residence time in a turbulent eddy of size l, where l can be calculated from the eddy diffusivity as

$$D_t = 0.045k^{1/2}\, l \tag{302}$$

and the residence time τ_R is given by the following expression:

$$l = \int_t^{t+\tau_R} |\mathbf{v}_d - \mathbf{u}|\, dt \tag{303}$$

Kuo and Yu [228] showed that the mixing rate increases with the mass-averaged injection pressure and with reduced injection duration. Poor end-of-injection can result in poor spray atomization and local poor fuel-air mixing. For a fixed injection rate, pulsed injection results in poor fuel-air mixing. However, gradual rises or falls in the injection rate reduce the mixing rate.

Chatwani and Bracco [229] extended a dense-spray model, using a line-source technique to account for the intact liquid core in full-cone sprays; they

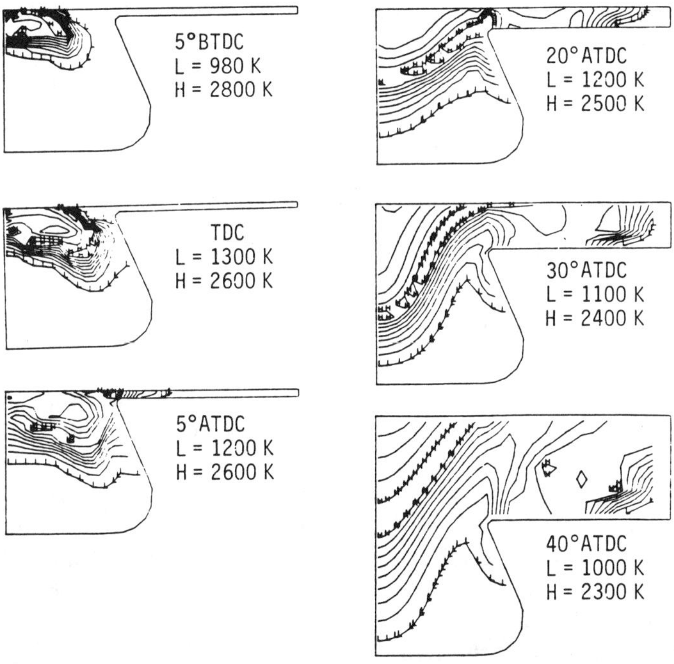

Figure 3.48 Temperature distributions in a reentrant bowl at selected crank angles at 2700 rpm [222].

showed that the intact core significantly modifies the computed structure of the near field. The far field was found to be insensitive to the near field, and the intact core length is given by

$$\frac{x_{\text{ic}}}{d_{\text{icn}}} = c\left(\frac{\rho_l}{\rho}\right)^{1/2} f^{-1}\left(\frac{\rho}{\rho_l}\,\frac{\mu_l^2\,V_T^2}{\sigma_l^2}\right) \tag{304}$$

where x_{ic} and d_{icn} are the intact core length and the intact core diameter at the nozzle exit, respectively; c is a constant; and V_T is the relative droplet-gas velocity for droplet formation.

Chehroudi et al. [230] imposed a voltage in sprays to measure the intact core length and found that current is carried not only by the intact core but also by atomized unconnected sprays. The shape of the intact core could be deduced only in the immediate vicinity of the nozzle and in the atomization regime $c=7$ in Eq. (304).

Chatwani and Bracco [229] estimated the initial spray angle 2α and the initial droplet diameter d_0 by means of the following expressions:

$$\tan\alpha = \frac{1}{A}\,4\pi\left(\frac{\rho}{\rho_l}\right)^{1/2} f\left(\frac{\rho}{\rho_l}\,\frac{\mu_l^2\,V_T^2}{\sigma_l^2}\right) \tag{305}$$

$$d_0 = B\,\frac{2\pi\sigma}{\rho\,V_T^2}\,g\left(\frac{\rho}{\rho_l}\,\frac{\mu_l^2\,V_T^2}{\sigma_l^2}\right) \tag{306}$$

where A and B are constants. For a nozzle having a length-to-diameter ratio equal to 4, Chatwani and Bracco [229] used the following values: $A = 4$; $B = (3/4\pi)^{1/3}$; $cf^{-1} = 7.5$; $V_T = V_a/4$, where $V_a = C_D(2\,\Delta p/\rho_l)^{1/2}$; C_D is a discharge coefficient; c and f are the constant and the function that appear in Eqs. (304) and (305), respectively; and Δp is the pressure drop in the nozzle.

Martinelli et al. [231] analyzed dense sprays at room temperature and showed that their thick spray model results in good agreement with the measured mean quantities, but the predicted standard deviation of the droplet velocity distribution or the droplet velocity fluctuation is smaller than the measured one. This may be caused by an inadequate modeling of volumetric displacement effects and injection-generated turbulence. Martinelli et al. [231] also showed that nonvaporizing sprays achieve a structure very similar to, but much farther than, that of fully developed incompressible jets. After development, the droplets exert very little influence on the gas phase, but the gas controls the motion of the droplets.

Duggal et al. [232] performed three-dimensional calculations in a DI diesel engine and showed that the initial air swirl velocity and the grid distribution have a strong effect on the predicted fuel penetration and spread rate and that the ratio of the injected fuel momentum to the initial swirl momentum has a significant influence on the fuel-air mixing.

Amsden et al. [233] developed the KIVA code to analyze two- and three-dimensional flows in DI engines. The KIVA code uses an ALE formulation, a subgrid-scale model (see Chapter 2), and a mean-reaction-rate model, which can be represented by the following reactions:

$$2C_8H_{18} + 25O_2 \rightarrow 16CO_2 + 18H_2O \tag{307}$$

$$O + N_2 \rightleftarrows NO + O \qquad N + O_2 \rightleftarrows NO + O \tag{308}$$

$$N + OH \rightleftarrows NO + H \tag{309}$$

Equations (308) and (309) correspond to the extended Zel'dovich mechanism. The following reactions are assumed to be in chemical equilibrium:

$$H_2 \rightleftarrows 2H \qquad O_2 \rightleftarrows 2O \qquad N_2 \rightleftarrows 2N \tag{310}$$

$$O_2 + H_2 \rightleftarrows 2\,OH \qquad O_2 + 2H_2O \rightleftarrows 4OH \tag{311}$$

$$O_2 + 2\,CO \rightleftarrows 2\,CO_2 \tag{312}$$

O'Rourke and Amsden [234] analyzed the spray penetration, vaporization, mixing, and combustion in the UPS-292 stratified-charge engine with the KIVA code and showed that combustion occurs in the wake of the spark plug. The predicted pressure diagrams compared well with the measured ones at the highest load conditions investigated, but the HC and CO emissions agreed poorly with experimental data. Their model accurately predicted the NO emissions, but the predicted wall heat transfer losses were approximately two-thirds of the measured values.

Figure 3.49 shows (left to right, top to bottom) the droplet locations, the velocity vectors, the isotherms, the turbulence kinetic energy, the equivalence ratio, and the octane mass fractions in a three-dimensional DISC engine model

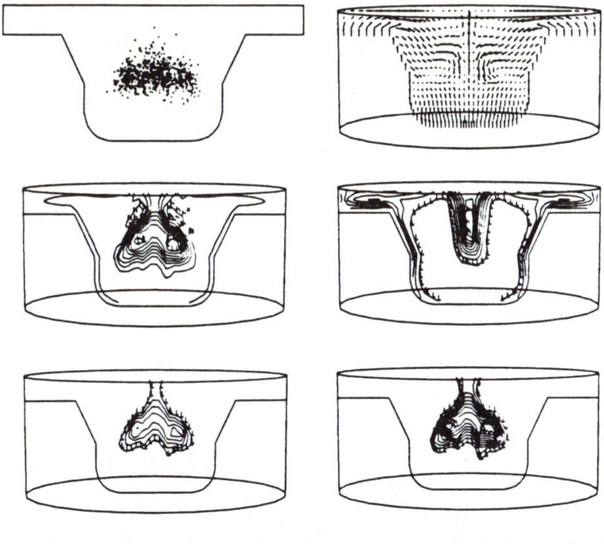

Figure 3.49 (From left to right, from top to bottom) Spray droplets, velocity vectors, isotherms, turbulence kinetic energy, equivalence ratio, and octane mass fraction at 28°BTDC in a three-dimensional DISC engine model calculated with the KIVA code [233].

predicted by the KIVA code [233]. Figure 3.50 shows a comparison between the measured and predicted pressure diagrams in the UPS-292 stratified-charge engine [234].

Reitz and Diwakar [235, 236] included droplet breakup in dense spray calculations. They used the KIVA code and two breakup regimes:

$$\text{Bag breakup} \qquad \text{We} = \rho\, |\mathbf{u}+\mathbf{u}'-\mathbf{v}_d|^2 \frac{r_d}{\sigma_l} > 6 \tag{313}$$

$$\text{Stripping breakup} \qquad \frac{\text{We}}{\text{Re}_d^{1/2}} > 0.5 \tag{314}$$

with the following lifetimes for the unstable droplets:

$$t_{\text{bag}} = c_1 \left(\frac{\rho_l\, r_d^3}{2\sigma_l}\right)^{1/2} \qquad t_{\text{strip}} = c_2\, r_d \left(\frac{\rho_l}{\rho}\right)^{1/2} |\mathbf{u}+\mathbf{u}'-\mathbf{v}_d|^{-1} \tag{315}$$

where c_1 and c_2 are constants. Reitz and Diwakar [235, 236] showed that droplet breakup influences the spray penetration, vaporization, and mixing in high-pressure sprays. The spray droplet size is controlled by breakup and coalescence; droplet breakup dominates in hollow-cone sprays because coalescence is minimized by the expanding spray geometry.

Reitz and Diwakar [236] injected droplets having a diameter equal to the nozzle exit diameter; these droplets break up due to interaction with the gas and yield a core region that contains relatively large droplets. The computed length, droplet size, droplet velocity, and spray-penetration data compared well with available measurements in high-pressure sprays [237].

Reitz and Diwakar [236] calculated the axial and radial velocities of the

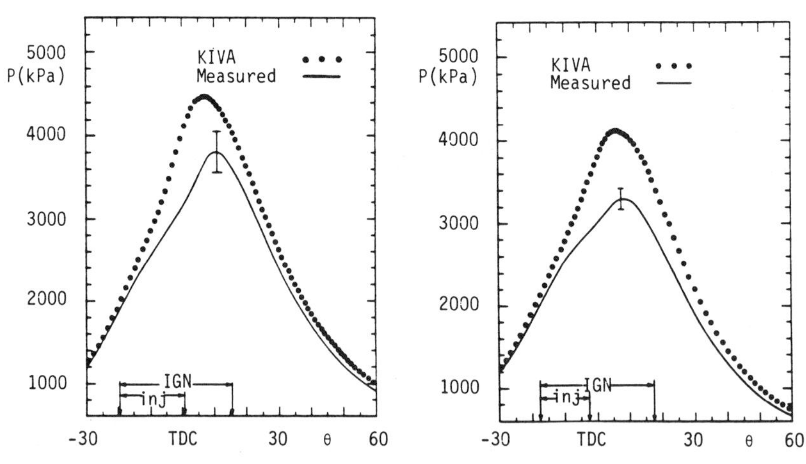

Figure 3.50 Measured and predicted pressure histories in the UPS-292 stratified-charge engine [234].

droplets as

$$V_{\text{axial}} = c_d \left(\frac{2\,\Delta p}{\rho_l} \right)^{1/2} \qquad V_{\text{radial}} = V_{\text{axial}} \tan \alpha \tag{316}$$

$$\tan \alpha = A \left(\frac{\rho}{\rho_l} \right)^{1/2} \qquad \text{A} = 0.7 \qquad \text{SMR} = B \frac{\sigma}{\rho\, V_{\text{axial}}^2} \tag{317}$$

where the axial velocity is equal to the injection velocity, B is a constant, α is the spray half angle, and c_d is the discharge coefficient across the nozzle.

The size of the unstable droplets was allowed to change with time according to the following equation:

$$\frac{dr}{dt} = -\frac{r - r_{st}}{\tau} \tag{318}$$

where $r_{st} = \sigma^2/(2\rho\, |\mathbf{u} + \mathbf{u}' - \mathbf{v}_d|^3\, \mu)$ is the radius given by the stripping breakup regime and τ is the characteristic droplet breakup time.

When droplet breakup occurs, the number of droplets in a parcel must be adjusted in order to conserve mass, momentum, and energy. Reitz and Diwakar [235, 236] found that the momentum and energy corrections that have to be introduced after droplet breakup were small. They also found that most breakups are of the stripping type and that bag breakup seldom occurs in high-pressure sprays.

3.5.6 Numerical Simulation of Fuel Injection

In the atomization regime, droplets are formed at the spray edge, and there is an intact core the length of which can be estimated from aerodynamic theories of liquid jet breakup [Eq. 304].

Multidimensional models of diesel engines are currently unable to deal with the liquid jet breakup into ligaments and droplets. Therefore, several approaches have been used to simulate liquid fuel injection.

3.5.6.1 Injection of small droplets. In the first approach droplets are injected at certain locations near the injector within the computational domain. The size of these droplets and the spray angle can be estimated from aerodynamic theories of liquid jet breakup [77, 81] as in Eqs. (305) and (306), whereas the droplet velocity can be estimated as a fraction of the injection velocity. The droplets may be injected at the end of the intact core length and on a circle having a radius equal to that of the injector. Droplets may also be injected at the end of the intact core length and at random axial and radial locations from the injector axis in order to simulate the unsteady characteristics of the fuel-injection process. The droplet size and spray angle may be estimated from Eqs. (305) and (306) or from Weber number stability criteria [212–215, 217]. At low Reynolds numbers the dynamic pressure tends to deform the droplets into oblate shapes, and analyses show that droplet breakup occurs for We = 5.33, where We denotes the Weber number [238]. The deformation, spreading, and breakup of droplets moving through gaseous streams can be obtained by asymptotic methods; the first term of the asymptotic expansion corresponds to spherical droplets, whereas the second term accounts for droplet spreading and deformation [239, 240].

Dukowicz [217], Nguyen et al. [212, 213], and Carpenter and Ramos [214, 215] used the following stability criterion to estimate the droplet size near the injector:

$$\rho \, |\mathbf{u} - \mathbf{v}_d|^2 \, \frac{d}{\sigma} = 20 \tag{319}$$

where $\mathbf{u}$ and $\mathbf{v}_d$ are the mean gas and droplet velocities, respectively; σ is the surface tension; d is the droplet diameter; and ρ is the gas density.

Equation (319) gives an upper bound for stable droplets, and $\mathbf{v}_d$ can be estimated as a fraction of the liquid injection velocity.

The mass of fuel injected at each time step can be allocated to a continuous distribution function of the form

$$f(r) = \frac{6}{\text{SMD}} \exp\left(-\frac{6r}{\text{SMD}}\right) \tag{320}$$

where SMD is the Sauter mean diameter, that is,

$$\text{SMD} = \frac{\displaystyle\int_0^\infty d^3 f(r)\, dr}{\displaystyle\int_0^\infty d^2 f(r)\, dr} \tag{321}$$

r is the droplet radius, and d is the droplet diameter ($d = 2r$).

Carpenter and Ramos [214, 215] used the Rosin-Rammler distribution function, that is,

$$\frac{d\text{Vol}}{dr} = \frac{2q}{d_{\text{ref}}} \left(\frac{d}{d_{\text{ref}}}\right)^{q-1} \exp\left[-\left(\frac{d}{d_{\text{ref}}}\right)^q\right] \tag{322}$$

where q is the distribution factor, dVol represents the volume of droplets with radii in the range r and $r + dr$, and d_{ref} is a reference diameter. The parameters q and d_{ref} are related to the location of the most postulated droplet size and the width of the droplet distribution function, respectively.

The mass injected in a time step is such that

$$\dot{m}_f \, \Delta t = \rho_l \int_0^\infty d\text{Vol} = \rho_l \int_0^\infty \frac{2q}{d_{\text{ref}}} \left(\frac{d}{d_{\text{ref}}}\right)^{q-1} \exp\left[-\left(\frac{d}{d_{\text{ref}}}\right)^q\right] dr \tag{323}$$

where $\dot{m}_f$ is the fuel injection rate and Δt is the time step.

In multidimensional calculations the droplet distributions given by Eqs. (320) and (322) are discretized into a finite number of droplet groups, each characterized by the same droplet velocity, radius, and temperature. For N droplet groups the mass of fuel injected at each time step must be equal to

$$\rho_l \int_0^\infty d\mathrm{Vol} = \rho_l \sum_{i=1}^{N} n_i \frac{\pi\, d_i^3}{6} \tag{324}$$

where n_i and d_i denote the number and diameter of the droplets in the ith group, respectively. Thus, at each time step, N groups of droplets are injected in the cylinder.

The injection of droplets into the cylinder directly from the nozzle exit is inadequate to reproduce the flow field near the injector because the exchange of momentum between the droplets and the gas is different from that between the liquid jet and the gas. For this reason the models mentioned in the previous paragraphs inject the droplets at some distance ($\sim 10\ d_{\text{inj}}$) downstream of the actual nozzle exit plane where the cross-sectional area of the spray can be estimated from Eqs. (304) and (305). Furthermore, these models inject a mixture of small droplets and entrained chamber gas moving at the same velocity in order to avoid spray overpenetration. However, these models do not account for the intact core length.

3.5.6.2 Line-source injection. Using a line-source droplet-injection technique that takes advantage of the fact that the intact core length is much longer than the nozzle radius, Chatwani and Bracco [229] injected droplets into the gas from a line along the axis of the spray. The intact core length, spray angle, and droplet diameter can be calculated from Eqs. (224), (226), and (229), respectively, which are applicable to full-cone sprays from single-hole nozzles in the immediate vicinity of the nozzle exit.

Chatwani and Bracco [229] assumed that the instantaneous length of the line source can be calculated by assuming that its tip moves at 70% of the injection velocity. The instantaneous droplet-generating surface is the surface of a truncated cone of base d_{inj} and height equal to the instantaneous length of the line source. The maximum length of the line source is L_{ic}. The initial droplet number flow rate can be determined from mass conservation using Eq. (229), the initial droplet axial velocity is the mass-mean injection velocity, and the initial droplet radial velocity can be calculated so that the computed initial spray angle is given by Eq. (226). The velocity that appears in Eqs. (224), (226), and (229) is the relative velocity between the liquid surface and the ambient gas and is a fraction of the injection velocity.

3.5.6.3 Droplet injection at the nozzle exit. Reitz and Diwakar [235, 236] injected droplets having a diameter equal to the nozzle exit diameter, and they assumed that the dynamics and breakup of these large droplets are similar to those of a liquid jet. Their method removes one of the major weaknesses of current spray calculations (i.e., the uncertainty in the specification of the upstream boundary conditions). However, their model uses a unique critical Weber number for droplet breakup, does not include the liquid viscosity, does not predict droplet oscillation and distortion, and requires the specification of the spray angle. Reitz and Diwakar [235, 236] considered bag and stripping breakup regimes [Eqs. (313)–(315)]. A similar model was developed by Watkins [241].

Recently, O'Rourke and Amsden [242] developed the Taylor analogy breakup (TAB) model, which predicts that there is no unique critical Weber number of droplet breakup, includes the liquid viscosity, and predicts droplet distortion and oscillation. The model also (1) predicts that the droplet breakup depends on the history of the droplet velocity relative to the gas; (2) considers the effects of the liquid viscosity on the oscillations of small droplets; (3) predicts droplet sizes that are consistent with experimentally determined mechanisms of liquid jet breakup; (4) predicts a velocity of the product droplets normal to the path of the original parent droplet (this normal velocity determines the initial spray angle, which means that there is no need to specify the spray angle); and (5) predicts and connects the bag and stripping breakup regimes. However, the TAB model can keep track of only one oscillation mode, whereas in reality there are many such modes.

The fundamental oscillation mode can be calculated using the following equation for a damped, harmonic, forced oscillator:

$$\frac{d^2x}{dt} = \frac{F}{m} - \frac{k}{m}x - \frac{d}{m}\frac{dx}{dt} \tag{325}$$

where

$$\frac{F}{m} = c_F \frac{\rho u^2}{\rho_l r} \qquad \frac{k}{m} = c_k \frac{\sigma}{\rho_l r^3} \qquad \frac{d}{m} = c_d \frac{\mu_l}{\rho_l r^2} \tag{326}$$

u is the relative velocity between the gas and liquid, r is the droplet radius, x is the displacement of the equator of the droplet from its equilibrium position, and c_k, c_F, and c_d are calculated by matching the fundamental oscillation frequency ($c_k = 8$, $c_d = 5$) and by postulating that breakup occurs if and only if the oscillation amplitude of the north and south poles equals the droplet radius ($c_F = 1/3$).

The model predicts

$$\mathrm{We}_{\mathrm{CRIT}} = \rho\, u^2 \frac{r}{\sigma} = 6 \tag{327}$$

$$t_{\mathrm{bu}} = \left(\frac{3\,\rho_l}{\rho}\right)^{1/2} \frac{r}{u} \qquad \mathrm{We} >> 1 \tag{328}$$

$$t_{\mathrm{bu}} = \pi \left(\frac{\rho_l\, r^3}{8\sigma}\right)^{1/2} \qquad (\mathrm{We} \sim \mathrm{We}_{\mathrm{CRIT}}, \text{ bag breakup}) \tag{329}$$

$$\tan\alpha = c_H \left(\frac{\rho}{3\rho_l}\right)^{1/2} \tag{330}$$

where c_H is a constant, We_{CRIT} denotes the critical Weber number for droplet breakup, t_{bu} is the breakup time, and α is the spray half angle.

Equation (327) is in remarkably good agreement with experimental data, and Eqs. (328)–(330) are in very good agreement with Eqs. (305) and (315).

O'Rourke and Amsden [242] calculated the droplet sizes after breakup by equating the energy of the parent droplet before breakup to the combined energies of the product droplets after breakup. The energy of the parent droplet is the sum of its surface energy and the energy in oscillation and distortion. In the bag breakup regime the Weber number based on the product droplet Sauter mean radius is 6, and to conserve mass, the number of droplets must be adjusted.

O'Rourke and Amsden [242] obtained good agreement with experimental data for the spray tip penetration and spray angle [216]. However, both the spray tip penetration and spray angle are fairly insensitive to many physical parameters, and the agreement between the theoretical and experimental results is not sufficient for model validation. Comparisons with as-yet-unavailable measured droplet sizes would be a more serious validation test of the TAB model.

The TAB model [242] of droplet breakup considers only one oscillation mode. Theoretical studies on inviscid droplet oscillations show that droplet oscillations are not axisymmetric and that there is a transfer of linear momentum and energy between modes [243]. Tsamopoulos and Brown [244, 245] indicated that mode coupling results in two-, three-, and four-lobed motions. They also showed that the frequency of oscillation decreases with the square of the amplitude. Natarajan and Brown [246] observed quadratic nonlinearities due to inertia, capillarity, and droplet deformation as well as stochastic motions in nonaxisymmetric droplet deformations. These motions redistribute the energy over all the degrees of freedom of the interacting modes.

Nguyen et al. [213], O'Rourke and Bracco [164], O'Rourke and Amsden [242], Kuo and Bracco [227], and Dukowicz [217] obtained good agreement between the measured and predicted spray angle and the spray tip penetration at different ambient pressures [216] using different spray models.

Nguyen et al. [213] extrapolated the experimental spray tip penetration rate at time $t = 0$ and used it as the initial droplet velocity, whereas Dukowicz [217] used initial droplet sizes that yielded good agreement with experimental data. Unfortunately, there are no droplet size measurements near the injector that can be used as boundary conditions in multidimensional models. These models are also sensitive to the initial levels of the turbulence kinetic energy and its rate of dissipation.

Some researchers have claimed that the flow field in reciprocating and rotary engines is sensitive mainly to the eddy diffusivity. This claim is misleading because eddy diffusivity is the product of a turbulent length scale and a fluctuating velocity (i.e., it involves time and length scales), and a constant value of the eddy diffusivity can be obtained with different time and length scales. Suppose, for example, that two flows have the same eddy diffusivity but different velocity and length scales. The flow with higher fluctuating velocities will result in larger droplet dispersion. This is a consequence of the k–ϵ model, where

$$\frac{\nu_t}{C_\mu} = \frac{k^2}{\epsilon} \qquad |\mathbf{u}'| \sim k^{1/2} \tag{331}$$

Equation (331) shows that for the same value of the eddy diffusivity ν_t, different levels of k and ϵ may yield different spray tip penetrations and spray angles. Thus, the background levels of k and ϵ play an important role in determining the flow field in reciprocating and rotary engines [247]. Unfortunately, experimental data for k and ϵ are not available at the present time.

Equation (331) suggests that even if accurate droplet size measurements were available at the nozzle exit, different levels of k and ϵ might give different spray tip penetrations and spray angles.

If droplet collisions, coalescence, and breakup occur, the number of droplet groups with different diameters may substantially increase. This may cause computer storage problems. Furthermore, if in a time step the mass, momentum, and energy exchanges between the gas and liquid phases are assigned to the computational cells and volumes in which the droplets are located at the end of the time step, the computations may not converge. This problem is particularly serious in the cells located near the nozzle exit. In order to avoid this problem, Nguyen et al. [212, 213] and Carpenter and Ramos [214, 215] apportioned the mass, momentum, and energy exchanges to the cells that the droplet groups crossed in each time step according to the fraction of time that the droplet groups spent in each cell. In the KIVA-II code [248] these exchanges are assigned randomly to the cells that the droplet groups cross in each time step so that, on the average, the droplets exchange mass, momentum, and energy with all the cells along their trajectories, even though an individual droplet group may cross many cells in one time step. These approximations permit the use of longer time steps. The KIVA-II code [248] also includes a two-equation (k–ϵ) model of turbulence, a stochastic spray model, an implicit-continuous Eulerian (ICE) method with conjugate residual iteration, a quasi-second-order-accurate convection scheme, and several other improvements over the original KIVA code [233, 234].

Figure 3.51 shows a comparison between the predicted and measured pressure diagrams in the UPS-292 stratified-charge engine and indicates that the KIVA-II code accurately predicts the cylinder pressure if the correct compression ratio is used in the calculations and if the wall heat transfer is increased 150% [242].

In the KIVA code [233] the time step was selected so that the droplet groups cannot travel more than one computational cell in one time step in order that the droplets exchange mass, momentum, and energy with all the cells along their trajectories.

Note that if the droplet groups were not allowed to cross more than one computational cell, the time step would be controlled by the flow near the nozzle exit. This would result in small time steps for which explicit methods may be more efficient than implicit ones. By way of comparison, Carpenter and Ramos

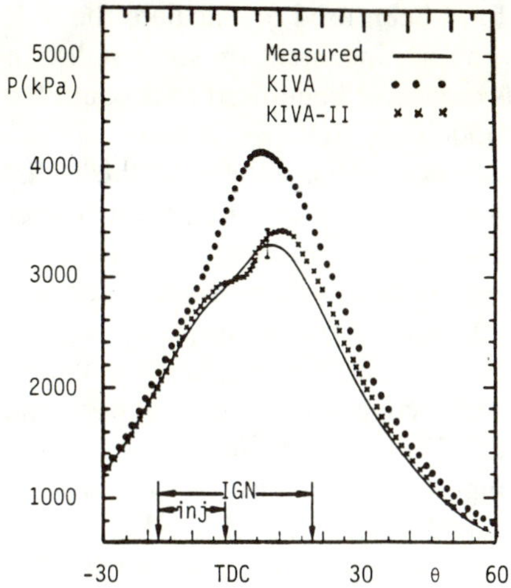

Figure 3.51 Measured and predicted pressure diagrams in the UPS-292 stratified-charge engine [242].

[214, 215] used between 5000 and 10,000 droplet groups and a time step equal to 0.1 μs in their studies of two-stroke diesel engines, whereas Watkins [241] and O'Rourke [249] used fewer than 5000 droplet groups and a time step of about 5 μs. Comparisons with the experimental data of Hiroyasu and Kadota [216] indicate that fewer than 5000 droplet groups are in general sufficient to predict the spray tip penetration and spray angle.

Watkins [241] developed the EPISO code, which includes (1) curvilinear-orthogonal coordinates in two directions and rectilinear coordinates in the third direction; (2) an implicit, noniterative, predictor-corrector method; (3) the thick spray model of O'Rourke and Bracco [164]; (4) no pressure gradient in the droplet equations; (5) cyclic boundary conditions; and (6) Stone's strongly implicit method for solving elliptic partial differential equations. Watkins [241] showed that quick and skew-upwind bounded methods result in little or no improvement over the standard donor cell or upwind method.

A disadvantage of the grid used by Watkins [241] is that it involves a small cylindrical surface along which periodic boundary conditions are imposed. The sensitivity of the numerical calculations to the radius of this small cylindrical surface should be established, since the velocity on the cylinder axis may not be calculated. The KIVA [233] and KIVA-II [248] codes do not suffer from this problem because they use arbitrary hexahedrons. On the other hand, the mesh generation technique employed by Watkins [241] can accommodate exactly the cylinder wall even though the cells located near this wall have large aspect ratios and may result in large numerical errors. The KIVA and KIVA-II codes approximate the cylinder head by locally planar surfaces.

3.5.7 Droplet Impingement on Solid Walls

Fuel sprays may impinge on the piston bowl in DI diesel engines and may have a great influence on the mixture formation and combustion processes. In Section

3.3 we presented some of the models that have been used to calculate the wall jets and air entrainment associated with spray impingement on solid walls. In this section we describe some of the models that have been used to account for droplet impingement in multidimensional models.

Fuel droplets impinging on solid walls may shatter, adhere to the walls, form a liquid film, or bounce off the walls. These phenomena are poorly understood at the present time and depend on the wall temperature, wall material, roughness, angle of incidence, etc. Droplet-wall impaction heat transfer depends on the droplet-wall contact heat transfer and on local convective and radiative heat transfer. The droplet-wall contact heat transfer depends on the Weber number and wall material, whereas the air convective heat transfer depends on the vaporization rate. At low Weber numbers (We $= \rho_l r_l U_n^2/\sigma$, where ρ_l is the liquid density, U_n is the droplet velocity component normal to the wall, r_l is the droplet radius, and σ is the surface tension) the droplets may rebound elastically from the wall. With hot walls a cushion of fuel vapor may form under the droplet, preventing it from wetting the wall. As the Weber number is increased, the normal velocity component of the rebounding droplet decreases, and the droplet may break up into smaller droplets. Beyond a critical value of the incident droplet Weber number, the liquid forms a layer on the wall surface. Whether the wall liquid fuel wets the surface or not depends on the wall temperature, wall material, roughness, angle of incidence, etc. At high wall temperatures, film boiling of the wall liquid may occur.

Liu and Yao [250] showed that droplet-wall contact heat transfer is larger than radiation, convection, and vaporization effects when single droplets impinge on stainless steel surfaces.

Fujimoto et al. [251] used a laser light extinction method to determine the droplet number density distribution function near solid walls; they showed that the Sauter mean diameter of the droplets in impinging diesel sprays is almost independent of the angle that the solid wall forms with the spray axis. They also showed that the droplet number density distribution function has a maximum at the spray edge near the wall.

Several approaches have been used in multidimensional models to account for droplet impingement on solid walls. In the KIVA code [158, 233, 234] the droplets impinging on the piston are assumed to attach to the wall, preserve their spherical symmetry, and vaporize as single, isolated droplets. This approach does not consider the formation of thin liquid films along the solid walls or droplet shattering and/or droplet bouncing off the walls. It concentrates the liquid fuel at the impingement site and reduces the fuel vaporization.

Nguyen et al. [212, 213] and Carpenter and Ramos [214, 215] assumed that the droplets impinging on solid walls are reflected specularly (i.e., the angle of incidence is equal to the angle of reflection), and the magnitudes of the velocity components parallel and perpendicular to the solid walls are preserved upon impingement. This approach does not consider the formation of liquid films along the walls and droplet shattering, and it may yield unrealistic results since the droplets that bounce off the solid walls may have large velocities and move toward the cylinder head after impingement.

The interaction between droplets and solid walls depends on the droplet linear momentum, surrounding gas velocity and temperature, turbulence levels, droplet temperature, wall temperature, Weber number, wall roughness, etc. The interaction between solid walls and droplets also depends on the droplet diameter, boundary layer thickness, and mean gas velocity parallel to the wall. These interactions are poorly understood even for single droplets impinging on solid walls.

The peak heat transfer for droplets impinging on hot surfaces depends on the droplet diameter, normal velocity of the droplets at impact, surface tension, and vapor density [252]. The heat transfer rates associated with droplet impingement are quite high, and as much as 50% of the droplet mass is evaporated during impact and bouncing. The case of zero-impact velocity is called the Leidenfrost phenomenon. The enhanced heat transfer from the wall to the liquid with wall wetting may have important effects on the droplet vaporization rate, which depends on the film thickness and droplet diameter.

Mizomoto and Ikai [253] studied the evaporation and ignition of spherical fuel droplets on hot surfaces having temperatures higher than the Leidenfrost temperature; they showed that droplet-wall contact heat transfer is very important for ignition and evaporation.

Wall-jet models similar to those presented in Section 3.3 can also be used in multidimensional calculations as follows. If the impinging liquid jet is assumed to be inviscid and the friction forces between the wall and the impinged jet are neglected, conservation of mass and linear momentum and a detailed analysis of the impact region yield the width and thickness of the wall jet [83]. Instead of using the results of the complete analysis of impinging inviscid jets [254], Naber and Reitz [255] postulated that the thickness of the wall jet depends on the conservation of mass and linear momentum and assumed that the droplets impinging on the solid walls follow trajectories tangent to the wall surface. Their approach neglects the friction forces between the liquid and the wall, breakup of the impinging jet as it impacts on the wall and the buildup of a continuous liquid film that wets the wall.

Wall wetting and vaporization phenomena affect the unburnt HC emissions and soot formation. At cold start, the liquid fuel may adhere to the cold intake manifold, and fuel droplets may be present in the combustion chamber and on the walls up to ignition. Smoke levels are also found in DI engines at cold start and are influenced by the accumulation of liquid fuel on the combustion chamber walls. On the other hand, wall wetting can be used to control combustion under certain circumstances. For example, the wall jet may burn as a diffusion process whereas the burning rate is controlled by the fuel vaporization off the wall; the vaporization rate is controlled by the air swirl and by the heat transfer from the walls and from the combustion gases.

Naber and Reitz [255] showed that their wall-jet model yields better agreement with experimental data than do the stick model used in the KIVA code [158, 233, 234] and the reflect model used by Carpenter and Ramos [214, 215]. Their results also show that the wall jet creates a head vortex and that small droplets are deflected away from the wall by the gas flow induced by the wall jet, whereas the larger droplets with correspondingly higher momentum impinge on the wall surface and then move along the wall.

The numerical results of Naber and Reitz [255] indicate that the width and height of the wall jet decrease as the gas density and the wall inclination angle are increased, and there is very little interaction between the impinging spray and the wall jet at high impingement angles. Their results also show that the details of the wall interaction model do not affect the spray shape, and the amount of liquid fuel that hits the wall decreases as the engine swirl or the gas density are increased.

3.6 OTHER MULTIDIMENSIONAL MODELS

The models presented in the previous sections account for the presence of the liquid fuel spray. In this section we discuss some models that have been applied to the DISC engine but that treat the spray as a gaseous fuel jet.

Diwakar [256] used an axisymmetric model, adiabatic walls, a constant eddy diffusivity, a mean-reaction-rate model for the oxidation of octane to carbon dioxide and water vapor, and the extended Zel'dovich mechanism for the NO formation to show that the fuel burns more rapidly and completely in the model than in the experiments. The predicted peak pressures, the NO concentrations, and the HC emissions were lower than the experimental data (Fig. 3.52). Diwakar [256] also showed that the exhaust gas emissions result from bulk quenching of the lean mixture emanating from the piston bowl and traveling outward during the expansion stroke.

Diwakar [257] used a two-equation model of turbulence and logarithmic laws of the wall to calculate the flow field and combustion in a DISC engine

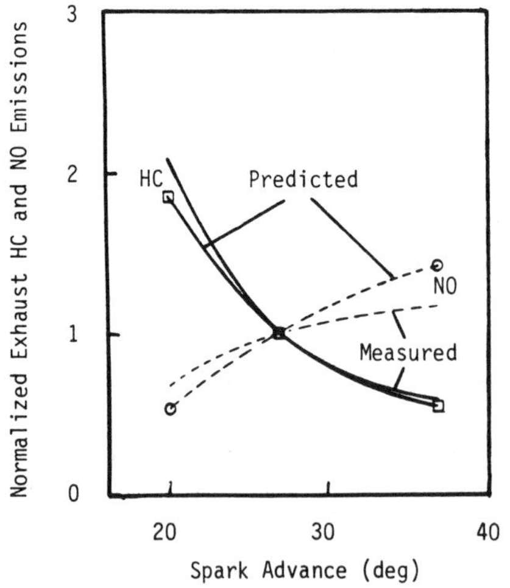

Figure 3.52 Normalized unburnt hydrocarbon and nitric oxide emissions as a function of the spark advance [256].

and showed that a two-equation model of turbulence yields better agreement with experimental data than does a constant eddy diffusivity model [256].

Benjamin et al. [258] developed a mathematical model to predict the flow field, heat transfer, and combustion in a stratified-charge engine. Their model tracks the flame front explicitly, and the turbulent flame speed is related to the laminar flame speed and turbulent velocity rms by means of an experimental correlation. Combustion behind the flame front was modeled by means of a one-step overall reaction. Their model uses transport equations for the fuel mass fraction and for the mixture fraction. Benjamin et al. [258] compared the numerical predictions for a divided-chamber engine operating under motored conditions and obtained good agreement with experimental data.

Gupta [259], Gupta et al. [260], Syed and Bracco [261], and Grasso and Bracco [262] used a mean-reaction-rate model and an eddy diffusivity proportional to engine rpm, a mean-reaction-rate model and a two-equation model of turbulence, and an eddy-breakup model and a two-equation model of turbulence to analyze the flow field and combustion in divided-chamber stratified-charge engines. Gupta et al. [260] showed that it was necessary to account for the turbulence generated by the prechamber jet in order to reproduce the combustion in the main chamber properly. Syed and Bracco [261] used an ignition model that burned the fuel contained in a computational cell in a finite time, adjusted the ignition time so that the predicted flame appears in the main chamber at the same time as the measured one, and obtained better agreement with experimental data than did Gupta et al. [260]. The mixing-controlled model employed by Grasso and Bracco [262] cannot account for combustion near cold walls, self-ignition and pollutant emissions, and rich and lean flammability limits because these processes are controlled by chemical kinetics. Grasso and Bracco [262] depleted the fuel in a computational cell in order to simulate ignition.

Gupta and Bracco [263] showed that if the prechamber pressure increases over a time that is short compared with the acoustic time in the main chamber, a pressure imbalance results that causes pressure oscillations. The decay of these oscillations depends on the throat Mach number and the strength and extent of the first compresion wave in the main chamber. The amplitude of the oscillations increases with the engine rpm and combustion rate in the prechamber and with decreasing throat areas.

Meintjes and Alkidas [264] performed theoretical and experimental studies of the flow field in diesel prechambers under motored conditions; they showed that the prechamber flow is dominated by the incoming jet, which forms a vortex initially confined to part of the prechamber. As the inflow velocity decreases toward the end of the compression stroke, the vortex tends to center itself, grows, and nearly fills the chamber.

Figure 3.53 shows a streak velocity representation of the flow field along the cylinder axis in a prechamber geometry [264], and Fig. 3.54 illustrates the flame propagation in a prechamber engine [260].

Boni et al. [265] used a turbulent diffusivity proportional to the cylinder bore and piston velocity to simulate combustion in a stratified-charge engine.

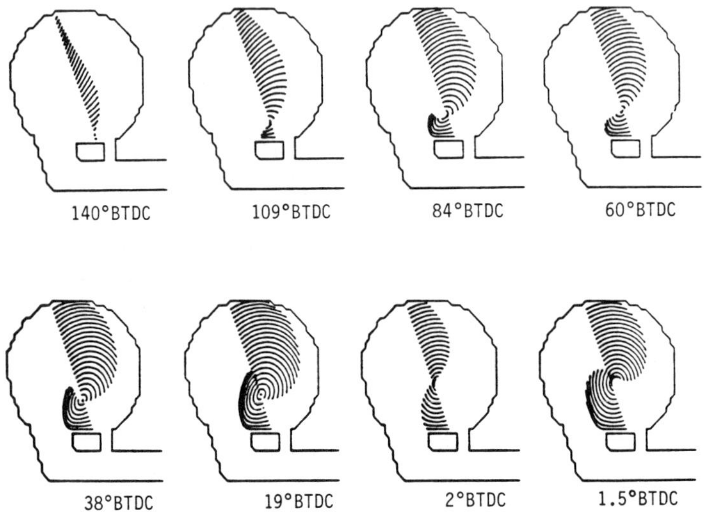

Figure 3.53 Velocity representation of the flow field along the injector axis in a prechamber geometry [264].

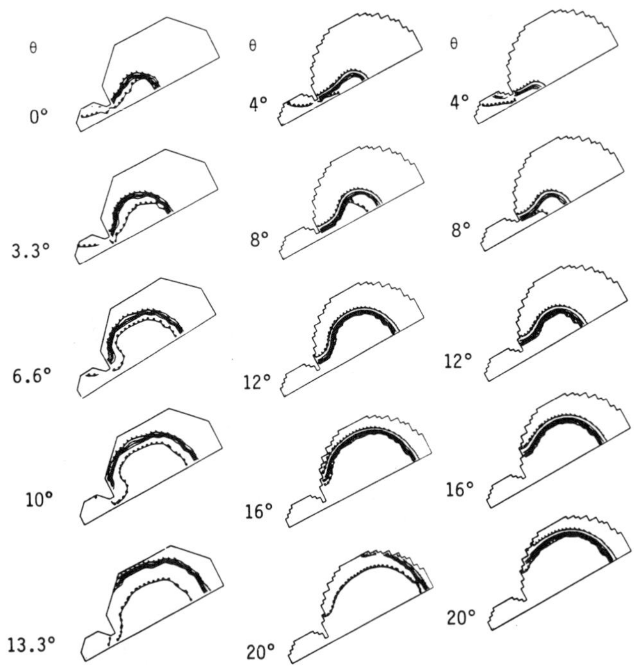

Figure 3.54 Flame propagation in a prechamber engine (left: 1030 rpm, overall equivalence ratio = 1.26, orifice area = 1.1 cm^2; center: 1110 rpm, equivalence ratio = 1.61, orifice area = 0.705 cm^2; right: 1520 rpm, equivalence ratio = 1.5, orifice area = 1.1 cm^2) [260].

Boni et al. [266] developed a two-dimensional axisymmetric model of stratified-charge engine combustion and used a global methane-air reaction and a constant eddy diffusivity. Their studies show that unburnt methane is found in the pre-chamber of a divided chamber engine at the end of combustion.

The propagation of flames through stratified mixtures exhibits very interesting physical phenomena. Ishikawa [267] showed that a flame propagating perpendicularly to concentration gradients is decelerated and may even be quenched. The ignition energy, local equivalence ratio, and concentration gradients play an important role in such quenching phenomena. Ishikawa [268] also showed that the induction distance in stratified mixtures depends on the amount of inerts in the fuel.

Spalding [269] developed the "shadow" method for two-phase combustion, which uses a local average fuel-particle size when the fuel and gas have different velocities. The shadow method does not seem to have been applied as yet to predict the flow field in diesel engines.

3.7 DIESEL ENGINE EMISSIONS

In this section we present a brief review of diesel engine emissions, which consist mainly of smoke, particulates, odor, NO_x, CO, and unburnt HC and are controlled by fuel-air mixing. Unfortunately, no current available model can satisfactorily predict diesel engine emissions [270].

Soot and carbonaceous particulates are formed in combustion processes and have carbon as the main constituent. Carbonaceous particulates that form from gas-phase processes are generally referred to as soot, and those formed from the pyrolysis of liquid hydrocarbon fuels are generally referred to as coke. Soot particulates in diesel engines absorb carcinogenic materials and increase the radiative power of the flame and the heat transfer rates.

The soot formation is related to the fuel type and flame in the combustion chamber; diesel engines are more smoky than spark-ignition engines. The soot emissions from a diesel engine are influenced by the spray atomization, combustion chamber geometry, air motion, turbulence, pressure, droplet size, swirl, spray cone angle, etc.

The soot propensity of premixed and diffusion flames can be classified as follows [271]:

Premixed flames: naphthalenes > benzenes > alkanes > alkenes > alkynes

Diffusion flames: naphthalenes > benzenes > alkynes > alkenes > alkanes

However, by controlling the temperature of sooting diffusion flames with the addition of additives, C_4 and C_5 olefins have the greatest tendency to soot, and the great sooting propensity of acetylene seems to be due entirely to its high flame temperature [271].

The fuel injected into the combustion chamber of a diesel engine undergoes pyrolysis and creates a precursor element from the pyrolysis components. These elements form precursors that could be aromatic. The aromatic structure grows due to nucleation reactions and forms condensed-phase materials, which subsequently absorb high-molecular-weight hydrocarbon compounds. The parent aromatic hydrocarbon can also undergo fragmentation which yields CH_x, C_2H_x, C_3H_x, polycyclic aromatic hydrocarbons (PAH), etc., which in turn form soot particles. Frenklach et al. [272] postulated that the pyrolysis of the aromatic hydrocarbons is caused by a free radical polymerization process initiated by the fragmentation of the aromatic hydrocarbon. The virgin soot particles undergo dehydrogenation and cyclization as well as simultaneous agglomeration, conglomeration, and thermal ionization.

The nucleation of soot particles in a flame can be described as a chemical nucleation process in the sense that small molecules add to molecules with increasingly large carbon numbers in a chain of reactions that leads to a macroscopic solid particle. Kennedy [273] studied the effects of strain (i.e., velocity gradients) on nucleation kinetics, in which the formation of condensable monomers is rate controlling. He also studied nucleation kinetics when the cluster growth reactions are rate controlling and showed that the fluid mechanics have a significant effect on the nucleation process.

The soot-formation process can be divided into two stages: soot particle formation and soot formation. Particle inception or formation can be defined as the first condensed-phase material that arises from the oxidation and/or pyrolysis of the fuel molecules. The pyrolysis products include unsaturated hydrocarbons, particularly acetylene, and PAH. These two types of molecules are considered the most likely precursors of soot in flames.

The time elapsed from the fuel decomposition to the appearance of the first soot particles in C_2 hydrocarbons can be written as

$$\tau = A[\mathrm{HC}]^{-n} \exp\left(\frac{E_A}{\tilde{R}T}\right) \tag{332}$$

where A and n are constants, [HC] denotes the concentration of the C_2 hydrocarbons, E_A is the activation energy, and τ denotes the induction time [274].

The soot particles can grow by surface condensation and coagulation. Coagulation is due to collisions and coalescence, and the coagulation rate depends on the soot volumetric fraction, particle size, density and temperature, and number of particles per unit volume. In turbulent flows where the average eddy size is much larger than the diameter of the coagulating particles, the rate of change of the particle number density depends on the dissipation rate of turbulence kinetic energy and on the kinematic viscosity of the fluid phase.

After the particle growth ceases, continued coalescence of the soot particles results in the formation of chains. The soot particles and chains are oxidized in those regions of the flame where oxidizing species are present, at a rate that depends on the temperature and partial pressure of oxygen, that is,

$$\dot{\omega}_o = M_c \left[\frac{xK_a p_{\mathrm{O}}}{1 + K_z p_{\mathrm{O}}} + K_b p_{\mathrm{O}}(1-x)\right] \tag{333}$$

where

$$x = \left[1 + \frac{K_t}{k_b\, p_O}\right]^{-1} \tag{334}$$

$\dot{\omega}_o$ is the soot-oxidation rate, M_c is the molecular mass of atomic carbon, p_O is the oxygen partial pressure, and K_a, K_b, K_t, and K_z are empirical constants that depend on the temperature [274, 275].

The soot-formation rate can be expressed as [94]

$$\dot{\omega}_f = A_f\, m_{fg}\, p^n \exp\left(-\frac{E_f}{\tilde{R}T}\right) \tag{335}$$

where A_f is the preexponential factor, m_{fg} is the mass of fuel vapor, p is the pressure, n is a constant, and E_f is the activation energy [276]. Typical values of A, n, and E_f are 8×10^9 s^{-1} atm^{-1}, 1, and 84,200 cal/mol · K, respectively.

The coagulation rate can be estimated by assuming Brownian motion and particle diameters that are small compared with the mean free path, by means of the following equation [274]:

$$\frac{dN}{dt} = -\frac{6}{5} K_{th}\, \phi^{1/6} N^{11/6} \tag{336}$$

where

$$K_{th} = \frac{5}{12}\left(\frac{3}{4\pi}\right)^{1/6}\left(\frac{6KT}{\rho_p}\right)^{1/2} G\alpha \tag{337}$$

ϕ is the particle volume fraction, ρ_p is the particle density, N is the number of particles, K is Boltzmann's constant, G is a factor that accounts for interparticle dispersion ($G \approx 2$ for spherical particles), and α depends on the collision rates of particles of different sizes.

Khan et al. [277] studied the coagulation of spherical soot particles by means of the following equations:

$$n^{-1} = n_0^{-1} + Ct \tag{338}$$

$$\frac{\rho_p\, d_0}{6m_0^{1/3}\, m^{2/3}} \frac{dm}{dt} = K \exp\left(-\frac{E}{\tilde{R}T}\right) p_O\, T^{-1/2} \tag{339}$$

where n is the particle number density, C is a coagulation constant, n_0 is the particle number density at time $t = 0$, ρ_p is the soot density, d_0 is the initial diameter of the soot particles or coagulates, m is the soot concentration, and m_0 is the initial soot concentration. The values of C, K, and E in Eqs. (338) and (339) are $C = 1.4 \times 10^{-4}$, $K = 1.085 \times 10^4$, and $E = 39{,}300$ cal/mol · K [276, 278].

Equations (338) and (339) correspond to the coagulation and surface reaction of soot. Khan et al. [277] assumed that soot is present at TDC and that 250×10^{-8} cm soot particles are uniformly distributed throughout the combustion-chamber.

D'Anna [276] used the following expression to determine the soot-oxidation rate:

$$\dot{\omega}_o = -S\, K_0\, p_{\mathrm{O}} \exp\left(-\frac{40{,}300}{\tilde{R}T}\right) \tag{340}$$

where $K_0 = 7.5 \times 10^4\, T^{-0.5}$ and S is the surface area of the soot particles.

Equations (335) and (340) can be used to calculate the soot-formation and -oxidation rates by assuming that soot coagulation results in spherical particles having diameters that affect the oxidation rate as indicated in Eq. (340).

In order to determine the diameters of the coagulated soot particles, it is necessary to specify their initial size and number. Electron micrographs indicate that the elementary soot particles, of which chains are formed, are spherical with diameters in the range of 50×10^{-8} cm $< d <$ 500×10^{-8} cm [274].

In Sections 3.7.1 and 3.7.2 we present a brief description of the particulates emitted by diesel engines. For the sake of convenience, carbonaceous particulate emissions are classified into two groups: soot and carbonaceous particulates. Soot is formed in gas-phase processes, whereas carbonaceous particulates are formed from the liquid hydrocarbon itself. For example, the impingement of liquid fuel sprays onto hot walls before they are substantially oxidized produces liquid-phase cracking, subsequent pyrolysis, and coking of the hydrocarbons. The resulting carbon particulate is called coke and is harder than the soot formed in gas-phase processes. Fuels containing aromatics form coke more readily than do other fuels.

3.7.1 Soot

Nondilute sprays produce fuel-rich vapor regions that undergo chemical reactions and may form soot. Soot is also formed on liquid films along the cylinder walls. If the fuel droplets are electrically charged, they will repel each other, increasing the fuel dispersion and reducing fuel-rich regions and soot formation and emissions [279].

Peterson and Alkidas [280] performed optical studies in simulated diesel engine prechambers and showed that the injected fuel is displaced by the flow and hits the wall opposite the injector. They also showed that ignition and combustion occur at times when liquid fuel and vapor exist in the prechamber and that liquid fuel is located within the volume and on the surfaces of the prechamber and produces luminous combustion associated with soot.

Flower [281] showed that the volumetric rate of soot formation in laminar diffusion flames increases with pressure. Rah et al. [282] correlated NO and soot formation with the ignition of liquid fuel droplets; they showed that soot formation can be suppressed by delaying ignition and by vaporizing as much fuel as possible prior to ignition. However, this favors the NO formation. Rah et al. [282] proposed the use of low oxygen concentrations in the oxidation gas to delay ignition and high fuel-air ratios to reduce NO_x as a possible method of controlling the NO_x and soot emissions.

Uyehara [283] showed that 2000 K $\leq T \leq$ 4000 K is the temperature range in which diesel engine flames contain soot and that the maximum soot formation occurs at 2150 K. He also showed that the flame contains negligible soot at 1950 K and 2400 K.

Soot and carbonaceous particulates may also be formed if there is extinction of the liquid fuel droplets [284]. The ignition of fuel droplets is kinetically controlled in cold environments, and it is droplet-heating controlled in hot environments [285]. The droplet heat-up period normally constitutes a large proportion of the total droplet lifetime. Therefore, calculations that ignore the heat-up period are prone to serious errors, particularly at high pressures [286], and may not accurately predict soot formation and emissions. In addition droplet ignition [287] and decomposition [288] influence the local equivalence ratio, the fuel-air mixing, and emissions from diesel engines. Some droplets may cross the flame envelope around diesel engine sprays and exhibit secondary diffusion flames, which can generate soot [289, 290].

Kagami et al. [291] studied the effects of the fuel viscosity, distillation range, and cetane number on the performance, gaseous and smoke emissions, knock, fuel consumption, startability, and cold performance of diesel engines and showed that fuels with low cetane numbers are associated with increased emissions and fuel consumption.

Tateishi et al. [292] developed a new combustion system for IDI diesel engines that reduces the fuel consumption, emissions, and noise. Engine noise is caused by the early stages of combustion, when there is rapid heat release [293].

Nitric oxide is formed in the postflame reactions that occur in the gas phase, whereas aldehydes are formed ahead of the flame and are associated with odor [294].

Callahan et al. [295] showed that the emissions of CO and particulates with torque transients increase with accelerations and decrease with decelerations. In steady state, rich mixtures produce high torques and increase the emissions of particulates, unburnt HC, and CO; however, the emissions of particulates and CO decrease with the engine speed because of better fuel-air mixing. Imperfect fuel-air mixing results in mixture stratifications and in rich and lean fuel regions; the rich fuel regions produce more soot than do the lean fuel zones [296].

Light-scattering methods and polarized argon-ion lasers can be used to measure the diameter and number density of soot particles in diesel engines [297]. Measurements in the exhaust port are subject to errors due to oxidation of the soot particles and unburnt HC. Sampling probes in the engine cylinder produce flow perturbations, decrease the mean velocity, increase the local mixing rates, and produce chemical effects if ionization probes are used to sample the cylinder charge [298].

Sinnamon and Cole [299] used sampling probes to analyze the emissions from a prechamber stratified-charge engine and showed that the prechamber jet induces turbulence that increases the fuel-air mixing rates and reduces the combustion interval. They also showed that an increase in the degree of stratification results in high NO and CO emissions for lean overall fuel-air ratios.

Turns and Borman [300, 301] showed that sampling probes induce flows around their tips and result in quenching bias, which is relatively unimportant in high-temperature regimes. They also showed that the flame temperature has the largest influence on the mass fraction of quench gases.

Yanagi [302] showed that changes in the flow and concentration gradients affect the composition of nonisokinetic sampling techniques. He also showed that large probes produce large errors and that the relative error decreases with an increase of the sampling rate within a finite range and with a decrease of the probe size.

Yamazaki et al. [303] showed that for injection temperatures above the boiling temperature, the mean droplet size decreases, the spray spreading rate increases, the thermal efficiency of the engine increases at high loads, the smoke density decreases at high loads, and the engine noise increases due to the long duration of the premixed combustion period and to shorter combustion duration. Oza and Sinnamon [304] showed that preheating of the fuel in the injector produces an expansion of the fuel vapor. At low temperatures, the liquid jet breaks up in the combustion chamber, whereas at high temperatures, there is boiling in the injector, and an underexpanded compressible two-phase flow is formed that results in a larger spray angle than that observed in liquid fuel sprays.

Soot formation in premixed gases is believed to be due to the formation of chemiions, which act as precursors on which free radicals, polyacetylenes, and polycyclic aromatic hydrocarbons (PAH) repeatedly add in fast ion-molecule reactions [305]. Olson and Pickens [306] studied the effects that the molecular structure of 55 pure hydrocarbons at atmospheric pressure have on soot formation; they showed that the soot thresholds depend on the equivalence ratio and the carbon-to-hydrogen atom ratio in the fuel. Olson and Madronich [307] showed that the soot volume fraction is a function of the flame temperature, but it is not a function of the fuel-to-oxygen and oxygen-to-nitrogen-plus-oxygen ratios.

Bernhardt [308] proposed use of the waste heat from the exhaust system to improve the preparation and distribution of the fuel-air mixture in order to decrease soot formation, exhaust odor, and NO_x and aldehyde emissions. The use of waste heat extends the lean-burning limit, particularly in alcohol-fueled and alcohol-gasoline-fueled engines. Shimoda et al. [309] used oil films to measure the flow direction near solid walls and LDV and schlieren photography to determine the mixture formation and burning in a DI diesel engine; they showed that squish increases the fuel-air mixing, decreases the ignition delay time, and results in better fuel consumption and lower smoke levels.

D'Anna [276] used a thermodynamic model of diesel engine combustion to determine the soot formation and oxidation rate. His model employs Eq. (141) and assumes that the liquid fuel droplets maintain their spherical symmetry throughout the vaporization and combustion processes. D'Anna's model [276] is based on a one-step overall chemical reaction and a d^2 law that accounts for forced-convection and droplet-interference effects. This model can be used to predict droplet ignition by solving the time-dependent one-dimensional conservation equations of mass, energy, and species mass fractions in the gas phase, and the energy equation in the liquid phase, and by defining ignition at the time at which $d^2T/dt^2 = 0$ in the gas phase [310]. Once ignition occurs, the gas-phase phenomena can be assumed to take place in a quasisteady manner.

The solution of the gas- and liquid-phase equations provides the distribution of the gas- and liquid-phase temperatures and the mass fractions of oxygen, fuel vapor, and combustion products as a function of the radial coordinate and time. These results can be used to determine the formation, coagulation, and oxidation of spherical soot particles by means of Eqs. (335), (338), and (340).

Experimental results on soot formation and oxidation in droplet burning [311] show that the instantaneous amount of soot follows the same trend as the instantaneous flame size. Almost complete oxidation of soot can be achieved by confining it within the closed flame, and weak convection promotes oxidation while early extinction can lead to substantial soot emissions. In view of these experimental observations, it can be postulated that soot formation occurs in the immediate neighborhood of the flames surrounding each individual liquid droplet. The flame location and temperature can be calculated by assuming that each droplet is spherically symmetric and burns as an isolated entity. The flame temperature and oxygen concentration can be used to determine the soot formation, coagulation, and oxidation by means of Eqs. (335), (338), and (340). Forced-convection effects and droplet interactions can be accounted for by modifying the constant that appears in the d^2 law, whereas the exchange of linear momentum between the liquid droplets and the gas phase can be calculated by means of, for example, Eq. (278). The effects of the gas-phase turbulence on the droplets can also be included by using models similar to those presented in Section 3.5.

3.7.2 Carbonaceous Particulates

Particulates are produced by fuel pyrolysis during incomplete combustion, and they contain polycyclic aromatic hydrocarbons (PAH), which may be mutagenic or carcinogenic. Particulate emissions can be reduced by increasing the temperature; however, as the temperature increases, the NO emissions increase.

Obuchi et al. [312] showed that the CO emissions and particulates in a diesel engine increase near the maximum rpm and maximum torque, whereas the NO emissions are independent of the engine rpm but increase as the torque is decreased.

The amount of liquid and solid particulates in diesel engines depends on the fuel-air mixing, advance of injection, load, etc. Iida et al. [313] observed that at light-loads, there was liquid particulate coagulation at the exhaust valve and at heavy loads, there was solid particulate coagulation.

Carbonaceous particulates have been observed in diesel engines for air-to-fuel ratios lower than 20; at higher air-to-fuel ratios, the volatile content increases [314]. Platinum-coated combustion chambers reduce the gaseous and particulate emissions from diesel engines, but they do not produce significant changes in the NO, CO, and unburnt HC emissions [315].

Veselinovic [316] showed that swirl in diesel engines may increase the amount of smoke. Van Gerpen et al. [317] showed that high swirl levels in diesel engines increase the amounts of unburnt HC emissions and particulates but have a small effect on ignition delay.

Van Gerpen et al. [317] also showed that the NO_x-particulate trade-off is favorably influenced by increases in the injection pressure. High swirl levels can increase the emissions of both unburnt HC and particulates for small-hole nozzles, while NO_x can reach a maximum at intermediate swirl levels. Van Gerpen et al. [317] also observed that the effects of swirl and injection parameters on ignition delay are small. These parameters can influence the temperature and pressure of the charge at the time of injection, but have little effect on the ignition delay time.

Nakanishi et al. [318] showed that the particulate concentration increases with increases in the swirl levels and air temperature, whereas the combustion duration decreases. Their experimental results also show that NO_x forms in the early stages of combustion, when particulates are also formed. The NO_x concentration does not change much in the late stages, where particulate oxidation is important.

The experimental results of Van Gerpen et al. [317] show that the NO_x, unburnt HC, and particulate emissions from diesel engines are determined by the fuel-air mixing, air motion, and injection parameters.

Hiroyasu et al. [319] showed that the swirl levels in a DI diesel engine are not uniform; they are highest in the piston bowl and almost constant between the piston and the cylinder head. The interaction between squish and swirl may result in good fuel-air mixing and reduce the combustion duration and exhaust smoke [320].

Norris-Jones et al. [321] showed that fuel burning from droplets and off-the-wall conditions yields more particulates than does fuel burning from fully evaporated conditions. They also showed that particulates are formed in the cylinder and that their number can be reduced by improving fuel atomization.

Du and Kittelson [322] showed that particulate coagulation and oxidation in diesel engines occur in the expansion stroke.

Figure 3.55 shows the Bosch smoke numbers in DI and IDI engines as a function of the diffusive combustion duration parameter θ_d. This figure indicates that the smoke levels increase with the duration of the diffusive burning [56]. Figure 3.56 shows that the NO emissions increase and the soot emissions decrease with increases in the swirl ratio, injection pressure, and injection timing [94].

3.7.3 Radiation Heat Transfer

In Sections 3.2.2 and 3.5.3 we described some of the correlations used to calculate convective and radiative heat transfer in reciprocating engines. These correlations [Eqs. (49), (50), and (55)] do not account for the spatial and temporal

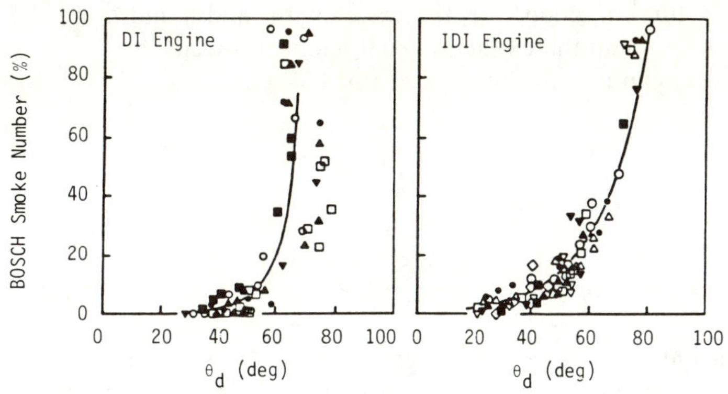

Figure 3.55 Bosch smoke number in direct-injection and indirect-injection engines as a function of the diffusive combustion duration parameter [56].

variations of radiative heat transfer and soot. In this section we describe some of the models that have been used to calculate radiative heat transfer. These models are thermodynamic in nature.

Zhou et al. [323] developed a multizone combustion model to predict the soot concentrations and emissivities. Gas emissivity mainly from carbon dioxide and water vapor was computed with a wide band model, whereas the soot emissivity was determined from a Rayleigh limit expression.

The spectral emittance of an isothermal luminous gas volume composed of soot in a nonradiating carrier gas can be written as

$$\epsilon_{\lambda s} = 1 - \exp(-K_{\lambda s} L) \tag{341}$$

where L is the optical length of the soot-laden gas and $K_{\lambda s}$ is the volumetric absorbance of soot, which for small particles can be expressed as [324]

$$K_{\lambda s} = 36nk\left(\frac{\pi}{\lambda}\right)\frac{f}{[n^2 - k^2 + 2]^2 + 4n^2k^2} \tag{342}$$

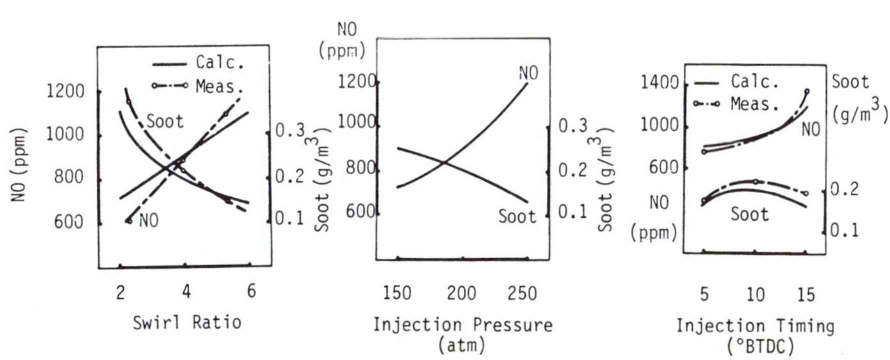

Figure 3.56 Effects of the swirl ratio (left), injection pressure (center), and injection timing (right) on nitric oxide and soot emissions [94].

where f is the soot volume fraction, λ is the wavelength, and n and k are the simple index of refraction and the extinction coefficient, respectively.

The soot emissivity can be calculated from Eq. (341) as

$$\epsilon_s = \int_0^\infty \epsilon_{\lambda s} E_b \frac{d\lambda}{\sigma T^4} \tag{343}$$

where

$$E_b = 2\pi h \frac{c^2}{\lambda^5}\left[\exp\left(\frac{hc}{\lambda KT}\right) - 1\right]^{-1} \tag{344}$$

h is the Planck constant, c is the speed of light, K is the Boltzmann constant, σ is the Stefan-Boltzmann constant, and the subscript b denotes blackbody.

Substitution of Eqs. (341), (342), and (344) into Eq. (343) yields

$$\epsilon_s = 1 - \frac{15}{\pi^4}\sum_{i=1}^{\infty}[s(3, a_i, x_i) - s(3, a_i, x_{i+1})] \tag{345}$$

where

$$s(3, a_i, x_i) - s(3, a_i, x_{i+1}) = \int_{x_i}^{x_{i+1}} e^{-a_i x}\frac{x^3}{e^x - 1}\,dx \tag{346}$$

$$x = \frac{hc}{\lambda KT} \qquad a = 36\pi n\, k\, f\, L \frac{KT}{hc}\{[n^2 - k^2 + 2]^2 + 4n^2 k^2\}^{-1} \tag{347}$$

If the gases contained in each spray zone are assumed to be CO, CO_2, O_2, H_2O, N_2, H_2, O, NO, H, OH, and N except in the air and liquid fuel zones, the dominant radiation characteristics are those of CO_2 and H_2O, which are located in bands. The emissivity of gas j in band i can be expressed as

$$\epsilon_{g_{j,i}} = \frac{15}{\pi^4}\int_{\text{band}_i}\frac{x^3}{e^x - 1}\,dx = \frac{15}{\pi^4}[s(3, 0, x_{l,i}) - x(3, 0, x_{u,i})] \tag{348}$$

where $x_{l,i}$ and $x_{u,i}$ are the lower and upper limits of the ith band.

The total gas emissivity can be obtained from Eq. (348) as

$$\epsilon_g = \sum_i^M \sum_j^N \epsilon_{j,i} \tag{349}$$

where M and N denote the number of bands and gases, respectively.

In order to calculate the soot and gas emissivities, it is necessary to estimate the optical mean length L_j and soot volumetric fraction f_j in each zone. The optical mean length can be calculated from the following expression:

$$L_j = 3.6\frac{V_j}{S_j} \tag{350}$$

where V_j and S_j are the volume and surface area of the jth zone, respectively.

The volumetric fraction of soot in the jth zone can be calculated as

$$f_j = \frac{c_{s_j}}{\rho_s V_j} \tag{351}$$

where ρ_s is the soot density, and the soot concentration in the jth zone c_{s_j} can be obtained from Eq. (126).

The radiative heat transfer from the walls can be estimated as

$$q_w = \epsilon_w \sigma (T^4 - T_w^4) \tag{352}$$

where ϵ_w is the wall emissivity. However, the walls absorb part of the energy and reflect the rest. In order to account for the reflected energy, an average wall temperature is used, $\langle T_w \rangle$, such that

$$\langle T_w \rangle = \frac{\sum_{j=1}^{J} A_{w_j} T_{w_j}}{\sum_{j=1}^{J} A_{w_j}} \tag{353}$$

where J is the number of zones and A_w and T_w are the surface area and wall temperature, respectively.

The total energy emitted by the jth zone is

$$\text{TE}_j = \epsilon_j \sigma T_j^4 \tag{354}$$

The wall absorbs part $(\epsilon_w \cdot \text{TE}_j)$ of this energy and reflects the rest $[(1-\epsilon_w)\ \text{TE}_j]$ so that the radiative heat exchange between the wall and the jth zone is

$$Q_{j,w} = \epsilon_w \text{TE}_j - \epsilon_j \cdot A_j \left[\sigma \epsilon_w T_w^4 + (1-\epsilon_w) \frac{\sum_{j=1}^{J} \text{TE}_j}{\sum_{j=1}^{J} A_j} \right] \tag{355}$$

Zhou et al. [323] showed that Eqs. (126) and (341)–(355) accurately predict the soot emissions as well as gas and soot radiation from low-heat-rejection diesel engines.

Whitehouse et al. [325] divided the cylinder into a large number of elements including the combustion chamber walls, and assigned the gas composition, soot concentration, and temperature to each zone according to equivalence ratio maps. Their model computes the transmittance along radiation paths and view factors, and it shows that radiative heat transfer peaks at a high rate during the main combustion period due to the presence of hot soot particles. Their model also indicates that radiative heat transfer may be important in reducing the maximum temperature of the hottest zones. However, radiative heat transfer calculations are seldom practicable because of the lack of information on soot concentrations.

Chen et al. [326] used a spherical flame radiation model to calculate radiative heat transfer in diesel engines, which accounts for the radiant energy emitted by each zone, the radiant energy emitted by the wall, which reaches the ith zone, and the radiant energy emitted by the surrounding zones, which reaches the ith zone.

Ovsjannikov [327] showed that radiative heat transfer is significant only at the beginning of visible combustion and is much less than convective heat transfer for most of the engine cycle.

The emissivity of the gases and soot in diesel engines depends on the optical properties, size, form, concentration, and temperature of the soot particles. The optical properties of the cylinder mixture can be characterized by scattering, absorption, and transmittance coefficients. Currently, no theory accounts for all these effects although some efforts in this direction have been made by Kostin and Rudnev [328].

Chapman et al. [111] used the Cummins engine model to study radiative heat transfer in diesel engines. They assumed that soot behaves as a gray body, employed a zonal model to calculate the view factors and heat transfer rates, assumed that the cylinder gases are optically thin, and showed that the primary radiation source is from soot formed in fuel-rich regions.

Figure 3.57 shows the radiative heat fluxes on the cylinder head and piston at 20°ATDC predicted by first-order P_1 and third-order P_3 spherical harmonics approximations to the radiative heat transfer equation [52].

3.8 DIESEL ENGINE NOISE

Diesel engine noise is caused by the initial rapid heat release in the early part of combustion [329]. Combustion and turbulence cause pressure fluctuations, which in turn may produce noise.

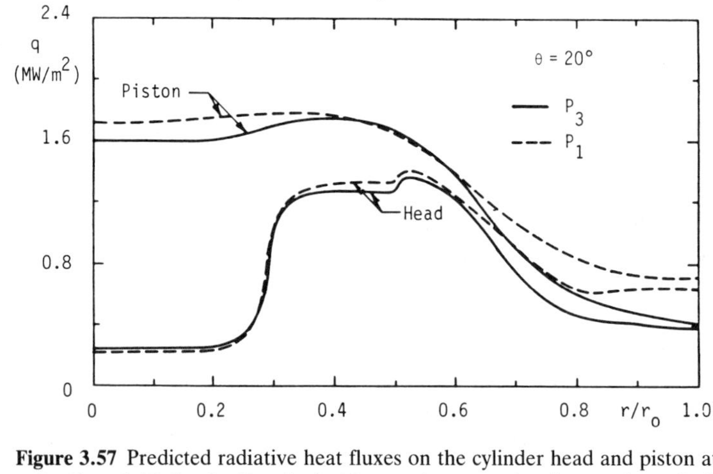

Figure 3.57 Predicted radiative heat fluxes on the cylinder head and piston at 20°ATDC with first-order P_1 and third-order P_3 spherical harmonics approximations to the radiative heat transfer equation [52].

Strahle et al. [330] studied the cetane rating and load effects on combustion noise in a single-cylinder air-cooled engine in order to determine the connection between combustion randomness and radiated noise. They found that the radiated noise is not related to the mean pressure-time history and that at high loads, the effect of the randomness (turbulence) on the total noise output is small.

Strahle [331] showed that engine turbulence and cycle-to-cycle variations account for a substantial amount of noise radiated from diesel engines. Strahle [331] used a three-dimensional time-dependent formulation and assumed that the heat released by combustion can be decomposed into periodic and random components. He found that randomness dominates the periodic part of the pressure-time history for frequencies above 1000 Hz.

Strahle et al. [332] showed that the combustion process in a diesel engine causes random pressure fluctuations in the cylinder pressure; these fluctuations are related to noise radiation.

3.9 MEASUREMENTS IN DIESEL ENGINES

Combustion in a DI diesel engine is closely related to the transient behavior of the fuel-air mixing processes, including atomization and vaporization at the spray edge. Accurate descriptions of the behavior of atomization and vaporization processes are very important if fuel-air mixing and combustion are to be understood.

In this section we describe some of the available measurements in diesel engines. For the sake of convenience these experimental data are classified in two groups: velocity measurements and droplet size measurements. Velocity data in diesel engines have been obtained by means of hot wires and laser Doppler anemometry in cold and reactive flows. Droplet size measurements and breakup lengths have been obtained with voltage, holography, laser light scattering, Mie scattering, and other techniques. However, very little information is available concerning the droplet size distribution near the nozzle exit because this region is nontransparent. This quality makes it difficult to obtain accurate velocity and droplet size measurements near the nozzle exit that can be compared with the predictions of multidimensional models.

Section 3.9.1, on velocity measurements, includes cold and reactive flows. In particular, it covers what is known about the effects of swirl and piston bowls on the engine flow field, turbulence, and mixing. It also describes the effects of the inlet port geometry on the in-cylinder flow field. Other velocity measurements in motored and firing engines are presented in Chapter 4.

3.9.1 Velocity Measurements

Wigley et al. [333] used laser Doppler anemometry to measure swirl velocities in a firing diesel engine provided with a quartz window, and they showed that the light scattered from the fuel plumes and the light emitted during the combustion process deteriorate the quality of the laser Doppler velocimetry (LDV) measurements. They also showed that combustion increases the velocity rms.

Johnston [334] obtained spray measurements prior to combustion in a diesel engine by means of a continuous wave (CW) laser, Mie scattering, and vibrational Raman scattering in order to distinguish between the liquid and vapor fuel. Mie scattering makes it possible to obtain time-averaged number density measurements near the spray boundary, but it is difficult to separate the effects of the liquid and fuel vapor.

Nishida et al. [335] showed that premixed combustion in the early stages of diesel engine combustion increases the swirl levels and that there is a double swirl structure in the piston bowl.

Arcoumanis et al. [336] performed LDV measurements in a four-stroke motored engine operating at 200 rpm and a compression ratio of 3.5. They showed that vortices are created during the intake stroke; these vortices disappear by the time that the intake valve closes. They also showed that the compression stroke is characterized by vortex formation and breakup and that the turbulence levels remain constant during the compression stroke (i.e., the compression-generated turbulence is small).

Arcoumanis et al. [336] observed that turbulence is generated during the exhaust stroke and that a bowl with reentrant lip results in more squish and higher turbulence levels near the piston than a bowl without lip.

Arcoumanis et al. [337] showed that the intake-generated vortices almost disappear in the compression stroke and that the turbulence kinetic energy decays in the intake stroke but that it is uniform and constant in the compression stroke. They also observed that the intake stroke is characterized by a solid body rotation, that the velocity rms decreases in the expansion stroke, and that reverse squish in the expansion stroke results in large levels of turbulence near the piston.

Arcoumanis et al. [338] showed that the interaction between squish and swirl in reciprocating engines results in solid body rotation and complex flow patterns in cylindrical and reentrant bowls, respectively. They also observed an increase in the turbulence levels in reentrant bowls and a decrease in the turbulence levels in cylindrical bowls.

The results of Fansler [339] show the presence of off-axis swirl with a moving center of rotation that varies with both depth and engine speed. Arcoumanis et al. [340] also showed that the swirl center oscillates during compression and that a shrouded valve produces a double vortex structure and more turbulence at TDC than a valve without shroud. They also showed that in engines equipped with an off-center valve, the vortical motion generated by the valve collapses in the absence of swirl or is transformed into a single vortex when swirl is present.

El Banhawy and Whitelaw [341] studied the interactions between fuel sprays and swirl; they showed that the combustion intensity decreases as the droplet diameter increases and as the swirl number decreases.

Ball et al. [342] obtained individual engine cycle measurements using a cutoff frequency of 28 kHz. They showed that large differences exist between different cycles; they also showed that the velocity rms may be 100% of the mean swirl velocity, but it is only about 30% in the piston bowl at TDC. Ball et al. [342] observed that the time scales in the piston bowl at TDC are 2°CA to 3°CA (crank shaft angle degrees), while the length scales are about 6 mm.

Catania [343] performed hot wire anemometry measurements in open-chamber diesel engines and showed that the swirl velocity profile has a solid-body rotation structure. The hot wire measurements of Wakisaka et al. [344] in motored engines showed that the turbulence at TDC is in an underrelaxed state. On the other hand, those of Tindall et al. [345] in a single-cylinder engine showed that the velocity rms peaks at half period during the induction stroke and decays during the compression stroke, indicating that in the compression stroke the dissipation rate of turbulence kinetic energy is larger than the production rate.

The LDV measurements of Morse et al. [346, 347] indicated that no pure solid-body rotation in motored four-stroke reciprocating engines is achieved until the piston reaches TDC. Rask and Saxena [348] observed that a penetrant bowl piston increases the swirl ratio and turbulence in DI diesel engines and that the throat diameter of the piston bowl is the major geometric parameter that affects the mean velocity, swirl, velocity rms, and squish.

Arcoumanis et al. [349] used laser Rayleigh scattering to measure the mean and fluctuating concentrations of a passive scalar in two-stroke motored engines. They observed that the laser Rayleigh system overestimates the mole fraction when the concentration fluctuations and the particle scattering are high; they also observed that the flow field is dominated by the high momentum of the transient jet generated during the intake stroke [350].

Arcoumanis et al. [350] observed that the flow around a valve is not uniform and that the mean axial velocity is very sensitive to the valve lift. They also observed that the turbulence intensity is almost uniform during compression but decays rapidly after 340°CA and that the mean swirl velocity exhibits axial variations in a disk chamber.

Bicen et al. [351] observed that the valve seat angle produces different flow regimes within the engine cylinder, that the flow angle depends less on the flow regimes than on the cylinder confinement, and that the in-cylinder flow characteristics are insensitive to the valve operation. Valve seat angles of 45° and 60° resulted in four and three flow regimes, respectively.

Matsuoka et al. [352] studied the interaction between squish and swirl in DI diesel engines and showed that the mean swirl velocity varies proportionally with the engine speed and that the bulk flow can be considered potential. Matsuoka et al. [353] discussed the errors introduced by velocity gradient broadening in LDV measurements and showed that the velocity rms in a diesel engine is about 20% of the mean flow velocity.

Coghe et al. [354] showed that the swirl velocities in an engine cylinder equipped with a helical port do not correspond to solid body rotation, because of the frequency response characteristics of hot wires, Coghe et al. used them to derive time scales. Their results indicate that the turbulent length scale is almost uniform in the radial direction and that large turbulence levels exist near the locations of the maximum velocity gradients. Dent and Salama [355] used Taylor's theory of frozen turbulence to calculate the Taylor microscale from hot wire measurements of microtime scales and showed that the microtime scale depends weakly on the engine speed and the combustion chamber geometry. They also showed that the integral length scale is three to four times larger than the Taylor microscale.

Tsuge et al. [356, 357] studied the decay of turbulence in closed vessels and showed the existence of two periods. In the first, or prerelaxation, period the decay is influenced by the way that turbulence was generated, whereas in the second, or homogeneous, period the decay is influenced by the vessel walls but very little by the way that turbulence was generated. Tsuge et al. [357] assumed that the turbulence is in steady state during a short interval and observed that the energy spectrum has a peak at low frequencies, indicating the presence of large-scale vortical structures. Willis et al. [358] showed that in a water model with induction swirl, the azimuthal velocity profile has a solid-body rotation near the centerline and behaves like a free potential vortex in the outer portions. The vortex becomes unstable and unsymmetrical with its center of rotation offset from the cylinder axis, and it forms a helix that extends the length of the cylinder. Willis et al. [358] also showed that the penetration of a jet in a swirling flow is a function of the injection characteristics, nozzle diameter, and swirl level.

The hot wire measurements of Huebner and McDonald [359] in piston engines with induction swirl, streamlined ports, and shrouded valves indicate that there is no solid-body rotation in the engine cylinder.

3.9.2 Droplet Size Measurements

Most of the droplet measurement systems in use today can be classified into one of three categories: extinction methods, diffraction sizing, and light-scattering techniques. In the extinction method the attenuation of a light beam is measured. This technique uses a high-intensity light source from which light is received and measured at a single detector. The extinction of the light by different droplets gives the droplet number concentration and an average droplet size. However, part of the light intensity is lost by scattering in different directions.

The presence of particles of various sizes within a laser beam causes light to be diffracted through different angles. The result is a pattern of diffraction rings of different radii; each ring is related to a particular droplet size. Diffraction techniques require a multiarray detector and a special data-reduction program to determine the droplet size distribution from the diffraction rings.

Light-scattering techniques are based on the measurement of the scattered light intensity as each individual droplet crosses a focused laser beam. The scattered light is received by a photomultiplier, and the light intensity is a measure of the droplet size. Light-scattering techniques provide the droplet number density as a function of the droplet diameter.

Holographic techniques for measuring the droplet size and fuel-air ratio in sprays were used by Polymeropoulos and Sernas [360]. Rao and Lefebvre [361] used a continuous sampling probe to determine the evaporation characteristics of kerosene sprays at atmospheric pressure, and they obtained empirical correlations for swirl atomizers. Some of these correlations may be used to predict the vaporization rate in thermodynamic models of diesel engine combustion.

Kamimoto et al. [362] used an incident-light-extinction technique to measure the droplet diameter and fuel concentration in nonevaporating diesel sprays; they developed a quasisteady one-dimensional jet model that gives good agreement with experimental data. Their experimental data show that the droplet number density function can be expressed as

$$\frac{dn}{n} = 62.5 \left(\frac{d}{\text{SMD}}\right)^2 \exp\left(-5\frac{d}{\text{SMD}}\right) d\left(\frac{d}{\text{SMD}}\right) \tag{356}$$

where n is the number of droplets with diameter d and SMD is the Sauter mean diameter.

Arai et al. [363] used an electrical-resistance technique to measure breakup lengths, photography to measure the spray angle and spray tip penetration, and laser-diffraction techniques to measure the droplet SMD. They observed that the spray angle decreases and the SMD increases if the liquid viscosity is increased, and they derived the following correlations for the breakup time and length:

$$t_{\text{bu}} = 28.65\, \rho_l \frac{d_{\text{inj}}}{(\rho_l\, \Delta p)^{1/2}} \tag{357}$$

$$l_{\text{bu}} = 15.8\, d_{\text{inj}} \left(\frac{\rho_l}{\rho}\right)^{1/2} \tag{358}$$

where the subscript bu denotes breakup and Δp is the pressure drop in the injector.

Varde and Popa [364] studied the penetration of diesel fuel sprays and derived the following correlation for the spray penetration, which accounts for the nozzle length-to-diameter ratio and yields good agreement with experimental data:

$$x = 1.1\, a_1^{0.3}\, a_2^{-0.008}\, a_3^{0.5} a_4^{0.16}\, t^{0.55} \tag{359}$$

where

$$a_1 = \Delta p\, \rho_l \frac{d_{\text{inj}}^2}{\mu_l^2} \qquad a_2 = \rho_l\, \sigma \frac{d_{\text{inj}}}{\mu_l^2} = \frac{\text{Re}^2}{\text{We}} \tag{360}$$

$$a_3 = \frac{\rho_l}{\rho} \qquad a_4 = \frac{L}{D} \tag{361}$$

Takeuchi et al. [365] used photography to determine the spray tip penetration as a function of time, pressure drop in the nozzle, and ambient density as well as to measure the droplet size distribution. Takeuchi et al. [366] used optical methods and microscopic photography to observe the distribution of fuel droplets and soot in diesel engine combustion chambers.

Kuniyoshi et al. [367] studied spray penetration in diesel engines and obtained the following correlation:

$$x = 418\, t^{0.375}\, (\rho_{ch})^{0.092 \ln t - 0.409}\, d_{inj}^{0.538} \tag{362}$$

where ρ_{ch} is the specific weight of the charge.

Chehroudi et al. [368] studied sprays from conical-oscillating-poppet injectors (COPI) and showed that the spray structure is affected by the chamber pressure. They observed that the injection duration, spray tip penetration, and initial axial velocity of the spray tip decreases as the chamber pressure increases. The use of COPI has been proposed for early-injection DISC engines.

Cavaliere et al. [369] used a two-dimensional laser-light-scattering technique to analyze sprays near the injector; they observed spray tip fragmentation at about 4 mm from the nozzle exit. Their technique uses coherent light and is not intrusive; however, there is substantial attenuation of the laser and scattered light if the laser sheet crosses the spray axis at small or large angles.

Hiroyasu [370] used a voltage technique to measure breakup lengths in diesel sprays; however, his measurements are not very accurate because both the intact core length and the liquid ligaments conduct electricity when voltage is applied to the injector [230]. Hiroyasu [370] also used photography (shadowgraphs and schlieren photographs), a helium-neon laser, and holography to measure the spray tip penetration, droplet size, and fuel vapor concentration. Both laser and holographic techniques are subject to errors caused by light attenuation and scattering, and they may not provide accurate and detailed information about the droplet sizes and velocity field near the injector. Highly penetrating X rays may be able to resolve the spray structure and the flow field near the injector.

Shimizu et al. [371] used light-scattering and holographic techniques to analyze the unsteady characteristics of fuel sprays and calculated the droplet size and volume distribution from the light-intensity distribution on photographs of fuel sprays.

Hess [372] developed a technique that combines the visibility of a laser Doppler signal and the intensity of the scattered light to measure the size and velocity of particles and droplets. Hess showed that using only visibility techniques may lead to large errors in dense sprays.

Reason [373] and Dibelius et al. [374] developed a laser optical probe to measure the droplet size distribution and velocity in sprays. The primary light source used in this technique is the beam of an argon-ion laser, which illuminates the droplets as they pass through an observation volume. This volume is formed by the intersection of two perpendicular fields of view from the objective lenses in the probe. A glass fiber transmits the laser light to one objective lens, and, depending on their size, the droplets scatter the light in different directions by reflection, refraction, and diffraction. Within a fixed range of scattered angles, the light is collected by the second objective lens and transmitted by an optical fiber to the photomultiplier. In the photomultiplier the light pulses are converted to voltage signals so that the number of pulses and the pulse height can be determined. The droplet velocity can be determined from the pulse widths [375]. Tardieu et al. [376] used the light intensity diffracted by a droplet cloud to measure the droplet distribution in sprays.

The main difficulty encountered in diffraction techniques is the determination of the droplet distribution function from the measured diffraction profile. This problem is ill conditioned since small measurement errors may lead to entirely different distributions. In previous studies this difficulty was overcome by imposing the shape of the size distribution at the outset. Using an array of 1024 photodiodes to provide a high-resolution sampling of the diffracted light intensity profile, Tardieu et al. [376] assembled successive diffraction patterns with partial overlap.

3.10 ADIABATIC ENGINES

Compared with conventional water-cooled engines, adiabatic or insulated diesel engines can considerably reduce the heat transfer losses to the cylinder walls. Eliminating the water coolant and insulating the combustion chamber walls and other critical engine components can reduce heat transfer losses by as much as 65% [377]. However, in adiabatic engines, radiative heat transfer plays a dominant role, with the cylinder temperature and the NO_x emissions being higher than in conventional water-cooled engines and the unburnt HC emissions being lower.

Due to the high temperatures of the combustion chamber in an adiabatic engine, the ignition delay is significantly reduced and the engine can burn low-cetane fuels [378]. However, part of the available energy is lost in the high-temperature exhaust gases, and the engine efficiency may not be higher than that of conventional engines unless the exhaust gases are used in a turbocompound system. Since the temperature of the cylinder becomes extremely high in adiabatic engines, a proper selection of the combustion chamber materials, lubricant, and lubrication method is necessary in order to have enough engine durability and lifetime under heavy-load conditions.

Ceramics have been used to coat the cylinder liner and walls with partially stabilized zirconia. Experimental studies with ceramic-coated valves and cylinder liners have shown that adiabatic engines can result in fuel improvements and lower HC, NO_x, and particulate emissions than conventional engines due to a shortened ignition delay and an increase in the burning rate during the diffusion stage of combustion [379]. These studies also show that adiabatic diesel engines have a multifuel capability and high fuel efficiency. However, thermal shocks can cause failure of the ceramics and valve coatings, and the coatings may be effective for only about 1000 hours of engine operation [380].

Morel et al. [381, 382] used a multizone model to provide a spatially and temporally resolved description of the heat transfer losses, gas velocity, and turbulence intensity in adiabatic engines. Their model uses a one-dimensional time-dependent submodel to analyze the heat transfer through layered or laminated walls, a steady-state heat-conduction submodel of the engine structure, and a thermodynamic cycle model. Their numerical results indicate that there are swing effects caused by the low thermal conductivity of ceramics. These temperature swings cause thermal stresses that may eventually result in failure of the ceramic coating.

Glance [383] also performed a thermal and structural analysis of adiabatic engine components. Hoag et al. [384] used a cycle-simulation model and performed a structural analysis of insulated diesel engines. Their model accounts for radiative heat transfer. Cole and Alkidas [385] used an air gap to reduce heat losses from an insulated piston and showed that the HC, CO, and smoke emissions decrease when an air gap is used. However, the NO emissions and brake-specific fuel consumption (bsfc) increased at low loads. Their experiments also showed that a low-heat-rejection piston reduces the piston-crown surface heat transfer by 3.5% and 5.5% at 1000 rpm under full and light loads, respectively. Elsbett et al. [386] also showed that, as compared to an aluminum engine, an insulated piston results in lower bsfc, higher torque, less heat transfer to the coolant, lower weight, lower HC emissions, and improved cold start and multifuel capability.

Solid lubrication instead of air gaps has also been used in adiabatic diesel engines [387]. Indig and Haman [388] employed turbocompounding in order to recover some energy from the exhaust gases. Kirloskar et al. [389] used an emptying and filling model for the induction and exhaust processes, Watson's heat release correlation, and Woschni's heat transfer correlation to analyze the performance of differentially cooled engines. Their studies indicate that for optimum mixing conditions, the liner temperature should be lower than 430 K and that the temperatures of the cylinder head and piston bear a reciprocal relationship to each other; a high piston temperature should be associated with a low cylinder head temperature, and vice versa, in order to promote efficient mixing.

3.11 SUMMARY

In this chapter single-zone, multizone, and multidimensional models of diesel engine combustion have been reviewed. It has been indicated that single-zone models, which assume that the cylinder mixture is homogeneous in both composition and temperature, may require a case-by-case adjustment of the Wiebe function parameters or burning rate law to accurately predict the cylinder pressure as a function of the engine rpm, combustion chamber geometry, load, and injection conditions. In general, the Wiebe function parameters are functions of the engine geometry and operating conditions.

Multizone models account for air entrainment and mixture inhomogeneities by dividing the fuel spray injected into the cylinder into parcels, the composition of which can be calculated as a function of time by using the first law of thermodynamics. Some of the multizone models discussed in this chapter consider a gaseous fuel jet and neglect the presence of fuel droplets. Other multizone models account for the droplet vaporization by dividing the injected liquid fuel into droplet groups, and they use experimental correlations for the spray penetration and air entrainment.

Multidimensional models of diesel engine combustion account for the engine geometry and the temporal and spatial variations of the flow field. In these models the mass of fuel injected at each time step is allocated to a continuous droplet distribution function, which in turn is subdivided into droplet groups in such a manner that all the droplets in a group have the same radius, velocity, and temperature.

Multidimensional models do not account for the liquid fuel jet breakup into ligaments and droplets. In most cases droplets are injected, not at the injector location but within the computational domain, and their interactions with the gas-phase turbulence are modeled by means of stochastic approximations, assuming that the turbulence is isotropic. This assumption is incorrect because the energy-containing eddies are anisotropic and depend on the engine geometry. Furthermore, most multidimensional models account only for the effects of the gas-phase turbulence on the droplets and void fraction, but neglect the effects of the droplets on turbulence; these effects depend on the eddy size, droplet diameter, eddy-turnover time, droplet residence time in a typical eddy, droplet loading, etc., and are poorly understood at the present time.

The large-scale eddies are responsible for the production of turbulence kinetic energy, whereas the smallest eddies are responsible for its dissipation. Therefore, in order to consider the effects of the droplets on the gas-phase turbulence, the length scales of the energy-containing eddies as well as those of the Kolmogorov eddies and the droplet diameter must be included in turbulence models of two-phase flows.

Accurate models are needed to represent the droplet-solid wall interactions. When droplets collide with solid walls, they may remain trapped in the boundary layer, bounce off, form a film along the walls, or shatter. Some of the multidimensional models discussed in this chapter assume that the collisions of the droplets with the solid walls are specular (i.e., the angle of incidence is equal to the angle of reflection) and that the magnitude of the droplet velocity is conserved upon collisions.

Aerodynamic theories of droplet breakup provide estimates of the droplet size and intact core length, but because of their linearity, they may be inadequate. Furthermore, the effects of nozzle geometry, turbulence, boundary layer instabilities, cavitation, and flow relaxation must be included in the determination of the intact core length and droplet size. The initial droplet size and velocity distribution are essential for accurate determination of the flow field, fuel vaporization, and combustion in diesel engines.

More experimental data are required to validate the predictions of multidimensional models. These data must include droplet size distributions and droplet velocities in diesel engines, and they can be used to determine the effects of turbulence and droplet collisions, coalescence, and breakup on the engine flow field and combustion.

REFERENCES

1. Krieger, R. B., and Borman, G. L., The Computation of Apparent Heat Release from Internal Combustion Engines. Paper No. 66-WA/DGP-4, ASME, New York, 1966.
2. Borman, G. L., Mathematical Simulation of Internal Combustion Engine Processes and Performance Including Comparison with Experiment, Ph.D. thesis, University of Wisconsin, Madison, 1964.
3. Watson, N., Pilley, A. D., and Marzouk, M., A Combustion Correlation for Diesel Engine Simulation, SAE International Congress and Exposition, Detroit, Mich., SAE Paper No. 800029, 1980.
4. Wolfer, H. H., Der Zunderzug im Diesel-motor, *CDE-Forschungsheft*, vol. 392, pp. 15–24, 1938.
5. Cheng, C.-Y., Computer Simulation for a Four-Stroke, Direct-Injection Diesel Engine, M.S. thesis, University of Rhode Island, Kingston, 1985.
6. Hurn, R. W., and Hughes, K. J., Combustion Characteristics as Measured in a Constant-Volume Bomb, *SAE Quarterly Transactions*, vol. 6, no. 1, pp. 24–35, 1952.
7. Baev, V. K., Bazhaikin, A. N., Buzukov, A. A., and Timoshenko, B. P., Induction Time Measurement for Ignition of Liquid Fuel Jets in Air at High Temperatures and Pressures, in *Dynamics of Flames and Reactive Systems*, eds. J. R. Bowen, N. Manson, A. K. Oppenheim, and R. I. Soloukhin, *Progress in Astronautics and Aeronautics*, vol. 95, pp. 554–562, AIAA, New York, 1984.
8. Ikegami, M., Miwi K., and Inada, M., A Study of Ignition and Combustion of a Diesel Spray by Means of a Rapid Compression Machine, *Bulletin of the JSME*, vol. 24, no. 195, pp. 1608–1615, 1981.
9. Tsao, K. C., Myers, P. S., and Uyehara, O. A., Gas Temperatures during Compression in Motored and Fired Diesel Engines, *SAE Transactions*, vol. 70, pp. 136–145, 1962.
10. Gerrish, H. C., and Ayer, B. E., Influence of Oil Temperature on the Combustion in a Prechamber Compression Ignition Engine, NACA Technical Note 565, 1936.
11. Spadaccini, L. J., Auto-Ignition Characteristics of Hydrocarbon Fuels at Elevated Temperatures and Pressures, *ASME Journal of Engineering for Power*, vol. 99, no. 1, pp. 83–87, 1977.
12. Walsh, G., and Cheng, W. K., Effects of Highly Heated Fuel on Diesel Combustion, SAE International Congress and Exposition, Detroit, Mich., SAE Paper No. 850088, 1985.
13. Scharnweber, D. H., and Hoppie, L. O., Hyperbolic Combustion in an Internal Combustion Engine, SAE International Congress and Exposition, Detroit, Mich., SAE Paper No. 850089, 1985.
14. Parker, T. E., Forsha, M. D., Stewart, H. E., Hom, K., Sawyer, R. F., and Oppenheim, A. K., Induction Period for Ignition of Fuel Sprays at High Temperatures and Pressures. SAE International Congress and Exposition, Detroit, Mich., SAE Paper No. 850087, 1985.
15. Shipinski, J., Uyehara, O. A., and Myers, P. S., Experimental Correlation between Rate-of-Injection and Rate-of-Heat-Release in a Diesel Engine, Paper No. 68-DGP-11, ASME, New York, 1968.
16. Sitkei, G., *Kraftstoffaufbereitung und Verbrennung bei Dieselmotoren*, Springer-Verlag, Berlin, 1964.
17. Hardenberg, H. O., and Hase, F. W., An Empirical Formula for Computing the Pressure Rise Delay of a Fuel from Its Cetane Number and from the Relevant Parameters of Direct-Injection Diesel Engines, SAE International Congress and Exposition, Detroit, Mich., SAE Paper No. 790493, 1979.

18. Dent, J. C., and Mehta, P. S., Phenomenological Combustion Model for a Quiescent Chamber Diesel Engine, SAE International Congress and Exposition, Detroit, Mich., SAE Paper No. 811235, 1981.
19. Assanis, D. N., A Computer Simulation of the Turbocharged Turbocompounded Diesel Engine System for Studies of Low Heat Rejection Engine Performance, Ph.D. thesis, Massachusetts Institute of Technology, Cambridge, 1985.
20. Gulder, O. L., and Glavincevski, B., Ignition Quality Determination of Diesel Fuels from Hydrogen Type Distribution of Hydrocarbons, *Combustion and Flame*, vol. 63, pp. 231–238, 1986.
21. Kirsch, L. J., Rosenfeld, J. L. J., and Summers, R., Studies of Fuel Injection into a Rapid Compression Machine, *Combustion and Flame*, vol. 43, pp. 11–21, 1981.
22. Belardini, P., Bertoli, C., Corcione, F. E., and Police, G., Ignition Delay Measurement in a Direct Injection Diesel Engine, *International Conference on Combustion Engineering*, vol. 2, pp. 1–7, I. Mech. E. Conference Publications 1983-3, London, 1983.
23. Wu, K.-J., Effects of Compression Ratio on Combustion Characteristics of a Direct-Injection Diesel Engine, SAE International Fuels and Lubricants Meeting and Exposition, Toronto, Ont., Canada, SAE Paper No. 872056, 1987.
24. Woschni, G., A Universally Applicable Equation for the Instantaneous Heat Transfer Coefficient in the Internal Combustion Engine, SAE International Congress and Exposition, Detroit, Mich., SAE Paper No. 670931, 1967.
25. Annand, W. J. D., Heat Transfer in the Cylinders of Reciprocating Internal Combustion Engines, *Proceedings of the Institution of Mechanical Engineers*, vol. 177, no. 36, pp. 973–990, 1963.
26. Kamel, M., and Watson, N., Heat Transfer in the Indirect Injection Diesel, SAE International Congress and Exposition, Detroit, Mich., SAE Paper No. 790826, 1979.
27. Trifunovic, R., Local Heat and Mass Transfer on the D.I. Diesel Engine Combustion Chamber Wall: Present Status of Investigation, XIX International Symposium on Heat and Mass Transfer in Gasoline and Diesel Engines, International Centre for Heat and Mass Transfer, Dubrovnik, Yugoslavia, 1987.
28. Dobovisek, Z., Hribernik, A., Cernej, A., and Humski, F., One-Dimensional Model for Prediction of Cylinder Convective Heat Transfer Coefficients, XIX International Symposium on Heat and Mass Transfer in Gasoline and Diesel Engines, International Centre for Heat and Mass Transfer, Dubrovnik, Yugoslavia, 1987.
29. Morel, T., Keribar, R., and Blumberg, P. N., Cyclical Thermal Phenomena in Engine Combustion Chamber Surfaces, SAE International Congress and Exposition, Detroit, Mich., SAE Paper No. 850360, 1985.
30. Morel, T., Fort, E. F., and Blumberg, P. N., Effect of Insulation Strategy and Design Parameters on Diesel Engine Heat Rejection and Performance, SAE International Congress and Exposition, Detroit, Mich., SAE Paper No. 850506, 1985.
31. Mansouri, S. H., and Heywood, J. B., Correlation for Viscosity and Prandtl Number of Hydrocarbon-Air Combustion Products, *Combustion Science and Technology*, vol. 23, pp. 251–256, 1980.
32. Svehla, R. A., and McBride, B. J., Fortran IV Computer Program Calculations of Thermodynamic and Transport Properties of Complex Chemical Systems, NASA Technical Note D-7056, 1973.
33. Mansouri, S. H., Heywood, J. B., and Radhakrishnan, K., Divided-Chamber Diesel Engines: Part 1. A Cycle-Simulation Which Predicts Performance and Emissions, SAE International Congress and Exposition, Detroit, Mich., SAE Paper No. 820273, 1982.
34. Poulos, S. G., and Heywood, J. B., The Effect of Chamber Geometry on Spark-Ignition Engine Combustion, SAE International Congress and Exposition, Detroit, Mich., SAE Paper No. 830334, 1983.
35. Tennekes, H., and Lumley, J. L., *A First Course in Turbulence*, MIT Press, Cambridge, Mass., 1972.
36. Reynolds, W. C., Modeling of Fluid Motions in Engines—An Introductory Overview, in

Combustion Modeling in Reciprocating Engines, eds. J. N. Mattavi and C. A. Amann, pp. 41–68, Plenum Press, New York, 1980.

37. Wong, V. W., and Hoult, D. P., Rapid Distortion Theory Applied to Turbulent Combustion, SAE International Congress and Exposition, Detroit, Mich., SAE Paper No. 790357, 1979.
38. Hoult, D. P., and Wong, V. W., The Generation of Turbulence in an Internal-Combustion Engine, in *Combustion Modeling in Reciprocating Engines*, eds. J. N. Mattavi and C. A. Amann, pp. 131–160, Plenum Press, New York, 1980.
39. Morel, T., and Keribar, R., A Model for Predicting Spatially and Time Resolved Convective Heat Transfer in Bowl-in-Piston Combustion Chambers. SAE International Congress and Exposition, Detroit, Mich., SAE Paper No. 850204, 1985.
40. Borman, G., and Nishiwaki, K., Internal Combustion Engine Heat Transfer, *Progress in Energy and Combustion Science*, vol. 13, pp. 1–46, 1987.
41. Oguri, T., and Shigewo, I., Radiant Heat Transfer in Diesel Engines, SAE International Congress and Exposition, Detroit, Mich., SAE Paper No. 720023, 1972.
42. Kunimoto, T., Matsuoka, K., and Oguri, T., Prediction of Radiative Heat Flux in a Diesel Engine, SAE International Congress and Exposition, Detroit, Mich., SAE Paper No. 750786, 1975.
43. Ebersole, G. D., Myers, P. S., and Uyehara, O. A., The Radiant and Convective Components of Diesel Engine Heat Transfer, SAE International Congress and Exposition, Detroit, Mich., SAE Paper No. 701C, 1963.
44. Sitkei, G., and Ramaniah, G. V., A Rational Approach for Calculation of Heat Transfer in Diesel Engines, SAE International Congress and Exposition, Detroit, Mich., SAE Paper No. 720027, 1972.
45. LeFeuvre, T., Myers, P. S., and Uyehara, O. A., Experimental Instantaneous Heat Fluxes in a Diesel Engine and Their Correlation, SAE International Congress and Exposition, Detroit, Mich., SAE Paper No. 690464, 1969.
46. Dent, J. C., and Suliaman, S. J., Convective and Radiative Heat Transfer in a High Swirl Direct Injection Diesel Engine, SAE International Congress and Exposition, Detroit, Mich., SAE Paper No. 770407, 1977.
47. Flynn, P., Mizusawa, M., Uyehara, O. A., and Myers, P. S., An Experimental Determination of the Instantaneous Potential Radiant Heat Transfer within an Operating Diesel Engine, SAE International Congress and Exposition, Detroit, Mich., SAE Paper No. 720022, 1972.
48. Overbye, V. D., Bennethum, J. E., Uyehara, O. A., and Myers, P. S., Unsteady Heat Transfer in Engines, *SAE Transactions*, vol. 69, pp. 461–494, 1961.
49. Annand, W. J. D., and Ma, T. H., Instantaneous Heat Transfer Rates to the Cylinder Head Surface of a Small Compression-Ignition Engine, *Proceedings of the Institution of Mechanical Engineers*, vol. 185, no. 72/71, pp. 976–987, 1970–1971.
50. Alkidas, A. C., and Cole, R. M., Transient Heat Flux Measurements in a Divided-Chamber Diesel Engine, *ASME Journal of Heat Transfer*, vol. 107, no. 2, pp. 439–444, 1985.
51. Alkidas, A. C., and Cole, R. M., The Effects of Coolant Temperature on the Performance and Emissions of a Single-Cylinder Divided-Chamber Diesel Engine, in *Diesel Engine Combustion and Emissions*, eds. R. A. Baranescu and T. Murayama, pp. 1–11, SAE, Warrendale, Pa., 1984.
52. Menguc, M. P., Viskanta, R., and Ferguson, C. R., Multidimensional Modeling of Radiative Heat Transfer in Diesel Engines, SAE International Congress and Exposition, Detroit, Mich., SAE Paper No. 850503, 1985.
53. Nakanishi, K., Arai, M., and Hiroyasu, H., The Effect of Air Swirl on the Time History of the Diesel Particulate in the Combustion Chamber, *ASME-JSME Thermal Engineering Joint Conference Proceedings*, vol. 4, pp. 241–246, ASME, New York, 1983.
54. Stull, D. R., and Prophet H., *JANAF Thermochemical Tables*, 2nd ed., National Bureau of Standards, Washington, D.C., 1971.
55. Wiebe, I., Halbempirische Formel dur die Verbrennungsgeschwindigkeit, in *Kraftstoffaufbereitung und Verbrennung bei Dieselmotoren*, ed. G. Sitkei, pp. 156–159, Springer-Verlag, Berlin, 1964.

56. Miyamoto, N., Chikahisa, T., Murayama, T., and Sawyer, R., Description and Analysis of Diesel Engine Rate of Combustion and Performance Using Wiebe's Functions. SAE International Congress and Exposition, Detroit, Mich., SAE Paper No. 850107, 1985.
57. Murayama, T., Miyamoto, N., Yamada, T., Kawashima, J.-I., and Itow, K., A Method to Improve the Solubility and Combustion Characteristics of Alcohol-Diesel Fuel Blends, SAE International Congress and Exposition, Detroit, Mich., SAE Paper No. 821113, 1982.
58. Ghojel, J. I., A Study of Combustion Chamber Arrangements and Heat Release in D.I. Diesel Engines, SAE International Congress and Exposition, Detroit, Mich., SAE Paper No. 821034, 1982.
59. Tinaut, F. V., Contribucion al Estudio del Proceso de Combustion en Motores de Encendido por Compresion de Inyeccion Directa, Dr. Eng. thesis, Polytechnic University of Valencia, Valencia, Spain, 1986.
60. Watson N., Transient Performance Simulation and Analysis of Turbocharged Diesel Engines, SAE International Congress and Exposition, Detroit, Mich., SAE Paper No. 810338, 1981.
61. Watson, N., and Kamel, M., Thermodynamic Efficiency Evaluation of an Indirect Injection Diesel Engine, SAE International Congress and Exposition, Detroit, Mich., SAE Paper No. 790039, 1979.
62. Winterbone, D. E., and Tennant, D. W. H., The Variation of Friction and Combustion Rates during Diesel Engine Transients, SAE International Congress and Exposition, Detroit, Mich., SAE Paper No. 810339, 1981.
63. Winterbone, D. E., and Loo, W. Y., A Dynamic Simulation of a Two-Stroke Turbocharged Diesel Engine, SAE International Congress and Exposition, Detroit, Mich., SAE Paper No. 810337, 1981.
64. Tomic, M., Personal communication, Faculty of Mechanical Engineering, University of Belgrade, Belgrade, Yugoslavia, September 8, 1987.
65. Brandstatter, W. R., Modeling of a Stratified-Charge Engine with an Unscavenged Prechamber, in *Combustion Modeling in Reciprocating Engines*, eds. J. N. Mattavi and C. A. Amann, pp. 509–536, Plenum Press, New York, 1980.
66. Whitehouse, N. D., and Way, R. J. B., A Simple Method for the Calculation of Heat Release Rates in Diesel Engines Based on Fuel Injection Rate, SAE International Congress and Exposition, Detroit, Mich., SAE Paper No. 710134, 1971.
67. Lyn, W. T., Study of Burning Rate and Nature of Combustion in Diesel Engines, *Ninth Symposium (International) on Combustion*, pp. 1069–1082, Academic Press, New York, 1963.
68. Carapanayotis, M., and Salcudean, M., Thermodynamic Simulation of a Two-Stroke Direct-Injection Turbocharged Diesel Engine and Comparison with Experimental Measurements, Department of Mechanical Engineering, University of British Columbia, Vancouver, B. C., Canada, 1986.
69. Kuo, K. K., *Principles of Combustion*, Wiley, New York, 1986.
70. Abramovich, G. N., *The Theory of Turbulent Jets*, MIT Press, Cambridge, Mass., 1963.
71. Whitehouse, N. D., and Sareen, B. K., Prediction of Heat Release in a Quiescent Chamber Diesel Engine Allowing for Fuel/Air Mixing, SAE International Congress and Exposition, Detroit, Mich., SAE Paper No. 740054, 1974.
72. Kumar, K., Babu, M. K. G., Gaur, R. R., and Garg, R. D., A Thermodynamic Simulation Model for a Four Stroke Medium Speed Diesel Engine, SAE International Congress and Exposition, Detroit, Mich., SAE Paper No. 840516, 1984.
73. Borman, G. L., and Johnson, J. H., Unsteady Vaporization Histories and Trajectories of Fuel Drops Injected into Swirling Air, SAE National Power Plant Meeting, Philadelphia, Pa., SAE Paper No. 598C, 1962.
74. Kuo, T.-W., and Bracco, F. V., On the Scaling of Impulsively Started Incompressible Turbulent Round Jets, *ASME Journal of Fluids Engineering*, vol. 104, pp. 191–197, 1982.
75. Kuo, T.-W., and Bracco, F.V., On the Scaling of Transient Laminar, Turbulent, and Spray Jets, SAE International Congress and Exposition, Detroit, Mich., SAE Paper No. 820038, 1982.
76. Taylor, G. I., Generation of Ripples by Wind Blowing over a Viscous Fluid, in *The Scientific*

Papers of Sir Geoffrey Ingram Taylor, ed. G. K. Batchelor, vol. 3, pp. 244–259, Cambridge University Press, Cambridge, U.K., 1963.

77. Reitz, R. D., and Bracco, F. V., Mechanism of Atomization of a Liquid Jet, *Physics of Fluids*, vol. 25, no. 10, pp. 1730–1742, 1982.
78. Reitz, R. D., and Bracco, F. V., Ultra-High-Speed Filming of Atomizing Jets, *Physics of Fluids*, vol. 22, no. 6, pp. 1054–1064, 1979.
79. Reitz, R. D., and Bracco, F. V., On the Dependence of Spray Angle and Other Spray Parameters on Nozzle Design and Operating Conditions, SAE International Congress and Exposition, Detroit, Mich., SAE Paper No. 790494, 1979.
80. Wu, K.-J., Reitz, R. D., and Bracco, F. V., Measurements of Drop Size at the Spray Edge near the Nozzle in Atomizing Liquid Jets, *Physics of Fluids*, vol. 29, no. 4, pp. 941–951, 1986.
81. Reitz, R. D., and Bracco, F. V., Mechanisms of Breakup of Round Liquid Jets, in *Encyclopedia of Fluid Mechanics*, ed. N. P. Cheremisinoff, vol. 3, pp. 233–249, Gulf Publishing, Houston, Tex., 1986.
82. Bracco, F. V., Modeling of Engine Sprays, SAE International Congress and Exposition, Detroit, Mich., SAE Paper No. 850394, 1985.
83. Idoum, A., Wallace, F. J., Packer, J. P., and Charlton, S. J., An Experimental and Analytical Study of Jet Impingement and Wall Jets in High Swirl D.I. Diesel Engines Using the Hydraulic Analogy, SAE International Congress and Exposition, Detroit, Mich., SAE Paper No. 850263, 1985.
84. Packer J. P., Wallace, F. J., Adler, D., and Karimi, E. R., Diesel Jet Mixing under High Swirl Conditions, *International Conference on Combustion Engineering*, vol. 2, pp. 9–22, I. Mech. E. Conference Publications 1983-3, London, 1983.
85. Mehta, P. S., and Gupta, A. K., Modelling of Spray-Swirl Interaction in Direct Injection Diesel Engine Combustion Chambers, *Proceedings of the Institution of Mechanical Engineers*, vol. 199, no. D3, pp. 187–199, 1985.
86. Ha. J.-H., Iida, N., Sato, G. T., Hayashi, A., and Tanabe, H., Experimental Investigation of the Entrainment into Diesel Spray, *Diesel Engine Combustion and Emissions*, pp. 73–79, SAE, Warrendale, Pa., 1984.
87. Sinnamon, J. F., Lancaster, D. R., and Stiener, J. C., An Experimental and Analytical Study of Engine Fuel Spray Trajectories, SAE International Congress and Exposition, Detroit, Mich., SAE Paper No. 800135, 1980.
88. Ricou, F. P., and Spalding, D. B., Measurements of Entrainment by Axisymmetric Turbulent Jets, *Journal of Fluid Mechanics*, vol. 9, pp. 21–32, 1961.
89. Hiraki, H., and Rife, J. M., Performance and NO_x Model of a Direct Injection Stratified Charge Engine, SAE International Congess and Exposition, Detroit, Mich., SAE Paper No. 800050, 1980.
90. Blizard, N., and Keck, J. C., Experimental and Theoretical Investigation of Turbulent Burning Model for Internal Combustion Engines, SAE International Congress and Exposition, Detroit, Mich., SAE Paper No. 740191, 1974.
91. Tabaczynski, R. J., Ferguson, C. R., and Radhakrishnan, K., A Turbulent Entrainment Model for Spark-Ignition Engine Combustion, SAE International Congress and Exposition, Detroit, Mich., SAE Paper No. 770647, 1977.
92. Lavoie, G. A., Heywood, J. B., and Keck, J. C., Experimental and Theoretical Study of Nitric Oxide Formation in Internal Combustion Engines, *Combustion Science and Technology*, vol. 1, pp. 313–326, 1970.
93. Komiyama, K., and Heywood, J. B., Predicting NO_x Emissions and Effects of Exhaust Gas Recirculation in Spark-Ignition Engines, SAE International Congress and Exposition, Detroit, Mich., SAE Paper No. 730475, 1973.
94. Kono, S., Nagao, A., and Motooka, H., Prediction of In-Cylinder Flow and Spray Formation Effects on Combustion in Direct Injection Diesel Engines, SAE International Congress and Exposition, Detroit, Mich., SAE Paper No. 850108, 1985.
95. Meguerdichian, M., and Watson, N., Prediction of Mixture Formation and Heat Release in

Diesel Engines, SAE International Congress and Exposition, Detroit, Mich., SAE Paper No. 780225, 1978.

96. Hiroyasu, H., Kadota, T., and Arai, M., Development and Use of a Spray Combustion Modeling to Predict Diesel Engine Efficiency and Pollutant Emissions: Part 1. Combustion Modeling, *Bulleting of the JSME*, vol. 26, no. 214, pp. 569–575, 1983.
97. Kobayashi, H., Kamimoto, T., and Matsuoka, S., A Photographic and Thermodynamic Study of Diesel Combustion in a Rapid Compression Machine, SAE International Congress and Exposition, Detroit, Mich., SAE Paper No. 810259, 1981.
98. Shahed, S. M., Flynn, P. F., and Lyn, W. T., A Model for the Formation of Emissions in a Direct-Injection Diesel Engine, in *Combustion Modeling in Reciprocating Engines*, eds. J. N. Mattavi and C. A. Amann, pp. 345–368, Plenum Press, New York, 1980.
99. Arai, M., Tabata, M., Hiroyasu, H., and Shimizu, M., Disintegrating Process and Spray Characterization of Fuel Jet Injected by a Diesel Nozzle, SAE International Congress and Exposition, Detroit, Mich., SAE Paper No. 840275, 1984.
100. Hiroyasu, H., Kadota, T., and Arai, M., Development and Use of a Spray Combustion Modeling to Predict Diesel Engine Efficiency and Pollutant Emissions: Part 2. Computational Procedure and Parametric Study, *Bulletin of the JSME*, vol. 26, no. 214, pp. 576–583, 1983.
101. Kuo, T.-W., Evaluation of a Phenomenological Spray-Combustion Model for Two Open-Chamber Diesel Engines, SAE International Fuels and Lubricants Meeting and Exposition, Toronto, Canada, SAE Paper No. 872057, 1987.
102. Hiroyasu, H., and Arai, M., Development and Use of a Spray Combustion Modeling to Predict Diesel Engine Efficiency and Pollutant Emissions, *Proceedings of the Fifth International Automotive Propulsion Systems Symposium*, pp. 264–288, SAE, Warrendale, Pa., 1980.
103. Hiroyasu, H., Kadota, T., and Arai, M., Supplementary Comments: Fuel Spray Characterization in Diesel Engines, in *Combustion Modeling of Reciprocating Engines*, eds. J. N. Mattavi and C. A. Amann, pp. 369–408, Plenum Press, New York, 1980.
104. Levich, V. I., *Physicochemical Hydrodynamics*, pp. 639–650, Prentice-Hall, Englewood Cliffs, N.J., 1962.
105. Ranz, W. E., and Marshall, W. R., Evaporation from Drops, *Chemical Engineering Progress*, vol. 48, pp. 141–146, pp. 173–180, 1952.
106. Shahed, S. M., Chiu, W. S., and Yumlu, V. S., A Preliminary Model for the Formation of Nitric Oxide in Direct Injection Diesel Engines and Its Application in Parametric Studies, SAE International Congress and Exposition, Detroit, Mich., SAE Paper No. 730083, 1973.
107. Chiu, W. S., Shahed, S. M., and Lyn, W. T., A Transient Spray Mixing Model of Diesel Combustion, *SAE Transactions*, vol. 85, pp. 502–512, 1977.
108. Shahed, S. M., Chiu, W. S., and Lyn, W. T., A Mathematical Model of Diesel Combustion, *Proceedings of the Institution of Mechanical Engineers*, C94/75, pp. 119–128, 1975.
109. Kuo, T.-W., Yu, R. C., and Shahed, S. M., A Numerical Study of the Transient Evaporating Spray Mixing Process in the Diesel Environment, SAE International Congress and Exposition, Detroit, Mich., SAE Paper No. 831735, 1983.
110. Elkotb, M. M., Elbahar, O. M. F., Salem, H., and Saad, M., Induced Air Flows in Diesel Fuel Sprays, in *International Symposium on Flows in Internal Combustion Engines—III*, eds. T. Uzkan, W. G. Tiederman, and J. M. Novak, pp. 97–105, ASME, New York, 1985.
111. Chapman, M., Friedman, M. C., and Aghan, A., A Time-Dependent Spatial Model for Radiant Heat Transfer in Diesel Engines, SAE International Congress and Exposition, Detroit, Mich., SAE Paper No. 831725, 1983.
112. Kuo, T.-W., and Bracco, F. V., On the Scaling of Impulsively Started Incompressible Turbulent Round Jets, *ASME Journal of Fluids Engineering*, vol. 104, pp. 191–197, 1982.
113. Gaillard, P., Multidimensional Numerical Study of the Mixing of an Unsteady Gaseous Fuel Jet with Air in Free and in Confined Situations, SAE International Congress and Exposition, Detroit, Mich., SAE Paper No. 840225, 1984.
114. Turns, S. R., and Ran, N., Transient Gas Jet Penetration and Growth in a Swirling Flow, in *Flows in Internal Combustion Engines*, ed. T. Uzkan, pp. 21–28, ASME, New York, 1982.

115. Varde, K. S., Popa, D. M., and Varde, L. K., Spray Angle and Atomization in Diesel Sprays, in *Diesel Engine Combustion and Emissions*, eds. R. A. Baranescu and T. Murayama, pp. 27–35, SAE, Warrendale, Pa., 1984.
116. Wu, K.-J., Su, C.-C., Steinberger, R. L., Santavicca, D. A., and Braco, F. V., Measurements of the Spray Angle of Atomizing Jets, *ASME Journal of Fluids Engineering*, vol. 105, pp. 406–413, 1983.
117. Kobayashi, H., Kamimoto, T., and Matsuoka, S., Prediction of the Rate of Heat Release of an Axisymmetrical Diesel Flame in a Rapid Compression Machine, SAE International Congress and Exposition, Detroit, Mich., SAE Paper No. 840519, 1984.
118. Kamimoto, T., Chang, Y. J., and Kobayashi, H., Rate of Heat Release and Its Prediction of a Diesel Flame in a Rapid Compression Machine, SAE International Congress and Exposition, Detroit, Mich., SAE Paper No. 841076, 1984.
119. Ikegami, M., Shioji, M., and Koike, M., Stochastic Approach to Model the Combustion Process in Direct-Injection Diesel Engines, *Twentieth Symposium (International) on Combustion*, pp. 217–224, The Combustion Institute, Pittsburgh, Pa., 1985.
120. Plee, S. L., and Ahmad, T., Relative Roles of Premixed and Diffusion Burning in Diesel Combustion, SAE International Congress and Exposition, Detroit, Mich., SAE Paper No. 831733, 1983.
121. Haenlein, A., On the Disruption of a Liquid Jet, NACA Technical Memorandum 659, 1932.
122. McCarthy, M. J., and Molloy, N. A., Review of Stability of Liquid Jets and the Influence of Nozzle Design, *Chemical Engineering Journal*, vol. 7, pp. 1–20, 1974.
123. Giffen, E., and Muraszew, A., *The Atomisation of Liquid Fuels*, Wiley, New York, 1953.
124. Putnam, A. A., Benington, F., Einbinder, H., Hazard, H. R., Kettelle, J. D., Levy, A., Miese, C. C., Pilcher, J. M., Thomas, R. E., Weller, A. E., and Landry, B. A., Injection and Combustion of Liquid Fuels, WADC Technical Report No. 56-344, ASTIA Document No. AD 118142, Battelle Memorial Institute, Wright-Patterson Air Force Base, Dayton, Ohio, 1957.
125. Ranz, W. E., Some Experiments on Orifice Sprays, *Canadian Journal of Chemical Engineering*, vol. 36, pp. 175–181, 1958.
126. Clark, C. J., and Dombrowski, N., Aerodynamic Instability and Disintegration of Inviscid Liquid Sheets, *Proceedings of the Royal Society of London, Series A*, vol. 329, pp. 467–478, 1972.
127. Asare, H. R., Takahashi, R. K., and Hoffman, M. A., Liquid Sheet Jet Experiments: Comparison with Linear Theory, *ASME Journal of Fluids Engineering*, vol. 103, pp. 595–604, 1981.
128. Hagerty, W. W., and Shea, J. F., A Study of the Stability of Plane Liquid Sheets, *ASME Journal of Applied Mechanics*, vol. 22, pp. 509–514, 1955.
129. Squire, H. B., Investigation of the Instability of a Moving Liquid Film, *British Journal of Applied Physics*, vol. 4, pp. 167–169, 1953.
130. Biswas, G., Som, S. K., and Gupta, A. S., Instability of a Moving Cylindrical Liquid Sheet, *ASME Journal of Fluids Engineering*, vol. 107, pp. 451–454, 1985.
131. Dombrowski, N., and Hooper, P. C., The Effect of Ambient Density on Drop Formation in Sprays, *Chemical Engineering Science*, vol. 17, pp. 291–305, 1962.
132. Dombrowski, N., and Johns, W. R., The Aerodynamic Instability and Disintegration of Viscous Liquid Sheets, *Chemical Engineering Science*, vol. 18, pp. 203–214, 1963.
133. Lin, S. P., and Kang, D. J., Atomization of a Liquid Jet, *Physics of Fluids*, vol. 30, pp. 2000–2006, 1987.
134. Gaster, M., A Note on the Relation between Temporally-Increasing and Spatially-Increasing Disturbances in Hydrodynamic Stability, *Journal of Fluid Mechanics*, vol. 14, pp. 222–224, 1962.
135. Nurick, W. H., Orifice Cavitation and Its Effect on Spray Mixing, *ASME Journal of Fluids Engineering*, vol. 98, pp. 681–687, 1976.
136. Lafrance, P., The Breakup Length of Turbulent Liquid Jets, *ASME Journal of Fluids Engineering*, vol. 99, pp. 414–415, 1977.

137. Ingebo, R. D., Acceleration Wave Breakup of Liquid Jets with Airstreams, in *Fluid Mechanics of Combustion Systems*, eds. T. Morel, R. P. Lohmann, and J. M. Rackley, pp. 101–105, ASME, New York, 1981.
138. Williams, F. A., *Combustion Theory*, 2nd ed., pp. 446–484, Benjamin-Cummings, Menlo Park, Calif., 1985.
139. Wallis, G. B., *One-Dimensional Two-Phase Flow*, McGraw-Hill, New York, 1969.
140. Soo, S. L., *Fluid Dynamics of Multiphase Systems*, Blaisdell, Waltham, Mass., 1967.
141. Ishii, M., *Thermo-Fluid Dynamic Theory of Two-Phase Flows*, Eyrolles, Paris, 1975.
142. Dobran, F., Constitutive Equations for Multiphase Mixtures of Fluids, *International Journal of Multiphase Flow*, vol. 10, no. 3, pp. 273–305, 1984.
143. Drew, D. A., Mathematical Modeling of Two-Phase Flow, *Annual Review of Fluid Mechanics*, vol. 15, pp. 261–291, 1983.
144. Dobran, F., Theory of Multiphase Mixtures: A Thermomechanical Formulation, *International Journal of Multiphase Flow*, vol. 11, no. 1, pp. 1–30, 1985.
145. Crowe, C. T., REVIEW—Numerical Models for Dilute Gas-Particle Flows, *ASME Journal of Fluids Engineering*, vol. 104, pp. 297–303, 1982.
146. Ransom, V. H., and Hicks, D. L., Hyperbolic Two-Pressure Models for Two-Phase Flows, *Journal of Computational Physics*, vol. 53, pp. 124–151, 1984.
147. Di Giacinto, M., Sabetta, F., and Piva, R., Two-Way Coupling Effects in Dilute Gas-Particle Flows, *ASME Journal of Fluids Engineering*, vol. 104, pp. 304–312, 1982.
148. Brenner, H., Rheology of Two-Phase Systems, *Annual Review of Fluid Mechanics*, vol. 2, pp. 137–176, 1970.
149. Batchelor, G. K., Transport Properties of Two-Phase Materials with Random Surface, *Annual Review of Fluid Mechanics*, vol. 6, pp. 227–255, 1976.
150. Sirignano, W. A., The Formulation of Spray Combustion Models: Resolution Compared to Droplet Spacing, *ASME Journal of Heat Transfer*, vol. 108, pp. 633–639, 1986.
151. Dobran, F., On the Consistency Conditions of Averaging Operators in 2-Phase Flow Models and on the Formulation of Magnetohydrodynamic 2-Phase Flow, *International Journal of Engineering Science*, vol. 19, pp. 1353–1368, 1981.
152. Dobran, F., On the Formulation of Conservation, Balance and Constitutive Equations for Multiphase Flow, *Proceedings of the Third Multi-Phase Flow and Heat Transfer Symposium*, eds. T. Veziroglu and A. E. Bergles, pp. 125–141, Elsevier, New York, 1983.
153. Dobran, F., A Two-Phase Fluid Model Based on the Linearized Constitutive Equations, *Advances in Two-Phase Flow and Heat Transfer*, vol. 1, pp. 41–59, Martinus Nijhoff, The Netherlands, 1982.
154. Elghobashi, S. E., and Abou-Arab, T. W., A Two-Equation Turbulence Model for Two-Phase Flows, *Physics of Fluids*, vol. 26, no. 4, pp. 931–938, 1983.
155. Gibeling, H. J., McDonald, H., and Banks, N. E., An Implicit Numerical Analysis for Two-Dimensional, Two-Phase Turbulent Interior Ballistic Flows, AIAA 21st Aerospace Sciences Meeting, Reno, Nev., Paper No. AIAA-83-0561, 1983.
156. Mostafa, A. A., and Elghobashi, S. E., A Study of the Motion of Vaporizing Droplets in a Turbulent Flow, in *Dynamics of Flames and Reactive Systems*, eds. J. R. Bowen, N. Manson, A. K. Oppenheim, and R. I. Soloukhin, *Progress in Astronautics and Aeronautics*, vol. 95, pp. 513–539, AIAA, New York, 1984.
157. O'Rourke, P. J., Collective Drop Effects on Vaporizing Liquid Sprays, Ph.D. thesis, Princeton University, Princeton, N.J., 1981.
158. Amsden, A. A., Ramshaw, J. D., O'Rourke, P. J., and Dukowicz, J. K., KIVA: A Computer Program for Two- and Three-Dimensional Fluid Flows with Chemical Reactions and Fuel Sprays, Report No. LA-10245-MS, Los Alamos National Laboratory, Los Alamos, N.M., 1985.
159. O'Rourke, P. J., Personal communication, Dubrovnik, Yugoslavia, September 1, 1987.
160. Modarress, D., Wuerer, J., and Elghobashi, S., An Experimental Study of a Turbulent Round Two-Phase Jet, *Chemical Engineering Communications*, vol. 28, pp. 341–354, 1984.

161. Modarress, D., Tan, H., and Elghobashi, S., Two-Component LDA Measurements in a Two-Phase Turbulent Jet, AIAA 21st Aerospace Sciences Meeting, Reno, Nev., Paper No. AIAA-83-0052, 1983.
162. Modarress, D., Wuerer, J., and Elghobashi, S., An Experimental Study of a Turbulent Round Two-Phase Jet, AIAA/ASME 3rd Joint Thermophysics, Fluids, Plasma and Heat Transfer Conference, St. Louis, Mo., Paper No. AIAA-82-0964, 1982.
163. Faeth, G. M., Mixing, Transport and Combustion in Sprays, *Progress in Energy and Combustion Science*, vol. 13, 293–345, 1987.
164. O'Rourke, P. J., and Bracco, F. V., Modelling of Drop Interactions in Thick Sprays and a Comparison with Experiments, *Stratified Charge Automotive Engines*, pp. 101–116, I. Mech. E. Conference Publications 1980-9, 1980.
165. Hinze, J. O., *Turbulence*, 2nd ed., pp. 427, 724–734, McGraw-Hill, New York, 1975.
166. Choi, Y. D., and Chung, M. K., Analysis of Turbulent Gas-Solid Suspension Flow in a Pipe, *ASME Journal of Fluids Engineering*, vol. 105, pp. 329–334, 1983.
167. Hinze, J. O., Turbulent Fluid and Particle Interaction, in *Progress in Heat and Mass Transfer*, eds. G. Hetsroni, S. Sideman, and J. P. Harnett, vol. 6, pp. 433–452, Pergamon Press, New York, 1972.
168. Tchen, C.-M., Mean Value and Correlation Problems Connected with the Motion of Small Particles Suspended in a Turbulent Fluid, Ph.D. thesis, University of Delft, Delft, The Netherlands, 1947.
169. Tchen, C.-M., Motion of Small Particles of Skew Shape Suspended in a Viscous Fluid, *Journal of Applied Physics*, vol. 25, pp. 463–473, 1954.
170. Ahmadi, G., and Goldschmidt, V. W., Kinematic Computer Simulation of the Turbulent Dispersion of Neutrally Buoyant Particles, in *Developments in Mechanics, Proceedings of the Eleventh Midwestern Mechanics Conference*, eds. H. J. Weiss, D. F. Young, W. F. Riley, and T. R. Roggie, vol. 5, pp. 201–214, Iowa State University Press, Ames, 1969.
171. Goldschmidt, V. W., Householder, M. K., Ahmadi, G., and Chuang, S. C., Turbulent Diffusion of Small Particles Suspended in Turbulent Jets, in *Progress in Heat and Mass Transfer*, eds. G. Hetsroni, S. Sideman, and J. P. Harnett, vol. 6, pp. 487–508, Pergamon Press, New York, 1972.
172. Hjelmfelt, A. T., and Mockros, L. F., Motion of Discrete Particles in a Turbulent Fluid, *Applied Scientific Research*, vol. 16, pp. 149–161, 1966.
173. Saffman, P. G., The Lift on a Small Sphere in a Slow Shear Flow, *Journal of Fluid Mechanics*, vol. 22, pp. 385–400, 1965.
174. Faeth, G. M., Evaporation and Combustion of Sprays, *Progress in Energy and Combustion Science*, vol. 9, no. 1/2, pp. 1–76, 1983.
175. Einstein, A., *Investigations on the Theory of the Brownian Motion*, Dover, New York, 1956.
176. Davies, C. N., Deposition of Aerosols from Turbulent Flow through Pipes, *Proceedings of the Royal Society of London, Series A*, vol. 289, pp. 235–246, 1966.
177. Cox, R. G. and Mason, S. G., Suspended Particles in Fluid Flow through Tubes, *Annual Review of Fluid Mechanics*, vol. 3, pp. 291–316, 1971.
178. Grasso, F., Wey, M.-J., Bracco, F. V., and Abraham, J., Three-Dimensional Computations of Flows in a Stratified-Charge Rotary Engine, SAE International Congress and Exposition, Detroit, Mich., SAE Paper No. 870409, 1987.
179. Tal, R., Lee, D. N., and Sirignano, W. A., Heat and Momentum Transfer around a Pair of Spheres in Viscous Flows, *International Journal of Heat and Mass Transfer*, vol. 27, no. 11, pp. 1953–1962, 1984.
180. Bellan, J., and Cuffel, R., A Theory of Nondilute Spray Evaporation Based upon Multiple Drop Interactions, *Combustion and Flame*, vol. 51, pp. 55–67, 1983.
181. Samson, R., Bedeaux, D., Saxton, M. J., and Deutch, J. M., A Simple Model of Fuel Spray Burning: I. Random Sprays, *Combustion and Flame*, vol. 31, pp. 215–221, 1978.
182. Samson, R., Bedeaux, D., and Deutch, J. M., A Simple Model of Fuel Spray Burning: II. Linear Droplet Systems, *Combustion and Flame*, vol. 31, pp. 223–229, 1978.
183. Bracco, F. V., Modeling of Two-Phase, Two-Dimensional, Unsteady Combustion for Internal

Combustion Engines, *Stratified Charge Engines*, pp. 167–187, I. Mech. E. Conference Publications 1976-11, 1976.

184. Bracco, F. V., Gupta, H. C., Krishnamurthy, L., Santavicca, D. A., Steinberger, R. L., and Warshaw, V., Two-Phase, Two-Dimensional, Unsteady Combustion in Internal Combustion Engines: Preliminary Theoretical-Experimental Results, SAE International Congress and Exposition, Detroit, Mich., SAE Paper No. 760114, 1976.
185. Gupta, H. C., and Bracco, F. V., Numerical Computations of Two-Dimensional Unsteady Sprays for Application to Engines, *AIAA Journal*, vol. 16, no. 10, pp. 1053–1061, 1978.
186. Haselman, L. C., and Westbrook, C. K., A Theoretical Model for Two-Phase Fuel Injection in Stratified Charge Engines, SAE International Congress and Exposition, Detroit, Mich., SAE Paper No. 780318, 1978.
187. Westbrook, C. K., Three-Dimensional Numerical Modeling of Liquid Fuel Sprays, *Sixteenth Symposium (International) on Combustion*, pp. 1517–1526, The Combustion Institute, Pittsburgh, Pa., 1977.
188. Clift, R., Grace, J. R., and Weber, M. E., *Bubbles, Drops, and Particles*, p. 30, Academic Press, New York, 1978.
189. Happel, J., and Brenner, H., *Low Reynolds Number Hydrodynamics*, p. 119, Prentice-Hall, Englewood Cliffs, N.J., 1962.
190. Renksizbulut, M., and Yuen, M. C., Experimental Study of Droplet Evaporation in a High-Temperature Air Stream, *ASME Journal of Heat Transfer*, vol. 105, no. 2, pp. 384–388, 1983.
191. Renksizbulut, M., and Yuen, M. C., Numerical Study of Droplet Evaporation in a High-Temperature Stream, *ASME Journal of Heat Transfer*, vol. 105, no. 2, pp. 389–397, 1983.
192. Crowe, C. T., Sharma, M. P., and Stock, D. E., The Particle-Source-In Cell (PSI-CELL) Model for Gas-Droplet Flow, *ASME Journal of Fluids Engineering*, vol. 99, pp. 325–332, 1977.
193. Prakash, S., and Sirignano, W. A., Theory of Convective Droplet Vaporization with Unsteady Heat Transfer in the Circulating Liquid Phase, *International Journal of Heat and Mass Transfer*, vol. 23, no. 3, pp. 253–268, 1980.
194. Prakash, S., and Sirignano, W. A., Liquid Fuel Droplet Heating with Internal Circulation, *International Journal of Heat and Mass Transfer*, vol. 21, pp. 885–895, 1978.
195. Law, C. K., Prakash, S., and Sirignano, W. A., Theory of Convective, Transient, Multicomponent Droplet Vaporization, *Sixteenth Symposium (International) on Combustion*, pp. 605–617, The Combustion Institute, Pittsburgh, Pa., 1977.
196. Sirignano, W. A., Fuel Droplet Vaporization and Spray Combustion Theory, *Progress in Energy and Combustion Science*, vol. 9, pp. 291–322, 1983.
197. Aggarwal, S. K., Tong, A. Y., and Sirignano, W. A., A Comparison of Vaporization Models in Spray Calculations, *AIAA Journal*, vol. 22, no. 10, pp. 1448–1457, 1984.
198. Jin, J. D., and Borman, G. L., A Model for Multicomponent Droplet Vaporization at High Ambient Pressures, SAE International Congress and Exposition, Detroit, Mich., SAE Paper No. 850264, 1985.
199. Tong, A. Y., and Sirignano, W. A., Multicomponent Droplet Vaporization in a High Temperature Gas, *Combustion and Flame*, vol. 66, pp. 221–235, 1986.
200. Tong, A. Y., and Sirignano, W. A., Multicomponent Transient Droplet Vaporization with Internal Circulation: Integral Equation Formulation and Approximate Solution, *Numerical Heat Transfer*, vol. 10, pp. 253–278, 1986.
201. Kerstein, A. R., Prediction of the Concentration PDF for Evaporating Sprays, *International Journal of Heat and Mass Transfer*, vol. 27, no. 8, pp. 1291–1309, 1984.
202. Marble, F. E., Dynamics of Dusty Gases, *Annual Review of Fluid Mechanics*, vol. 2, pp. 397–446, 1970.
203. Peskin, R. L., Turbulent Fluid-Particle Interaction, in *Handbook of Multiphase Systems*, ed. G. Hetsroni, pp. 1.184–1.204, Hemisphere, Washington, D.C., 1982.
204. Boysan, F., Ayers, W. H., Swithenbank, J., and Pan, Z., Three-Dimensional Model of Spray Combustion in Gas Turbine Combustors, AIAA 19th Aerospace Sciences Meeting, St. Louis, Mo., Paper No. AIAA-81-0324, 1981.

205. Boysan, F., and Binark, H., Predictions of Induced Air Flows in Hollow Cone Sprays, *ASME Journal of Fluids Engineering*, vol. 101, pp. 312–318, 1979.
206. Gosman, A. D., and Ioannides, E., Aspects of Computer Simulation of Liquid-Fuelled Combustors, AIAA 19th Aerospace Sciences Meeting, St. Louis, Mo., Paper No. AIAA-81-0323, 1981.
207. Gosman, A. D., and Johns, R. J. R., Computer Analysis of Fuel-Air Mixing in Direct-Injection Engines, SAE International Congress and Exposition, Detroit, Mich., SAE Paper No. 800091, 1980.
208. Gosman, A. D., and Harvey, P. S., Computer Analysis of Fuel-Air Mixing and Combustion in an Axisymmetric D.I. Diesel, SAE International Congress and Exposition, Detroit, Mich., SAE Paper No. 820036, 1982.
209. Elkotb, M. M., Fuel Atomization for Spray Modelling, *Progress in Energy and Combustion Science*, vol. 8, pp. 61–91, 1982.
210. Abramzon, B., and Elata, C., Unsteady Heat Transfer from a Single Sphere in Stokes Flow, *International Journal of Heat and Mass Transfer*, vol. 27, no. 5, pp. 687–695, 1984.
211. Abramzon, B., Numerical Analysis of Steady Convective Heat Transfer from a Sphere with Non-Uniform Surface Temperature, Paper No. 84-HT-56, ASME, New York, 1984.
212. Nguyen, H. L., Carpenter, M. H., and Ramos, J. I., The Effects of Engine Speed and Injection Characteristics on the Flow Field and Fuel/Air Mixing in Motored Two-Stroke Diesel Engines, AIAA 25th Aerospace Sciences Meeting, Reno, Nev., Paper No. AIAA-87-0227, 1987.
213. Nguyen, H. L., Schock, H. J., Carpenter, M. H., Ramos, J. I., and Stegeman, J. D., Numerical Simulation of the Flow Field and Fuel Sprays in an IC Engine, SAE International Congress and Exposition, Detroit, Mich., SAE Paper No. 870599, 1987.
214. Carpenter, M. H., and Ramos J. I., Mathematical Modelling of Uniflow-scavenged Two-Stroke Diesel Engines, in *Modelado en Motores de Combustion Interna*, eds. F. Payri, J. M. Desantes, and F. Boada, pp. 321–332, Universidad Politecnica de Valencia, Valencia, Spain, 1987.
215. Carpenter, M. H., and Ramos, J. I., Spray Penetration, Vaporization and Mixing in Diesel Engines, XIX International Symposium on Heat and Mass Transfer in Gasoline and Diesel Engines, International Centre for Heat and Mass Transfer, Dubrovnik, Yugoslavia, 1987.
216. Hiroyasu, H., and Kadota, T., Fuel Droplet Size Distribution in Diesel Combustion Chamber, SAE Congress and Exposition, Detroit, Mich., SAE Paper No. 740715, 1974.
217. Dukowicz, J. K., A Particle-Fluid Numerical Model for Liquid Sprays, *Journal of Computational Physics*, vol. 35, pp. 229–253, 1980.
218. Cloutman, L. D., Dukowicz, J. K., and Ramshaw, J. D., Numerical Simulation of Reactive Flow in Internal Combustion Engines, in *Seventh International Conference on Numerical Methods in Fluid Dynamics*, eds. W. C. Reynolds and R. W. MacCormack, *Lecture Notes in Physics*, vol. 141, pp. 119–124, Springer-Verlag, New York, 1981.
219. Butler, T. D., Cloutman, L. D., Dukowicz, J. K., and Ramshaw, J. D., Multidimensional Numerical Simulation of Reactive Flow in Internal Combustion Engines, *Progress in Energy and Combustion Science*, vol. 7, pp. 293–315, 1981.
220. Cloutman, L. D., Dukowicz, J. K., Ramshaw, J. D., and Amsden, A. A., CONCHAS-SPRAY: A Computer Code for Reactive Flows with Fuel Sprays, Report No. LA-9294-MS, Los Alamos National Laboratory, Los Alamos, N.M., 1982.
221. Bassoli, C., Biaggini, G., Bodritti, G., and Cornetti, G. M., Two-Dimensional Combustion Chamber Analysis of Direct Injection Diesel, SAE International Congress and Exposition, Detroit, Mich., SAE Paper No. 840228, 1984.
222. Bassoli, C., Bodritti, G., and Cornetti, G. M., Optimum Air Momentum and Spray Formation for D.I. Diesel, SAE International Congress and Exposition, Detroit, Mich., SAE Paper No. 850501, 1985.
223. Butler, T. D., Cloutman, L. D., Dukowicz, J. K., Ramshaw, J. D., and Krieger, R. B., Toward a Comprehensive Model for Combustion in a Direct-Injection Stratified-Charge Engine, in *Combustion Modeling in Reciprocating Engines*, eds. J. N. Mattavi and C. A. Amann, pp. 231–264, Plenum Press, New York, 1980.
224. Ramshaw, J. D., and Dukowicz, J. K., APACHE: A Generalized-Mesh Eulerian Computer

Code for Multicomponent Reactive Fluid Flow, Report No. LA-7427, Los Alamos National Laboratory, Los Alamos, N.M., 1979.

225. Hirt, C. W., An Arbitrary Lagrangian-Eulerian Computing Technique, in *Proceedings of the Second International Conference on Numerical Methods in Fluid Dynamics*, ed. M. Holt, *Lecture Notes in Physics*, vol. 8, pp. 350–355, Springer-Verlag, New York, 1971.
226. Rivard, W. C., Butler, T. D., and Farmer, O. A., The Transient Dynamics of Chemically Reactive Gaseous Mixtures with Turbulence, in *Proceedings of the Fourth International Conference on Numerical Methods in Fluid Dynamics*, ed. R. D. Richtmyer, *Lecture Notes in Physics*, vol. 35, pp. 334–340, Springer-Verlag, New York, 1975.
227. Kuo, T.-W., and Bracco, F. V., Computations of Drop Sizes in Pulsating Sprays and of Liquid-Core Length in Vaporizing Sprays, SAE International Congress and Exposition, Detroit, Mich., SAE Paper No. 820133, 1982.
228. Kuo, T.-W., and Yu, R. C., Modeling of Transient Evaporating Spray Mixing Processes—Effect of Injection Characteristics, SAE International Congress and Exposition, Detroit, Mich., SAE Paper No. 840226, 1984.
229. Chatwani, A. U., and Bracco, F. V., Computation of Dense Spray Jets in *Proceedings of the 1985 International Conference on Liquid Atomisation and Spray Systems, ICLASS-85*, eds. P. Eisenklam and A. Yule, vol. 1, pp. IB/1/1-IB/1/12, Institute of Energy, London, 1985.
230. Chehroudi, B., Chen, S.-H., Bracco, F. V., and Onuma, Y., On the Intact Core of Full-Cone Sprays, SAE International Congress and Exposition, Detroit, Mich., SAE Paper No. 850126, 1985.
231. Martinelli, L., Bracco, F. V., and Reitz, R. D., Comparisons of Computed and Measured Dense Spray Jets, in *Dynamics of Flames and Reactive Systems*, eds. J. R. Bowen, N. Manson, A. K. Oppenheim, and R. I. Soloukhin, *Progress in Astronautics and Aeronautics*, vol. 95, pp. 484–512, AIAA, New York, 1984.
232. Duggal, V. K., Kuo, T.-W., Mukerjee, T., Przekwas, A. J., and Singhal, A. K., Three-dimensional Modeling of In-Cylinder Processes in DI Diesel Engines, SAE International Congress and Exposition, Detroit, Mich., SAE Paper No. 840227, 1984.
233. Amsden, A. A., Butler, T. D., O'Rourke, P. J., and Ramshaw, J. D., KIVA—A Comprehensive Model for 2-D and 3-D Engine Simulations, SAE International Congress and Exposition, Detroit, Mich., SAE Paper No. 850554, 1985.
234. O'Rourke, P. J., and Amsden, A. A., Three Dimensional Numerical Simulations of the UPS-292 Stratified Charge Engine, SAE International Congress and Exposition, Detroit, Mich., SAE Paper No. 870597, 1987.
235. Reitz, R. D., and Diwakar, R., Effect of Drop Breakup on Fuel Sprays, SAE International Congress and Exposition, Detroit, Mich., SAE Paper No. 860469, 1986.
236. Reitz, R. D., and Diwakar, R., Structure of High-Pressure Fuel Sprays, SAE International Congress and Exposition, Detroit, Mich., SAE Paper No. 870598, 1987.
237. Wu, K.-J., Santavicca, D. A., Bracco, F. V., and Coghe, A., LDV Measurements of Drop Velocity in Diesel-Type Sprays, *AIAA Journal*, vol. 22, no. 9, pp. 1263–1270, 1984.
238. Frankel, I., and Weihs, D., Shape and Stability of a Liquid Drop Moving at Low Reynolds Numbers, *Applied Scientific Research*, vol. 40, pp. 279–294, 1983.
239. Gonor, A. L., and Zolotova, N. V., Spreading and Break-up of a Drop in a Gas Stream, *Acta Astronautica*, vol. 11, no. 2, pp. 137–142, 1984.
240. Sundarajan, T., and Ayyaswamy, P. S., Heat and Mass Transfer Associated with Condensation on a Moving Drop: Solutions for Intermediate Reynolds Numbers by a Boundary Layer Formulation, *ASME Journal of Heat Transfer*, vol. 107, no. 2, pp. 409–416, 1985.
241. Watkins, A. P., Three-Dimensional Modelling of Gas Flow and Sprays in Diesel Engines, Advanced Course on Computer Simulation of Fluid Flow, Heat and Mass Transfer and Combustion in Reciprocating Engines, International Centre for Heat and Mass Transfer, Dubrovnik, Yugoslavia, 1987.
242. O'Rourke, P. J., and Amsden, A. A., The TAB Method for Numerical Calculation of Spray Droplet Breakup, SAE International Fuels and Lubricants Meeting and Exposition, Toronto, Ont., Canada, SAE Paper No. 872089, 1987.

243. Natarajan, R., and Brown, R. A., Third-Order Resonance Effects and Nonlinear Stability of Drop Oscillations, *Journal of Fluid Mechanics*, vol. 183, pp. 95–121, 1987.
244. Tsamopoulos, J. A., and Brown, R. A., Nonlinear Oscillations of Inviscid Drops and Bubbles, *Journal of Fluid Mechanics*, vol. 127, pp. 519–537, 1983.
245. Tsamopoulos, J. A., and Brown, R. A., Resonant Oscillations of Inviscid Charged Drops, *Journal of Fluid Mechanics*, vol. 147, pp. 373–395, 1984.
246. Natarajan, R., and Brown, R. A., Quadratic Resonance in the Three-Dimensional Oscillations of Inviscid Drops with Surface Tension, *Physics of Fluids*, vol. 29, no. 9, pp. 2788–2797, 1986.
247. Karlsson, A., Personal communication, Dubrovnik, Yugoslavia, September 1, 1987.
248. Amsden, A. A., Butler, T. D., and O'Rourke, P. J., The KIVA-II Computer Program for Transient Multidimensional Chemically Reactive Flows with Sprays, SAE International Fuels and Lubricants Meeting and Exposition, Toronto, Ont., Canada, SAE Paper No. 872072, 1987.
249. O'Rourke, P. J., Personal communication, Dubrovnik, Yugoslavia, September 1, 1987.
250. Liu, L., and Yao, S.-C., Heat Transfer Analysis of Droplet Flow Impinging on a Hot Surface, Department of Mechanical Engineering, Carnegie-Mellon University, Pittsburgh, Pa., 1983.
251. Fujimoto, H., Saitou, M., Minoura, A., Cho, I. Y., and Senda, J., Characteristics of a Diesel Spray Impinging on a Flat Surface, XIX International Symposium on Heat and Mass Transfer in Gasoline and Diesel Engines, International Centre for Heat and Mass Transfer, Dubrovnik, Yugoslavia, 1987.
252. Holman, P. J., *Heat Transfer*, 6th ed., pp. 511–512, McGraw-Hill, New York, 1986.
253. Mizomoto, M., and Ikai, S., Evaporation and Ignition of a Fuel Droplet on a Hot Surface: Part 4. Model of Evaporation and Ignition, *Combustion and Flame*, vol. 51, pp. 95–104, 1983.
254. Taylor, G. I., Oblique Impact of a Jet on a Plane Surface, *Philosophical Transactions of the Royal Society of London, Series A*, vol. 260, pp. 96–100, 1966.
255. Naber, J. D., and Reitz, R. D., Modeling Engine Spray/Wall Impingement, Report No. GMR-6097, General Motors Research Laboratories, Warren, Mich., 1987.
256. Diwakar, R., Multidimensional Modeling Applied to the Direct-Injection Stratified-Charge Engine-Calculation versus Experiment, SAE International Congress and Exposition, Detroit, Mich., SAE Paper No. 810225, 1981.
257. Diwakar, R., Direct-Injection Stratified-Charge Engine Computations with Improved Submodels for Turbulence and Heat Transfer, SAE International Congress and Exposition, Detroit, Mich., SAE Paper No. 820039, 1982.
258. Benjamin, S. F., Weaving, J. H., Glynn, D. R., Markatos, N. C., and Spalding, D. B., Development of a Mathematical Model of Flow, Heat Transfer and Combustion in a Stratified Charge Engine, *Stratified Charge Automotive Engines*, pp. 91–99, I. Mech. E. Conference Publications 1980-9, London, 1980.
259. Gupta, H. C., REC (Reciprocating Engine Combustion): A Computer Program for the Computation of Combustion in Open and Divided Chamber Reciprocating Engines with Uniform and Stratified Charges, AMS Report No. 1365, Department of Aerospace and Mechanical Sciences, Princeton University, Princeton, N.J., 1978.
260. Gupta, H. C., Steinberger, R. L., and Bracco, F. V., Combustion in a Divided Chamber, Stratified Charge, Reciprocating Engine: Initial Comparisons of Calculated and Measured Flame Propagation, *Combustion Science and Technology*, vol. 22, pp. 27–61, 1980.
261. Syed, S. A., and Bracco, F. V., Further Comparisons of Computed and Measured Divided-Chamber Engine Combustion, SAE International Congress and Exposition, Detroit, Mich., SAE Paper No. 790247, 1979.
262. Grasso, F., and Bracco, F. V., Evaluation of a Mixing-controlled Model for Engine Combustion, *Combustion Science and Technology*, vol. 28, pp. 185–210, 1982.
263. Gupta, H. C., and Bracco, F. V., The Origin of Pressure Oscillations in Divided Chamber Engines, *Combustion and Flame*, vol. 48, pp. 33–49, 1982.
264. Meintjes, K., and Alkidas, A. C., An Experimental and Computational Investigation of the Flow in Diesel Prechambers, SAE International Congress and Exposition, Detroit, Mich., SAE Paper No. 820275, 1982.

265. Boni, A. A., Chapman, M., and Schneyer, G. P., Computer Simulation of Combustion Processes in a Stratified Charge Engine, *Acta Astronautica*, vol. 3, pp. 293–307, 1976.
266. Boni, A. A., Chapman, M., Cook, J. L., and Schneyer, G. P., Computer Simulation of Combustion in a Stratified Charge Engine, *Sixteenth Symposium (International) on Combustion*, pp. 1527–1544, The Combustion Institute, Pittsburgh, Pa., 1977.
267. Ishikawa, N., A Diffusion Combustor and Methane-Air Flame Propagation in Concentration Gradient Fields, *Combustion Science and Technology*, vol. 30, pp. 185–203, 1983.
268. Ishikawa, N., Combustion of Stratified Methane/Air Layers, *Combustion Science and Technology*, vol. 30, pp. 311–325, 1983.
269. Spalding, D. B., The "Shadow" Method of Particle-Size Calculation in Two-Phase Combustion, *Nineteenth Symposium (International) on Combustion*, pp. 941–951, The Combustion Institute, Pittsburgh, Pa., 1982.
270. Shahed, S. M., The Role of Fuel-Air Mixing in Diesel Combustion and Emissions, in *International Symposium on Flows in Internal Combustion Engines—III*, eds. T. Uzkan, W. G., Tiederman, and J. M. Novak, pp. 159–164, ASME, New York, 1985.
271. Glassman, I., and Yaccarino, P., The Temperature Effect in Sooting Flames, *Eighteenth Symposium (International) on Combustion*, pp. 1175–1183, The Combustion Institute, Pittsburgh, Pa., 1981.
272. Frenklach, M., Taki, S., and Matula, R. A., A Conceptual Model for Soot Formation in Pyrolysis of Aromatic Hydrocarbons, *Combustion and Flame*, vol. 49, pp. 275–282, 1983.
273. Kennedy, I. M., Flow Field Effects on Nucleation in a Reacting Mixing Layer, *Physics of Fluids*, vol. 28, no. 12, pp. 3515–3524, 1985.
274. Haynes, H. S., and Wagner, H. Gg., Soot Formation, *Progress in Energy and Combustion Science*, vol. 7, pp. 229–273, 1981.
275. Smith, O. I., Fundamentals of Soot Formation in Flames with Application to Diesel Engine Particulate Emissions, *Progress in Energy and Combustion Science*, vol. 7, pp. 275–291, 1981.
276. D'Anna, A., Combustione nei Motori Diesel, Istituto di Ricerche sulla Combustione, CNR, University of Naples, Naples, Italy, 1987.
277. Khan, I. M., Wang, C. H. T., and Langridge, B. E., Coagulation and Combustion of Soot Particles in Diesel Engines, *Combustion and Flame*, vol. 17, pp. 409–419, 1971.
278. Lee, K. B., Thring, M. W., and Beer, J. M., On the Rate of Combustion of Soot in a Laminar Soot Flame, *Combustion and Flame*, vol. 6, pp. 137–145, 1962.
279. Bellan, J. R., A New Approach to Soot Control in Diesel Engines by Fuel-Drop Charging, *Combustion and Flame*, vol. 51, pp. 117–119, 1983.
280. Peterson, R. C., and Alkidas, A. C., A Visual Study of a Divided-Chamber Diesel Combustion using a Rapid Compression Machine, *Combustion and Flame*, vol. 53, pp. 65–81, 1983.
281. Flower, D. L., The Effect of Elevated Pressure on the Rate of Soot Production in Laminar Diffusion Flames, *Combustion Science and Technology*, vol. 48, pp. 31–43, 1986.
282. Rah, S.-C., Sarofim, A. F., and Beer, J. M., Ignition and Combustion of Liquid Fuel Droplets: Part I. Impact on Pollution, *Combustion Science and Technology*, vol. 48, pp. 273–284, 1986.
283. Uyehara, O. A., Diesel Combustion Temperature on Soot, SAE International Congress and Exposition, Detroit, Mich., SAE Paper No. 800969, 1980.
284. Buchholz, E. K., and Tapper, M. L., Time to Extinction of Liquid Hydrocarbon Fuel Droplets Burning in a Transient Diesel-like Environment, *Combustion and Flame*, vol. 31, pp. 161–171, 1978.
285. Law, C. K., Theory of Thermal Ignition in Fuel Droplet Burning, *Combustion and Flame*, vol. 31, pp. 285–296, 1978.
286. Chin, J. S., and Lefebvre, A. H., The Role of the Heat-up Period in Fuel Drop Evaporation, AIAA 21st Aerospace Sciences Meeting, Reno, Nev., Paper No. AIAA-83-0068, 1983.
287. Aggarwal, S. K., and Sirignano, W. A., Ignition of Polydisperse Sprays: Importance of D_{20}, *Combustion Science and Technology*, vol. 46, pp. 289–300, 1986.
288. Choudhury, P. R., and Gerstein, M., Effect of Liquid Phase Decomposition on Fuel Droplet Distribution Function, AIAA 21st Aerospace Sciences Meeting, Reno, Nev., Paper No. AIAA-83-0069, 1983.

289. Seth, B., Aggarwal, S. K., and Sirignano, W. A., Flame Propagation through an Air-Fuel Spray Mixture with Transient Droplet Vaporization, *Combustion and Flame*, vol. 39, pp. 149–168, 1980.
290. Aggarwal, S. K., and Sirignano, W. A., Unsteady Spray Flame Propagation in a Closed Volume, *Combustion and Flame*, vol. 62, pp. 69–84, 1985.
291. Kagami, M., Akasaka, Y., Date, K., and Maeda, T., The Influence of Fuel Properties on the Performance of Japanese Automobile Diesels, SAE International Congress and Exposition, Detroit, Mich., SAE Paper No. 841082, 1984.
292. Tateishi, M., Imoto, K., Inenaga, N., Ishikawa, H., and Harada, S., New Combustion System of the IDI Diesel Engine, SAE International Congress and Exposition, Detroit, Mich., SAE Paper No. 841081, 1984.
293. Heywood, J. B., Engine Combustion Modeling—An Overview, in *Combustion Modeling in Reciprocating Engines*, eds. J. N. Mattavi and C. A. Amann, pp. 1–38, Plenum Press, New York, 1980.
294. Henein, N. A., Analysis of Pollutant Formation and Control and Fuel Economy in Diesel Engines, in *Energy and Combustion Science*, ed. N. A. Chigier, student ed. 1, pp. 283–325, Pergamon Press, New York, 1979.
295. Callahan, T. J., Ryan, T. W., Dietzmann, H., and Waytulonis, R., The Effect of Discrete Transients in Speed and Load on Diesel Engine Exhaust Emissions, SAE International Congress and Exposition, Detroit, Mich., SAE Paper No. 850109, 1985.
296. Takagi, T., Okamoto, T., and Uyehara, O. A., Estimation of Imperfect Mixing in Turbulent Diffusion Flames, SAE International Congress and Exposition, Detroit, Mich., SAE Paper No. 850114, 1985.
297. Nishida, O., and Mukohara, S., Optical Measurement on Soot Particle in a Nonsteady Diffusion Flame, SAE International Congress and Exposition, Detroit, Mich., SAE Paper No. 841079, 1984.
298. Johnston, S. C., and Green, R. M., Engine Hydrocarbon Concentration Measurements: A Comparison between a Sampling Probe and Raman Spectroscopy, SAE International Congress and Exposition, Detroit, Mich., SAE Paper No. 820499, 1982.
299. Sinnamon, J. F., and Cole, D. E., The Influence of Overall Equivalence Ratio and Degree of Stratification on the Fuel Consumption and Emissions of a Prechamber, Stratified-Charge Engine, SAE Congress and Exposition, Detroit, Mich., SAE Paper No. 790438, 1979.
300. Turns, S. R., and Borman, G. L., Internal Combustion Engine Sampling Probes: An Experimental Study of Sample-Gas Origins and Flow Disturbances, *Combustion and Flame*, vol. 38, pp. 313–327, 1980.
301. Turns, S. R., and Borman, G. L., An Estimation of Quench-Gas Bias in Internal Combustion Engine In-Cylinder Sampling, *Combustion and Flame*, vol. 39, pp. 97–109, 1980.
302. Yanagi, T., Effect of Probe Sampling Rates on Sample Composition, *Combustion and Flame*, vol. 28, pp. 33–44, 1977.
303. Yamazaki, N., Miyamoto, N., and Murayama, T., The Effects of Flash Boiling Fuel Injection on Spray Characteristics, Combustion, and Engine Performance in DI and IDI Diesel Engines, SAE International Congress and Exposition, Detroit, Mich., SAE Paper No. 850071, 1985.
304. Oza, R. D., and Sinnamon, J. F., An Experimental and Analytical Study of Flash-boiling Fuel Injection, SAE International Congress and Exposition, Detroit, Mich., SAE Paper No. 830590, 1983.
305. Calcote, H. F., Mechanisms of Soot Nucleation in Flames—A Critical Review, *Combustion and Flame*, vol. 42, pp. 215–242, 1981.
306. Olson, D. B., and Pickens, J. C., The Effects of Molecular Structure on Soot Formation: I. Soot Thresholds in Premixed Flames, *Combustion and Flame*, vol. 57, pp. 199–208, 1984.
307. Olson, D. B., and Madronich, S., The Effect of Temperature on Soot Formation in Premixed Flames, *Combustion and Flame*, vol. 60, pp. 203–213, 1985.
308. Bernhardt, W., Combustion Technology for the Improvement of Engine Efficiency and Emission Characteristics, *Sixteenth Symposium (International) on Combustion*, pp. 223–232, The Combustion Institute, Pittsburgh, Pa., 1977.

309. Shimoda, M., Shigemori, M., and Tsuruoka, S., Effect of Combustion Chamber Configuration on In-Cylinder Air Motion and Combustion Characteristics of D.I. Diesel Engine, SAE International Congress and Exposition, Detroit, Mich., SAE Paper No. 850070, 1985.
310. Rah, S.-C., Sarofim, A. F., and Beer, J. M., Ignition and Combustion of Liquid Fuel Droplets: Part II. Ignition Studies, *Combustion Science and Technology*, vol. 49, pp. 169–184, 1986.
311. Randolph, A. L., and Law, C. K., Influence of Physical Mechanisms on Soot Formation and Destruction in Droplet Burning, *Combustion and Flame*, vol. 64, pp. 267–284, 1986.
312. Obuchi, A., Ohi, A., Aoyama, H., and Obuchi, H., Evaluation of Gaseous and Particulate Emission Characteristics of a Single Cylinder Diesel Engine, *Combustion and Flame*, vol. 70, pp. 215–224, 1987.
313. Iida, N., Sato, G. T., and Hasebe, J., Study on Characteristics of Particulate Emissions from a Direct Injection Diesel Engine Using a Particulate Method in Sampling Process, SAE International Congress and Exposition, Detroit, Mich., SAE Paper No. 841077, 1984.
314. Roessler, D. M., and Faxvog, F. R., Changes in Diesel Particulates with Engine Air/Fuel Ratio, *Combustion Science and Technology*, vol. 26, pp. 225–231, 1981.
315. Siegla, D. C., and Plee, S. L., Heterogeneous Catalysis in the Diesel Combustion Chamber, *Combustion Science and Technology*, vol. 27, pp. 97–102, 1982.
316. Veselinovic, B., Some More Notes on the Amount of Diffusion Burning Causing Smoke Formation, SAE International Congress and Exposition, Detroit, Mich., SAE Paper No. 850050, 1985.
317. Van Gerpen, J. H., Huang, C.-W., and Borman, G. L., The Effects of Swirl and Injection Parameters on Diesel Combustion and Heat Transfer, SAE International Congress and Exposition, Detroit, Mich., SAE Paper No. 850265, 1985.
318. Nakanishi, K., Arai, M., and Hiroyasu, H., The Effect of Air Swirl on the Time History of the Diesel Particulate in the Combustion Chamber, *ASME-JSME Thermal Engineering Joint Conference Proceedings*, vol. 4, pp. 241–246, ASME, New York, 1983.
319. Hiroyasu, H., Arai, M., Nishida, K., and Yoshikawa, S., Behaviour of Combustion Swirl in Direct Injection Diesel Engine, *ASME-JSME Thermal Engineering Joint Conference Proceedings*, vol. 4, pp. 233–239, ASME, New York, 1983.
320. Shimoda, M., Shigemori, M., and Tsuruoka, S., Effect of Combustion Chamber Configuration on In-Cylinder Air Motion and Combustion Characteristics of D.I. Diesel Engine, SAE International Congress and Exposition, Detroit, Mich., SAE Paper No. 850070, 1985.
321. Norris-Jones, S. R., Hollis, T., and Waterhouse, C. N. F., A Study of the Formation of Particulates in the Cylinder of a Direct Injection Diesel Engine, SAE International Congress and Exposition, Detroit, Mich., SAE Paper No. 840419, 1984.
322. Du, C. J., and Kittelson, D. B., Total Cylinder Sampling from a Diesel Engine: Part III. Particle Measurements, SAE International Congress and Exposition, Detroit, Mich., SAE Paper No. 830243, 1983.
323. Zhou, Q.-B., Wei, X.-Y., Zhu, T.-Z., and Wang, R.-S., The Study of the Heat Radiation Transfer Process in D.I. Diesel Engine, XIX International Symposium on Heat and Mass Transfer in Gasoline and Diesel Engines, International Centre for Heat and Mass Transfer, Dubrovnik, Yugoslavia, 1987.
324. Siegel, R., and Howell, J. R., *Thermal Radiation Heat Transfer*, 2nd ed., Hemisphere, Washington, D.C., 1981.
325. Whitehouse, N. D., Fu, H., Dannish, S. A. M., Murakami, T., and Clough, E., An Estimate of Local Instantaneous Conditions in a Diesel Engine Based on Sampled Data for Local Constituents and on Radiative and Convective Heat Transfer, XIX International Symposium on Heat and Mass Transfer in Gasoline and Diesel Engines, International Centre for Heat and Mass Transfer, Dubrovnik, Yugoslavia, 1987.
326. Chen, B., Wang, R., and Hu, G., Computer Simulation of Three Types of Combustion Systems for D.I. Diesel Engines, XIX International Symposium on Heat and Mass Transfer in Gasoline and Diesel Engines, International Centre for Heat and Mass Transfer, Dubrovnik, Yugoslavia, 1987.
327. Ovsjannikov, M. K., Measuring of Convective and Radiant Heat Flow to the Walls of a Diesel

Engine Compression Chamber, XIX International Symposium on Heat and Mass Transfer in Gasoline and Diesel Engines, International Centre for Heat and Mass Transfer, Dubrovnik, Yugoslavia, 1987.

328. Kostin, A. K., and Rudnev, B. I., Radiative Heat Transfer in a Diesel Cylinder: An Experiment and Method of Calculation, XIX International Symposium on Heat and Mass Transfer in Gasoline and Diesel Engines, International Centre for Heat and Mass Transfer, Dubrovnik, Yugoslavia, 1987.
329. Kidin, N., Librovich, V., Roberts, J., and Vuillermoz, M., On Sound Sources in Turbulent Combustion, in *Dynamics of Flames and Reactive Systems*, eds. J. R. Bowen, N. Manson, A. K. Oppenheim, and R. I. Soloukhin, *Progress in Astronautics and Aeronautics*, vol. 95, pp. 343–355, AIAA, New York, 1984.
330. Strahle, W. C., Handley, J. C., and Varma, M. S., Cetane Rating and Load Effects on Combustion Noise in Diesel Engines, *Combustion Science and Technology*, vol. 17, pp. 51–61, 1977.
331. Strahle, W. C., Combustion Randomness and Diesel Engine Noise: Theory and Initial Experiments, *Combustion and Flame*, vol. 28, pp. 279–290, 1977.
332. Strahle, W. C., Muthukrishnan, M., and Handley, J. C., Turbulent Combustion and Diesel Engine Noise, *Sixteenth Symposium (International) on Combustion*, pp. 337–346, The Combustion Institute, Pittsburgh, Pa., 1977.
333. Wigley, G., Patterson, A. C., and Renshaw, J., Swirl Velocity Measurements in a Firing Production Diesel Engine by Laser Anemometry, in *Fluid Mechanics of Combustion Systems*, eds. T. Morel, R. P. Lohmann, and J. M. Rackley, pp. 29–39, ASME, New York, 1981.
334. Johnston, S. C., An Experimental Investigation into the Application of Spontaneous Raman Scattering to Spray Measurements in an Engine, in *Fluid Mechanics of Combustion Systems*, eds. T. Morel, R. P. Lohmann, and J. M. Rackley, pp. 107–118, ASME, New York, 1981.
335. Nishida, K., Murakami, A., and Hiroyasu, H., Measurements of Swirling Flow in a Direct Injection Diesel Engine under Motored and Fired Conditions, *Twentieth Symposium (International) on Combustion*, pp. 201–208, The Combustion Institute, Pittsburgh, Pa., 1985.
336. Arcoumanis, C., Bicen, A. F., and Whitelaw, J. H., Measurements in a Motored Four-Stroke Reciprocating Model Engine, in *Fluid Mechanics of Combustion Systems*, eds. T. Morel, R. P., Lohmann, and J. M. Rackley, pp. 21–27, ASME, New York, 1981.
337. Arcoumanis, C., Bicen, A. F., and Whitelaw, J. H., Measurements in a Motored Four-Stroke Reciprocating Model Engine, *ASME Journal of Fluids Engineering*, vol. 104, pp. 235–241, 1982.
338. Arcoumanis, C., Bicen, A. F., and Whitelaw, J. H., Squish and Swirl-Squish Interaction in Motored Model Engines, *ASME Journal of Fluids Engineering*, vol. 105, pp. 105–112, 1983.
339. Fansler, T. D., Laser Velocimetry Measurements of Swirl and Squish Flows in an Engine with a Cylindrical Piston Bowl, SAE International Congress and Exposition, Detroit, Mich., SAE Paper No. 850124, 1985.
340. Arcoumanis, C., Bicen, A. F., Vafidis, C., and Whitelaw, J. H., Three-dimensional Flow Field in Four-Stroke Model Engines, SAE International Congress and Exposition, Detroit, Mich., SAE Paper No. 841360, 1984.
341. El Banhawy, Y., and Whitelaw, J. H., Experimental Study of the Interaction between a Fuel Spray and Surrounding Combustion Air, *Combustion and Flame*, vol. 42, pp. 253–275, 1981.
342. Ball, W. F., Rettifer, H. F., and Waterhouse, C. N. F., Laser Doppler Velocimetry Measurements in a Direct-Injection Diesel Combustion Chamber, *International Conference on Combustion Engineering*, vol. I, pp. 163–174, I. Mech. E. Conference Publications 1983-3, London, 1983.
343. Catania, A. E., 3-D Swirling Flows in an Open-Chamber Automotive Diesel Engine with Different Induction Systems, in *Flows in Internal Combustion Engines*, ed. T. Uzkan, pp. 53–66, ASME, New York, 1982.
344. Wakisaka, T., Hamamoto, Y., and Shimamoto, Y., Turbulence Structure of Air Swirl in Reciprocating Engine Cylinders, in *Flows in Internal Combustion Engines*, ed. T. Uzkan, pp. 93–100, ASME, New York, 1982.

345. Tindal, M. J., Williams, T. J., and Aldoory, M., The Effect of Inlet Port Design on Cylinder Gas Motion in Direct-Injection Diesel Engines, in *Flows in Internal Combustion Engines*, ed. T. Uzkan, pp. 101–111, ASME, New York, 1982.
346. Morse, A. P., Whitelaw, J. H., and Yianneskis, M., The Influence of Swirl on the Flow Characteristics of a Reciprocating Piston-Cylinder Assembly, *ASME Journal of Fluids Engineering*, vol. 102, pp. 478–480, 1980.
347. Morse, A. P., and Whitelaw, J. H., Measurements of the In-Cylinder Flow of a Motored Four-Stroke Reciprocating Engine, *Proceedings of the Royal Society of London, Series A*, vol. 377, pp. 309–329, 1981.
348. Rask, R. B., and Saxena, V., Influence of Geometry on Flow in the Combustion Chamber of a Direct-Injection Diesel Engine, in *International Symposium on Flows in Internal Combustion Engines—III*, eds. T. Uzkan, W. G. Tiedermann, and J. M. Novak, pp. 19–28, ASME, New York, 1985.
349. Arcoumanis, C., Green, H. G., and Whitelaw, J. H., The Application of Laser Rayleigh Scattering to a Reciprocating Model Engine, SAE International Congress and Exposition, Detroit, Mich., SAE Paper No. 840376, 1984.
350. Arcoumanis, C., Vafidis, C., and Whitelaw, J. H., Valve and In-Cylinder Flow Field Generated by a Helical Port in a Production Diesel Engine, in *International Symposium on Flows in Internal Combustion Engines—III*, eds. T. Uzkan, W. G. Tiedermann, and J. M. Novak, pp. 29–37, ASME, New York, 1985.
351. Bicen, A. F., Vafidis, C., and Whitelaw, J. H., Steady and Unsteady Airflow through the Intake Valve of a Reciprocating Engine, *ASME Journal of Fluids Engineering*, vol. 107, pp. 413–420, 1985.
352. Matsuoka, S., Kamimoto, T., Urushihara, T., Mochimaru, Y., and Morita, H., LDA Measurement and a Theoretical Analysis of the In-Cylinder Air Motion in a DI Diesel Engine, SAE International Congress and Exposition, Detroit, Mich., SAE Paper No. 850106, 1985.
353. Matsuoka, S., Nagakura, K., Tawai, T., Kamimoto, T., and Aoyagi, Y., Application of Laser Doppler Anemometry to a Motored Diesel Engine, SAE International Congress and Exposition, Detroit, Mich., SAE Paper No. 800965, 1980.
354. Coghe, A., Gamma, F., Mauri, M., Brunello, G., Calderini, F., and Ferri degli Antoni, F., In-Cylinder Air Motion Measurements by Laser Velocimetry under Steady-State Flow Conditions, SAE International Congress and Exposition, Detroit, Mich., SAE Paper No. 850123, 1985.
355. Dent, J. C., and Salama, N. S., The Measurement of the Turbulence Characteristics in an Internal Combustion Engine Cylinder, SAE International Congress and Exposition, Detroit, Mich., SAE Paper No. 750886, 1975.
356. Tsuge, M., Kido, H., and Yanagihara, H., Decay of Turbulence in a Closed Vessel (1st Report, Decay of Turbulence Intensity), *Bulletin of the JSME*, vol. 16, no. 92, pp. 244–251, 1973.
357. Tsuge, M., Kido, H., Kato, K., and Nomiyama, Y., Decay of Turbulence in a Closed Vessel (2nd Report, Time-Scales of Turbulence), *Bulletin of the JSME*, vol. 17, no. 107, pp. 587–594, 1974.
358. Willis, D. A., Meyer, W. E., and Birnie, C., Mapping of Airflow Patterns in Engines with Induction Swirl, SAE International Congress and Exposition, Detroit, Mich., SAE Paper No. 660093, 1966.
359. Huebner, K. H., and McDonald, A. T., Experimental Determination of Airflow Patterns in Piston Engines with Induction Swirl, SAE International Congress and Exposition, Detroit, Mich., SAE Paper No. 720026, 1972.
360. Polymeropoulos, C. E., and Sernas, V., Measurement of Droplet Size and Fuel-Air Ratio in Sprays, *Combustion and Flame*, vol. 29, pp. 123–131, 1977.
361. Rao, K. V. L., and Lefebvre, A. H., Evaporation Characteristics of Kerosine Sprays Injected into a Flowing Air Stream, *Combustion and Flame*, vol. 26, pp. 303–309, 1976.
362. Kamimoto, T., Ahn, S. K., Chung, Y. J., Kobayashi, H., and Matsuoka, S., Measurement of Droplet Diameter and Fuel Concentration in a Non-evaporating Diesel Spray by Means of an Image Analysis of Shadow Photographs, SAE International Congress and Exposition, Detroit, Mich., SAE Paper No. 840276, 1984.

363. Arai, M., Tabata, M., Hiroyasu, H., and Shimizu, M., Disintegrating Process and Spray Characterization of Fuel Injected by a Diesel Nozzle, SAE International Congress and Exposition, Detroit, Mich., SAE Paper No. 840275, 1984.
364. Varde, K. S., and Popa, D. M., Diesel Fuel Spray Penetration at High Injection Pressures, SAE International Congress and Exposition, Detroit, Mich., SAE Paper No. 830448, 1983.
365. Takeuchi, K., Senda, J., and Shikuya, M., Transient Characteristics of Fuel Atomization and Droplet Size Distribution in Diesel Fuel Spray, SAE International Congress and Exposition, Detroit, Mich., SAE Paper No. 830449, 1983.
366. Takeuchi, K., Senda, J., Ito, Y., and Ishida, K., Distribution of Fuel Droplets, Hydrocarbon and Soot in Diesel Combustion Chamber, SAE International Congress and Exposition, Detroit, Mich., SAE Paper No. 830456, 1983.
367. Kuniyoshi, H., Tanabe, H., Sato, G. T., and Fujimoto, H., Investigations on the Characteristics of Diesel Fuel Spray, SAE International Congress and Exposition, Detroit, Mich., SAE Paper No. 800968, 1980.
368. Chehroudi, B., Lombardi, P., Felton, P. G., and Bracco, F. V., Spray Photographs, Poppet Lift, and Injection Pressure of an Oscillating Poppet Injection, *ASME Journal of Fluids Engineering*, vol. 109, pp. 289–296, 1987.
369. Cavaliere, A., Ragucci, R., D'Alessio, A., and Noviello, C., Analysis of Diesel Spray Characteristics through Two-dimensional Laser Light Scattering Technique, XIX International Symposium on Heat and Mass Transfer in Gasoline and Diesel Engines, International Centre for Heat and Mass Transfer, Dubrovnik, Yugoslavia, 1987.
370. Hiroyasu, H., Measurements of Spray Characteristics and Fuel Vapor Concentration in a Diesel Spray, XIX International Symposium on Heat and Mass Transfer in Gasoline and Diesel Engines, International Centre for Heat and Mass Transfer, Dubrovnik, Yugoslavia, 1987.
371. Shimizu, I., Shimoda, M., Suzuki, T., and Emori, Y., A Light Scattering and Holographic Technique for Determining Droplet Size and Volume Density Distribution in Diesel Fuel Sprays, SAE International Congress and Exposition, Detroit, Mich., SAE Paper No. 820355, 1982.
372. Hess, C. F., A Technique Combining the Visibility of a Doppler Signal with the Peale Intensity of the Pedestal to Measure the Size and Velocity of Droplets in a Spray, AIAA 22nd Aerospace Sciences Meeting, Reno, Nev., Paper No. AIAA-84-0203, 1984.
373. Reason, J., Laser Probe Measures Droplet Size in Wet Steam and Gases, *Power Magazine*, p. 120, October 1986.
374. Dibelius, G., Renz, U., Schmitz, R. M., Voβ, H. and Wilhelmi, J., Instationare Zerstaubung des Brennstoffstrahles bei der Dieselmotorischen Einspritzung, Sonderforschungsbereich 224, Motorische Verbrennung, Kolloquium, Rheinish-Westfalische Technische Hochschule Aachen, F.R.G., November, 18–19, 1985.
375. FEV of America, Inc., Laser Optical Probe, Palo Alto, Calif., December 1986.
376. Tardieu, A., Candel, S. M., and Esposito, E., Droplet Size Distributions from Diffracted Light Intensities, in *Dynamics of Flames and Reactive Systems*, eds. J. R. Bowen, N. Manson, A. K. Oppenheim, and R. I. Soloukhin, *Progress in Astronautics and Aeronautics*, vol. 95, pp. 736–749, AIAA, New York, 1984.
377. Sudhakar, V., Performance Analysis of Adiabatic Engine, SAE International Congress and Exposition, Detroit, Mich., SAE Paper No. 840431, 1984.
378. Yoshimitsu, T., Toyama, K., Sato, F., and Yamaguchi, H., Capabilities of Heat Insulated Diesel Engine, SAE International Congress and Exposition, Detroit, Mich., SAE Paper No. 820431, 1982.
379. Wade, W. R., Havstad, P. H., Ounsted, E. J., Trinkler, F. H., and Garwin, I. J., Fuel Economy Opportunities with an Uncooled DI Diesel Engine, SAE International Congress and Exposition, Detroit, Mich., SAE Paper No. 841286, 1984.
380. Asnani, M., and Kuonen, F. L., Ceramic Valve and Seat Insert Performance in a Diesel Engine, SAE International Congress and Exposition, Detroit, Mich., SAE Paper No. 850358, 1985.
381. Morel, T., Keribar, R., and Blumberg, P. N., Cyclical Thermal Phenomena in Engine Com-

bustion Chamber Surfaces, SAE International Congress and Exposition, Detroit, Mich., SAE Paper No. 850360, 1985.

382. Morel, T., Fort, E. F., and Blumberg, P. N., Effect of Insulation Strategy and Design Parameters on Diesel Engine Heat Rejection and Performance, SAE International Congress and Exposition, Detroit, Mich., SAE Paper No. 850506, 1985.
383. Glance, P. C., Computer Aided Engineering Analysis of Adiabatic Engine Components, SAE International Congress and Exposition, Detroit, Mich., SAE Paper No. 850361, 1985.
384. Hoag, K. L., Brands, M. C., and Bryzik, W., Cummins/TACOM Adiabatic Engine Program, SAE International Congress and Exposition, Detroit, Mich., SAE Paper No. 850356, 1985.
385. Cole, R. M., and Alkidas, A. C., Evaluation of an Air-Gap-Insulated Piston in a Divided-Chamber Diesel Engine, SAE International Congress and Exposition, Detroit, Mich., SAE Paper No. 850359, 1985.
386. Elsbett, K., Elsbett, G., Elsbett, L., and Behrens, M., The Ferrum Piston for Diesel Engines—A Two-Piece Articulate Piston Design, SAE International Congress and Exposition, Detroit, Mich., SAE Paper No. 850505, 1985.
387. Moorhouse, P., Mortimer, B., and Kamo, R., Solid Lubrication Studies for Adiabatic Diesel Engines, SAE International Congress and Exposition, Detroit, Mich., SAE Paper No. 850508, 1985.
388. Indig, H., and Haman, A. C., Experimental Analysis of an Inwardly-opposed Piston Engine, SAE International Congress and Expositon, Detroit, Mich., SAE Paper No. 850362, 1985.
389. Kirloskar, C. S., Chandorkar, S. B., and Narayan Rao, N. N., Investigation on the Differentially Cooled System, SAE International Congress and Exposition, Detroit, Mich., SAE Paper No. 840427, 1984.

CHAPTER

FOUR

MATHEMATICAL MODELS OF SPARK–IGNITION ENGINES

4.1 INTRODUCTION

The objective of this chapter is to present and review some of the mathematical models that have been used to analyze the flow field and combustion in spark-ignition engines. Mathematical models for spark-ignition engines can be divided into two main groups—thermodynamic and dimensional models. Thermodynamic models can in turn be classified in two subgroups—single- and multizone models—whereas dimensional models can be divided into one- and multidimensional models.

In single-zone models the cylinder charge is assumed to be uniform in pressure, temperature, and composition. These models can be used as diagnostic (heat release analysis) or predictive tools. In heat release analyses experimentally determined pressure diagrams are used as an input to predict the heat release rate or the mass burning rate. Because of their simplicity, single-zone models can account for mass flows into and out of crevices. However, they ignore the flame propagation and combustion chamber geometry.

Single-zone models can also be used as predictive tools if the mass burning rate is specified. The mass burning rate depends on the combustion duration, ignition angle, engine geometry, equivalence ratio, residual mass, etc. Therefore, tuning may be required to predict the pressure diagrams in different engines or in the same engine operating under different conditions.

Multizone models attempt to resolve the combustion phenomena in a more physical manner than do single-zone models. The combustion chamber is generally divided into burnt and unburnt regions; sometimes thermal boundary layers in the burnt and unburnt gases are also considered. The cylinder charge is frequently assumed to be composed of ideal gases (frozen in the unburnt-gas region and in chemical equilibrium in the burnt-gas region), and the first law of thermodynamics, equation of state, and conservation of mass and volume are applied to the burnt and unburnt gases. The pressure is assumed to be uniform throughout the cylinder charge and n first-order ordinary differential equations can be obtained for the pressure, mass, volume, composition, and temperature of the burnt and unburnt gases. Since there are only $n-1$ conservation equations, the system of equations cannot be solved unless the burning rate or the turbulent flame speed is calculated or specified as a function of, for example, the laminar flame speed, engine rpm, etc.

The turbulent flame speed does not determine the mass burning rate unless the flame front surface area is specified. Frequently, the flame is assumed to propagate spherically throughout the combustion chamber.

Most of the multizone models that specify the turbulent flame speed account for the effects of the turbulent intensity and laminar flame speed on the flame propagation and combustion but do not consider the effects of the turbulent length scales on flame propagation. These effects can be accounted for by considering a turbulence kinetic energy cascade process and by solving appropriate equations for the turbulence kinetic energy and its dissipation rate. Molecular transport effects can be introduced by assuming that, for example, the Taylor eddies burn in a laminar manner. In this way the effects of the laminar flame speed, turbulence intensity, and turbulence length scales on combustion can be included. The effects of the intake-generated turbulence, swirl, squish, shear, and compression can also be included in the equations for the turbulence kinetic energy and its dissipation rate.

Multizone models assume that the compression stroke is characterized by uniform levels of turbulence throughout the combustion chamber and neglect the mean flow field. (The mean flow can be considered by means of an equation for the mean flow kinetic energy.) Experimental data and multidimensional calculations have shown that the large-scale structures (eddies) generated during the intake stroke may persist up to top dead center (TDC) of the compression stroke and that the turbulence relaxation time is on the order of 120°CA (crank angle degrees). Therefore, during such a time interval multizone models of engine turbulence are not accurate. Furthermore, these models consider an infinitesimally thin flame, neglect the flame structure, and must be supplied with heat transfer correlations unless an integral method is used to calculate the heat transfer losses to the solid walls. Heat transfer correlations are presented in Chapters 2 and 3.

One-dimensional combustion models were developed before multidimensional models and assumed planar or cylindrical flame propagation. In planar flame propagation models the distance between the piston and the cylinder head must be much larger than the cylinder bore; this is clearly unrealistic in conventional reciprocating engines. Cylindrical flame propagation models can simulate combustion phenomena after the flame reaches the piston; however, they are not accurate for predicting the initial stages of combustion.

Multidimensional models are based on the time-dependent three-dimensional conservation equations of mass, momentum, energy, and species. Because of the impossibility of resolving all the flow scales with current and future computers, modeling is necessary; the instantaneous flow variables are decomposed into the sum of mean flow and fluctuating quantities, which are substituted into the governing equations. Time or mass (density) averaging is introduced and results in a number of correlations, which are frequently modeled as terms proportional to the gradients of the mean flow variables (see Chapter 2).

Averaging the conservation equations is a delicate subject. Time average is strictly valid for statistically stationary turbulence. Engine turbulence is not stationary, and flow relaxation times may be larger than cycle times. Therefore, time average may not be valid for engine flow analyses (see Chapter 2). In addition, most of the turbulence models used to analyze engine turbulence are based on those developed for statistically stationary, incompressible, thin shear flows. Engine flows are unsteady and compressible, and they involve complex recirculation zones characterized by streamline curvature.

In spite of their limitations, multidimensional models have been used to predict the flow fields in motored reciprocating engines at low rpm. Comparisons between predicted and measured mean velocities and turbulence intensities have been encouraging. However, most of the experimental data obtained to date include cycle-to-cycle variations in the determination of the turbulent intensity and may not be accurate enough to assess the validity of multidimensional models. In fact, cycle-resolved measurements have shown that non-cycle-resolved data may overestimate the turbulence intensity in motored engines.

In this chapter an attempt is made to present and review mathematical models of spark-ignition engines and their limitations. Experimental data are also discussed in order to emphasize that mathematical models of spark-ignition engine combustion need further refinement in order to predict the engine performance and emissions and to show that our knowledge of engine turbulence is very poor.

It may be safely stated that many mathematical models are available that have been used to analyze the flow field in reciprocating engines. However, only multizone models have been the subject of intensive validation studies with experimental data. Multidimensional models have been validated mainly with motored engine data, but they cannot as yet predict the effects that such factors as the engine geometry, spark plug location, and bore-to-stroke ratio have on the engine performance, combustion, and emissions under firing conditions. Furthermore, accurate and detailed experimental data on engine turbulence, including turbulence root-mean-square (rms) velocities and length scales, are unavailable, although they may be forthcoming.

In this chapter a brief review of multidimensional models of spark-ignition engines is presented. Turbulence and combustion models have been discussed at length in Chapter 2, which also contains a detailed discussion of the assumptions and limitations of the turbulence and combustion models used to predict the flow field and flame propagation phenomena in reciprocating and rotary engines.

4.2 SINGLE–ZONE MODELS

In single-zone models the pressure, temperature, and composition of the cylinder charge are assumed to be uniform. These models define the state of the cylinder charge in terms of average properties, do not distinguish between burnt and unburnt gases, and assume that the cylinder charge is homogeneous. Multizone models permit a more accurate treatment of the thermodynamic properties of the cylinder mixture; the burnt and unburnt gases are considered as separate thermodynamic systems that are uniform in composition and state. However, the geometry of each zone must be tracked in order to calculate the heat transfer and composition of the burnt and unburnt gases and flows into crevices. Single-zone models have the advantage of including heat transfer and gas flow phenomena more simply. Combustion in single-zone models can be considered as a heat addition process, and the chamber charge is regarded as a simple fluid.

The first law of thermodynamics applied to an open system (Fig. 4.1) can be written as (see Chapter 2)

$$\frac{d(me)}{d\theta} = -p\frac{dV}{d\theta} - \frac{dQ}{d\theta} + \Sigma\, \dot{m}_i h_i \qquad e = e^0 + \int_{T_0}^{T} C_v\, dT \tag{1}$$

where p, T, and m are the pressure, temperature, and mass of the cylinder charge, respectively; e is the mixture specific internal energy; C_v is the specific heat at constant volume; V is the combustion chamber volume; $dQ/d\theta$ represents the heat losses; h_i is the specific enthalpy of the gases flowing into the cylinder with a mass flow rate equal to $\dot{m}_i$; T_0 is a reference temperature; e^0 is the internal energy of formation at the reference temperature T_0; and θ is the crank shaft angle.

In the absence of injection and flows into crevices, $dm/d\theta = 0$. In premixed charge engines, there are flows into and out of crevices (i.e., volumes between the piston, cylinder wall, and piston rings and the spark plug threads). The crevices can be modeled as a single volume at the cylinder pressure or as a series of volumes, connected by restrictions to simulate the piston-ring-cylinder wall region and blowby.

Equation (1) can be written as

$$\frac{dQ_{CH}}{d\theta} = mC_v\frac{dT}{d\theta} + p\frac{dV}{d\theta} + \frac{dQ}{d\theta} + (h - e)\frac{dm_{CR}}{d\theta} \tag{2}$$

where $dQ_{CH}/d\theta$ represents the heat released by combustion, $dm_{CR}/d\theta$ represents the mass flow rate into crevices, and h is the specific enthalpy.

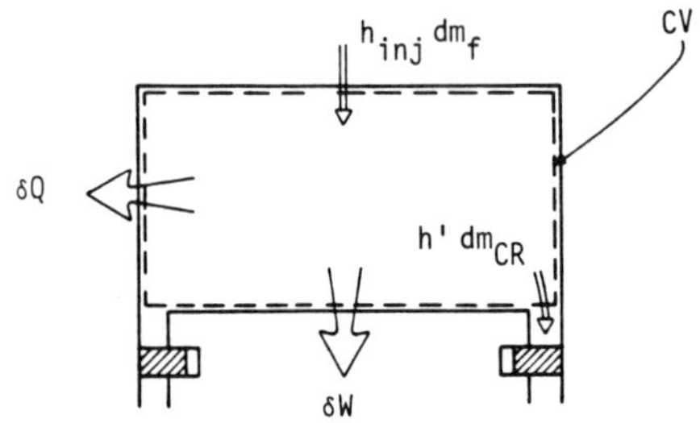

Figure 4.1 Schematic of a thermodynamic system (h_{inj} = enthalpy of injected fuel; m_f = fuel mass; m_{CR} = mass in crevice; CV = control volume; δW = piston work; δQ = heat transfer losses; h' = enthalpy of the cylinder charge).

Conservation of mass applied to the combustion chamber yields

$$\frac{dm}{d\theta} = -\frac{dm_{CR}}{d\theta} \tag{3}$$

When the cylinder pressure is high, $dm_{CR}/d\theta > 0$ and the value of h corresponds to that in the combustion chamber:

$$h = e + \frac{pV}{m} \tag{4}$$

However, during the expansion stroke, $dm_{CR}/d\theta < 0$ and the value of h is that of the gases contained in the crevices.

If the crevice volume and temperature are assumed constant and the crevice pressure is equal to that of the cylinder charge, the mass flow rate into the crevices can be written as (see Chapter 2)

$$\frac{dm_{CR}}{d\theta} = V_{CR}\,\frac{dp/d\theta}{RT_w} \tag{5}$$

where the crevice temperature was set equal to the temperature of the cylinder wall T_w, and V_{CR} is the crevice volume. Equation (5) can be substituted into Eqs. (2) and (3) to obtain an equation for the heat released by combustion once the heat transfer losses $dQ/d\theta$ are specified. Heat transfer correlations have been presented in Chapters 2 and 3 and are not repeated here.

Borman and Nishiwaki [1] reviewed heat transfer correlations based on bulk gas temperatures and models that account for thermal boundary layers and bulk gas motion. Most of the available correlations correspond to cycle-averaged heat fluxes and may not be accurate due to the lag between the temperature and heat flux; this lag may result in infinite values of the film heat transfer coefficients. One-dimensional analyses of heat transfer through the solid walls have shown that the average wall temperature fluctuates approximately 30°C. Furthermore, the heat transfer losses are affected by soot formation, radiation, and deposits on the walls.

A new formula for the film heat transfer coefficient in reciprocating engines, which accounts for the turbulence intensity of the charge, has been derived by Rao and Bardon [2], assuming that the angular momentum of the large eddies is conserved. This formula can be written as $h = 0.058\,\rho C_p u'$ where h, C_p, and u' are the film heat transfer coefficient, specific heat at constant pressure, and the rms of the velocity fluctuations, respectively.

Alkidas and Myers [3] measured heat fluxes in the cylinder head and liner of a single-cylinder engine and showed that the heat flux is greatest near stoichiometry. They also showed that heat fluxes increase the cylinder volumetric efficiency (load), and they proposed correlations that agree with experimental data from intake valve closing (IVC) to exhaust valve opening (EVO).

A derivation of Eq. (2) that accounts for the internal energies of the burnt and unburnt gases has been presented in Chapter 2. In addition to Eqs. (2)–(5) the cylinder charge can be considered as an ideal gas, that is,

$$pV = m\,RT \tag{6}$$

where the gas constant R is a function of the mixture pressure, temperature, equivalence ratio, and mass fraction of the residual gases, that is, $R = R(T,p,\phi,f)$ where f is the residual gas mass fraction.

Equations (3)–(6) can be substituted into Eq. (2) to obtain an equation that relates the heat release rate to the combustion chamber pressure and volume, mass fraction of residual gases, wall temperature, and crevice volume. The resulting equation can be used in heat release analyses if experimentally determined pressure diagrams are used to predict the heat release rate. Alternatively, the resulting equation can be used as a predictive tool if the heat release rate $dQ_{CH}/d\theta$ is specified as a function of the crank shaft angle to yield the mixture pressure and temperature. The mass burning rate can be specified by means of, for example, a Wiebe function (see Chapters 2 and 3).

Gatowski et al. [4] showed that the heat release rate varies substantially from cycle to cycle and that small negative heat release rates may be obtained under some operating conditions as combustion is completed, particularly in fast-burning cycles. These errors were attributed to the heat transfer correlation, the method used to represent the thermodynamic properties, and thermal effects in the pressure transducer.

If the mass flow rates into crevices are neglected, the mass of the cylinder charge is constant, $dm/d\theta = dm_b/d\theta + dm_u/d\theta = 0$, where the subscripts b and u denote burnt and unburnt gases, respectively. The mass fraction of burnt gases x_b can be specified as [5]

$$x_b = \frac{m_b}{m} = \frac{1 - \cos\left[(\theta - \theta_0)\,\pi/\Delta\theta_b\right]}{2} \tag{7}$$

$$x_b = \frac{m_b}{m} = 1 - \exp\left[-a\left(\frac{\theta - \theta_0}{\Delta\theta_b}\right)^{m+1}\right] \tag{8}$$

where θ is the crank shaft angle, $\Delta\theta_b$ is the duration of combustion, θ_0 denotes the angle at which burning starts, and a and m are parameters (e.g., $a = 5$, $m = 2$).

The heat release rate $dQ_{CH}/d\theta$ can be written as

$$\frac{dQ_{CH}}{d\theta} = C\frac{dm_b}{d\theta} \tag{9}$$

where C is the heat of combustion and $dm_b/d\theta$ can be calculated from Eq. (7) or Eq. (8). The heat of combustion is approximately $e_u^0 - e_b^0$ (see Chapter 2).

The burnt gases in the cylinder are close to thermodynamic equilibrium, and their properties are usually specified through curve fits to thermodynamic equilibrium calculations. Curve fits are used to reduce the computation time [6]. Blumberg and Kummer [5] showed that the NO emissions decrease as the inlet temperature and compression rate are decreased. Blumberg et al. [7] reviewed phenomenological models for spark-ignition engines, where the NO formation is calculated from the extended Zel'dovich mechanism with O, OH, H, O_2, and N_2 assumed to be in equilibrium, with a steady-state assumption for N, and

$$N + NO \rightleftarrows N_2 + O \qquad N + O_2 \rightleftarrows NO + O \tag{10}$$

$$N + OH \rightleftarrows NO + H \tag{11}$$

Although the burnt gases are almost in equilibrium, the NO formation is not an equilibrium process. Furthermore, the NO formation is affected by the temperature gradients in the burnt-gas region and depends on the temperature and O_2 concentration histories [8]. Nitric oxide concentration gradients are expected throughout the combustion chamber, and NO freezes in the expansion stroke. The major source of unburnt hydrocarbons (HC) are the oil and quench layers and crevices. The HC and CO emissions increase with the equivalence ratio ϕ for rich mixtures. The HC emissions also increase as the equivalence ratio is decreased for lean mixtures because of the slow flame propagation [9]. Nitric oxide seems to form in the postflame reactions, whereas aldehydes are formed ahead of the flame front and are associated with knock and odor [10]. The emissions of unburnt HC, CO, and NO from spark-ignition engines are described in Sections 4.6 and 4.7.

Single-zone models do not account for the combustion chamber geometry except in a global manner through V in Eqs. (2), (4), and (6). Furthermore, these models do not consider flame propagation phenomena and do not account for the presence of burnt and unburnt gases in the combustion chamber. In addition, the parameters a, θ_0, $\Delta\theta_b$, and m, which appear in the mass fraction of the burnt gases [Eqs. (7) and (8)], may be a function of the engine rpm, air-to-fuel ratio, and combustion chamber geometry.

4.3 MULTIZONE MODELS

Multizone models account for the combustion chamber geometry and for the presence of burnt and unburnt gases (Fig. 4.2). In these models the cylinder mixture is divided into burnt and unburnt zones, which are separated from each other by a surface of discontinuity (the flame). The composition and temperature

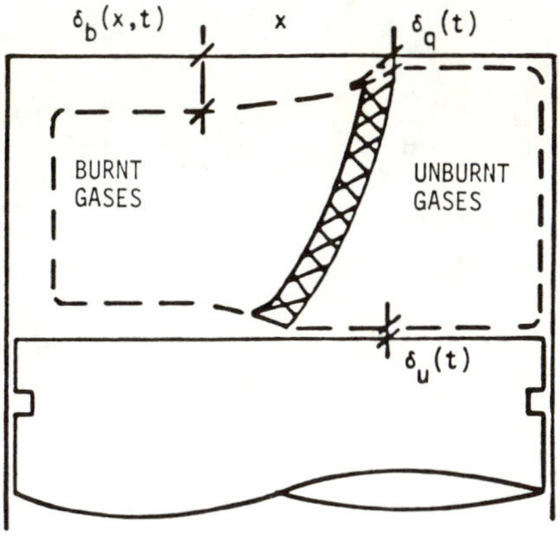

Figure 4.2 Two-zone thermodynamic model showing the burnt- and unburnt-gas regions and the thermal boundary layers (δ_b = boundary layer thickness in the burnt gases; δ_u = boundary layer thickness in the unburnt gases; δ_q = quenching distance; t = time; x = distance from the flame front).

of the burnt and unburnt gases are different, and the pressure is uniform throughout the combustion chamber. Although the assumption of uniform temperature in the unburnt gases is reasonable, significant temperature gradients may exist in the burnt gases due to the differences between first burning and then compressing the burnt gas, as compared to first compressing and then burning the fresh mixture.

Multizone models do not consider the flame structure but may account for quenching.

The equations governing the burnt- and unburnt-gas zones can be written as (see Chapter 2)

$$m = m_u + m_b \qquad \frac{dm}{d\theta} = \frac{dm_u}{d\theta} + \frac{dm_b}{d\theta} = -\frac{dm_{u,CR}}{d\theta} - \frac{dm_{b,CR}}{d\theta} \tag{12}$$

$$V = V_u + V_b \tag{13a}$$

$$p\,V_u = m_u\,R_u\,T_u \tag{13b}$$

$$p\,V_b = m_b\,R_b\,T_b \tag{13c}$$

$$\frac{d}{d\theta}(m_u\,e_u) = -p\frac{dV_u}{d\theta} - \frac{dQ_u}{d\theta} + h_u\frac{dm_{u,R}}{d\theta} - h_u\frac{dm_{u,CR}}{d\theta} \tag{14}$$

$$\frac{d}{d\theta}(m_b\,e_b) = -p\frac{dV_b}{d\theta} - \frac{dQ_b}{d\theta} + h_u\frac{dm_{b,R}}{d\theta} - h_b\frac{dm_{b,CR}}{d\theta} \tag{15}$$

$$e_u = e_u^0 + \int_{T_0}^{T_u} C_{vu}\,dT \qquad e_b = e_b^0 + \int_{T_0}^{T_b} C_{vb}\,dT \tag{16}$$

where the subscripts u and b denote the unburnt and burnt gases, respectively; the third terms on the right-hand sides of Eqs. (14) and (15) denote the enthalpy fluxes associated with the chemical reaction R; and the fourth terms on the right-hand sides of Eqs. (14) and (15) denote the enthalpy fluxes associated with flows into crevices CR.

Note that

$$\frac{dm_{u,R}}{d\theta} = -\frac{dm_{b,R}}{d\theta} = -\frac{dm_c}{d\theta} \tag{17}$$

where $dm_c/d\theta$ is the combustion or mass burning rate.

In the absence of leakage and flows into crevices,

$$\frac{dm_{u,R}}{d\theta} = \frac{dm_u}{d\theta} \qquad \frac{dm_{b,R}}{d\theta} = \frac{dm_b}{d\theta} \tag{18}$$

If there is leakage,

$$\frac{dm}{d\theta} = -\frac{dm_{u,CR}}{d\theta} - \frac{dm_{b,CR}}{d\theta} \tag{19}$$

$$\frac{dm_b}{d\theta} = -\frac{dm_{b,CR}}{d\theta} + \frac{dm_{b,R}}{d\theta} \tag{20}$$

$$\frac{dm_u}{d\theta} = -\frac{dm_{u,CR}}{d\theta} + \frac{dm_{u,R}}{d\theta} \tag{21}$$

where the first terms on the right-hand sides of Eqs.(20) and (21) denote the flows of burnt and unburnt gases into the crevices, respectively.

Equations (14), (15), and (17)–(21) can be written as

$$m_u C_{v_u} \frac{dT_u}{d\theta} = -p\frac{dV_u}{d\theta} - \frac{dQ_u}{d\theta} + (h_u - e_u)\frac{dm_{u,R}}{d\theta} + (e_u - h)\frac{dm_{u,CR}}{d\theta} \tag{22}$$

$$m_b C_{v_b} \frac{dT_b}{d\theta} = -p\frac{dV_b}{d\theta} - \frac{dQ_b}{d\theta} + (h_b - e_b)\frac{dm_{b,R}}{d\theta} + (e_b - h)\frac{dm_{b,CR}}{d\theta} \tag{23}$$

where

$$h_u = e_u + p\frac{V_u}{m_u} \qquad h_b = e_b + p\frac{V_b}{m_b} \tag{24}$$

and the specific enthalpy h, which appears in Eqs. (14), (15), (22), and (23), is that of the unburnt and burnt gases if $dm_{u,CR}/d\theta > 0$ and $dm_{b,CR}/d\theta > 0$, respectively.

The leakage of burnt and unburnt gases can be calculated from Eq. (5), and the combustion chamber pressure can be calculated by adding Eqs. (13*b*) and (13*c*), that is,

$$p(V_u + V_b) = m_u R_u T_u + m_b R_b T_b \tag{25}$$

where $R_u = R_u(\phi, f, p, T_u)$, $R_b = R_b(\phi, f, p, T_b)$, f is the residual gas mass fraction, and ϕ is the equivalence ratio.

The mass fractions of the burnt and unburnt gases can be written as

$$x_u = \frac{m_u}{m} \tag{26a}$$

$$x_b = \frac{m_b}{m} \tag{26b}$$

Equation (26*b*) can be differentiated with respect to the crank shaft angle to yield

$$\frac{dm_b}{d\theta} = x_b \frac{dm}{d\theta} + m \frac{dx_b}{d\theta} = \frac{dm_c}{d\theta} - \frac{dm_{b,CR}}{d\theta} \tag{27}$$

or

$$\frac{dm_c}{d\theta} = \frac{dm_{b,CR}}{d\theta} + x_b \frac{dm}{d\theta} + m \frac{dx_b}{d\theta} \tag{28}$$

Equations (5), (12), (13), and (16)–(26) yield first-order ordinary differential equations for m_u, m_b, V_b, T_u, T_b, m_c, and p. Note that V_u can be calculated from Eq. (13*a*) once V_b is known.

This system of equations is not closed (i.e., there are more unknowns than equations). Closure can be achieved by specifying the mass burning rate $dm_c/d\theta$ and the geometry of the flame front. Frequently, the flame is assumed to propagate spherically from the spark plug, and the mass burning rate (or equivalently, the turbulent flame speed) is either specified [Eqs. (7) and (8)] or calculated by means of phenomenological models.

The assumption of a spherical flame front with center at the ignition location can lead to a large underprediction of the flame surface area because this assumption constrains the flame surface area-to-volume ratio to be an absolute minimum and because the spherical flame assumption may be prematurely truncated by the combustion chamber walls. Furthermore, the interaction of the flame front with the combustion-generated flow field may result in highly curved flames, aerodynamic and geometric stretchings, and flame quenching.

In the next sections, three different approaches to spark-ignition engine combustion are described. The first approach involves phenomenological combustion models that calculate the mass burning rate from physical considerations. In the second approach the turbulent flame speed is specified, whereas in the third approach the mass burning rate is specified.

4.3.1 Calculation of the Mass Burning Rate

Some multizone models attempt to predict the burning rate from fundamental physical quantities such as the turbulence intensity, integral length scale, Kolmogorov eddy size, and kinetics of the fuel-air oxidation process. The purpose of these models is to predict the ignition delay and combustion rate as a function of the engine design and operating conditions. In spark-ignition engines the burning process can be modeled as a flame of surface area A_f, frequently assumed to be a sphere propagating through the unburnt-gas mixture of density ρ_u with a turbulent flame speed S_T such that

$$\frac{dm_b}{dt} = \rho_u A_f S_T \tag{29}$$

Figure 4.2 is a schematic of a flame propagating through the cylinder charge, the burnt- and unburnt-gas regions, and the quench layer.

Experimental data have shown that S_T is proportional to the turbulent intensity u'. Some researchers have assumed that S_T is proportional to the laminar flame speed S_L but with a proportionality constant that is a function of u'. One of the major drawbacks of these approaches is that the effects of the turbulence length scale on S_T are not included. The turbulence length scale is very important during ignition and extinction, where quenching due to velocity gradients is important [11]. If $S_T = gS_L$ where g is a function of the engine rpm, the resulting model is unable to predict the combustion duration $\Delta\theta_d$ as a function of the ignition delay time and equivalence ratio (see Section 4.3.3).

4.3.1.1 Blizard-Keck (BK) model. The Blizard-Keck (BK) model [12] considers the effect of u', S_L, and l (turbulence length scale) on the burning rate and assumes that the large eddies entrain the fresh mixture whereas the small eddies burn in a laminar manner with a characteristic time $\tau_b = l/S_L$, and the mass entrainment of the fresh mixture into the flame front is

$$\frac{dm_e}{dt} = \rho_u A_f u_e \tag{30}$$

where u_e is the entrainment velocity.

Blizard and Keck [12] also assumed that the entrained eddies have an exponential distribution of burning times that yields the following mass burnt at a time t after ignition:

$$\frac{dm_b}{dt} = \frac{m_e - m_b}{\tau_b} \tag{31}$$

Blizard and Keck [12] assumed that the entrainment velocity is proportional to the inlet gas speed and that l is proportional to the valve lift.

Zeleknik [13] generalized the BK model to the case of time-dependent burning times and showed that unburnt hydrocarbons always appear in the thermodynamic kinetic modeling of exhaust gases. Zeleknik and McBride [14, 15] used Zeleknik's model [13] to predict combustion in spark-ignition engines with finite rate chemistry in the burnt gases; the unburnt gases were assumed to be frozen.

The equations governing the composition of the burnt- and unburnt-gas zones can be written as

$$\frac{dn_u}{d\theta} = 0 \qquad \frac{dn_{b,i}}{d\theta} = R_{b,i} + (n^e_{b,i} - n_{b,i}) \frac{dm_b/d\theta}{m_b} \tag{32}$$

where n represents the number of moles per unit mass, the superscript e denotes chemical equilibrium, and $R_{b,i}/\rho_b$ is the volumetric production rate of species i. When combustion is completed, $dn_{b,i}/d\theta = R_{b,i} = 0$ and $n_{b,i} = n^e_{b,i}$.

The BK model [12] is an algebraic model since the length scale and the entrainment velocity are specified as functions of the inlet gas speed and valve lift, respectively. In the next sections algebraic ϵ models are presented. These models use a conservation equation for the turbulence kinetic energy in order to calculate the entrainment velocity but either specify the turbulent length scale algebraically or calculate it from equilibrium considerations.

4.3.1.2 Algebraic ϵ models I. In this section we present algebraic ϵ models that use a specified length scale proportional to the valve lift or to the distance between the cylinder head and the piston.

Poulos and Heywood [16] used the BK model but calculated the entrainment velocity $u_e = S_L + u'$, where $u' = (2k/3)^{1/2}$, from the following conservation equations for the mean flow and turbulence kinetic energies:

$$\frac{dK}{dt} = \dot{m}_{\text{int}} \frac{V^2_{\text{int}}}{2} - P - K \frac{\dot{m}_{\text{exh}}}{m} \qquad P = C\left(\frac{KL}{l^2}\right)\left(\frac{k}{m}\right)^{1/2} \tag{33}$$

$$\frac{dk}{dt} = P - m\epsilon - k \frac{\dot{m}_{\text{exh}}}{m} \qquad \epsilon = \frac{(2k/3m)^{3/2}}{L} \tag{34}$$

where K and k are the mean flow and turbulence kinetic energies, respectively; l is the length scale of the energy-containing eddies; $\dot{m}$ denotes mean flow rates; ϵ is the dissipation rate of turbulence kinetic energy; L is a length scale; C is a constant; and the subscripts int and exh indicate flows into and out of the cylinder, respectively. The length scale of the energy-containing eddies was assumed to be equal to L and given by the following expression:

$$L = l = \frac{V}{\pi B^2/4} \tag{35}$$

where B is the cylinder bore.

At the start of combustion the initial values of L and K are supplied by Eqs. (33)–(35), and the turbulence flow field is assumed homogeneous and isotropic (i.e., the flame-generated turbulence is neglected). Once combustion begins, the unburnt gases are assumed to be compressed by the flame at a rate sufficiently high so that the dissipation of turbulence kinetic energy in the large eddies can be neglected on the time scale of the flame propagation. This implies that the angular momentum of the large eddies is conserved, that is,

$$\frac{L}{L_0} = \left(\frac{\rho_{u_0}}{\rho}\right)^{1/3} \qquad \frac{u'}{u'_0} = \left(\frac{\rho_u}{\rho_{u_0}}\right)^{1/3} \tag{36}$$

where the subscript 0 denotes the time of spark ignition. The burning time can be calculated by assuming that combustion is laminar in eddies of dimensions on the order of the Taylor microscale λ, which can be calculated from isotropic turbulence as

$$\frac{\lambda}{L} = \left(15 \frac{\nu}{u'L}\right)^{1/2} \qquad \tau_b = \frac{\lambda}{S_L} \tag{37}$$

where ν is the kinematic viscosity.

If the turbulence is isotropic, the fluctuating rms velocity is given by the following expression:

$$u' = \left(\frac{2k}{3m}\right)^{1/2} \tag{38}$$

The solution of Eqs. (33)–(38) can yield a characteristic velocity and length scale that can be used to calculate the heat transfer rates by means of a Nusselt-Reynolds-Prandtl correlation (see Chapters 2 and 3).

The combustion process can be modeled as two simultaneous phenomena due to the entrainment of the unburnt mixture by the flame and combustion within the flame. Burning on the Taylor microscale occurs at the laminar flame speed since diffusion is laminar on that scale. These phenomena can be represented by the following equations

$$\frac{dm_e}{dt} = \rho_u A_f (u' + S_L) \tag{39a}$$

$$\frac{dm_b}{dt} = \frac{m_e - m_b}{\tau_b} \tag{39b}$$

$$\tau_b = \frac{\lambda}{S_L} \tag{40a}$$

$$\frac{S_L}{S_{L_{\text{ref}}}} = \left(\frac{T_u}{T_{\text{ref}}}\right)^{\alpha} \left(\frac{p}{p_{\text{ref}}}\right)^{\beta} (4.706 f^2 - 4.062 f + 1) \tag{40b}$$

where f is the residual mass fraction, α and β are constants, the subscript ref indicates reference conditions, and λ can be calculated from Eqs. (35)–(38).

Figure 4.3 shows a schematic of concentrated vorticity regions, and Fig. 4.4 illustrates a turbulent eddy structure.

Poulos and Heywood [16] and Poulos [17] showed that their model underscores the importance of the combustion chamber geometry in determining the burning duration. Significant differences were observed between combustion chambers with side ignition and with central ignition.The spark plug location was found to be more important than the chamber geometry in determining the burning duration.

Tabaczynski et al. [18] assumed that l is proportional to the chamber height and developed a second-order ordinary differential equation for the complete

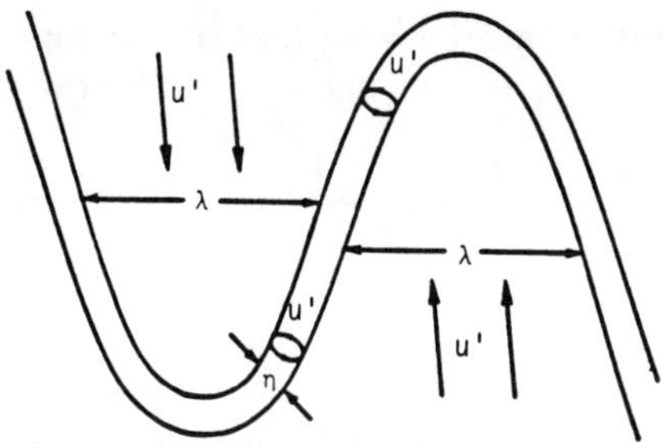

Figure 4.3 Model of concentrated vorticity regions (λ = Taylor microscale; η = Kolmogorov eddy size; u' = fluctuating velocity rms).

eddy-burning process. Belli et al. [19] solved this differential equation for a range of turbulence intensities, eddy sizes, and equivalence ratios; they showed that

$$\frac{\tau_{bu}}{\tau_e} = 1.8\left(\frac{\tau_b}{\tau_e}\right)^{2/3} \qquad \tau_e = \frac{l}{u'} \qquad l = c_1\left(\frac{m}{\rho_u}\right)^{1/3} \tag{41}$$

$$\tau_{bu} = 4.44c_2\left(\frac{l^2\,\nu}{u'^2\,S_L^2}\right)^{1/3} \tag{42}$$

where τ_e and τ_{bu} are the eddy-turnover and -burnup times, respectively, and c_1 and c_2 are constants.

The entrainment volume and the mass burning rate can be calculated from

$$V_e = \frac{m_e - m_b}{\rho_u} + \frac{m_b}{\rho_b} \qquad \frac{dm_b}{dt} = \frac{m_e - m_b}{\tau_b} \tag{43}$$

Belli et al. [19] calculated the turbulent intensity necessary to determine the entrainment velocity from the following expressions:

$$m\,k = \int_0^t \left(\frac{\dot{m}\,V_{\text{int}}^2}{2} - m\,\epsilon\right) dt \tag{44}$$

$$u' = \left(\frac{2k}{3}\right)^{1/2} \qquad \epsilon = \frac{(2k/3)^{3/2}}{l} \tag{45}$$

where $\dot{m}\,V_{\text{int}}^2/2$ is the kinetic energy associated with the incoming flow, and the

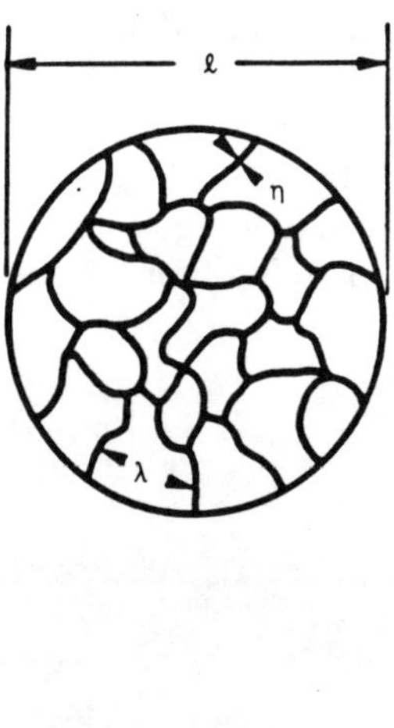

Figure 4.4 Turbulent eddy structure (l = length of the large scale eddies).

length scale of the turbulence dissipation rate coincides with that of the combustion model. Belli et al. [19] also accounted for the quench layer as follows:

$$\frac{dm}{d\theta} = \frac{dm_b}{d\theta} + \frac{dm_u}{d\theta} + \frac{dm_q}{d\theta} \qquad m_q = \int_0^t \rho_u \, \delta_q \, \frac{dA_q}{dt} \, dt \tag{46}$$

$$\delta_q = \alpha_u \frac{\text{Pe}}{S_L} \tag{47}$$

where δ_q is the quenching distance and α_u, Pe, and A_q are the thermal diffusivity, Péclet number, and area wetted by the quenched gases, respectively. Ferguson and Keck [20] showed that quenching in an engine is laminar, depends on the upstream free stream unburnt-gas temperature rather than the wall temperature, and the characteristic length for heat loss is the flame thickness. Similar results were obtained by Goolsby and Haskell [21] in a Cooperative Fuels Research (CFR) engine.

Hires et al. [22] used the model of Tabaczynski et al. [18] to predict the ignition delay time and combustion duration as a function of the combustion chamber geometry. Lucas and Brunt [23] showed that the flame speed is largely independent of the combustion chamber geometry. However, the geometry has an important effect on the burning rate, cylinder pressure, and cycle-to-cycle variations.

Tabaczynski et al. [18] simulated the combustion period in two stages: the ignition delay period and the turbulent flame propagation period. In the ignition delay period the initial burning takes place over a small spatial region, so that the turbulence is not at work. Their model postulates the burning of an individual eddy having a size proportional to the instantaneous chamber height. The eddy-burnup time can be obtained by assuming that combustion is laminar in eddies on the order of the Taylor microscale, which can be calculated from isotropic turbulence. These assumptions lead to the following expression:

$$\frac{t_d \, u_p}{l} = c \left(\frac{\nu}{l \, u_p}\right)^m \left(\frac{u_p}{S_L}\right)^n \tag{48}$$

where t_d is the ignition delay time, u_p is the mean piston velocity, l is the average combustion chamber height during t_d, c is an empirical constant, ν is the kinematic viscosity, and S_L is the average laminar flame speed during t_d. Tabaczynski et al. [18] suggested $m = \frac{1}{3}$ and $n = \frac{2}{3}$, and used Eq. (48) to obtain the time required to burn the first 1% of the cylinder charge.

Tabaczynski et al. [18] applied rapid-distortion theory [Eq. (36)] to predict the variation of the turbulence root-mean-square velocity u' and integral length scale L caused by changes in density during burning, and they obtained the following flame speed ratio:

$$\text{FSR} = \frac{S_T}{S_L} = D \left(\frac{u'}{S_L}\right)^{1/3} \left(\frac{\rho_u}{\rho_0}\right)^{1/9} \left(\frac{u'L}{\nu}\right)^{1/3} \tag{49}$$

where D is a constant, ρ_0 is the initial density, and S_T is the turbulent flame speed.

Groff and Matekunas [24] used pressure diagrams, a spherical flame propagation model, and Woschni's heat transfer correlation to predict the burning velocity in homogeneous-charge spark-ignition engines; they showed that

$$\text{FSR} = 3.2L^{-0.34} \tag{50}$$

Rapid-distortion theory applied to the model of Tabaczynski et al. [18] indicates that

$$L = L_0 \left(\frac{\rho}{\rho_0}\right)^{1/3} \qquad u' = u'_0 \left(\frac{\rho}{\rho_0}\right)^{1/3} \tag{51}$$

which, for a polytropic compression with exponent equal to 1.3, yield

$$L = L_0 \left(\frac{p}{p_0}\right)^{-0.26} \qquad u' = u'_0 \left(\frac{p}{p_0}\right)^{0.26} \tag{52}$$

Equations (50) and (52) imply that

$$\text{FSR} \sim L^{-0.34} \sim \left(\frac{p}{p_0}\right)^{0.09} \tag{53}$$

whereas the experimental data of Groff and Matekunas [24] indicate that

$$\text{FSR} = a + b\,\frac{u'}{S_L}\left(\frac{p}{p_{\max}}\right)^{n} S \tag{54}$$

where a, b, and n are constants, $p_{\max}$ is the maximum pressure, and S is a spark advance factor.

Groff [25] used the Blizard-Keck model [12] with $L = 25$ mm and showed that this model underestimates the experimental flame acceleration unless the burning velocity is reduced at small flame radii by an empirical term that is a function of the flame radius and thickness. Groff [25] obtained the following flame speed ratio:

$$\frac{S_T}{S_L} = 2.273 + 1.681\,\frac{u'}{S_L} \tag{55}$$

Tabaczynski et al. [26] refined the Blizard-Keck-Tabaczynski model represented by Eqs. (30) and (31) and showed that if the burning is instantaneous, $m_e = m_b$ and the turbulent flame speed $S_T = u' + S_L$, which agrees with Damkohler's original expression. They also used a length scale proportional to the chamber height and showed that

$$\frac{S_T}{S_L} = C\left(\frac{u'}{S_L}\right)^{1/3}\left(\frac{\rho_u}{\rho_{u_0}}\right)^{1/9}\left(\frac{u'L}{\nu}\right)^{1/3} \tag{56}$$

where C is a constant.

Equation (56) shows that S_T depends on L and u' Tabaczynski et al. [26] showed that if the turbulence intensity is assumed proportional to the engine rpm, their model cannot predict the combustion dependence on the spark advance. Equations (36)–(40) were found to predict the experimental data variations with the engine rpm, load, dilution, spark advance, and equivalence ratio. However, in these equations the length scale of the energy-containing eddies is assumed to be proportional to the valve lift or given by Eq. (35) (i.e., the variations of the turbulence length scales with k and ϵ are neglected).

Experimental data on flame propagation in spark-ignition engines have shown that a laminarlike burning process immediately follows the spark discharge for about 15°CA compared with a total burning time of 80°CA. Tagalian and Heywood [27] observed laminar flame growth in the early stages of flame propagation and deduced that during those stages only a small fraction of the charge mass was burnt. Kalghatgi [28] showed that the initial flame is spherical and that as the compression ratio is decreased, the flame speed decreases but the residual mass fraction increases. He also observed that the flame speed increases with the turbulence levels, that the initial burning velocity is not equal to the laminar flame speed, and that cyclic variations decrease as the laminar flame speed increases. Kalghatgi [28] also showed that ignition influences the early flame development and cyclic variations. Gatowski et al. [29] obtained schlieren photographs in a square cross-section engine and showed that both bulk gas motion and small-scale turbulence affect the flame kernel. Large- and small-scale influences were observed as the flame grew. However, the reaction zone surfaces were seen to be smooth for a radius of curvature of about 1 mm. Hanson and Thomas [30] showed that the strain rate is more important than turbulence in enhancing combustion in a closed volume. Heywood and Vilchis [31] obtained schlieren photographs of flame propagation in spark-ignition engines and showed that faster flames result in fewer cycle-to-cycle variations and that the ratio of the turbulence intensity to the laminar flame speed plays an important role in determining the burning rate and cyclic variations.

Recently, Zur Loye and Bracco [32] used Mie scattering of TiO_2 seeding particles and pulsed laser sheets to visualize ignition kernels in ported premixed-charge spark-ignition engines fueled with stoichiometric propane-air mixtures at 300, 1200, and 3000 rpm; they showed that the turbulent flow field affects the ignition kernels as small as 2 mm and at times after ignition as short as 100 μs. Zur Loye and Bracco [32] also showed that distortion and cycle-to-cycle variations in the ignition kernel increase with the engine rpm and that the flame front thickness is smaller than 200 μm (i.e., the flame is in the reaction-sheet regime).

Keck [33] showed by means of schlieren, shadowgraphs, and laser-scattering measurements, that flames in spark-ignition engines are highly wrinkled and that the burning rate is controlled by eddy entrainment and laminar flame stretching and wrinkling. In the initial stages the flame in a spark-ignition engine is subject to geometric and aerodynamic stretches; the geometric stretch is due to the flame surface area increase after ignition, whereas the aerodynamic stretch is due to tangential velocity gradients at the flame front. Flame stretching can result in flame extinction when the Karlovitz number, defined as $u'\delta_L/lS_L$, where δ_L is the laminar flame thickness, exceeds some critical value [34, 35]. A theory of flame development in spark-ignition engines based on geometric and aerodynamic stretchings at the flame front has been proposed by Thomas [36].

Beretta et al. [37] modified the Blizard-Keck-Tabaczynski (BKT) model to account for the initial expansion of the flame front, which is close to that of a laminar flame. Their model can be expressed as

$$\frac{dm_b}{dt} = \rho_u A_f S_L + \frac{M}{\tau_b} \tag{57a}$$

$$\frac{dM}{dt} = \rho_u A_f S_T - \frac{M}{\tau_b} \tag{57b}$$

and uses experimentally determined correlations for S_L, l, and S_T. Equation (57*a*) indicates that the burning rate is laminar if $M = 0$. Beretta and Keck [38] established energy and entropy balances in an open-chamber internal combustion engine with detailed models of wall heat transfer and thermal boundary layers, and they obtained analytical solutions.

4.3.1.3 Algebraic ϵ models II. The algebraic ϵ models presented in this section calculate the length scale of the energy-containing eddies by assuming that the production of turbulence kinetic energy is equal to its dissipation rate.

Arici et al. [39] used Eqs. (37)–(40) to predict the lean misfire limit in spark-ignition engines with $u' = k^{1/2}$ and the following equation for the turbulence kinetic energy:

$$\frac{dk}{dt} = \left(\frac{2k}{3\rho}\right)\frac{d\rho}{dt} - \epsilon \tag{58a}$$

$$\epsilon \sim \frac{k^{3/2}}{l} \tag{58b}$$

If the angular momentum of the large eddies is conserved, the eddy viscosity is given by the following expression:

$$\nu_t \sim k^{1/2}\, l = A \tag{59}$$

where A is a constant.

Equations (58*b*) and (59) imply that

$$\epsilon \sim \frac{k^2}{A} \tag{60}$$

that is, ϵ is related to k. This result is unrealistic since the effects of the turbulence length scale on k and ϵ appear only indirectly. Furthermore, Eqs. (58*b*) and (60) indicate that $l \sim k^{-1/2}$ (i.e., the length scale of the energy-containing eddies is inversely proportional to the square root of the turbulence intensity).

Anderson and Lin [40] showed that the use of a fully developed value for the eddy diffusivity [Eq. (59)] during ignition is not valid in spark-ignition engines; their model was unable to satisfactorily predict the effect of the engine speed on misfire.

4.3.1.4 k–ϵ model. Ordinary differential equations for k and ϵ can be obtained by integrating the corresponding multidimensional equations presented in Chapter 2 throughout the combustion chamber volume. The resulting equations can be written as

$$\frac{dk}{dt} = \frac{2}{3}\frac{k}{\rho}\frac{d\rho}{dt} - \epsilon \tag{61}$$

$$\frac{d\epsilon}{dt} = \frac{4}{3}\frac{\epsilon}{\rho}\frac{d\rho}{dt} - c_2\frac{\epsilon^2}{k} \tag{62}$$

where c_2 is a constant.

Equations (61) and (62) neglect the fluxes of k and ϵ associated with combustion; these fluxes will be considered later in this section. The first terms in Eqs. (61) and (62) correspond to the production of k and ϵ by compression and have been modeled according to rapid-distortion theory, in which the turbulence interacts more rapidly with the mean flow than with itself (see Chapter 2).

In equilibrium, a balance between the production and dissipation of turbulence kinetic energy yields

$$\epsilon = B\nu\frac{u'^2}{\lambda^2} = C\frac{u'^3}{l} \tag{63}$$

where ν is the kinematic viscosity and B and C are constants; $B = 15$ for isotropic homogeneous turbulence.

For isotropic homogeneous turbulence, Eq. (63) yields

$$\frac{\lambda}{l} = \left(\frac{B}{C\,\mathrm{Re}}\right)^{1/2} \tag{64a}$$

$$\mathrm{Re} = \frac{u'l}{\nu} \tag{64b}$$

$$u' = \left(\frac{2k}{3}\right)^{1/2} \tag{64c}$$

Equations (59), (63), and (64c) and conservation of mass in the large eddies, that is,

$$\rho\, l^3 = D \tag{65}$$

where D is a constant, yield

$$l \sim \rho^{-1/3} \tag{66a}$$

$$u' \sim \rho^{1/3} \tag{66b}$$

$$k \sim \rho^{2/3} \tag{66c}$$

$$\epsilon \sim \rho^{4/3} \tag{66d}$$

Equations (66*a*) and (66*b*) have been derived from Eqs. (59) ($k^{1/2}l = A$ is equivalent to $u'l$ = constant), (64*c*), and (65).

Equations (61) and (62) include rate processes whereas Eq. (63) assumes equilibrium between the production and the dissipation rate of turbulence kinetic energy. This equilibrium assumption may explain why the model used by Anderson and Lin [40] was unable to satisfactorily predict the effect of the engine speed on misfire.

Turbulent flows are spotty and intermittent in the sense that the dissipation of turbulence kinetic energy does not take place at all locations at the same time. The spotty character of ϵ can be considered as in the model proposed by Tennekes [41], which accounts for the spatial nonuniform distributions of viscous dissipation. Tennekes [41] assumed that the small-scale structure of turbulence can be modeled as vortex tubes of diameter η (the Kolmogorov eddy size) that are stretched by eddies of size λ (the Taylor microscale).

For locally isotropic turbulence [42]

$$\epsilon = C\frac{u'^3}{l} = 15\nu\frac{u'^2}{\lambda^2} = D\nu\frac{v^2}{\eta^2} \tag{67}$$

where C and D are constants, ν is the kinematic viscosity, and v is the velocity of the Kolmogorov eddies, which is such that

$$\frac{v\eta}{\nu} = 1 \tag{68}$$

that is, the Reynolds number based on the velocity and length scale of the Kolmogorov eddies is equal to 1.

Equations (67) and (68) yield

$$\frac{\lambda}{l} = \left(\frac{15}{D\mathrm{Re}}\right)^{1/2} \tag{69a}$$

$$\eta = \left(\frac{D\,\nu^3}{\epsilon}\right)^{1/4} \tag{69b}$$

$$v = \left(\epsilon\,\frac{\nu}{D}\right)^{1/4} \tag{69c}$$

$$\frac{\eta}{l} = \left(\frac{D}{C\mathrm{Re}^3}\right)^{1/4} \qquad \mathrm{Re} = \frac{u'l}{\nu} \tag{70}$$

$$\frac{\lambda}{\eta} = 15^{1/2}\,\frac{\mathrm{Re}^{1/4}}{D^{3/4}} \tag{71}$$

The strain rate within the vortex tube can be approximated as u'/η, and the dissipation rate per unit mass is

$$\epsilon = E\,\nu\left(\frac{u'}{\eta}\right)^2 \tag{72}$$

where E is a constant.

The ratio of the mass contained in the vortex tube to the total mass is proportional to

$$\eta^2\,\frac{\lambda}{\lambda^3} = \frac{\eta^2}{\lambda^2} \tag{73}$$

so that the average dissipation rate within the vortex tube can be obtained by multiplying Eqs. (72) and (73). The result is

$$\epsilon = E\,\nu\left(\frac{u'}{\eta}\right)^2 \frac{\eta^2}{\lambda^2} = E\,\nu\,\frac{u'^2}{\lambda^2} \tag{74}$$

which is consistent with Eq. (67) if $E = 15$.

For a vortex tube of length L_v, the rate at which it is being stretched can be written as

$$\frac{1}{L_v}\frac{dL_v}{dt} = 2\,\frac{u'}{\lambda} \tag{75}$$

and the rate at which mass is being added to the length L_v can be estimated from the following equation:

$$\frac{d}{dt}\left(\rho\,\frac{\pi}{4}\,\eta^2\,L_v\right) = F\,\rho\pi\eta\,L_v\,v = F\,\rho\pi\eta\,L_v\,\frac{\nu}{\eta} \tag{76}$$

where we have used Eq. (68) and F is a constant.

If the density is constant, Eq. (76) can be written as

$$\frac{d\eta}{dt} = \eta\left(2F\,\frac{\nu}{\eta^2} - \frac{1}{2L_v}\frac{dL_v}{dt}\right) \tag{77}$$

which for $d\eta/dt = 0$ (i.e., if the viscous and stretching terms are equal) yields

$$\frac{2F\nu}{\eta^2} = \frac{1}{2L_v}\frac{dL_v}{dt} = \frac{u'}{\lambda} \tag{78}$$

where we have used Eq. (75).

Equation (78) can be written as

$$\frac{\eta}{\lambda} = \left(\frac{2F}{\mathrm{Re}_\lambda}\right)^{1/2} \tag{79a}$$

$$\mathrm{Re}_\lambda = \frac{u'\lambda}{\nu} \tag{79b}$$

Equations (79*a*) and (69*a*) can be combined to yield

$$\frac{\lambda}{\eta} = 15^{-1/4} D^{1/4} (2F)^{-1/2} \mathrm{Re}^{-1/4} \tag{80}$$

Equations (80) and (71) are identical if

$$F = 15^{-1/2} \left(\frac{D}{2}\right) \tag{81}$$

Therefore, Tennekes's model is compatible with Eq. (71).

Annand [43] showed that no model can accurately predict the effect of the bore-to-stroke ratio on flame propagation in spark-ignition engines, but that the Blizard-Keck-Tabaczynski (BKT) model is best overall. The BKT model has been modified by Tabaczynski [44] to account for conservation of circulation around a vortex tube of diameter η (the Kolmogorov eddy size), entrained mass, local equilibrium and control of the large-scale eddies by volume change as follows:

$$k^{1/2} \eta = c_1 \tag{82a}$$

$$\frac{d(\rho \eta^2 \lambda)}{dt} = \rho c_2 \eta \lambda \frac{\nu}{\eta} \tag{82b}$$

$$\eta = c_4 \lambda \left(\frac{k^{1/2} l}{\nu}\right)^{-1/4} \tag{83a}$$

$$\frac{\lambda}{l} = c_5 \left(\frac{k^{1/2} l}{\nu}\right)^{-1/2} \tag{83b}$$

where η is the Kolmogorov eddy size and c_1, c_2, c_4, and c_5 are constants.

Equation (82*b*) is identical to Eq. (76) if $L_\nu = \lambda$, whereas Eq. (83) is based on local equilibrium between the production of turbulence kinetic energy and its dissipation rate. Equations (83*a*) and (83*b*) are identical to Eqs. (71) and (69*a*), respectively.

Equations (82) and (83) can be combined, differentiated with respect to time, and written as

$$\frac{dk}{dt} = \frac{2}{3} \frac{k}{\rho} \frac{d\rho}{dt} - \frac{2}{3} k \left(\frac{1}{\nu} \frac{d\nu}{dt} + c_2 \frac{\nu}{\eta^2}\right) \tag{84}$$

$$\frac{d\eta}{dt} = -\frac{1}{3} \frac{\eta}{\rho} \frac{d\rho}{dt} + \frac{\eta}{3} \left(\frac{1}{\nu} \frac{d\nu}{dt} + c_2 \frac{\nu}{\eta^2}\right) \tag{85}$$

$$\frac{d\lambda}{dt} = -\frac{\lambda}{3\rho} \frac{d\rho}{dt} + \frac{\lambda}{3} \left(c_2 \frac{\nu}{\eta^2} - \frac{2}{\nu} \frac{d\nu}{dt}\right) \tag{86}$$

$$\frac{dl}{dt} = -\frac{l}{3\rho} \frac{d\rho}{dt} + \frac{l}{3} \left(c_2 \frac{\nu}{\eta^2} - \frac{8}{\nu} \frac{d\nu}{dt}\right) \tag{87}$$

where ν is a function of the mixture pressure and temperature.

Equation (69b) can be substituted into Eq. (85) to yield

$$\frac{d\epsilon}{dt} = \frac{4\epsilon}{3\rho}\frac{d\rho}{dt} + \frac{5\epsilon}{12\nu}\frac{d\nu}{dt} - \frac{4\epsilon}{3}c_2\frac{\nu}{\eta^2} \tag{88}$$

Equation (72) is identical to the third term on the right-hand side of Eq. (88) if

$$2\,k\,\frac{c_2}{3} = E\,u'^2 \tag{89}$$

and $d\nu/dt = 0$.

If the turbulence is assumed to be isotropic $u'^2 = 2k/3$, Eq. (89) results in $c_2 = E$. For $c_2 = E$ and isotropic turbulence, Eq. (72) can be written as

$$\epsilon = 2\,c_2\frac{\nu k}{3\eta^2} \tag{90}$$

which can be substituted into Eqs. (84) and (88) to yield

$$\frac{dk}{dt} = \frac{2}{3}\frac{k}{\rho}\frac{d\rho}{dt} - \frac{2}{3}\frac{k}{\nu}\frac{d\nu}{dt} - \epsilon \tag{91}$$

$$\frac{d\epsilon}{dt} = \frac{4\epsilon}{3\rho}\frac{d\rho}{dt} + \frac{5\epsilon}{12\nu}\frac{d\nu}{dt} - 2\frac{\epsilon^2}{k} \tag{92}$$

For a fluid with constant kinematic viscosity, $d\nu/dt = 0$ and the value of the constant that multiplies the third term on the right-hand side of Eq. (92) is in remarkably good agreement with the value of 1.92 used in multidimensional models (see Chapter 2). It is also in good agreement with the value of 1.833 obtained by Tabaczynski [44]. Note that if the production of turbulence by compression is equal to its dissipation rate, that is, $dk/dt = d\epsilon/dt = 0$, Eqs. (91) and (92) yield the result $\epsilon = (2k/3\rho)\,d\rho/dt$ if $d\nu/dt = 0$. This is not the case in multidimensional models (see Chapter 2) and in the model of Tabaczynski [44].

In the early part of compression the first terms on the right-hand sides of Eqs. (91) and (92) are smaller than the third ones. If the second terms are also neglected, Eqs. (91) and (92) can be solved to yield $\epsilon = Ak^2$, where A is a constant. This solution combined with $\epsilon \sim k^{3/2}/l$ yields $k^{1/2}l$ = constant (i.e., during the early parts of compression the angular momentum of the energy-containing eddies is conserved).

The k–ϵ equations [Eqs. (91) and (92)] can be used to calculate the Taylor microscale [Eq. (83b)], the burning time [Eq. 40(a)], and the turbulence intensity [Eq. (64c)]. Note that for isotropic turbulence, $c_5 = 15^{1/2}$, $l = k^{3/2}/\epsilon$, and $u' = (2k/3)^{1/2}$. However, in order to account for the effects of combustion, the value of u' that appears in the burning rate should be modified by the density ratio across the flame front [45]. Tabaczynski [45] and Daneshyar and Hill [46] suggested that the average axial propagation velocity induced by the combustion of the vortex tubes should be

$$u' \sim u'_{ff} \left(\frac{2}{3} \frac{\rho_u}{\rho_b} \right)^{1/2} \tag{93}$$

where u'_{ff} is the turbulence intensity in front of the flame front given by $u'_{ff} = (2k/3)^{1/2}$, where k is the turbulence kinetic energy of the unburnt gases.

Equations (39*a*) and (93) imply that the turbulent flame speed is

$$S_T = u' + S_L = S_L + c\, u'_{ff} \tag{94}$$

where c depends on the density ratio between the unburnt and burnt gases [Eq. (93)].

Equation (94) can also be written as

$$\frac{S_T}{S_L} = 1 + c \frac{u'_{ff}}{S_L} \tag{95}$$

which is in good agreement with available experimental data for $u'_{ff}/S_L > 10$ and $\text{Re} = u'l/\nu > 1000$ [46]. Equation (95) was proposed by Damkohler in 1947 (see Section 4.5).

Equation (94) accounts for the effects of the turbulent velocity and length scales through u'_{ff} or k and ϵ [Eqs. (91) and (92)].

The derivation of Eqs. (91) and (92) is based on equilbrium considerations for the length scales of the energy-containing eddies, Kolmogorov eddies, and Taylor microscale. These equilibrium conditions may be applicable at high Reynolds numbers when there is an inertial subrange and the small-scale eddies can be considered isotropic. If the Reynolds number is not large enough, the large-scale eddies interact with the small ones, and Eqs. (69)–(71) are no longer valid. Experimental data are needed to establish the validity of Eqs. (69)–(71) in engine flows.

At present, there is no theory that can predict Eq. (93) based on physical principles. It is remarkable, however, that Eq. (95) with $\rho_u/\rho_b = 6$ and $c = 2$ is in very good agreement with experimental data [45].

4.3.1.5 Two-zone k–ϵ model. If the k–ϵ model [Eqs. (91) and (92)] can be applied to the burnt- and unburnt-gas zones, the resulting equations can be written as [46]

$$\frac{dk_u}{dt} = \frac{2k_u}{3\rho_u} \frac{d\rho_u}{dt} - \frac{2}{3} \frac{k_u}{\nu_u} \frac{d\nu_u}{dt} - \epsilon_u \tag{96}$$

$$\frac{d\epsilon_u}{dt} = \frac{4\epsilon_u}{3\rho_u} \frac{d\rho_u}{dt} + \frac{5\epsilon_u}{12\nu_u} \frac{d\nu_u}{dt} - 2 \frac{\epsilon_u^2}{k_u} \tag{97}$$

$$\frac{dk_b}{dt} = \frac{2k_b}{3\rho_b} \frac{d\rho_b}{dt} - \frac{2}{3} \frac{k_b}{\nu_b} \frac{d\nu_b}{dt} - \epsilon_b + F_k^b \tag{98}$$

$$\frac{d\epsilon_b}{dt} = \frac{4\epsilon_b}{3\rho_b}\frac{d\rho_b}{d} + \frac{5}{12}\frac{\epsilon_b}{\nu_b}\frac{d\nu_b}{dt} - 2\frac{\epsilon_b^2}{k_b} + F_\epsilon^b \tag{99}$$

where F_k^b and F_ϵ^b denote the fluxes of k and ϵ across the flame front, which can be expressed as

$$F_k^b = \frac{1}{m_b}\frac{dm_b}{dt}(k_b^* - k_b) \tag{100}$$

$$F_\epsilon^b = \frac{1}{m_b}\frac{dm_b}{dt}(\epsilon_b^* - \epsilon_b) \tag{101}$$

and k_b^* and ϵ_b^* can be calculated using rapid-distortion theory, assuming that the transition from the unburnt to burnt gases occurs in a very short time.

If the mass and angular momentum of a fluid element are conserved upon crossing the flame front,

$$\rho_b\, l_b^3 = \rho_u\, l_u^3 \tag{102}$$

$$l_b\, k_b^{*1/2} = l_u\, k_u^{1/2} \tag{103}$$

Eliminating l_b from Eqs. (102) and (103) yields

$$k_b^* = \left(\frac{\rho_b}{\rho_u}\right)^{2/3} k_u \tag{104}$$

The dissipation rate ϵ is proportional to $k^{3/2}/l$. Therefore, Eqs. (102) and (103) yield

$$\epsilon_b^* = \epsilon_u\left(\frac{\rho_b}{\rho_u}\right)^{4/3} \tag{105}$$

Rapid-distortion theory may not be applicable in internal combustion engines, particularly near TDC. Near TDC the large-scale eddies generated during the intake stroke have been damped by compression, the mean flow field is almost unidirectional in the absence of swirl and squish, and the turbulence intensity is very high. Therefore, the interaction of the turbulence with itself may be comparable to, if not larger than, the interaction between the turbulence and the mean flow. If this occurs, rapid-distortion theory is not applicable.

Equations (104) and (105) imply that $k_b^* < k_u$ and $\epsilon_b^* < \epsilon_u$, and for isotropic turbulence, Eq. (93) can be written as

$$u' \sim \left(\frac{2k_u}{3}\right)^{1/2}\left(\frac{2\rho_u}{3\rho_b}\right)^{1/2} \tag{106}$$

Tabaczynski [45] used Eqs. (39), (40), and (96)–(105) and showed that the burning time decreases with the compression ratio and with the mixture equivalence ratio for lean mixtures, but increases with the equivalence ratio for rich mixtures. The burning time also decreases as the turbulence intensity is increased. Tabaczynski [45] also showed that the burning duration increases with the amount of exhaust gas recirculation (EGR), but decreases as the equivalence ratio is increased in lean mixtures. These results are physically plausible and indicate that an entrainment-burning model is able to predict the experimentally observed trends.

Milane et al. [47] used Eqs. (96) and (98) and a stochastic model of turbulence to study ignition and combustion in a stratified-charge spark-ignition engine. Their model accounts for molecular and turbulent diffusion, assumes that combustion is controlled by turbulent diffusion, uses a Monte Carlo technique, and correctly predicts the variations of the burning duration and burning rate with the spark advance, EGR, mixture stratification, and stroke.

Lavoie et al. [48] developed an unmixed thermodynamic model (see Section 4.3.3) that assumes that the elements of the charge that burn at different times do not mix but retain their identity as the burnt gases are compressed and expanded. Lavoie et al. [48] showed that the NO concentrations are rate controlled in the burnt gases behind the flame front and that they freeze early during the expansion stroke. Measurements at two different locations from the spark plug showed that a gradient in the NO concentration existed and that this gradient was of the same order of magnitude as their model predictions. Lavoie et al. [48] assumed that each gas element is isentropically compressed after combustion and used a six-reaction mechanism for the NO formation where O, O_2, H, OH, and N_2 were considered to be in equilibrium. Komiyama and Heywood [49] used the model developed by Lavoie et al. [48] and showed that the NO concentrations freeze early in the expansion stroke for lean mixtures. For rich mixtures freezing occurs later, after all the charge has been burnt, and substantial NO decomposition takes place. Komiyama and Heywood [49] used the extended Zel'dovich mechanism and showed that their model matches experimental data for NO emissions over a wide range of engine-operating conditions and that NO emissions are generally less sensitive in rich mixtures than in lean mixtures to changes in engine-operating conditions. Exhaust gas recirculation (EGR) increases the time required to establish a flame kernel, the time for the flame to propagate across the cylinder charge, and the cycle-to-cycle pressure variations. Komiyama and Heywood [49] stated that the time required to establish a stable flame kernel appears to be the most significant effect since a turbulent flame front model suggested that this is the cause of increased cycle-to-cycle pressure fluctuations.

Rashidi [50] showed that the main cause of cycle-to-cycle variations is the small-scale turbulence and the mixing it produces. Small- and medium-scale turbulence influences the mixing of the reactive charge in the small, unstable flame kernel and causes variations in the reaction rate and in the ignition delay time. The large-scale turbulence was found to affect the flame front surface area. Akindele et al. [51] observed that the early stages of flame development are controlled by molecular processes and that the flame kernel moves away from the spark plug with a velocity approximately equal to the turbulent rms velocity. Hansel [52] linked the cycle-to-cycle variations to turbulent combustion, used a flame propagation model with a turbulent flame speed equal to the sum of the turbulent rms velocity and the laminar flame speed, and showed that the variations in the peak pressure are functions of the turbulent rms and equivalence ratio when the flame kernel is formed and interacts with the turbulent flow field. Hansel [52] attributed the cycle-to-cycle variations to swirl and turbulence; velocity gradients at the flame front distort and wrinkle the flame. Tromans [53] stated that the cycle-to-cycle variations are due to cyclic variations in the strain rate of the turbulent flow field around the spark, particularly for lean mixtures.

Chomiak [54] argued that turbulence causes random variations in the flame kernel growth rate and that the interaction of the ignition kernel with the roll-up vortex formed by the piston sliding over the cylinder wall is the major cause of quenching and cycle-to-cycle variations. It is doubtful that the roll-up vortex can be the cause of cycle-to-cycle variations since its size is small. Hot wire anemometry has shown evidence of the roll-up vortex [55]; however, laser Doppler anemometry with single photon counting and digital correlation has shown that the intake-generated turbulence relaxes during the compression stroke, although large-scale structures have been observed [56]. Hot wire and laser Doppler anemometry have shown that large-scale structures are responsible for the cycle-to-cycle variations [57]. Cole and Swords [58] showed a correlation between the large-scale velocity fluctuations in the spark at ignition and the ultimate combustion efficiency. Their conditionally sampled laser Doppler velocimetry data indicate a weak correlation between the pressure variations and the high-frequency turbulence, but a strong correlation between the low-frequency, large-scale turbulence and the pressure fluctuations. Wakisaka and Hamamoto [59] showed that the inflammability limits become narrower with increases in the turbulent intensity and that flames are disintegrated into small flamelets by strong turbulence and then extinguished due to heat losses. Hamamoto et al. [60] observed that the flame shape and size in the initial stages of combustion fluctuate from cycle to cycle even if the mixture composition is kept fixed. They also observed that a reduction of the combustion period suppresses the cyclic variations. Martin et al. [61] showed that the combustion-induced mean flow field produces a small increase in the bulk turbulent kinetic energy ahead of the flame front and that the turbulence in the preflame gases close to the flame front is amplified in the direction of the flame propagation.

Davis et al. [62] developed a quasidimensional model based on the concentrated vorticity regions proposed by Tennekes [41]. In this model the vorticity is concentrated in vortex tubes of diameter on the order of the Kolmogorov length scale with a spacing equal to the Taylor microscale (Fig. 4.3). Tennekes's model explains intermittency and the dissipation of turbulence kinetic energy [63]. The following assumptions are made in the model: (1) conservation of circulation around a vortex tube (this is similar to the assumption of a constant eddy diffusivity); (2) entrainment of the mass into the individual vortex tubes at a rate governed by ν and η; and (3) local equilibrium (i.e., production of turbulence kinetic energy equal to its dissipation rate). These assumptions yield the following expressions:

$$k^{1/2}\,\eta = \text{constant} \qquad \frac{d(\rho\eta^2\lambda)}{dt} = c_2\,\rho\eta\lambda\left(\frac{\nu}{\eta}\right) \tag{107}$$

$$\frac{\lambda^3}{\eta^2 l} = \text{constant} \qquad \frac{\eta}{\lambda} = c\left(k^{1/2}\,\frac{\eta}{\nu}\right)^{-1} \tag{108}$$

from which first-order ordinary differential equations for η, λ, l, and k can be deduced. Equations (107) and (108) are similar to Eqs. (82) and (83). Davis et al. [62] showed that the model represented by Eqs. (107) and (108) appears to have the proper sensitivity to velocities past the intake valve with regard to predicting the turbulence level near TDC of the compression stroke. Belaire et al. [64] specified the length scale as a function of the chamber shape and piston location relative to TDC and included the effects of combustion-generated turbulence in the squish region. Belaire et al. [64] also showed that compact high-squish-area combustion chambers have significantly faster combustion rates than open chambers when a high-swirl intake port is used in a homogeneous-charge engine with nearly central spark plug location. They also showed that the combustion rate is strongly influenced by the combustion chamber geometry.

Borgnakke et al. [65] used the model represented by Eqs. (82)–(87), analyzed the turbulent boundary layers in the burnt and unburnt gases, and developed an instantaneous heat transfer model. The influence of the turbulent flow field on the heat transfer rate was expressed as a function of the Reynolds number and the integral length scale. Their model is sensitive to geometry changes such as bore-to-stroke ratio, accurately predicts the heat transfer and burning rates, and is more sensitive to geometry changes and to changes in operating conditions than the correlations of Annand and Woschni (see Chapters 2 and 3).

Borgnakke et al. [66] used a swirl model based on an integral formulation of the angular momentum equation, which was solved with an assumed tangential velocity profile that enables prediction of a non-solid-body rotation as a function of the inlet flow, wall shear, and squish motion during the engine cycle. The intake-generated turbulence was assumed to be proportional to the velocity past the intake valve, and the production of turbulence due to the boundary layer was also included. Their model was found to be superior to a solid-body rotation model and showed the expected trends in the swirl velocity and turbulence intensity with changes in the combustion chamber geometry. Borgnakke et al. [66] divided the combustion chamber into two concentric cylinders; one cylinder contained the bowl and the other one extended to the cylinder wall. Energy and angular momentum transfers between the cylinders were considered.

Davis and Borgnakke [67] extended the model proposed by Borgnakke et al. [66], used the BKT model of turbulent combustion, and accounted for turbulence generation by compression, squish, and shear. Their results indicate that the swirl, squish, and turbulence levels can be varied to produce a minimum in fuel consumption.

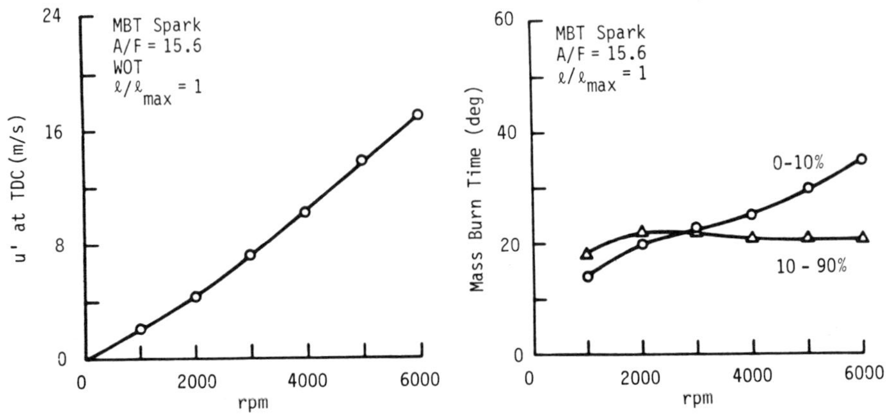

Figure 4.5 Turbulence intensity at TDC (left) and burning time (right) as a function of engine rpm [62] (MBT = mean best torque; A/F = air-fuel ratio; WOT = wide-open throttle; l = valve lift).

Witze et al. [68] used a two-zone thermodynamic model to predict the pressure history in spark-ignition engines and the results from motored tests to establish the initial conditions for their thermodynamic model. The model accurately predicted the pressure diagram for a range of engine speeds and ignition locations. Witze et al. [68] showed that turbulence was generated in the compression stroke and that a two-equation $k-\epsilon$ model accurately predicted the decay of turbulence after the production phase was completed.

Figures 4.5–4.17 show some sample results that have been obtained with the one- and two-equation models presented in this section. Figure 4.5 indicates that the turbulence intensity at TDC increases with the engine rpm, whereas the burn time for burnt-gas mass fractions between 0.10 and 0.90 is almost

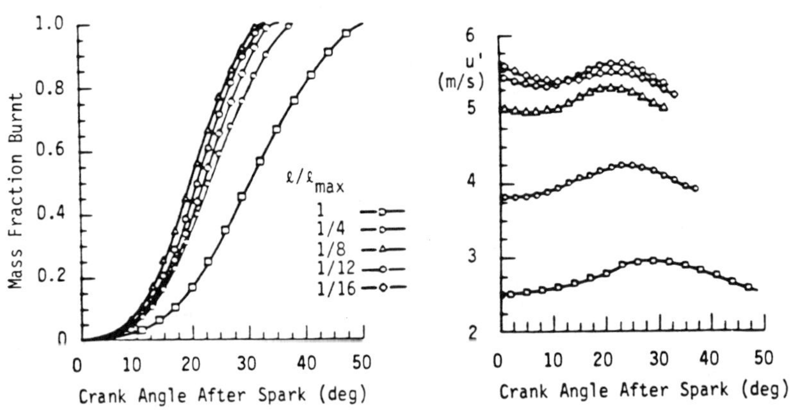

Figure 4.6 Mass fraction burnt (left) and turbulence intensity (right) during combustion in the unburnt gases as functions of the valve lift and crank shaft angle [62].

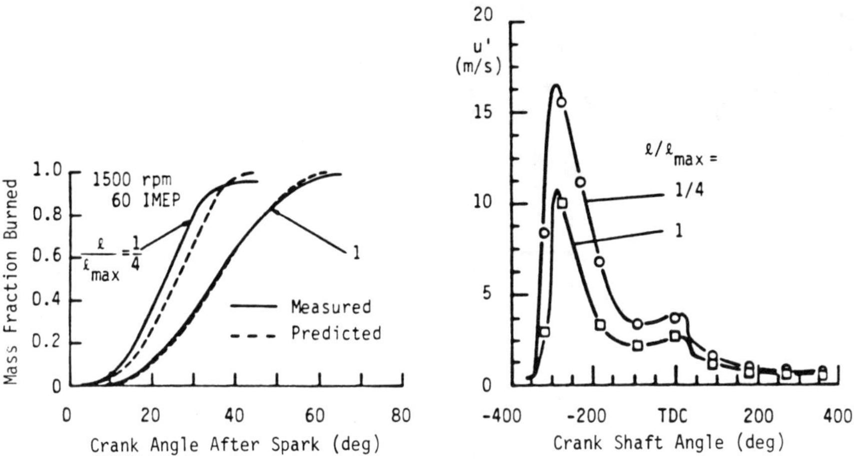

Figure 4.7 Mass fraction burnt (left) and turbulence intensity (right) as functions of the valve lift and crank shaft angle [62].

independent of the engine rpm. Figure 4.6 shows the mass fraction burned as a function of the crank shaft angle and valve lift and indicates that the burning time first decreases and then increases as the valve lift is decreased. Figure 4.6 also shows that the turbulence intensity in the unburnt gases during combustion decrease as the valve lift is increased. Figure 4.7 indicates that the burning time and the turbulence intensity at TDC decrease as the valve lift is decreased.

Figure 4.8 shows the turbulence intensity predicted by a two-equation $k-\epsilon$ model and an algebraic ϵ model as a function of the crank shaft angle; the algebraic model predicts higher turbulence intensities than the two-equation model. Figure 4.9 illustrates an idealized combustion chamber with swirl and squish.

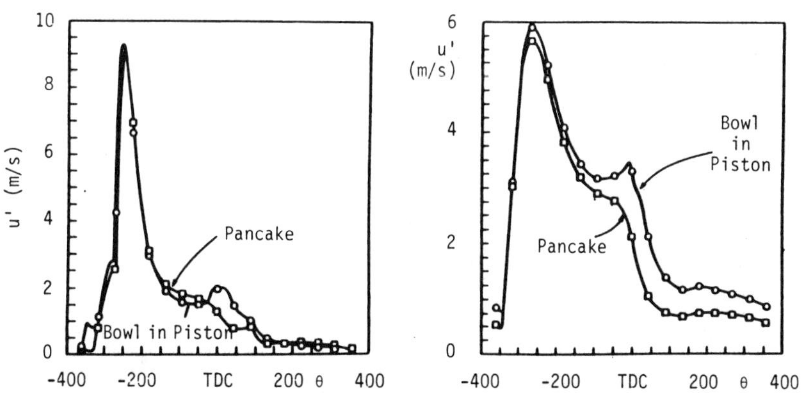

Figure 4.8 Turbulence intensity as a function of the crank shaft angle for full $k-\epsilon$ model (left) and algebraic ϵ model (right) [44].

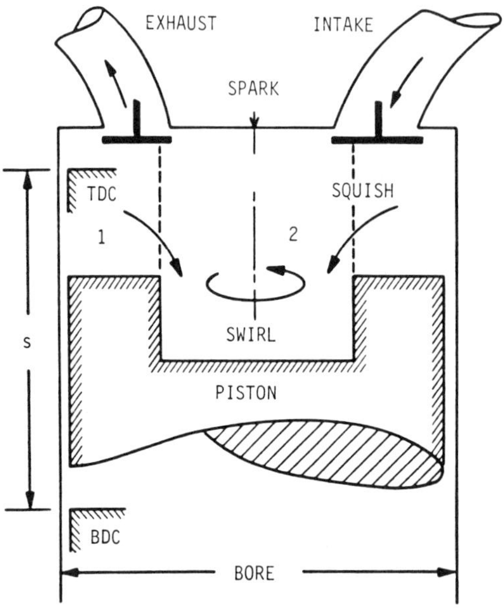

Figure 4.9 Idealized combustion chamber geometry with squish and swirl [66].

Figure 4.10 shows that in a bowl-in-piston combustion chamber the burning time increases with the amount of EGR. Figure 4.10 also shows that a two-zone model can accurately predict the pressure diagram in a bowl-in-piston combustion chamber.

Figure 4.11 shows that the turbulence intensity and swirl ratio increase monotonically with the squish area, but the burning time decreases monotonically. Figure 4.12 indicates that the turbulence intensity increases monotonically with the swirl ratio.

In Fig. 4.13 the mass fraction burned, which was predicted by the models proposed by Blizard and Keck (BK) [12] and Borgnakke, Arpaci, and Tabac-

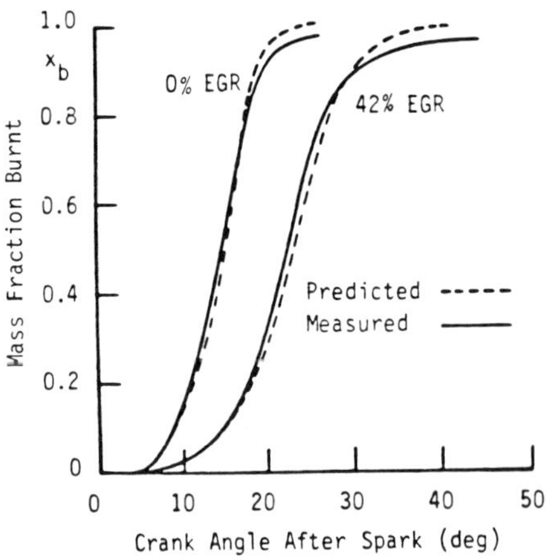

Figure 4.10 Mass fraction burnt as a function of the crank shaft angle and percentage of exhaust gas recirculation [67].

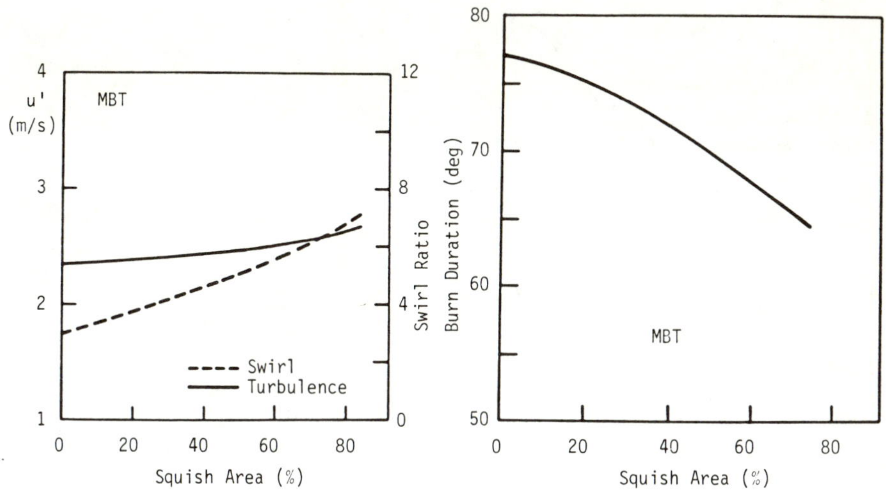

Figure 4.11 Turbulence intensity (left) and burn duration (right) as functions of the squish area [67].

zynski (BAT) [65], is presented. Figure 4.14 also compares the heat transfer rates calculated using Annand's and Woschni's correlations with the BAT model [65], which uses a von Karman-Pohlhausen method to calculate the thermal boundary layer thickness.

Figure 4.15 shows that the 0–10% and 10–90% burning times increase and decrease respectively with the spark advance. Figure 4.15 also shows that the combustion times in combustion chambers with squish are smaller than those in disk chambers. Figure 4.16 indicates that the burning time in a combustion chamber provided with a single central spark is smaller than the burning times in combustion chambers equipped with dual wall sparks and single wall sparks. Figure 4.17 shows that the burning time is a monotonically increasing function of the bore-to-stroke ratio.

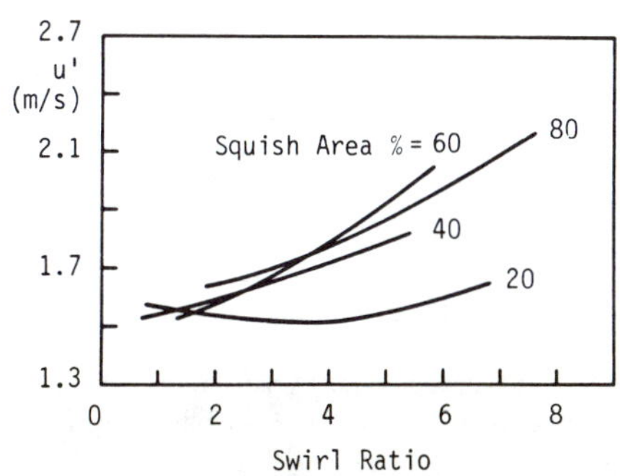

Figure 4.12 Turbulence intensity as a function of the swirl ratio and percentage of squish area [66].

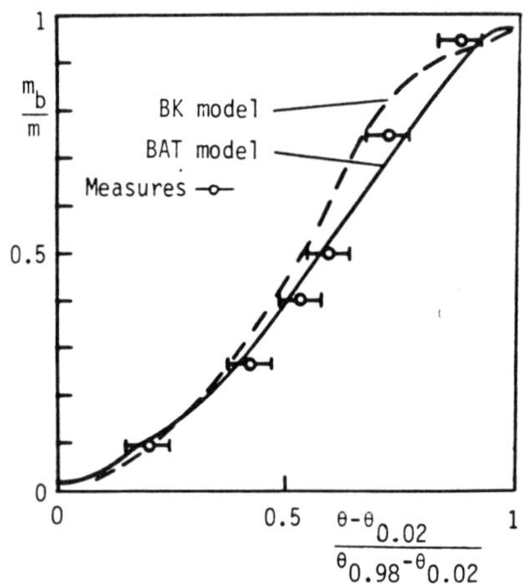

Figure 4.13 Mass fraction burnt as a function of the crank shaft angle [65]: BK model, Blizard and Keck; BAT model, Borgnakke, Arpaci, and Tabaczynski.

4.3.2 Specification of the Turbulent Flame Speed

In Section 4.3.1 models based on the calculation of the mass burning rate were presented. In this section we describe models in which the turbulent flame speed is specified as a function of the turbulence intensity, laminar flame speed, and engine rpm. These models also assumed that the flame propagates spherically throughout the combustion chamber.

Mathur et al. [69] used Eq. (29) where the turbulent flame speed is given by the following expression:

$$\text{FSR} = \frac{S_T}{S_L} = 1 + 4.01\,\frac{u'}{S_L} \qquad (109)$$

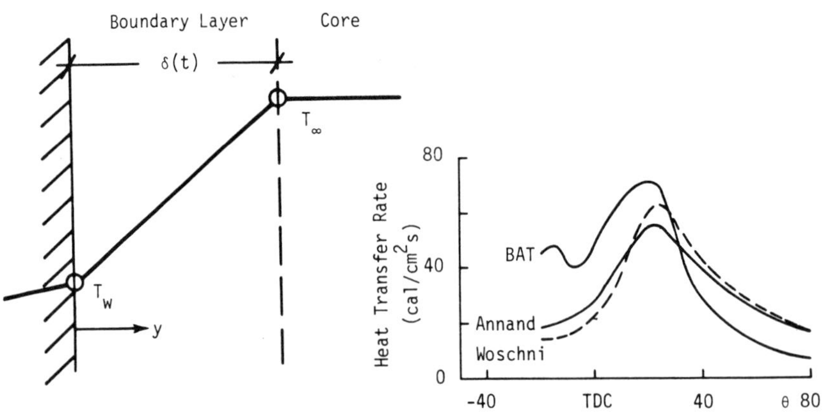

Figure 4.14 Instantaneous local thermal boundary layer and averaged heat transfer rates [65] (y = distance from the wall; T_W = wall temperature; T_∞ = core temperature; δ = boundary layer thickness).

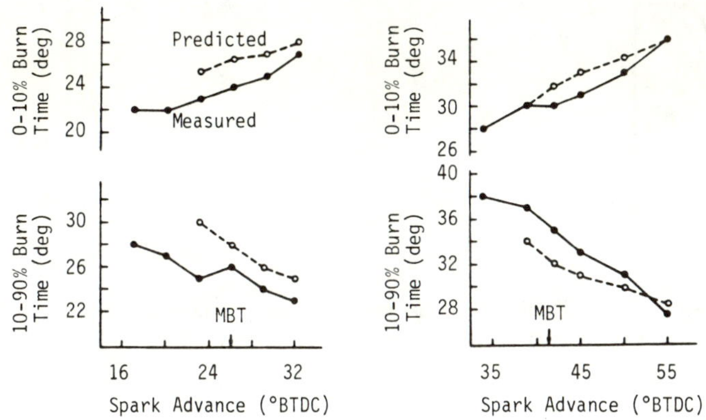

Figure 4.15 Burn time as a function of the spark advance for squish (left) and disk (right) combustion chambers [64].

During the flame development period, which extends up to the point when the flame radius is approximately 0.03 m, the flame speed ratio (FSR) was modified as follows:

$$\text{FSR}_{fd} = \left(\frac{R_f}{0.03}\right)^{0.5} \text{FSR} \tag{110}$$

where R_f is the flame front radius.

Equations (109) and (110) account only for the effects of the turbulent intensity on the flame speed. The expansion velocity can be obtained from the following expression:

$$\frac{S_T}{S_F} = \frac{\rho_u}{\rho_b} - 1 \tag{111}$$

Equation (109) is more physical than

$$S_T = f\,S_L \tag{112}$$

where f is a turbulence-enhancement factor that can be considered proportional to the engine rpm, for example, $f = 1 + 0.00197$ rpm. Equation (112) assumes that the turbulence intensity is constant during the compression of the unburnt

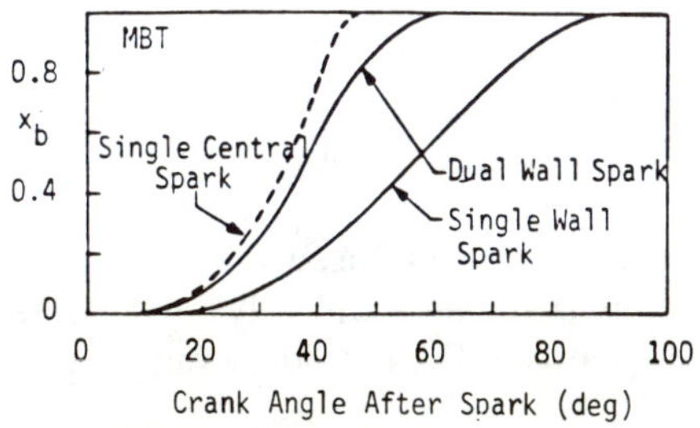

Figure 4.16 Mass fraction burnt as a function of the crank shaft angle for single central, single wall, and dual wall spark configurations [67].

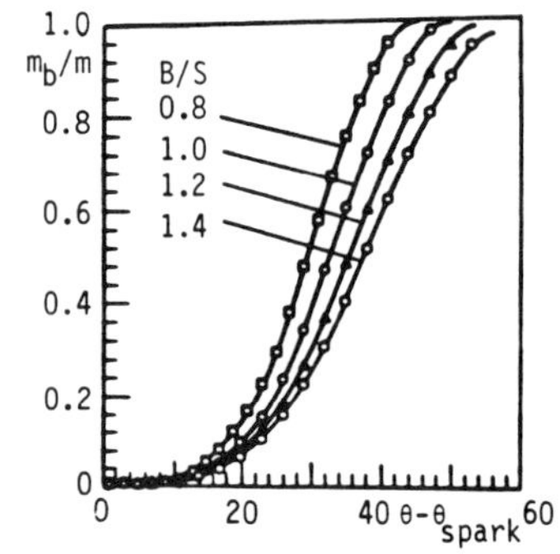

Figure 4.17 Mass fraction burnt as a function of the bore-to-stroke (B/S) ratio and crank shaft angle [62].

charge and fails to predict the variation of the combustion duration and ignition delay time with the equivalence ratio and spark timing.

Hiroyasu and Kadota [70] specified a burning velocity given by Eq. (112) with $f = 1 + 0.002$ rpm and used partial equilibrium for 11 species and a nonequilibrium NO model that considers that the NO formation occurs in post-flame reactions. Hiroyasu and Kadota [70] assumed a spherical flame front, uniform pressure throughout the combustion chamber, and a heterogeneous temperature distribution. The cylinder charge was assumed to consist of numerous elements; these elements burnt during successive time increments.

Fagelson et al. [71] and Rubin and McLean [72] used a multizone model and a semiempirical turbulent flame speed given by

$$S_T = A\ \mathrm{Re}^B\ S_L \tag{113}$$

where Re is the Reynolds number and A and B are constants. They also used the extended Zel'dovich mechanism to predict the NO formation in hydrogen-fueled and methanol-fueled engines. In hydrogen-fueled engines Fagelson et al. [71] observed fast burning rates, high pressures and temperatures, and fast NO formation and destruction. The maximum NO concentration was observed at an equivalence of 0.8. For richer mixtures, NO decomposition lowers the exhaust emissions; however, NO_x control by EGR was found to be more efficient in lean mixtures. Near-stoichiometric mixtures resulted in excessively high pressure rises, and EGR moderated the burning rate and reduced the power output. Rubin and McLean [72] showed that their semiempirical turbulent flame speed model results in good agreement with the experimentally determined power, efficiency, and emissions. The maximum efficiency was obtained for $0.7 \leq \phi \leq 0.8$, whereas the maximum NO concentration occurred at $\phi = 0.9$. Rubin and McLean [72] also observed that addition of water to methanol reduces the NO_x emissions and engine efficiency.

Experimental data show that at Reynolds numbers based on the Taylor microscale higher than 100, a wrinkled laminar flame is unlikely, and flame propagation is associated with the small dissipative eddies. Andrews et al. [73]

proposed the following flame speed ratio correlation:

$$\frac{S_T}{S_L} = fcn\ (\mathrm{Re}_\lambda) \tag{114}$$

where $\mathrm{Re}_\lambda = \rho\, u'\lambda/\mu$ is the Reynolds number based on the Taylor microscale λ and $fcn\ (\mathrm{Re}_\lambda)$ denotes a function of the Reynolds number Re_λ.

Haghgooie et al. [74] used hot wires to measure turbulent time scales in spark-ignition engines, and they showed that high frequencies and narrow distributions are observed at high rpm and at high inlet port velocities. They also scaled their results with those of isotropic turbulence and found good agreement with experimental data when the Taylor microscale was determined from the most probable frequency.

On the other hand, the hot wire anemometry data of Kido et al. [75] indicate that the length scales predicted by isotropic turbulence are not strictly valid in engine flows. Kido et al. [75] found the following correlations to their experimental data:

$$\frac{l}{\eta} = 1.96\left(\frac{u'l}{\nu}\right)^{0.60} \qquad \lambda^3 = 2\, l^2\, \eta \tag{115}$$

$$l = A^{1/2}\, C^{-5/2}\, \delta\left(\frac{u'\delta}{\nu}\right)^{-1/4} \qquad \lambda = \frac{15^{1/2}}{C^2}\, \delta\left(\frac{u'\delta}{\nu}\right)^{-2/5} \tag{116}$$

where δ is the combustion chamber height, $C = 2.1$, and $A = 28.9$.

Kido et al. [75] also found that

$$l \sim \mathrm{CR}^{-0.93} \qquad l \sim \mathrm{rpm}^{-0.28} \qquad l \sim \mathrm{CR}^{-0.82} \qquad l \sim \mathrm{rpm}^{-0.45} \tag{117}$$

where CR is the compression ratio.

During the compression stroke, u' and l decrease monotonically, but the turbulent time scale remains almost constant. The swirl velocity and the turbulence intensity decay monotonically during the compression stroke in swirling flows, and the turbulent time and length scales are slightly smaller and larger, respectively, than those without swirl.

The experimental results of Kneen et al. [76] and Nagao and Tanaka [77] indicate that large-scale vortex motions and swirl increase the turbulent flame speed, the combustion rate, and the engine noise. However, the engine volumetric efficiency decreases with increases in the swirl levels at the inlet port. In order to improve the combustion efficiency in spark-ignition engines, Nagao and Tanaka [77] suggested a control of the swirl levels at the intake port.

Figure 4.18 shows that the microtime scale decreases with the engine speed [44].

Smith [78] used a pulsed microschlieren technique, varied the engine rpm, and used different shrouded valves in order to vary the in-cylinder turbulent flow field. His results show that the Reynolds number based on the Taylor microscale correlates with the scale of the flame structure and the total burn

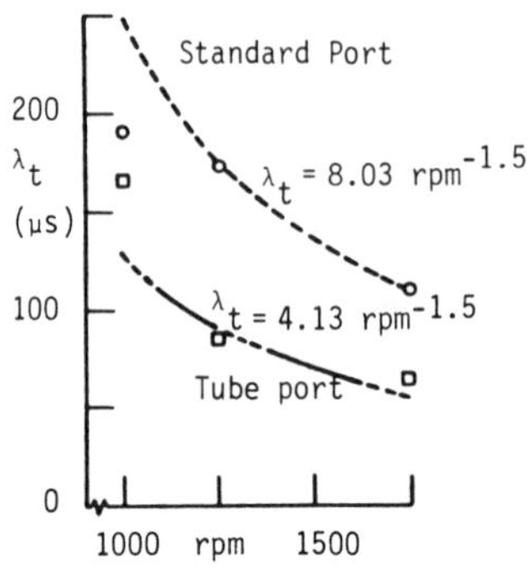

Figure 4.18 Microtime scales for two inlet port configurations as a function of the engine speed [44]. Symbols denote experimental data.

duration in spark-ignition engines and that the flame develops a wrinkled structure soon after ignition.

The flame speed in spark-ignition engines can be measured by means of optical techniques [79–82] and correlated with the combustion chamber geometry, ignition timing, rpm, etc. The resulting empirical correlations can be used to predict the mass burning rate.

The laminar flame speed that appears in Eqs. (109)–(114) can be correlated with the combustion chamber pressure, temperature, and residual mass fraction [83, 84].

The schlieren visualization studies of Namazian et al. [81] in a square cross-section piston-cylinder assembly show that (1) the intake flow enters the cylinder and interacts with the cylinder wall; (2) a recirculation zone at the corner between the cylinder head and the cylinder wall is created in the intake stroke and persists throughout compression, combustion, and expansion; (3) the turbulent flame propagates approximately spherically and consists of an irregular, thick front; (4) only a fraction of the mixture contained behind the propagating flame front is fully burnt; (5) the flame front shows large- and small-scale irregularities; (6) the small-scale structure at the flame front appears to be burnt gases surrounded by a thin flame zone and then unburnt mixture; (7) on the back side of the flame, the small-scale structure consists of unburnt mixture within a thin flame zone with burnt mixture outside; (8) differences in flame development and propagation can be related to the occurrence of the peak pressure and can cause cycle-to-cycle variations; (9) gases from the crevice volumes behind and between the piston rings issue from the corners of the gap between the piston crown and the walls during the expansion stroke and contain a high concentration of unburnt HC; (10) the blowdown flow entrains the denser thermal boundary layer gas off the cylinder head and upper part of the cylinder wall; and (11) a roll-up vortex is formed in the exhaust stroke near the corner between the cylinder wall-piston-crown-surface.

Figure 4.19 shows the FSR as a function of u'/S_L at the 50% mass burnt point for open and wedge combustion chambers. The effect of the engine speed on the turbulence intensity for open and wedge combustion chambers is shown

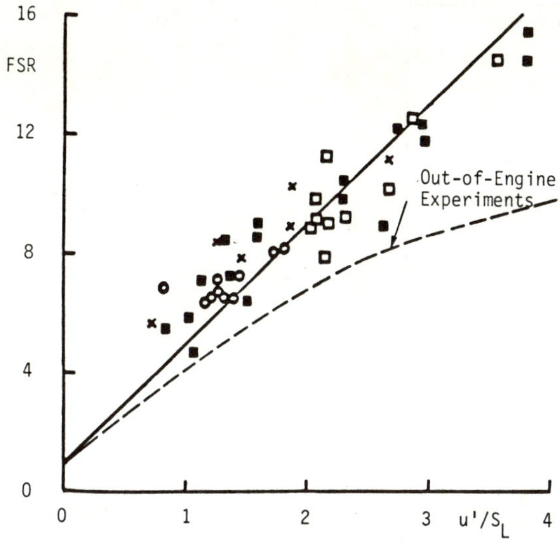

Figure 4.19 Flame speed ratio (FSR) as a function of u'/S_L at the 50% mass burnt point for open and wedge combustion chambers [86]. Symbols denote experimental data.

in Fig. 4.20. A comparison between the measured and predicted mass burning rates using an FSR model is shown in Fig. 4.21.

4.3.3 Specification of the Mass Burning Rate

In some models the mass fraction of burnt gases is specified by means of a Wiebe function [Eqs. (7) and (8)], and the flame is assumed to propagate spherically throughout the combustion chamber. Equation (29) can be used to calculate the

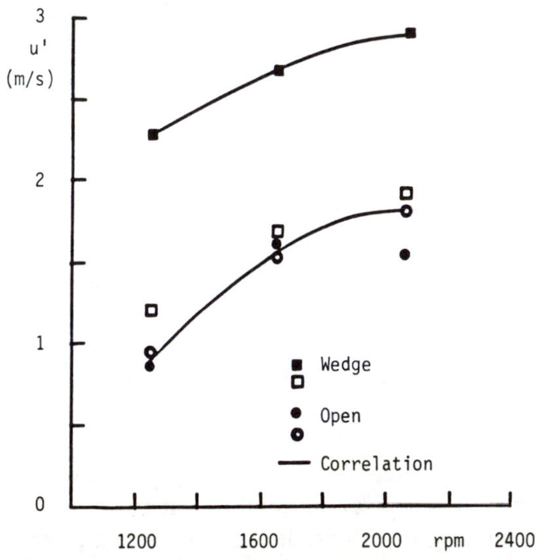

Figure 4.20 Effect of engine speed on turbulence intensity in open and wedge combustion chambers [86]. Symbols denote experimental data.

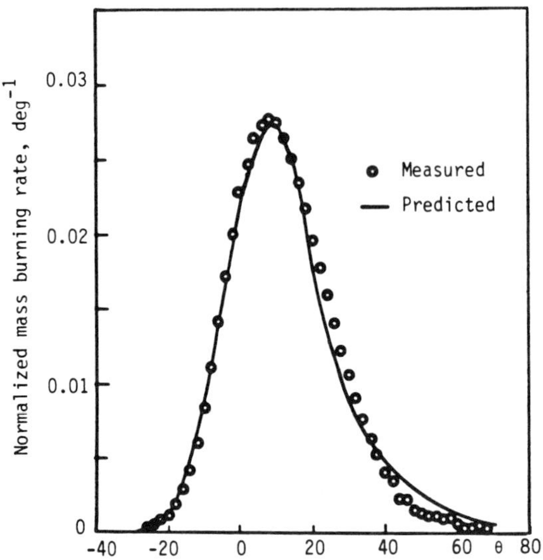

Figure 4.21 Normalized mass burning rate as a function of the crank shaft angle [86].

mass burning rate $dm_c/d\theta$, which can be written as

$$\frac{dm_c}{d\theta} = \rho_u A_f S_T \tag{118}$$

In the absence of leakage and flows into crevices $dm_c/d\theta = dm_b/d\theta$ and Eqs. (7) or (8) and Eq. (118) can be used to calculate the turbulent flame speed, provided that the flame front geometry is known.

Lavoie and Blumberg [85] used Eq. (7), inferred the combustion angle $\Delta\theta_b$ from measured pressure diagrams, and showed the importance of the flame-generated NO in premixed combustion. They also showed that stratification for NO control depends on the overall equivalence ratio; for equivalence ratios less than or equal to 0.80, mixture stratification results in a loss of NO control compared with premixed charges.

Mattavi et al. [86] used a two-zone thermodynamic model as a diagnostic tool (heat release analysis) where the pressure diagram is specified, and calculated the mass, temperature, and volume of the burnt and unburnt gases and the turbulent flame speed, assuming a spherical flame propagation model. They also obtained an experimentally determined flame speed ratio, which was used to predict the cylinder pressure and the mass, volume, and temperature of the burnt- and unburnt-gas regions as a function of time.

Lavoie and Blumberg [87] developed a model for spark-ignition engines that accounts for thermal boundary layers, quench layers, adiabatic core, scrapping vortices, exhaust process, and ring crevices. The model results in good agreement with measured NO_x emissions and fuel consumption. Blumberg [88] assumed that the liquid fuel is completely vaporized prior to combustion in stratified-charge engines, used Eq. (7), accounted for air humidity and EGR, employed the extended Zel'dovich mechanism, and showed that stratified charges result in lower NO emissions than homogeneous charges.

Daneshyar and Watfa [89] developed mixed and unmixed models to predict the NO formation by means of the extended Zel'dovich mechanism in spark-ignition engines. The mixed model assumes that the burnt gases mix instantaneously and that the temperature and composition are uniform in the burnt-gas region. The unmixed model considers small elements of burnt gases with different temperature and composition and predicts the NO formation more accurately because it accounts for the history of the burnt gases.

Hires et al. [90] used Eq. (7) to predict the performance and NO_x emissions of a prechamber stratified-charge engine; they showed that in the absence of EGR, the prechamber is the major source of NO emissions for overall air-fuel ratios leaner than 22 : 1.

Heywood and Watts [91] used Eq. (7) and divided the combustion chamber into three regions—the unburnt gases, the adiabatic core, and the boundary layer—and accounted for the rate of change of NO concentrations in the adiabatic core and in the boundary layer. They also used the extended Zel'dovich mechanism for NO formation [Eqs. (10) and (11)] and a steady-state approximation for N, and they assumed that O, O_2, OH, and H are in chemical equilibrium in the adiabatic core. They also assumed that NO freezes in the boundary layer and used one-dimensional quasi-steady flow equations to calculate the flow through the intake and exhaust valves, assuming that the intake and exhaust manifolds behave as plenum chambers. Flow reversals in the intake valve were modeled as plug flows.

The energy equations of the unburnt gases u, adiabatic core a, and boundary layer bl can be written as:

$$\frac{dE_u}{dt} = -\frac{dQ_u}{dt} - p\frac{dV_u}{dt} - \dot{m}_b h_u \tag{119}$$

$$\frac{dE_a}{dt} = -p\frac{dV_a}{dt} + \dot{m}_b h_u - \dot{m}_{bl} h_a \tag{120}$$

$$\frac{dE_{bl}}{dt} = -p\frac{dV_{bl}}{dt} - \frac{dQ_{bl}}{dt} + \dot{m}_{bl} h_a \tag{121}$$

where the internal energy of the burnt gases is

$$m_b \bar{u}_b = m_{bl} u_{bl} + m_a u_a \tag{122}$$

and the boundary layer temperature is assumed equal to the film temperature, that is,

$$T_{bl} = \frac{T_a - T_w}{ln(T_a/T_w)} \tag{123}$$

where T_w is the wall temperature and u is the specific internal energy.

Heywood and Watts [91] assumed that the unburnt-gas region, boundary layer, and adiabatic core consist of ideal gases and showed that the NO concentration peaks at $\phi = 0.94$.

Watts [92] and Watts and Heywood [93] used Woschni's heat transfer correlation (see Chapter 2) and accounted for compression, expansion, scavenging, and combustion effects to analyze the effects of turbocharging and reduced heat transfer in spark-ignition engines.

Figure 4.22 illustrates the two-zone thermodynamic model of Heywood and Watts [91] and the division of the burnt-gas zone into an adiabatic core A and a thermal boundary layer BL; Fig. 4.23 shows a comparison between measured and predicted cylinder pressure diagrams.

4.4 ONE-DIMENSIONAL MODELS

One-dimensional models of spark-ignition engines are based on the time-averaged conservation equation of mass, momentum, species, and energy presented in Chapter 2. These equations are then averaged across the combustion chamber from the cylinder head to the piston or from the symmetry axis to the cylinder wall. If the cross-sectional average is performed in the axial direction, the fluid properties are functions only of time and radial coordinate, whereas if the average is performed in the radial direction, the fluid properties are functions of time and axial coordinate. The first case corresponds to cylindrical flame propagation, whereas the second case results in a planar flame propagation. Both

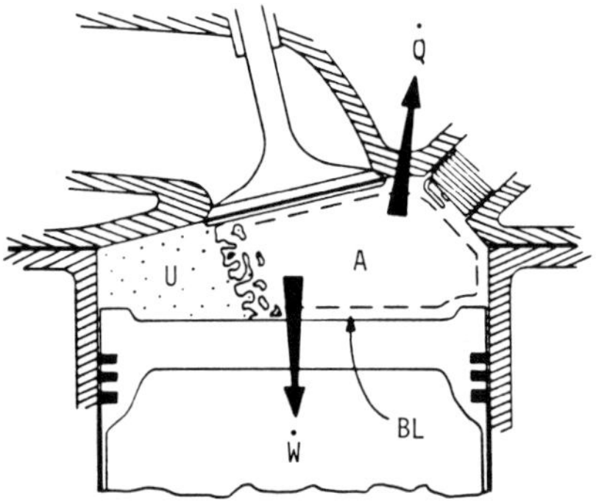

Figure 4.22 Two-zone thermodynamic model illustrating the division of the burnt gases into an adiabatic core A and thermal boundary layers BL [91] (U = unburnt gases; $\dot{W}$ = work; $\dot{Q}$ = heat transfer losses).

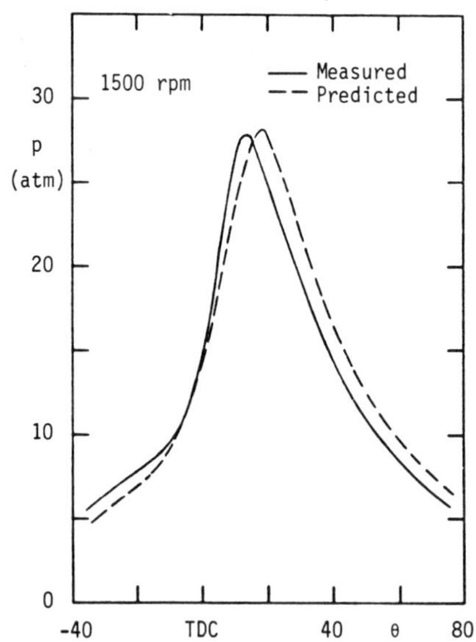

Figure 4.23 Predicted and measured pressure diagrams as a function of the crank shaft angle [91].

cases are unrealistic. However, the cylindrical flame propagation is more realistic, particularly after the flame reaches the piston, assuming, of course, that the the fluid velocity is radial and that ignition occurs along a line source that coincides with the cylinder symmetry axis.

Sirignano [94] proposed the first one-dimensional model of spark-ignition engines with a planar flame propagation. This approximation is strictly valid only if the bore-to-stroke ratio is much smaller than 1 and the primary direction of flame propagation is axial. This is clearly unrealistic since the ratio between the bore and the distance between the piston and cylinder head is greater than 1 for most of the combustion process.

Payri et al. [95] and Ramos [96] developed one-dimensional models that consider cylindrical flame propagation, a constant eddy diffusivity proportional to the engine rpm, and a one-step overall chemical reaction for the oxidation of octane to carbon dioxide and water vapor; they showed that large eddy diffusivities result in high pressures and short combustion angles. However, two-dimensional calculations have shown that there is an upper limit to the TDC turbulence [97–99].

Hayder et al. [97] presented comparisons between the predictions of a two-equation k–ϵ model and experimental data in a ported engine with and without swirl and in a valved engine without swirl; they showed that there is an upper limit to the magnitude of TDC turbulence intensity in reciprocating engines with pancake combustion chambers. This limit is independent of the characteristics of the flow field at intake valve closing (IVC) and is due to the chamber walls. Below this limit, the TDC turbulence intensity is influenced mainly by the turbulence diffusivity at IVC. Hayder et al. [97] also used a zero-dimensional k–ϵ model. They showed that the predictions of the zero-dimensional model agree with those of the two-dimensional calculations as long as diffusive effects are small and that near TDC, the turbulence dissipation rate dominates over the compression-generated turbulence production.

Grasso and Bracco [98] used a two-dimensional axisymmetric model to predict turbulence in motored engines; they showed that the TDC turbulence intensity increases with the engine speed, is insensitive to the compression ratio, and depends linearly on the engine load in the absence of swirl, but is insensitive to the load in the presence of swirl. Grasso and Bracco [98] also showed that in bowl-in-piston combustion chambers, turbulence near the edge of the bowl is dominated by production due to squish and its intensity is proportional to the engine speed but insensitive to the engine load and compression ratio. The turbulence near the bottom of the bowl behaves as in a pancake combustion chamber.

Hayder et al. [97] and Grasso and Bracco [98, 99] assumed that the initial levels of the turbulence kinetic energy and its dissipation rate are proportional to the volumetric efficiency and rpm to the second and third powers, respectively, whereas the initial mean velocities are equal to zero. These assumptions seem questionable in view of the available theoretical and experimental work that shows that the intake-generated eddies persist up to TDC and that the intake-generated turbulence is clearly nonuniform throughout the combustion chamber at IVC. Two-dimensional calculations in ported (uniflow-scavenged) engines have indicated that the turbulence becomes uniform throughout the cylinder before the piston reaches TDC; however, nonuniform velocity fields are observed beyond TDC [100–103]. Two-dimensional axisymmetric calculations in four-stroke motored engines have shown that the velocity field at IVC depends on the intake valve seat angle. For an angle of 45°, the intake-generated eddies break down into two new eddies by the end of the intake stroke; these eddies play an important role in determining the nonuniform levels of turbulence kinetic energy at IVC, and their lifetime depends on the engine rpm and compression ratio [104–106].

In the compression stroke, turbulence is generated by compression and destroyed by viscous dissipation. The calculations performed by Hayder et al. [97] indicate that viscous dissipation in the bulk flow and diffusion from the solid walls may overcome the compression-generated turbulence. However, the k–ϵ model of turbulence assumes that the smallest scales of turbulence respond instantaneously to changes in the mean flow. This may not be the case in internal combustion engines, where cycle times are short compared with relaxation times for important flow adjustments and prediction methods based on statistically stationary turbulent flows may not be valid for engine flow analysis [107]. Furthermore, most of the modifications introduced in the standard k–ϵ model to account for compression-generated turbulence are based on rapid-distortion theory, which is based on local linearization of the full nonlinear equations and assumes that the turbulence is distorted more rapidly by changes in the mean flow than by interaction with itself [108, 109].

Westbrook [110] used a one-dimensional planar model to study flame propagation in homogeneous- and stratified-charge engines. He accounted for the temporal and spatial variations of the pressure field and used detailed and one-step reaction mechanisms for the oxidation of methane. He showed that in stratified-charge engines, the fuel can be completely burnt before it reaches the combustion chamber wall, effectively eliminating wall quenching as a source of unburnt HC emissions.

Singh and Surakomol [111] developed a one-dimensional cylindrical flame propagation model to analyze the heat release rate in spark-ignition engines. They used a k–ϵ model of turbulence and a one-step reaction and assumed that the pressure is uniform throughout the combustion chamber. Their results indicate that the flame speeds in engines with spark plugs located at the cylinder axis are higher than those with eccentric sparks and that the NO concentration peaks at fuel-air ratios about 10% leaner than stoichiometric conditions.

Bellan and Sirignano [112] developed a cylindrical flame propagation model that uses a time-dependent eddy diffusivity proportional to the instantaneous piston speed and that depends on the average velocity at the intake port. Their model uses a Nusselt number correlation to calculate the heat transfer losses through the solid walls, a one-step overall chemical reaction mechanism, and the Zel'dovich mechanism to calculate the NO formation, assuming that the atomic nitrogen is in steady state and that the atomic and molecular oxygen are in equilibrium. Bellan and Sirgnano [112] simulated ignition by assuming that a certain amount of fuel was burnt, calculated the initial temperature that corresponded to the heat released by the burnt fuel, and showed that mixture stratification results in NO reductions.

Bellan and Sirignano [113] used a two-equation k–ϵ model of turbulence to analyze NO formation in homogeneous- and stratified-charge engines and showed that ignition in the fuel-rich region results in a premixed flame that propagates toward the cylinder wall. Diffusion-type flames were observed in the wake of the premixed flame, due to the mixing of excess air and fuel. Bellan and Sirignano [113] showed that the turbulence kinetic energy at ignition plays an important role in determining the burning rate and that as the overall mixture ratio becomes leaner, the burning angle increases.

Westbrook [114] studied the propagation of a one-dimensional flame through a stratified charge by means of a modified RICE technique, which accounts for density variations [115]. [The ICE (implicit-continuous fluid Eulerian) technique accounts for pressure changes that are due mainly to density changes; this can result in systematic errors in combustion systems where energy deposition may produce large pressure changes.] Westbrook showed that the CO and NO_x emissions are lower than in homogeneous-charge engines. However, significant amounts of unburnt HC were observed at the solid walls.

O'Rourke [116] indicated that the original ICE technique yields satisfactory results, even when energy deposition produces large pressure changes, and that the modifications introduced by Westbrook [115] are not necessary to calculate chemically reacting flows. These modifications may, however, accelerate the convergence of the ICE technique.

Sorenson and Pan [117] used a one-dimensional model to analyze combustion in a dual-chamber stratified-charge engine equipped with three valves. They also used a one-step overall chemical reaction, accounted for ignition delay times, and modeled the NO formation with the Zel'dovich mechanism. The atomic and molecular oxygen were assumed to be in equilibrium, and a steady-state assumption was used for the atomic nitrogen, that is,

$$\frac{d[\mathrm{N}]}{dt} = 0 \qquad \mathrm{O} \rightleftarrows \frac{1}{2}\,\mathrm{O}_2 \tag{124}$$

where the brackets denote concentrations.

Sorenson and Pan [117] assumed that the fuel was completely vaporized before ignition, used the two-step explicit predictor-corrector method of MacCormack [118] and an effective diffusivity ν_{eff}, which depends on the mean piston speed u_p and mean flow velocity u, that is,

$$\nu_{\mathrm{eff}} = K(u_p + |u|) \tag{125}$$

where K is a constant.

The one-dimensional time-averaged conservation equations for cylindrical flame propagation can be written as [96]

$$(r \, \delta \, \rho)_{,t} + (\rho r \delta u)_{,r} = 0 \tag{126}$$

$$(r\delta\rho Y_i)_{,t} + (\rho\delta ru Y_i)_{,r} = \left[r\delta\left(\frac{\mu}{\text{Sc}} + \frac{\mu_t}{\text{Sc}_t}\right) Y_{i,r} \right]_{,r} + r\dot{\omega}_i, i = 1, 2, \ldots, N-1 \tag{127}$$

$$(r\delta\rho H)_{,t} + (\rho\delta ru H)_{,r} = \delta p_{,t} + \left[\delta r C_p \left(\frac{\mu}{\text{Pr}} + \frac{\mu_t}{\text{Pr}_t}\right) T_{,r} \right]_{,r} - r\delta \sum_{i=1}^{N} h_i^0 \, \dot{\omega}_i - r(Q_P + Q_H) \tag{128}$$

$$p = \rho \, \tilde{R} T \sum_{i=1}^{N} \left(\frac{Y_i}{M_i}\right) \qquad p = p(t) \qquad H = \sum_{i=1}^{N} Y_i \int_{T_0}^{T} C_{p_i} \, dT \tag{129}$$

$$Q_p = h(T - T_p) \qquad Q_H = h(T - T_H) \qquad C_p = \sum_{i=1}^{N} C_{p_i} Y_i \tag{130}$$

$$\frac{hr}{k} = 0.023 \, \text{Re}^{0.8} \, \text{Pr}^{0.4} \qquad \sum_{i=1}^{N} Y_i = 1 \tag{131}$$

where ρ, p, T, and u are the density, pressure, temperature, and mean velocity, respectively; r and t are the radial coordinate and time, respectively; δ is the distance between the piston and the cylinder wall; $\dot{\omega}_i$, Y_i, and C_{p_i} are the mean reaction rate, mass fraction, and specific heat at constant pressure of species i, respectively; μ and μ_t are the molecular and turbulent dynamic viscosities, respectively; Pr and Pr_t are the molecular and turbulent Prandtl numbers, respectively; Sc and Sc_t are the molecular and turbulent Schmidt numbers, respectively; $\tilde{R}$ is the universal gas constant; N is the number of species; Q_P and Q_H denote the heat transfer losses through the piston and cylinder head, respectively; T_P and T_H are the temperatures of the piston and cylinder head, respectively; k is the thermal conductivity; Re is the Reynolds number; and M_i is the molecular mass of species i.

Equations (126)–(128) can be transformed into a system of integrodifferential reaction-diffusion equations by means of the mapping

$$(t, r) \rightarrow (\tau, \eta) \qquad t = \tau \tag{132}$$

$$\rho r \delta = M \frac{\partial \eta}{\partial r} \qquad -\delta r \rho u = M \frac{\partial \eta}{\partial t} \qquad M = \int_0^R \rho r \delta \, dt \tag{133}$$

where M denotes the cylinder charge mass per radian.

Equations (132) and (133) can be substituted into Eqs. (126)–(128) to yield

$$\frac{\partial}{\partial \eta}(ru) = M \frac{\partial}{\partial \tau}[(\rho\delta)^{-1}] \tag{134}$$

$$\frac{\partial Y_i}{\partial \tau} = \frac{\delta^2}{M^2}\frac{\partial}{\partial \eta}\left[r^2\rho\left(\frac{\mu}{\mathrm{Sc}} + \frac{\mu_t}{\mathrm{Sc}_t}\right)\frac{\partial Y_i}{\partial \eta}\right] + \dot{\omega}_i \qquad i = 1, 2, \ldots, N-1 \qquad (135)$$

$$\begin{aligned}\frac{\partial T}{\partial \tau} &= \frac{1}{\rho C_p}\frac{dp}{d\tau} + \frac{\delta^2}{M^2}\frac{\partial}{\partial \eta}\left[r^2\rho\left(\frac{\mu}{\mathrm{Pr}} + \frac{\mu_t}{\mathrm{Pr}_t}\right)\frac{\partial T}{\partial \eta}\right] \\ &\quad - \frac{1}{\rho C_p}\left[\sum_{i=1}^{N} \dot{\omega}_i h_i^0 + \frac{Q_P + Q_H}{\delta}\right]\end{aligned} \qquad (136)$$

$$r^2 = \frac{2M}{\delta}\int_0^{\eta}\frac{d\eta}{\rho} \qquad (137a)$$

$$p = \frac{2\delta}{R^2}\tilde{R}M\int_0^1 T\sum_{i=1}^{N}\left(\frac{Y_i}{M_i}\right)d\eta \qquad (137b)$$

where R is the cylinder radius.

Equations (134)–(137) can be solved by means of, for example, a Crank-Nicolson method to obtain the values of u, Y_i, and T [119]. The pressure can then be calculated from Eq. (137*b*). Some sample results are presented in Fig. 4.24, which shows the temperature and flame-surface-to-piston-area ratio predicted by a two-zone thermodynamic model and a one-dimensional model [96]. The temperature predicted by a single-zone model is also shown in Fig. 4.24 [96].

4.5 MULTIDIMENSIONAL MODELS AND EXPERIMENTAL DATA

Multidimensional models of reciprocating engine combustion have been presented in Chapter 2, where both turbulence and combustion models are dis-

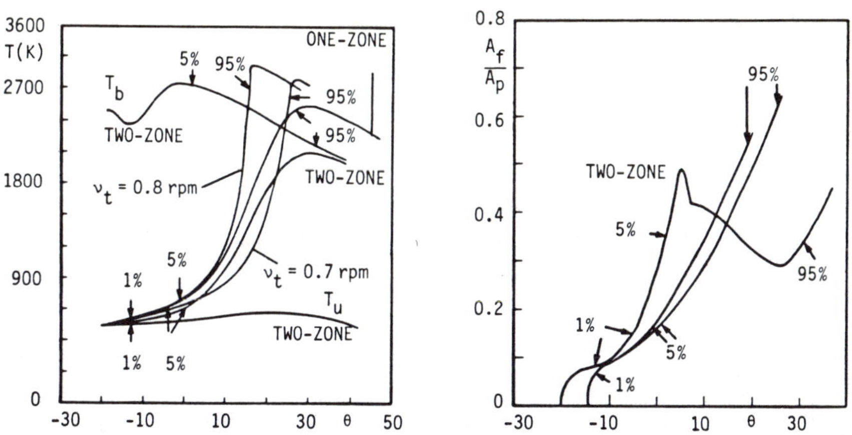

Figure 4.24 Mass-averaged temperatures (left) and flame-surface-to-piston-area ratio (right) calculated with one-zone, two-zone, and one-dimensional models as a function of the crank shaft angle [96] (ν_t = eddy diffusivity; T_u = unburnt gas temperature; T_b = burnt gas temperature; A_f = flame surface area; A_p = piston area).

cussed. Briefly, most of the combustion models of spark-ignition engines can be classified in three groups: mean-reaction-rate models [120–123], eddy-breakup or mixing-controlled models [124], and hybrid models [125, 126]. In mean-reaction-rate models the contributions of the fluctuating temperature and species mass fractions to the time-averaged reaction rate are neglected. This may result in substantial underestimates in premixed reactive flows [127].

In eddy-breakup models, the conversion of fuel to combustion products is controlled by the mixing of the reactants and hot products, and the chemical kinetics time is assumed to be shorter than the mixing time. Abraham et al. [126] used a hybrid chemical kinetics-mixing-controlled combustion model that accounts for the initial chemistry-controlled reactions until the flame travels a distance of the order of the length scale of the energy-containing eddies; thereafter, combustion is mixing controlled. The main advantage of the formulation proposed by Abraham et al. [126] is that the conversion rates are calculated with respect to their equilibrium values, and their model is able to predict local chemical equilibrium (see Chapter 2).

Gilaber [128] and Pinchon and Baritaud [129] used the KIVA code [130] and an eddy-breakup model of turbulent combustion. Gilaber [128] studied the combustion of propane-air mixtures by means of the following chemical reactions:

$$2C_8H_{18} + 25O_2 \rightarrow 16CO_2 + 18H_2O \tag{138}$$

$$CO_2 \rightleftarrows CO + \frac{1}{2}O_2 \tag{139}$$

where Eq. (139) is assumed to be in chemical equilibrium

Pinchon and Baritaud [129] used Eq. (139)

$$2C_3H_8 + 10(O_2 + 3.75N_2) \rightarrow 6CO_2 + 8H_2O + 37.5N_2 \tag{140}$$

and modeled the mean reaction rate as

$$\langle \dot{w}_f \rangle = C_{\mathrm{EBU}}\, Y_f \frac{Y_{f_0} - Y_f}{\tau_m} \tag{141}$$

where C_{EBU} is a constant, the subscript f denotes the fuel, Y_f and Y_{f_0} are the local and initial fuel mass fractions, respectively, and τ_m is the mixing time. Gilaber [128] assumed a constant $\tau_m = 0.67 \times 10^{-3}$ s, and Pinchon and Baritaud [129] calculated τ_m from the turbulence kinetic energy equation as

$$\tau_m = L\, k^{-1/2} \tag{142}$$

where L is a length scale of the order of the grid spacing.

Eddy-breakup models can predict, after appropriate tunings, the correct trends of the flame speed, heat release, and pressure as a function of the engine load and rpm [124, 126]. However, these models are inadequate to study ignition, knock, pollutant formation, rich and lean misfire limits, and wall quenching because these phenomena are controlled by chemical kinetics.

Rivard et al. [131] developed the RICE technique, which solves the energy equation explicitly, whereas the species equations are treated in three steps: explicit advection, explicit diffusion, and implicit reaction. Rivard et al. [131] introduced a pressure correction in order to satisfy the continuity equation. Butler and O'Rourke [132] used the implicit continuous-fluid Eulerian formulation of the RICE technique and artificially thickened the flame front so that the flame speed and jumps across the flame front are preserved. Artificial thickening of the flame front may suppress or alter the growth rate of fluid dynamic instabilities with wavelengths comparable to the flame front thickness, and it can be achieved by increasing the diffusivities, which in turn require a decrease in the reaction rate in order to predict the flame speed.

Bracco [133] applied scaling to flow fields with flames and increased the flame speed and thickness in order to reduce computational times in low-Mach-number flows. These flow fields are insensitive to the flame thickness except at the flame front. They also depend on the initial conditions and geometry and are difficult to implement in practice. Bracco and O'Rourke [134] applied α and β transformations, a mean-reaction-rate model, and a constant eddy diffusivity proportional to the engine rpm to predict the flow field in reciprocating engines. The α transformation is independent of the flame speed if the engine rpm is used for scaling and the Mach number of the flow is much smaller than unity. The β transformation, however, is such that the flow fields in front of and behind the flame front are independent of the flame thickness if this thickness is much smaller than the length of the computational domain.

Diwakar [135] proposed a modified one-step global combustion rate expression that provided good agreement with experimental data. However, the modified expression required case-to-case tuning when the volumetric efficiency or the fuel-air equivalence ratio were changed. The preexponential factor required for such a tuning was found to correlate with the residual gas mass fraction and with the equivalence ratio. Diwakar [135] also showed that his modified rate expression provided good agreement with experimental data for the wall heat transfer and emissions.

Spalding [136] proposed a two-fluid model based on the analogy between turbulent combustion and two-phase flows, which are characterized by patchiness Abraham et al. [137] indicated that the combustion regime in premixed-charge engines is not clear, but up to and including idle speed, combustion occurs in the reaction-sheet regime. At high engine speeds the parameters fail the conditions known to be sufficient for occurrence of either the reaction-sheet regime or the distributed-reaction regime, and the flame structure is not clear at this time.

Multidimensional models have attempted to reproduce the burning velocity without attempting to reproduce the flame structure, because the burning velocity and the flame area are the principal parameters that affect the pressure history. If the reaction sheet prevails, it may be difficult to accurately resolve the flame structure unless adaptive or Lagrangian methods are employed to track the flame front.

Gosman [138] and Smith [139] showed theoretically and experimentally that the flame front in homogeneous-charge engines is thin (of thickness comparable to the size of the Kolmogorov eddies) and highly wrinkled at low engine speeds but multiply connected at higher speeds. Hakberg and Gosman [140] determined the turbulent flame speed from combustion models. They found that the flame speed is proportional to the turbulence intensity, where the proportionality constant depends on the fuel composition according to the Bray-Libby-Moss (BLM) model [141], but is independent of the fuel composition according to the Magnussen-Hjertager model [142]. Hakberg and Gosman [140] showed that experimental data favor the Magnussen-Hjertager model. Benjamin et al. [143] used a turbulent flame speed proportional to the turbulence intensity. However, experimental data [144–147] indicate that the turbulent-to-laminar flame speed ratio depends on the Reynolds number, the ratio between the laminar flame speed and the turbulence rms, and the ratio between the Kolmogorov eddy size and the laminar flame thickness. Chomiak [148] showed that at high Reynolds numbers, the reaction is not uniformly distributed but is concentrated in small, isolated regions having a volume that is a small fraction of the flow field. Chomiak proposed a model that consists of permanent random vortex tubes having diameters equal to the Kolmogorov eddy size and separated from each other by a distance of the order of the Taylor microscale. This model yields good agreement with experimental data and is valid for the distributed-combustion regime.

Ballal and Lefebvre [146] and Ballal [147] varied the turbulence intensity and length scale in enclosed premixed flames and showed that the turbulent flame speed increases with the turbulent length scale for weak turbulence (i.e., $u' < 2S_L$, where u' and S_L are the turbulent rms velocity and the laminar flame speed, respectively). However, the turbulent burning velocity decreases with the turbulent length scale for strong turbulence (i.e., $u' > 2S_L$) and is independent of the length scale in the transition region, where $u' = 2S_L$. The burning velocity increased with the turbulence rms velocity for both weak and strong turbulence. Ballal [147] showed that there is a correlation between turbulent flame propagation and flame-generated turbulence. For weak turbulence and $\eta > \delta_L$, where η and δ_L are the Kolmogorov eddy size and laminar flame thickness, the flame-generated turbulence increases with the turbulent burning velocity and rms velocity; a similar behavior was observed for strong turbulence and $\eta < \delta_L$, whereas in the transition regime, the flame-generated turbulence increased with the turbulent burning velocity. Experimental results have shown that the flame generates turbulence; however, these results did not separate the cycle-to-cycle variations and turbulence [149].

Bray et al. [150] analyzed a one-dimensional time-averaged turbulent flame and showed that with sufficient heat release, turbulence increases upon passage through the flame because of the self-induced mean pressure gradient. This mechanism was found to overwhelm the dilatation effect that decreases turbulence behind the flame. The same mechanism (i.e., the pressure-velocity correlation) was found to be responsible for countergradient diffusion. Present multidimensional models account for turbulence generation by shear and compression but do not consider the pressure-velocity correlations and may result in countergradient diffusion (see Chapter 2).

Other experimental data have shown that the TDC turbulence is isotropic, with both intensity and length scales determined by the intake process [151]. It has also been shown that the turbulent-to-laminar flame speed ratio is a linear function of the turbulence rms velocity and volumetric flow rate at the intake [152].

Witze [153–156] used hot wire anemometry and showed that the intake-generated turbulence decays rapidly; high levels of turbulence were observed at TDC. However, Witze [153] was not able to conclude whether the measured TDC turbulence was shear- or compression-generated turbulence. He also showed that it is difficult to interpret the hot wire signals and that most of the turbulence kinetic energy was contained in low-frequency (<10 Hz) eddies. Witze [155] showed that the mean velocity and turbulence intensity vary linearly with the engine speed, whereas the turbulence length scales are only a function of the engine geometry. However, accurate hot wire turbulence intensity measurements appear to be possible only for the intake and exhaust strokes, for low turbulence levels and compression ratios, and when the flow direction is known due to the directional sensitivity of hot wires [156]. Witze and Vilchis [157] used laser Doppler velocimetry (LDV) to characterize the precombustion fluid motion; they showed that cycle-to-cycle variations are not necessarily decreased by increasing the burning rate, which increases with the swirl level when the flame remains attached to the spark plug. Cycle-to-cycle variations were enhanced by random detachments of the flame from the spark plug.

Witze et al. [158] used laser Doppler anemometry and ionization probes to identify the time of flame arrival. Phase-averaged measurements from many cycles were conditionally sampled according to flame arrival time. The measurements showed that piston compression increases turbulence and suggested that cyclic variations in burn duration are caused by cyclic variations in the bulk turbulence intensity ahead of the flame front. Martin et al. [159] showed that combustion appears to amplify the turbulence in the preflame gases, although the measured changes may be due, in part, to flame-induced convection of nonhomogeneous turbulence.

Spicher and Velji [160] used fiber optics to deduce the flame front shape, flame volume, and flame front velocity. Their LDV data indicate that the intake swirl (generated by a shrouded valve) results in high mean velocities and flame speeds as well as distorting the flame front.

Wakisaka et al. [161, 162] showed that in a motored engine equipped with a shrouded valve, the mean velocity and the turbulence intensity increase with the engine rpm. They also showed that there is strong turbulence with relatively low frequency and that the ratio of the integral length scale to the Taylor microscale is almost constant throughout the combustion chamber. However, both the integral length scale and the Taylor microscale increase with the engine rpm.

Wakisaka et al. [161] used Taylor's theory of frozen turbulence to calculate the integral length scale from the integral time scale, which was measured with hot wires. However, Taylor's theory of frozen turbulence is based on the assumption that the rms velocity is smaller than the mean flow velocity. This assumption may not be valid in engines.

Wakisaka et al. [163] showed that the intake-generated turbulence contributes significantly to the turbulence levels at TDC. Their results indicate that TDC turbulence in an engine with swirl is in an underrelaxed state.

Most of the engine turbulence measurements performed to date have been based on cycle-ensembled quantities and include cycle-to-cycle variations. Fraser et al. [164] proposed the following definitions: (1) velocity = local gas instantaneous velocity; (2) bulk velocity = low-frequency components (below a cutoff frequency) of the gas velocity in each cycle; (3) turbulence = high-frequency components (above the cutoff frequency) of the gas velocity in each cycle (i.e., the difference between the velocity and the bulk velocity at the same crank angle and in the same cycle); (4) ensemble-averaged mean velocity = ensemble average of the velocity in small crank angle interval windows; (5) fluctuation = difference between the velocity and the ensemble-averaged velocity at the same crank angle; (6) fluctuation intensity = ensemble-averaged rms of the fluctuation in small crank angle interval windows = non-cycle-resolved turbulence intensity; (7) cyclic variation in the bulk velocity = ensemble-averaged rms fluctuation of the bulk velocity; (8) turbulence intensity = ensemble-averaged rms of the turbulence in small crank angle interval windows = cycle-resolved turbulence intensity; (9) turbulence integral length scale = integral length scale based on the spatial correlation coefficient of turbulence; and (10) fluctuation integral length scales = integral length scales based on the spatial correlation coefficient of the fluctuation = fluctuation length scale.

Liou et al. [165–167] showed that cyclic bulk velocity fluctuations are smaller than 200% and that ensemble averages give rms that are 50% to 100% larger than cycle-resolved rms. Liou et al. [165] also showed that the cyclic variations increase with the rpm but decrease with swirl. If the cyclic variations are included, the turbulence is overestimated (100% in ported engines without swirl, 300% in valved engines without swirl, 10% in ported engines with swirl) compared with cycle-resolved measurements. Liou et al. [165] also showed that the rms velocity is homogeneous and proportional to the engine rpm in ported engines with and without swirl. In both valved and ported engines the turbulence intensity is isotropic with ±20%, whereas in open combustion chambers both the turbulence rms and length scale are influenced by the intake, but their homogeneity, isotropy, or proportionality to the engine speed are not affected by the intake process.

James and Lucas [168] used hot wire anemometry, which has a very good frequency response, to measure the flow field in spark-ignition engines with squish. Their results show that the turbulence energy is contained in frequencies lower than 700 Hz, and that the low-frequency fluctuations (<200 Hz) increase linearly with the engine speed.

Catania and Mittica [169] showed that the induction process in a reciprocating engine is anisotropic and nonhomogeneous; however, the flow tends to become isotropic and homogeneous in the compression stroke. Catania and Mittica [169] showed that the turbulence microtime scale is virtually uniform during induction and compression, particularly at points away from the cylinder wall. Fraser and Bracco [170] measured directly integral length scales, using a two-point, single-probe-volume laser Doppler velocimeter in a motored, ported, single-cylinder internal combustion engine with a pancake combustion chamber. They showed that the measured lateral fluctuation integral length scales are not linearly proportional to the TDC clearance height, whereas multidimensional k–ϵ models predict that the length scale is proportional to the clearance height. Fraser and Bracco [170] attributed this discrepancy to cyclic variations in the bulk velocity and anisotropy in the turbulence field. Another possible reason for the discrepancy may be due to the number of data points used in the determination of the length scale.

Cycle-resolved measurements are subject to several limitations. First, there may not be adequate seeding in the flow, and very few statistical data points may be available to obtain accurate velocity and turbulence measurements. Second, laser Doppler velocimeters have a poorer frequency response than do hot wires, and there has not as yet been a detailed and careful evaluation of the effects of the cutoff frequency on the measured cycle-resolved velocities. Third, even when two-point probes are used to measure Eulerian length scales, it is difficult to obtain the Lagrangian time scales. These scales are frequently calculated by means of Taylor's theory of frozen turbulence, which is based on the assumption that the fluctuating velocities are much smaller than the mean flow velocities; this assumption is generally violated near TDC, where the mean flow is almost unidirectional and the turbulent intensity is very high, even in the absence of swirl.

Even if accurate and detailed cycle-resolved measurements were available, it would be difficult and misleading to compare them with the numerical predictions of turbulence models. It is unclear what multidimensional models of reciprocating engines predict because they are based on time- or mass-averaging techniques, which are strictly applicable to statistically stationary turbulence, a situation that is not present in reciprocating engines.

Most reciprocating engine models are based on those developed for incompressible, thin shear, statistically stationary flows (see Chapter 2). The predictions of some of these models have been compared with available experimental data, and in some cases good agreement has been obtained after proper tuning of the inlet boundary conditions, grid distribution, turbulence model, and time steps. However, it is not clear that these comparisons should even be made because the numerical predictions may not correspond to what is measured.

Some researchers, including this writer, have blamed the disagreement between the numerical predictions and experimental data on the deficiencies of the k–ϵ model, inadequate simulation of the intake and exhaust valves, and lack of adequate experimental data at the inlet and exhaust ports. However, even if the k–ϵ model were appropriate for engine flows and accurate experimental data were available at inlet and exhaust ports, any comparison between predictions and experimental data would be misleading and prone to erroneous conclusions about the accuracy of multidimensional models of reciprocating engines.

Wong and Hoult [108] used rapid-distortion theory, which is valid when the turbulence is distorted more rapidly by the changes in the mean flow than by interaction with itself, a cylindrical flame propagation model, where the burning rate is given by

$$\frac{dm_b}{dt} = 2\pi\, r_f\, \rho_u\, S_T \tag{143}$$

where r_f, ρ_u, and S_T are the flame radius, unburnt-gas density, and turbulent flame speed; and they used the following correlations to calculate the flame speed:

$$\text{Damkohler [171]: } \frac{S_T}{S_L} = 1 + \frac{u'}{S_L} \tag{144}$$

Shchelkin and Troshin [172] and Karlovitz et al. [173]:

$$\frac{S_T}{S_L} = 1 + \left(\frac{2u'}{S_L}\right)^2 \tag{145}$$

Equations (144) and (145) account only for the increase in the turbulent burning rate due to flame wrinkling through the rms velocity, do not consider the effects of the turbulence length scale on the turbulent flame speed [146, 147, 174], and are not applicable to distributed chemical reactions. Therefore, they may not be valid at high rpm.

Equations (144) and (145) can be used to predict the burning rate in two-zone thermodynamic models. Some sample results are shown in Figs. 4.25 and

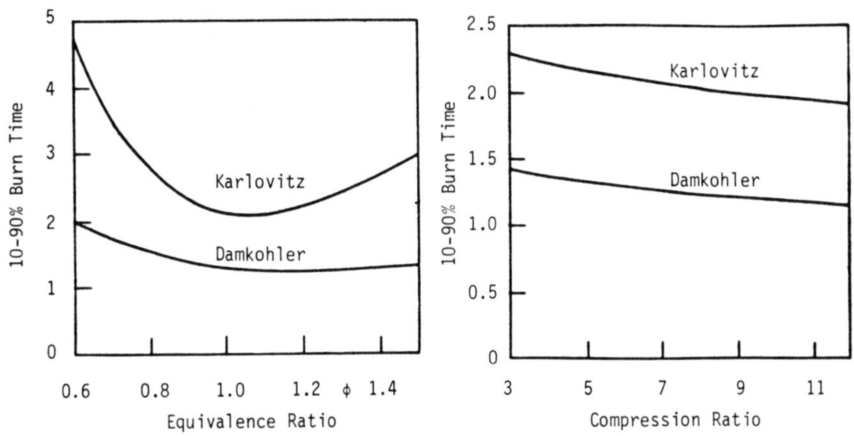

Figure 4.25 Burning time as a function of the equivalence ratio (left) and compression ratio (right) [108].

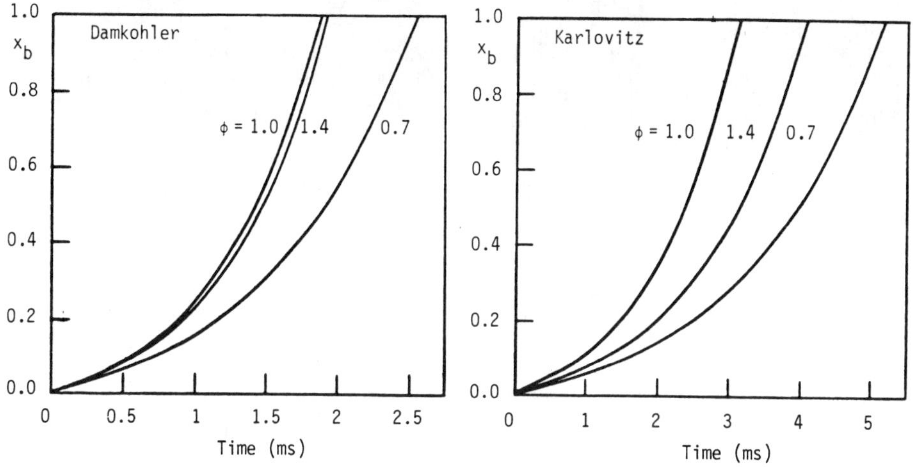

Figure 4.26 Mass fraction burnt as a function of time and mixture equivalence ratio [108].

4.26. Figure 4.25 indicates that Eqs. (144) and (145) predict combustion durations that first decrease and then increase as the equivalence ratio is increased. The burning times predicted by Eqs. (144) and (145) increase as the mixture equivalence ratio is increased or decreased beyond stoichiometry (Fig. 4.26).

Basso [175] and Basso and Rinolfi [176] used the REC code [177] to analyze combustion in spark-ignition engines and showed that for a given displacement, the indicated efficiency is insensitive to the chamber geometry because those changes that lead to faster combustion also lead to larger wall heat transfer losses. Basso [175] obtained a linear correlation between the indicated efficiency and the chamber surface-to-volume ratio. Some sample results are shown in Fig. 4.27, which presents the isotherms and velocity vectors at selected crank angles.

Markatos and Mukerjee [178] presented three-dimensional calculations of the flow field in internal combustion engines throughout the complete four-stroke cycle using a mean-reaction-rate model. Griffin et al. [179, 180] used MacCormack's explicit predictor-corrector method [118] to calculate the flow field in a four-stroke engine cycle. Some of the three-dimensional calculations were based on inviscid flow theory, and combustion was simulated as a constant-volume heat addition process. Two-dimensional inviscid calculations for a planar geometry with finite rate chemistry were also performed by considering the oxidation of octane to carbon monoxide and using partial equilibrium for (N_2, O), (NO, O), (O, O, X), (CO, O_2), and (H, H, X) where X denotes a third body. Griffin et al. [179, 180] showed that the combustion products freeze in the power stroke.

Figures 4.28–4.30 show the velocity field, fuel mass fraction profiles, and pressure diagram computed with a two-dimensional axisymmetric engine model [122].

Oppenheim and Rotman [181] used a vortex dynamics method based on low-Mach-number equations to analyze the flow field in a spark-ignition engine.

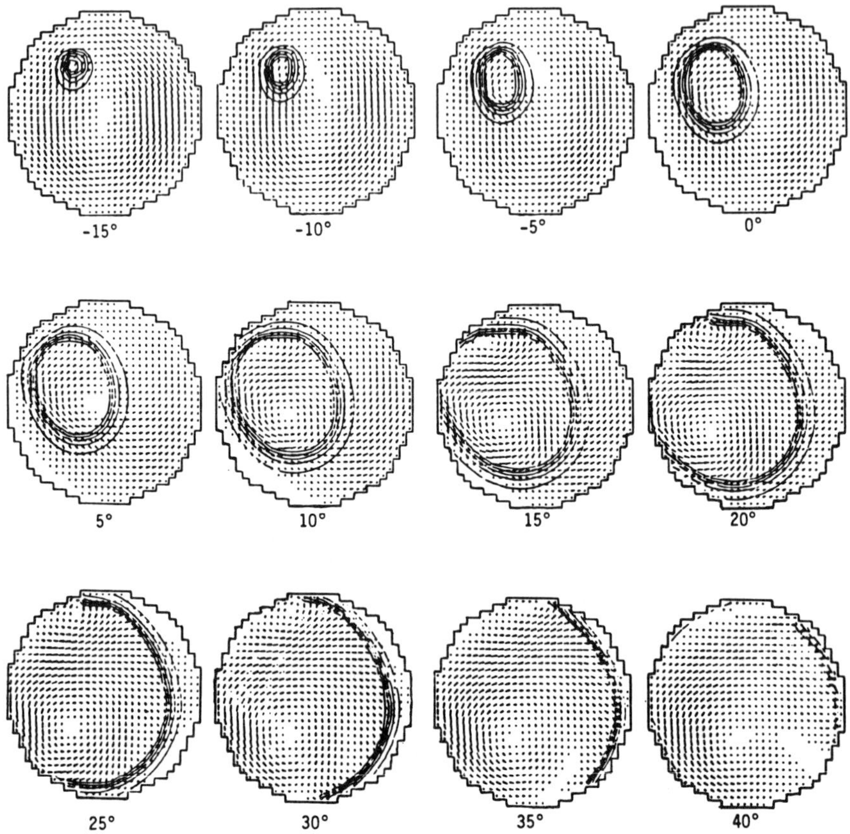

Figure 4.27 Isotherms and velocity vectors in a two-dimensional engine model at selected crank shaft angles [175].

The flame front was endowed with four different properties: (1) it is advected by the flow field, (2) it self-advances at a normal burning speed, (3) it is a source of velocity due to the expansion of the burnt gases, and (4) it is a source of vorticity. Vortex dynamics methods are free-grid algorithms valid for high-Reynolds-number flows. They do not require a turbulence model, and they divide the combustion chamber into burnt- and unburnt-gas regions, separated by an infinitesimally thin flame that propagates with the laminar flame speed.

4.6 UNBURNT HYDROCARBON EMISSIONS

The main emissions from spark-ignition engines are unburnt hydrocarbons (HC), nitric oxide (NO), and carbon monoxide (CO). Unburnt HC emissions are described in this section, whereas those of NO and CO are treated in Section 4.7.

The main sources of unburnt HC from spark-ignition engines are the spark plug, the flame (slow/partial burns), the bulk gas (rich mixtures), deposits, crevice outgassing, exhaust port burnup, oil layers, crevice HC burnup, exhaust manifold burnup, tailpipe burnup, and quench layers. These sources of unburnt HC emissions are described in the next sections.

4.6.1 Flame Quenching

Adamczyk et al. [182] used a single-pulse sampling valve to measure the unburnt HC concentration in the wall layers of a combustion bomb and showed that the

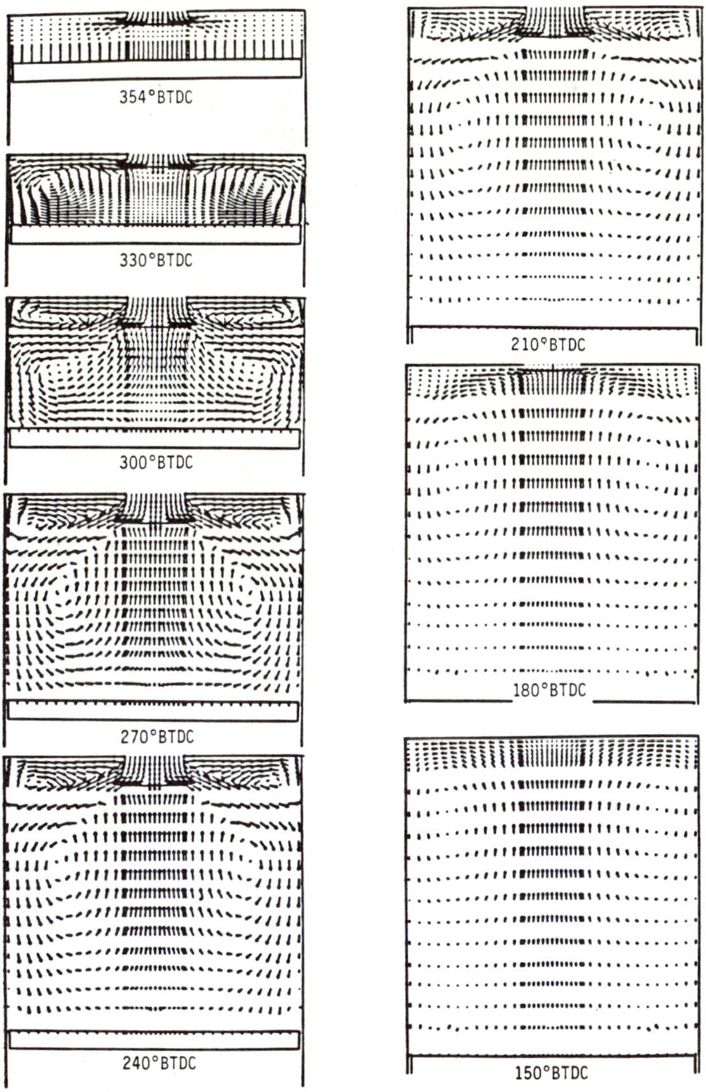

Figure 4.28 Velocity field in a two-dimensional axisymmetric engine model at selected crank shaft angles [122].

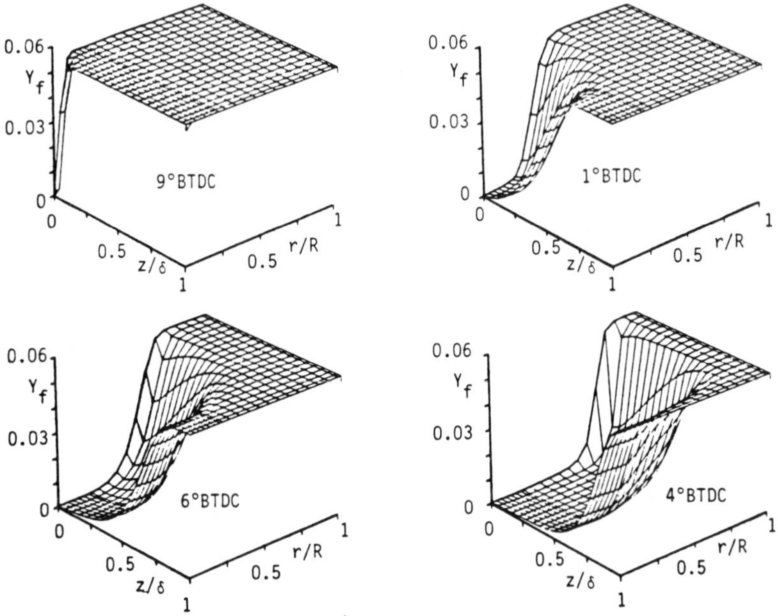

Figure 4.29 Fuel mass fraction profiles in a two-dimensional axisymmetric engine model at selected crank shaft angles [122] (z = axial distance from the cylinder head; δ = distance between piston and cylinder head; r = radial distance measured from the cylinder axis; R = cylinder radius).

HC present in the quench layer are substantially oxidized within 4 ms of flame arrival for head-on flame quenching.

Sloane and Schoene [183] studied the end-wall flame quenching of lean methane-oxygen mixtures at low pressures and showed that heterogeneous radical destruction at cooled walls has a minor effect on quenching compared with homogeneous destruction in the cooled flame gases near the wall surface. They also showed that cooled crevices have a strong effect on quenching and that the crevices are very important as sources of unburnt HC. Sloane and Ratcliffe [184] used molecular beam mass spectrometry to study side-wall flame quenching at low pressures and showed that H, OH, and O radicals may recombine at the solid walls.

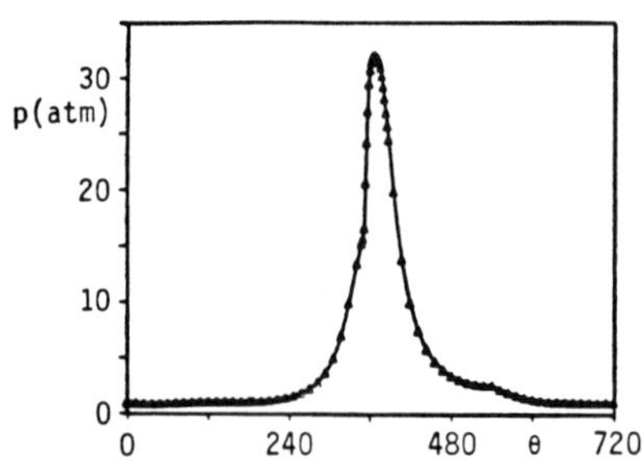

Figure 4.30 Pressure diagram calculated with a two-dimensional axisymmetric engine model [122].

The numerical studies of Westbrook et al. [185] on flame propagation perpendicular to noncatalytic solid walls indicate that radical recombination, chain-branching, and chain-propagation reactions are responsible for quenching. Westbrook et al. [185] also showed that thermal wall quenching may not be the major source of unburnt HC emissions at or near stoichiometric conditions. Fendell et al. [186] showed that the residual near-wall hydrocarbons are almost completely oxidized prior to blowdown and that head-on wall quenching is not the major source of unburnt HC. Experimental data [187] show that a cold wall begins to influence the propagation of a flame at a distance approximately equal to 10 to 18 times the thermal flame thickness.

Fendell et al. [186] conjectured that perhaps the main source of unburnt HC are the crevices. Similar conclusions were achieved by LoRusso et al. [188], who used a sampling valve and showed that unburnt HC in the quench layer are not the major source of exhaust HC emissions. LoRusso et al. [188] demonstrated that unburnt HC are rapidly oxidized after flame arrival and quenching at the walls, and they developed a model showing that the HC in the quench layer contribute approximately 3%–12% to the HC emissions in the exhaust gases. They also showed that the major sources of HC emissions are the ring crevices, absorption and desorption of the fuel in the oil layers along the cylinder wall, and surface deposits. The experimental data of LoRusso et al. [188] show evidence of incomplete combustion in the bulk gas under homogeneous fuel-rich conditions.

Adamczyk et al. [189] studied propane combustion in a bomb and showed that the ring crevices, head gaskets and spark plugs contribute 80.5%, 12.5%, and 5% to the unburnt HC emissions, respectively.

LoRusso et al. [190] also showed that the quench layers are not an important source of unburnt HC, and they conjectured that HC emissions may proceed from the ring crevices and fuel absorption/desorption in oil layers. Kaiser et al. [191] studied unburnt HC emissions in a single-cylinder Cooperative Fuels Research (CFR) engine; they showed that HC emissions increase as the compression ratio is increased and decrease as the spark timing is retarded relative to mean best torque (MBT) at a constant compression ratio. They also showed an increase in HC emissions from crevices and oil layers at higher compression ratios because of the higher peak pressures. Adamczyk et al. [192] used gas chromatography and Fourier transform infrared spectroscopy to analyze combustion in a closed bomb; they showed that the principal exhaust effluent is unburnt HC absorbed in and desorbed from the oil layer.

Hocks et al. [193] studied flame quenching in front of a cold wall under two-step kinetics of the form

$$A + B \rightarrow 2B \qquad \text{chain branching} \tag{146}$$

$$2B \rightarrow C \qquad \text{chain breaking} \tag{147}$$

and showed that if the chain-breaking reaction is fast, the concentration of B is low and quenching is fast. However, if the chain-breaking reaction is slow, B

sustains the chain-branching reaction, the flame remains in front of the wall for a long time, and the fuel is consumed. Hocks et al. [193] also showed that a one-step chemical reaction cannot predict flame quenching because it does not account for chain-branching and chain-breaking reactions.

Vosen et al. [194] studied premixed methane-air flames at atmospheric pressure in constant-volume bombs and for equivalence ratios in the range 0.7–1.2; they showed that the maximum heat release is related to the quenching distance. Their numerical data were successfully correlated, using heat release data prior to flame quenching. Vosen et al. [194] also showed that a one-step reaction predicts the heat transfer as well as a detailed kinetics model within 15% of the experimental data and that the thermal diffusion and chemical reaction rate are the dominant factors that determine the heat flux during quenching.

Adamczyk et al. [195] studied the combustion of lean and rich propane-air mixtures in constant-volume bombs; they showed that in lean mixtures, charging and outgassing of fuel from storage volumes into cold bulk gases are the main sources of unburnt HC. They also found that the flame-quench HC rapidly diffuse and oxidize. For fuel-rich mixtures ($\phi > 1.2$) processes other than wall-quenching storage were found responsible for unburnt HC in clean vessels.

Clendening et al. [196] used Raman scattering to measure side-wall quench layers. Arpaci and Tabaczynski [197] showed that the contribution of radiation to heat transfer and laminar flame quenching in small diesel engines can be as much as 35%.

The interactions between a planar premixed flame and parallel or perpendicular end walls have been studied by Fendell and co-workers [198–203] by means of one-dimensional time-dependent or steady-state two-dimensional equations. Carrier et al. [198] analyzed the steady, isobaric two-dimensional interaction between a fuel-lean hydrocarbon-air premixed flame stabilized on a heat-sink burner and a perpendicular wall by means of a one-step irreversible Arrhenius-type chemical reaction for large values of the activation energy. Their numerical calculations show an abrupt cessation of flame propagation through a premixed mixture when the flame is in too close proximity to cold parallel walls. This study is important for understanding of flame quenching in parallel walls and in parallel-plate crevice-type ducts with cold walls.

Carrier et al. [199] analyzed the unsteady one-dimensional interaction of a planar flame propagating through a fuel-lean mixture with a parallel impervious, noncatalytic wall; they showed that the flame speed decreases near an isothermal wall and increases near an adiabatic wall. Isothermal walls are pertinent to cooled engines, whereas adiabatic walls are of importance for proposed adiabatic engines. Carrier et al. [200] used a one-dimensional unsteady formulation to analyze the interaction of a planar, premixed, laminar, isobaric flame propagating through a fuel-lean mixture with a parallel adiabatic end wall; they showed that the flame speed increases as the wall is approached.

Bush et al. [201] studied unsteady one-dimensional, laminar, isobaric flame propagation phenomena through homogeneous fuel-lean premixtures characterized by large activation energies; they showed that large gas-phase temperatures are achieved at intermediate wall temperatures for a thermal diffusivity greater than the mass diffusivity. In such cases the heat release by the chemical reaction leads to gas temperatures appreciably higher than the bulk-gas adiabatic flame temperature. Futhermore, thermal shocks (i.e., exposures to large thermal gradients within one engine cycle) constitute a hazard for the ceramic caps of piston crowns.

Fink et al. [202] and Carrier et al. [203] studied unsteady, one-dimensional, laminar, nonisobaric propagation of a flame through a gaseous fuel-lean combustible premixture characterized by a large Arrhenius activation energy. The premixture was confined between planar, parallel, impervious, noncatalytic, isothermal walls. Fink et al. [202] showed the great importance of three phenomena that are not present in isobaric flame propagation: (1) the variation of the diffusion coefficient with the pressure, (2) the variation of the reaction rate with the pressure, and (3) the contribution of the temporal dependence of the pressure to the energy equation.

Fink et al. [202] simulated ignition by means of two models: (1) simulation of spark ignition with minimum enthalpy increase of a given duration at one wall and (2) simulation of hot-kernel ignition with a minimum enthalpy increase of a given volume. Their model assumes that the pressure is only a function of time and is unable to predict the conditions under which preheating of the residual charge is excessive and results in the breakdown of a spatially homogeneous pressure field. This phenomenon may result in an almost spatially uniform conversion of the residual charge to combustion products (i.e., an explosion rather than flame propagation) and engine knock. (Engine knock is discussed in Section 4.9.)

Kooker [204] showed that flame-generated pressure disturbances lead to pulsating flames having velocities that may be twice as high as those observed in the absence of pressure fluctuations. Kooker [204] also showed that flame quenching at adiabatic walls involves oscillations of the internal flame structure and that quenching at cold walls is dominated by heat losses that decrease the flame speed and eventually force the flame to retreat.

Goolsby and Haskell [205] used ion probes to detect flames, and they observed that the flame-quench distance depends strongly on the equivalence ratio in lean mixtures.

Mori et al. [206] studied catalytic combustion on a platinum plate, and they observed that low-temperature surface combustion ($<$1100 K) is possible at low equivalence ratios and that the catalytically active regions are not uniformly distributed on the platinum plate. Fakheri and Buckius [207] studied the heterogeneous catalytic combustion of fuel on an adiabatic flat plate and observed that radiation heat losses increase the ignition distance and the ignition delay time and that they lower the plate temperature.

Jarosinski [208] studied downward and upward flame propagation in narrow channels. His studies may be used to analyze flame quenching in narrow passages. Henningsen [209] studied the propagation and quenching of an isobaric one-dimensional two-reaction flame, considered the following species C_nH_m,

O_2, CO_2, CO, H_2O, and N_2, and obtained good agreement with experimental data for fuel-air ratios above stoichiometry.

4.6.2 Oil Layers

Shyy and Adamson [210] used Crocco's integral, which relates the velocity and temperature in boundary layers, and a two-layer structure for the turbulent boundary layer along the cylinder wall to analyze the unburnt HC emissions from spark-ignition engines. They showed that in the quench layer, the temperature is lower than the ignition temperature, that during the compression stroke, unburnt HC are absorbed in the oil layers along the cylinder liner, and that during the expansion stroke, unburnt HC are desorbed.

Shyy and Adamson [210] also showed that outgassing from the first ring crevice is important and that their one-dimensional absorption-desorption model predicts the experimental trends.

Adamczyk and Kach [211] studied the effect of oil layers on unburnt HC emissions in a combustion bomb by means of gas chromatography. They used five different oils and studied the combustion of propane, ethane, and butane at an equivalence ratio of 0.9. Adamczyk and Kach [211] showed that the oil layer may become saturated with unburnt fuel and that the unburnt HC emissions are caused mainly by the initial fuel. They also showed that the unburnt HC emissions vary proportionally to the amount of oil present in the combustion bomb, the initial fuel concentration, and the solubility of the fuel in the oil. Further, they showed that the consumption of the oil layer can have an important influence on the unburnt HC emissions.

Kaiser et al. [212] studied the effects of the oil layers on the HC emissions from spark-ignition engines in a CFR engine and showed that the exhaust HC emissions increase proportionally to the amount of oil added to the engine. Their gas chromatography studies showed that the fuel and fuel oxidation species (not the fuel itself) are responsible for most of the increase in HC emissions that result when oil is added to the engine.

Kaiser et al. [212] analyzed the combustion of isooctane-air and propane-air mixtures, showed that the oil does not increase HC emissions at the exhaust, and concluded that the main source of the increase in HC emissions is the dissolution of the fuel in the oil layer during compression and the subsequent release into the cooling burnt gases during the expansion stroke.

Carrier et al. [213] developed a one-dimensional unsteady model to analyze the absorption and desorption of gases in a constant-thickness isothermal oil layer and showed that the unburnt HC emissions decrease with an increase in the engine rpm.

Kaiser et al. [214] studied the effect of oil layers on the HC emissions generated during the combustion of hydrogen, methane, ethane, propane, and *n*-butane in closed bombs; they showed that for lean mixtures ($\phi \leq 1$), the fuel is absorbed in the oil layers prior to ignition and then desorbed into the cooled burnt gases after combustion is complete, yielding large amounts of HC emissions. For rich mixtures ($\phi > 1$), the oil layer is degraded to light HC during the passage of the flame. This degradation is very sensitive to the mixture equivalence ratio and type of fuel used.

4.6.3 Deposit Layers

Adamczyk and Kach [215] studied the effect of engine deposit layers on HC emissions from a closed combustion bomb. They analyzed the combustion of nine fuels by gas chromatography and used two deposit layers in their experiments. Adamczyk and Kach [215] showed that in thin deposit layers, unburnt fuel is the principal exhaust effluent and that as the solubility of the fuel in the deposit layer increases, the HC emissions increase. In thick deposit layers the HC decreases as loading and sampling times are reduced.

Lavoie et al. [216] developed a comprehensive HC emission model that accounts for ring crevices, exhaust gas scavenging, exhaust port burnup, oil layer absorption and desorption, and wall layer burnup. The fraction of unburnt charge remaining in the cylinder was assumed to be

$$F = c_1 \exp\left[-\frac{\theta_{EVO} - \theta_{90}}{c_2\,(\theta_{90} - \theta_0)}\right] \tag{148}$$

where c_1 and c_2 are constants, the subscript EVO denotes exhaust valve opening, θ is the crank shaft angle, and the subscripts 90 and 0 correspond to the crank angles at which 90% and 0% of the cylinder charge is burnt, respectively.

Lavoie et al. [216] showed that with high exhaust gas recirculation (EGR), partial burning is responsible for significant changes in the trends of HC emissions. Their simple semiempirical model also shows that the flame does not propagate completely across the combustion chamber when the HC emissions are high.

4.6.4 Crevices

Wentworth [217] showed that sizable HC reductions are possible when the piston-bore-ring crevice volume is essentially eliminated.

Carrier et al. [218] developed an isobaric, two-dimensional, steady-state model to analyze laminar flame propagation through and quenching in lean mixtures, and they showed that the flame speed decreases as the distance from the flame to the wall decreases. They also showed that the flame fails to propagate into narrow piston crevices due to the heat losses to the cold walls and that warm piston crowns and crevices discharge high-temperature fuel vapor into the hot oxygen-rich bulk cylinder in the power stroke.

Fairchild et al. [219] used Raman spectroscopy to analyze flame quenching in a duct-type crevice at atmospheric pressure with premixed propane-air mixtures; they showed that as the wall separation is reduced, the thermal boundary layers interact. This effect is more pronounced for lean mixtures where the flame speed is reduced. With adiabatic hot walls, the reduced thermal boundary layer thickness is associated with the flame front moving closer to the wall.

Driscoll et al. [220] used high-speed schlieren photography to analyze the flow field produced by a wall crevice in a constant-volume combustion chamber. They showed that the most important parameter is the orifice area-to-volume ratio; as this parameter is increased, rapid filling and emptying of the wall crevice is obtained. Driscoll et al. [220] assumed that the crevice temperature is constant and that the flow through the orifice is choked. They also assumed that the density of the main chamber is constant, and they used a lumped energy equation that accounts for the heat transfer losses from the main chamber by means of a Nusselt number correlation for pipe flow.

4.6.5 Exhaust Port Emissions

Lavoie et al. [221] used a thermodynamic model in which the mass fraction of burnt gases is given by a cosine-type Wiebe function [Eq. (7)] to analyze the HC emissions from spark-ignition engine models. They used an integral formulation of the von Karman-Pohlhausen type to analyze the thermal boundary layer growth along the cylinder wall and the evolution of the quench layer. They also used a plug flow reactor concept to analyze HC burnup in the exhaust port, the extended Zel'dovich mechanism with a correction that accounts for prompt NO, and the Reynolds analogy between momentum and mass transfer in the cylinder and in the exhaust port; they accounted for the vortex-ring crevice and quench layer. Their results show that HC emissions increase with the compression ratio. The model correctly predicted the effects of the equivalence ratio, spark timing, and exhaust gas recirculation, but yielded poor agreement with experimental data for variations in the engine rpm.

Herrin et al. [222] used a continuous stirred-tank reactor (CSTR) to model the HC disappearance in the exhaust port. Caton et al. [223] performed a thermodynamic analysis as a function of time and axial distance in the exhaust port to analyze the HC oxidation in spark-ignition engine exhaust systems. They accounted for the port heat transfer, reaction rates, and instantaneous and overall levels of HC as a function of the engine-operating conditions.

Experimental data [223] show that the fraction of HC reacted in the exhaust port ranges from 2% to 70% due to changes in the gas temperature, port residence times, and oxygen concentration. The HC oxidation in the exhaust port depends on the cylinder gas temperature, port wall temperature, exhaust port length, and exhaust port cross-sectional area. Mendillo and Heywood [224] used a model similar to that of Caton et al. [223] to analyze HC oxidation in the exhaust port of a spark-ignition engine.

Caton et al. [223] and Caton and Heywood [225, 226] divided the exhaust port into several control volumes of length Δx along the port and established energy and mass balances in each zone as follows (Fig. 4.31):

$$m_i C_p \frac{dT_i}{dt} = \dot{m}_{i-1} C_p (T_{i-1} - T_i) - \dot{m}_{i+1} C_p (T_i - T_{i+1}) - \dot{Q}_i \quad (149)$$

$$m_i \frac{dT_i}{dt} = T_i (\dot{m}_{i+1} - \dot{m}_{i-1}) \quad (150)$$

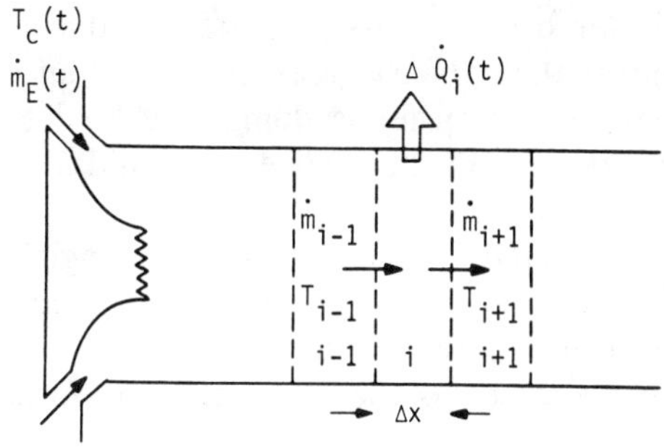

Figure 4.31 Schematic of control volumes in the exhaust port [225] (T_c = temperature of the cylinder charge; $\dot{m}_E$ = mass flow rate through the exhaust port; $\Delta\dot{Q}_i$ = heat transfer losses from the ith control volume).

where m_i and T_i are the mass and temperature of the ith zone, the subscripts $i - 1$ and $i + 1$ denote the zones adjacent to the ith zone, the dots denote differentiation with respect to time, and $\dot{Q}_i$ is the heat loss from the ith zone.

Equation (150) has been derived assuming that the pressure in each zone is constant and that the gas constant is also constant, that is,

$$\frac{p_i V_i}{R_i} = m_i\, T_i = \text{constant} \tag{151}$$

where V_i is the volume of the ith zone.

Differentiation of Eq. (151) with respect to time yields

$$m_i \frac{dT_i}{dt} = -T_i \frac{dm_i}{dt} = -T_i\,(\dot{m}_{i-1} - \dot{m}_{i+1}) \tag{152}$$

which is identical to Eq. (150).

The heat transfer loss $\dot{Q}_i$ was calculated from a Nusselt number correlation as

$$\dot{Q}_i = h_i\, P_i\, (T_i - T_w) \tag{153}$$

where h_i is the film transfer coefficient, P_i is the wetted perimeter of the ith zone, T_w is the wall temperature, and

$$\text{Nu}_i = \frac{h_i L_i}{\lambda_i} = c_1\, c_2\, 0.037\, \text{Re}^{0.8}\, \text{Pr}^{0.33} \tag{154}$$

$$L_i = \frac{V_i}{S_i} \qquad \text{Re} = \frac{\dot{m}_i L_i}{\mu_i S_i} \qquad \text{Pr} = \mu_i \frac{C_p}{\lambda_i} \tag{155}$$

S_i is the cross-sectional area, μ is the dynamic viscosity, and the constants c_1 and c_2 account for developing flow and roughness effects, respectively.

The HC oxidation in the exhaust port was modeled by means of the following global reaction rate:

$$\frac{d[\text{HC}]}{dt} = -1.36 \times 10^{10} \exp\left(-\frac{31000}{\tilde{R}T}\right) [\text{HC}][\text{O}_2]^{0.25} \tag{156}$$

where the brackets denote concentrations, and the reaction can be expressed as

$$C_xH_{2x} + 1.5xO_2 \rightarrow xCO_2 + xH_2O \tag{157}$$

Caton and Heywood [226] also considered the large-scale motion induced by flow separation and the impingement of the exhaust gases on the exhaust pipe. They showed that the exhaust port heat transfer is dominated by jet-induced large-scale fluid motions as opposed to wall shear-generated fluid motions.

Figure 4.32 shows that the fraction of reacted HC decreases as the engine load is increased but increases as the engine speed is increased. The fraction reacted first decreases and then increases as the spark advance is decreased; the fraction reacted first increases and then decreases as the equivalence ratio is increased.

Kaiser et al. [227] studied the effects of the fuel (octane and unleaded gasoline) and engine-operating variables (rpm, equivalence ratio, spark timing, exhaust gas recirculation, and coolant temperature) on the HC species in the exhaust. They showed that the concentrations of CH_4 and C_2H_2 increase as the mixture becomes richer, indicating incomplete combustion in the cylinder. With no EGR, increasing the engine rpm and retarding the spark advance result in lower HC emissions. With EGR, the HC emissions increase as the spark timing is retarded due to partial fuel oxidation.

Annand [228] analyzed heat transfer in engine ports and showed that adiabatic ports contribute to the oxidation of HC and NO. He also showed that radiation may be important in spark-ignition engines.

A review of exhaust gas modeling has been presented by Newhall [229], to which the reader is referred for more details on the history of engine modeling.

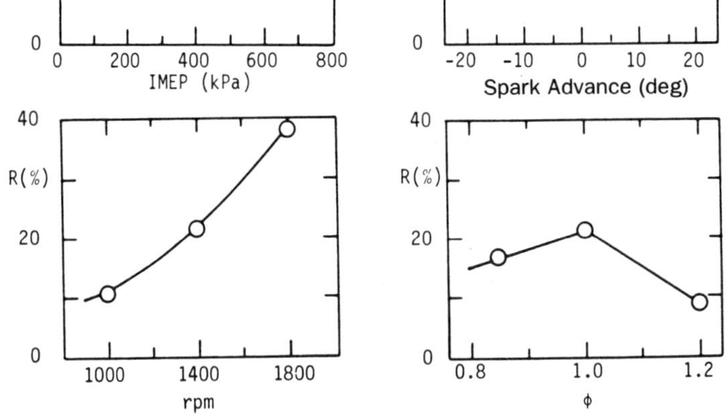

Figure 4.32 Fraction of reacted hydrocarbons in the exhaust port as a function of the engine load (IMEP: indicated mean effective pressure), spark timing, speed, and equivalence ratio [225].

4.6.6 Bulk Flow Quenching and Cycle-to-Cycle Variations

Quenching in the bulk flow is another source of unburnt hydrocarbons. Chomiak and Jarosinski [35] showed that flame quenching by turbulence occurs when $10 \leq \mathrm{Ka} \leq 20$, where Ka is the Karlovitz-Kovasznay number, defined as $\mathrm{Ka} = (u'/l)\ (\delta_L/S_L)$, where u' is the velocity rms, l is the integral length scale, δ_L is the laminar flame thickness, and S_L is the laminar flame speed. The Karlovitz-Kovasznay number can be interpreted as the ratio between the chemical reaction time and a typical mixing time, and it accounts for the effects of flame stretching. DeSoete [34] showed that aerodynamic stretch due to velocity gradients tangent to the flame front is more important than geometrical stretch in spark-ignition engines and may cause flame extinction.

Chomiak [54] showed that turbulence causes a random variation in the ignition kernel growth rate and that the interaction between the ignition kernel and the roll-up vortex formed by the piston sliding over the cylinder wall causes flame quenching and is the main source of cycle-to-cycle variations. Tromans [53] showed that the random strain field produced by the turbulent motion around the spark plug inhibits flame propagation and may be a major source of cycle-to-cycle variations in lean mixture engines. Hansel [52] attributed cycle-to-cycle variations to velocity fluctuations, which interact, distort, and may inhibit flame propagation.

Komiyama and Heywood [49] indicated that the cycle-to-cycle variations are closely related to the time required to establish a stable flame kernel. Rashidi [50] showed that the main cause of cycle-to-cycle variations is the small-scale turbulence and the mixing it produces. Small and medium scales affect the mixing of the reactive charge in the small, unstable flame kernel and cause variations in the reaction rate and in the ignition delay time. However, the large-scale turbulence affects only the flame front surface area.

Daneshyar et al. [230] used high-activation-energy asymptotics to show that flame propagation in internal combustion engines may be inhibited by the turbulent strain field. Their analysis also shows that fluctuations in the strain rate field may be responsible for cycle-to-cycle variations.

Turbulent fluctuations affect the flame front (e.g., wrinkling) and the combustion rate. The flame front distortion affects the turbulent flow field due to the expansion of the burnt gases and results in geometrical and aerodynamical stretches that may inhibit flame propagation [231, 232]. Furthermore, the flame front distortion and curvature may result in large pressure-velocity correlations in premixed charges [233], which may increase the turbulence and contribute to flame inhibition.

Nohira et al. [234] analyzed cycle-to-cycle variations from pressure measurements [235] and showed that variations in the indicated mean effective pressure (IMEP) are caused by variations in the heat release rate (fuel supply system), variations in the charge internal energy (heat release pattern), and their covariance. Smith [236] studied the effects of the ignition source location on the IMEP of a homogeneous-charge high-swirl engine and observed that the IMEP does not vary significantly with the ignition source location in a high-swirl engine.

With no swirl, the best IMEP is obtained if the ignition source is located near the cylinder centerline. Witze [237] obtained laser shadowgraphs of the flow field in a homogeneous-charge engine, and indicated that without swirl, the minimum burning time is obtained with a spark plug located at the cylinder centerline so that the flame travel distance is minimized. Central ignition is also preferable for low swirl. However, for high swirl the burning rate is increased by moving the spark toward the cylinder wall. Witze [237] also observed that cycle-to-cycle variations increase with the burning duration and decrease with the swirl level.

Beretta et al. [238] used photography to deduce a new burning rate and developed a thick-flame model, which is governed by the following equations:

$$m = m_b + m_u = m_f + m_f^- = m_f + \rho_u V_f^- \tag{158}$$

$$V = \frac{m_b}{\rho_b} + \frac{m_u}{\rho_u} = V_f + V_f^- \tag{159}$$

$$\rho_u A_f u_e = \dot{m}_f = -\dot{m}_f^- = -\rho_u \dot{V}_f^- - \dot{\rho}_u V_f^- \tag{160}$$

$$\rho_u A_f u_{gf} = \dot{\rho}_u V_f^- - \rho_u A_{pf}^- u_p \tag{161}$$

$$u_e = u_f - u_{gf} \qquad \dot{V} = -A_p u_p \tag{162}$$

$$\dot{x}_b = \frac{x_f - x_b}{\tau} \tag{163}$$

where m_f and m_f^- are the mass inside and outside the flame front, respectively; V_f^- is the volume of the mixture excluding the flame volume; u_e is the entrainment velocity or turbulent flame speed; u_{gf} is the sum of the combustion- and piston-induced compression velocities; A_{pf}^- is the piston surface area ahead of the thick flame; u_p is the piston velocity; V is the combustion chamber volume; A_p is the piston surface area ($A_p = \pi B^2/4$); B is the cylinder bore; τ is the burning time; x_f and x_b denote the mass fractions of the unburnt and burnt gases, respectively; u_f is the flame velocity in a laboratory frame of reference; and the dots denote differentiation with respect to time.

The solution of Eqs. (158)–(163) yields the mass burning rate $\dot{m}_b$, which can be associated with an equivalent laminar flame front of surface area

$$A_b = \frac{\dot{m}_b}{\rho_u S_L} \tag{164}$$

so that A_b/A_f can be interpreted as a flame stretch factor.

Beretta et al. [238] assumed that the flame was spherical and observed that u_e increases at the beginning of the flame propagation phenomenon and then remains constant. They also assumed that unburnt gases are inside the flame region during flame propagation and observed that the mass burning rate can be expressed as

$$\theta(\eta) = \theta_s + \frac{g(\eta)}{(1 - a)\, C(a,b)} \Delta\theta_b \tag{165}$$

where

$$g(\eta) = b\eta^{1-a} - (1 - b)(1 - \eta)^{1-a} + (1 - b) \qquad \eta = \frac{x_b}{x_{b_{\max}}} \tag{166}$$

$a = 0.84$, $b = 0.42$, $C(a,b)$ is a function of a and b, $\Delta\theta_b$ is the burning angle (i.e., $\Delta\theta_b = \theta_{90} - \theta_{10}$, where θ_{90} and θ_{10} are the crank shaft angles at which $x_b = 0.90$ and 0.10, respectively), and $x_{b_{\max}}$ denotes the maximum mass fraction of the burnt gases.

Beretta et al. [238] showed that

$$\Delta\theta_b = 6 \text{ rpm} \frac{x_{b_{\max}}}{(\dot{x}_b)_{\max}} \tag{167}$$

and that the variations in $\Delta\theta_b$ and $\Delta\theta_d = \theta_{10} - \theta_s$, where θ_s is the spark time and $\Delta\theta_d$ is the ignition delay time, are the reasons for cycle-to-cycle variations.

Al-Alousi and Karim [239] examined the effects of cycle-to-cycle variations on the ignition delay time and combustion duration in a spark-ignition engine fueled with methane. Their studies showed that the ignition delay time is much more affected by the cycle-to-cycle variations than by the combustion duration. Lowry and Devoto [240] studied the exhaust emissions from a single-cylinder engine fueled with gasoline, methanol, and ethanol; they showed that gasoline yields more unburnt HC than methanol but about the same as ethanol. The NO emissions were small for gasoline-fueled engines but nearly identical for methanol and ethanol.

Aiman [241] showed that the EGR can be increased without misfire by increasing the spark duration. Misfire, partial burning, and slow combustion may be due to high EGR and can be avoided with advanced spark timing.

The lean misfire limit can be extended with high-swirl and high-combustion turbulence by use of a helical port control with multipoint controlled-fuel injection and with a feedback control system of the air-fuel ratio by means of a lean mixture sensor [242]. The lean misfire limit, NO_x emissions, and torque fluctuations depend strongly on the fuel-injection timing; the NO_x emissions increase for the leanest misfire system. In order to meet the NO_x emissions standards and improve the engine drivability (engine torque fluctuations), Matsushita et al. [242] proposed use of a homogeneous mixture throughout the engine cylinder.

Matsushita et al. [243] demonstrated that a helical port with a swirl control valve (SCV) improves the combustion efficiency and maintains a high volumetric efficiency. (Swirl generally reduces the valve discharge coefficient and the mass flow rate through the intake valve.)

Matsushita et al. [243] obtained pressure-volume diagrams, photographs, and hot wire measurements with and without SCV and showed that the swirl center has a more active movement with SCV, indicating higher turbulence levels in the cylinder. Their SCV system resulted in higher flame speeds and higher combustion rates for the same swirl number than did swirl without valve control. The SCV reduced the combustion duration, improved the lean misfire limit and the EGR tolerance, improved the combustion at partial loads and the engine performance at full load, and produced swirling motions with an eccentric axis that rotated around the cylinder axis.

Experimental [244] and theoretical [245] studies have also shown that the turbulence levels within the engine cylinder are sensitive to both squish and swirl and that they increase with the swirl levels.

A mathematical engine model that includes the intake manifolds, combustion dynamics, exhaust emissions, and air-fuel ratio variations at the inlet port has been developed by Yuen and Servati [246].

4.7 NITRIC OXIDE AND CARBON MONOXIDE EMISSIONS

The emissions of nitric oxide and carbon monoxide from spark-ignition engines are described in Sections 4.7.1 and 4.7.2. The modeling of NO emissions has also been discussed throughout the text (see Sections 4.2 and 4.3).

The CO emissions depend to a large extent on the following reaction [247]:

$$CO + OH \rightleftarrows CO_2 + H \tag{168}$$

The CO emissions are generated in fuel-rich regions and increase as the mixture equivalence ratio is increased.

4.7.1 Nitric Oxide Emissions

The NO formation is a nonequilibrium phenomenon, which depends on the temperature gradients in the burnt gases. These temperature gradients arise because the elements of the unburnt mixture that burn first are compressed as combustion products, whereas the last part of the charge to burn is compressed as an unburnt mixture. The thermodynamic properties and equilibrium composition of the gas elements that burn at different times during the combustion process can be computed from the pressure diagram.

At the high pressures found in spark-ignition engines, the hydrocarbon oxidation reactions go rapidly to completion. In addition, the reaction zone is thin, although the turbulent flame front may extend across several burning eddies. Therefore, it is a reasonable approximation to assume that the burnt gases are almost in chemical equilibrium [248, 249]. The NO concentrations can be calculated by means of the extended Zel'dovich or thermal mechanism, that is,

$$N_2 + O \underset{k_2}{\overset{k_1}{\rightleftarrows}} NO + N \tag{169}$$

$$N + O_2 \underset{k_4}{\overset{k_3}{\rightleftarrows}} NO + O \tag{170}$$

$$N + OH \underset{k_6}{\overset{k_5}{\rightleftarrows}} NO + H \tag{171}$$

The rate of formation of NO can be expressed as

$$\frac{d[\mathrm{NO}]}{dt} = k_1[\mathrm{N_2}][\mathrm{O}] - k_2[\mathrm{NO}][\mathrm{N}] + k_3[\mathrm{N}][\mathrm{O_2}] - k_4[\mathrm{NO}][\mathrm{O}] + k_5[\mathrm{N}][\mathrm{OH}] - k_6[\mathrm{NO}][\mathrm{H}] \tag{172}$$

where the brackets denote concentrations, and k_1, k_2, k_3, k_4, k_5, and k_6 are reaction-rate constants.

Equation (172) indicates that in order to calculate the NO formation, the concentrations of O, OH, H, and N must be known. Frequently, a steady-state approximation for the N atom is used, that is,

$$\frac{d[\mathrm{N}]}{dt} = k_1[\mathrm{N_2}][\mathrm{O}] - k_2[\mathrm{NO}][\mathrm{N}] - k_3[\mathrm{N}][\mathrm{O_2}] + k_4[\mathrm{NO}][\mathrm{O}] - k_5[\mathrm{N}][\mathrm{OH}] + k_6[\mathrm{NO}][\mathrm{H}] = 0 \tag{173}$$

The N concentration can be written as

$$[\mathrm{N}] = \frac{k_1[\mathrm{N_2}][\mathrm{O}] + k_4[\mathrm{NO}][\mathrm{O}] + k_6[\mathrm{NO}][\mathrm{H}]}{k_2[\mathrm{NO}] + k_3[\mathrm{O_2}] + k_5[\mathrm{OH}]} \tag{174}$$

Substitution of Eq. (174) into Eq. (172) yields

$$\frac{d[\mathrm{NO}]}{dt} = 2\{k_1[\mathrm{N_2}][\mathrm{O}] - k_2[\mathrm{NO}][\mathrm{N}]\} \tag{175}$$

The concentrations of O, OH, and H can be calculated using the following partial equilibrium reactions

$$\frac{1}{2}\,\mathrm{O_2} \rightleftarrows \mathrm{O}\ (k_{e_\mathrm{O}}) \tag{176}$$

$$\mathrm{H} + \mathrm{O_2} \rightleftarrows \mathrm{OH} + \mathrm{O}\ (k_{e_\mathrm{H}}) \tag{177}$$

where k_{e_O} and k_{e_H} denote equilibrium constants and

$$k_{e_\mathrm{O}} = \frac{[\mathrm{O}]}{[\mathrm{O_2}]^{1/2}} \tag{178}$$

$$k_{e_\mathrm{H}} = \frac{[\mathrm{OH}][\mathrm{O}]}{[\mathrm{H}][\mathrm{O_2}]} \tag{179}$$

Substitution of Eqs. (178) and (179) into Eq. (175) yields

$$\frac{d[\mathrm{NO}]}{dt} = R_1\,\frac{1 - ([\mathrm{NO}]/[\mathrm{NO}]_e)^2}{1 + k\,[\mathrm{NO}]/[\mathrm{NO}]_e} \tag{180}$$

where

$$R_1 = k_1[\mathrm{O}]_e[\mathrm{N_2}]_e = k_2[\mathrm{N}]_e[\mathrm{NO}]_e \tag{181}$$

$$R_2 = k_3[\mathrm{N}]_e[\mathrm{O_2}]_e = k_4[\mathrm{NO}]_e[\mathrm{O}]_e \tag{182}$$

$$R_3 = k_5[\mathrm{N}]_e[\mathrm{OH}]_e = k_6[\mathrm{NO}]_e[\mathrm{H}]_e \tag{183}$$

$$k = \frac{R_1}{R_2 + R_3} \tag{184}$$

and the subscript e denotes equilibrium.

If the NO concentration is well below equilibrium, Eq. (175) can be approximated by

$$\frac{d[\mathrm{NO}]}{dt} = 2k_1[\mathrm{N_2}][\mathrm{O}] \tag{185}$$

Substitution of Eq. (178) into Eq. (185) yields

$$\frac{d[\mathrm{NO}]}{dt} = 2k_1\, k_{e_\mathrm{O}}\, [\mathrm{N_2}][\mathrm{O_2}]^{1/2} \tag{186}$$

The concentration of N atoms is very small; therefore, a steady-state approximation for N atoms is justified [Eq. (173)]. Furthermore, at the high pressures and temperatures in spark-ignition engines, the hydrocarbon oxidation reactions go rapidly to completion, and it is reasonable to assume that O, O_2, OH, and H are in equilibrium [Eqs. (176) and (177)]. This equilibrium is maintained well into the expansion stroke beyond the crank angle at which the NO chemistry freezes.

Spectral absorption measurements have shown that NO formation occurs primarily in postflame gases and can be accurately predicted by Eqs. (169) and (170) with the oxygen- and nitrogen-atom concentrations set equal to their equilibrium values [250]. Starkman [251] showed that the production of CO, NO, and unburnt HC is governed by nonequilibrium processes, particularly in fuel-lean mixtures.

Newhall [252] showed that during the expansion stroke, the atomic hydrogen and oxygen and the hydroxyl radical fail to recombine at rates sufficiently high to maintain shifting chemical equilibrium concentrations. The resulting persistence of excess atomic species and hydroxyl radicals and the accompanying partial equilibrium established by the fast bimolecular reactions involving these species yields a hydrogen-atom-to-hydroxyl-radical ratio much greater than the ratio for total equilibrium.

Figures 4.33–4.35 show that the equilibrium concentrations of hydroxyl radicals, NO, and CO during the expansion stroke are smaller than those obtained by accounting for finite rate, nonequilibrium chemical kinetics processes [252].

The NO freezes in the expansion stroke, and there are NO concentration gradients throughout the combustion chamber. Moreover, NO formation is strongly dependent on the temperature and O_2 concentration histories.

The activation energy of the forward reaction in Eq. (169) is higher than those of Eqs. (170) and (171); therefore, the characteristic times for the forward reaction in Eq. (169) always exceed those of the hydrocarbon oxidation, and it

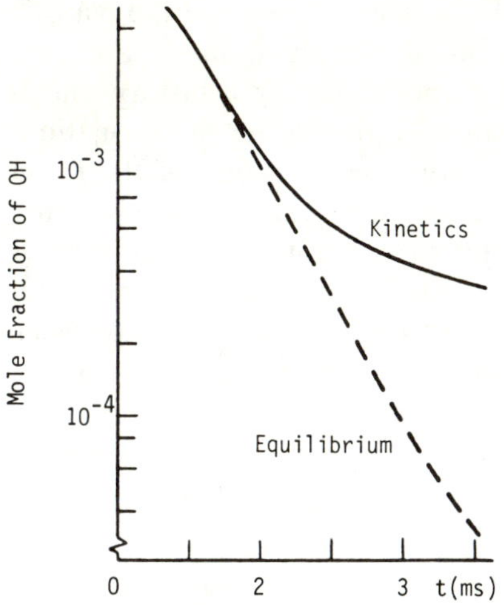

Figure 4.33 Hydroxyl radical concentration during the expansion stroke (compression ratio = 9, engine speed = 4000 rpm, fuel = C_8H_{18}, equivalence ratio = 1) [252].

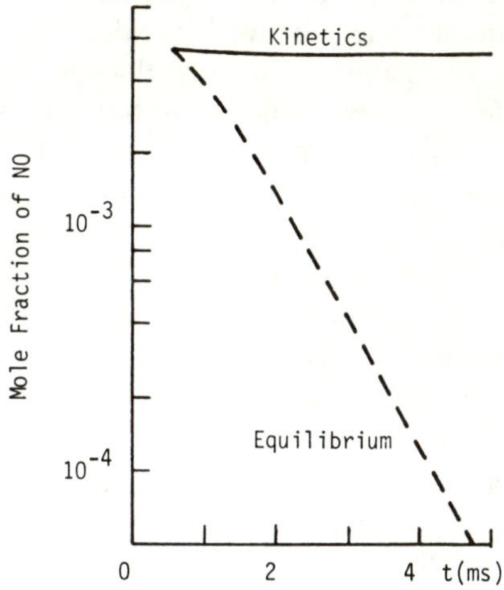

Figure 4.34 Nitric oxide concentration during the expansion stroke (compression ratio = 9, engine speed = 4000 rpm, fuel = C_8H_{18}, equivalence ratio = 1) [252].

is feasible to decouple the thermal NO formation process from that of the hydrocarbon combustion.

The strong dependence of the forward reaction in Eq. (169) with the temperature implies that in order to accurately predict NO formation in spark-ignition engines, it is necessary to consider the temperature gradients that arise because the unburnt mixture that burns first is compressed as combustion products, whereas the part of the mixture that burns last is compressed only as unburnt gases. The effects the temperature gradients throughout the combustion chamber have on NO formation can be considered by dividing the burnt gases into several elements or groups; the thermodynamic and equilibrium composition of elements that burn at different times during the combustion process can be computed

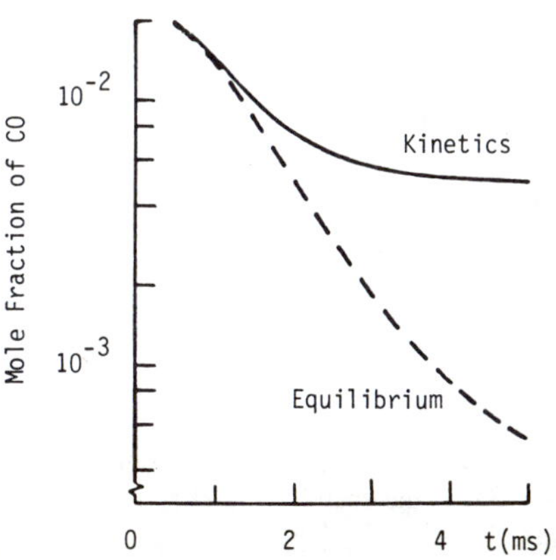

Figure 4.35 Carbon monoxide concentration during the expansion stroke (compression ratio = 9, engine speed = 4000 rpm, fuel = C_8H_{18}, equivalence ratio = 1) [252].

from the cylinder pressure and temperature. Thus, in a two-zone thermodynamic model, the mixture is divided into two regions—unburnt and burnt gases; the burnt gases are in turn divided into several elements having a history that is followed as a function of time in order to accurately predict the NO formation.

The thermodynamic properties and equilibrium composition of the burnt gases can be calculated as follows. Consider a gas element that burns at pressure p_i. For $p < p_i$ this element is compressed as unburnt mixture to the state p_i, h_i along an isentropic line s_u, where h_i is the specific enthalpy of the element. After isenthalpic constant-pressure combustion, its entropy is s_{b_i}. If mixing between elements and heat transfer from the element are neglected, this element is further compressed along the isentropic line s_{b_i}. Thus, given p_i and h_i, where h_i is equal to the enthalpy of the unburnt-gas mixture at p_i and s_{u_i}, then s_{b_i} can be calculated, and the time history of the element can be determined as it is further compressed and expanded [49].

The NO formation is kinetically controlled, and its value lags behind the local equilibrium level while the pressure and temperature continue to rise. The NO concentration may reach equilibrium levels at the maximum pressure and temperature in the gas elements that burn early, but the NO concentration does not reach the peak equilibrium levels in the gas elements that burn late. Furthermore, as the burnt gases are expanded, some decomposition occurs before the temperature falls low enough for the NO concentration to freeze.

James [253] showed that the assumptions of uniform temperature and composition in the burnt gases are inadequate to predict NO emissions. Improvements can be achieved by assuming one-dimensional temperature gradients and partial mixing in the burnt gases. James [253] also showed that in premixed flames, prompt NO is important, and mixing behind the flame increases the NO emissions from the combustion chamber.

Figure 4.36 shows the NO concentration in the burnt gases as a function of the crank shaft angle for five elements that burn at different times [49]. Figure

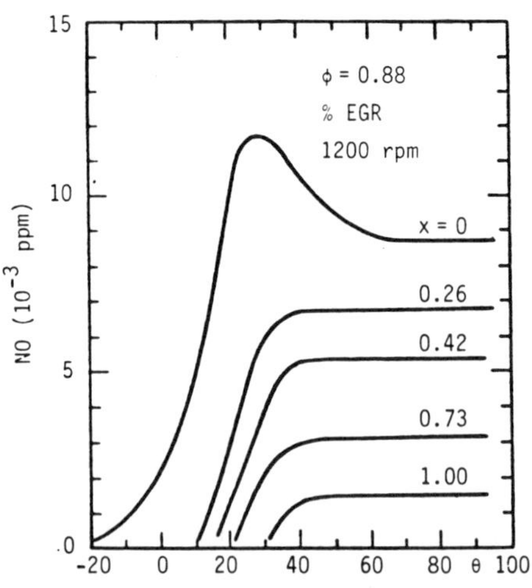

Figure 4.36 Nitric oxide concentration in the burnt gases as a function of crank shaft angle for five elements that burn at different times (ϕ = 0.88, engine speed = 1200 rpm, WOT = wide open throttle) [49].

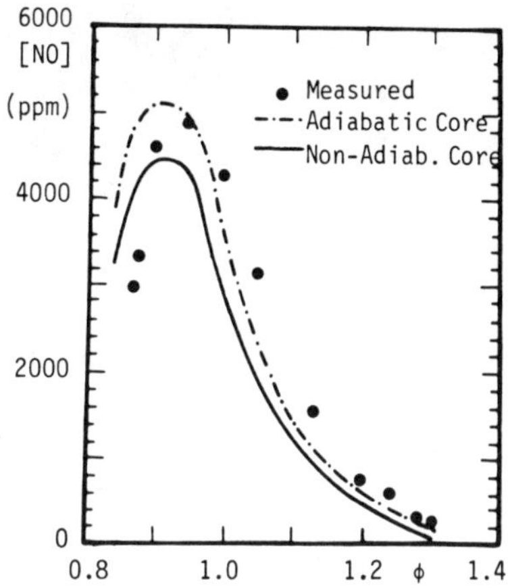

Figure 4.37 Nitric oxide concentrations predicted by adiabatic core (dashed-dotted line) and nonadiabatic core (solid line) models [19].

4.37 shows that the NO concentration peaks at equivalence ratios on the lean side of stoichiometry [19]. Figure 4.38 indicates that the exhaust NO concentrations decrease with the percentage of EGR for lean, stoichiometric, and rich mixtures, but increase with the engine speed for rich mixtures (Fig. 4.39).

The equilibrium concentrations of O, NO, N, O_2, H, and OH that appear in Eqs. (181)–(183) depend on the mixture equivalence ratio, burnt mass fraction, pressure, and temperature. For fuel-lean and stoichiometric mixtures, the main combustion products are unburnt fuel, H_2O, CO_2, O_2, and N_2, whereas for fuel-rich mixtures, the main combustion products are unburnt fuel, CO_2, CO, H_2, H_2O, O_2, and N_2. The calculation of equilibrium concentrations is described in Chapters 1 and 2.

Krieger [254] and Aiman [255] showed that in rich mixtures NO is controlled by decomposition, whereas in lean mixtures NO is controlled by formation

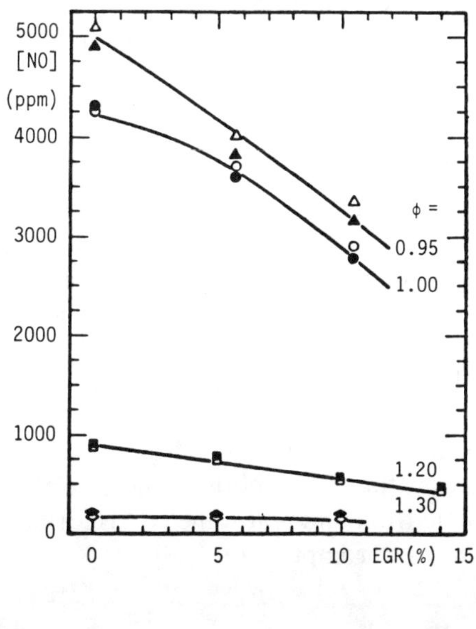

Figure 4.38 Exhaust NO concentration as a function of the amount of exhaust gas recirculation and the equivalence ratio [49]. Symbols denote experimental data.

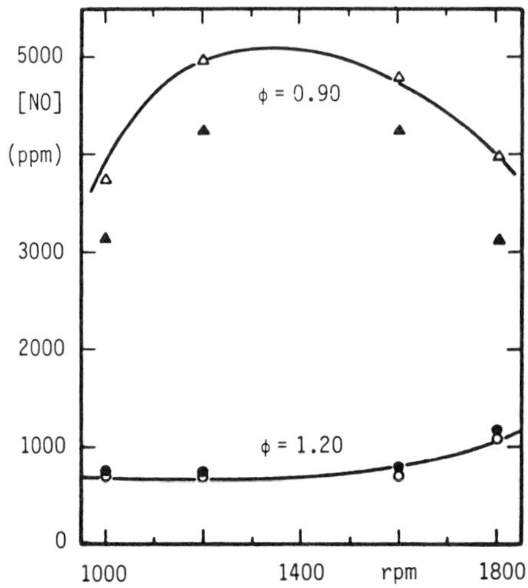

Figure 4.39 Nitric oxide concentration as a function of the engine speed and the equivalence ratio [49]. Symbols denote experimental data.

processes. They also showed that there is a trade-off between the NO emissions and fuel consumption. Aiman [255] added NO to the intake manifold of a spark-ignition engine and observed that predictions based on formation-rate-controlled models are not adequate to predict NO emissions. He also observed that both formation and decomposition processes are important, that the NO is not in equilibrium at the peak temperature, and that there seems to be a change in the NO formation mechanism at stoichiometry.

In fuel-lean mixtures the NO formation is rate controlling in the postflame gases, little NO decomposes after the crank shaft angle at which the maximum pressure is achieved, and the NO freezes in the expansion stroke. In fuel-rich mixtures NO is almost in equilibrium at the maximum pressure and temperature, and NO decomposition is important in the expansion stroke.

Aiman [255] showed that temperature gradients are important in the formation and decomposition of NO. He also showed that the order of burning or burning history is important for accurately predicting the NO emissions.

Brandstatter and Decker [256] showed that a combustion prechamber is desirable to achieve low NO emissions, that the injection time can be kept constant over the entire operating range of the engine, and that with a prechamber, improved control of mixture stratification can be achieved over a wide range of engine speeds. Fujimoto et al. [257] showed that impinging jet mixing and combustion reduces the NO_x emissions at stoichiometry, whereas standard engines yield large NO_x concentrations at high loads (stoichiometry).

The extended Zel'dovich mechanism [Eqs. (169)–(171)] for NO formation indicates that NO is formed mainly in the postflame or burnt gases and is based on the idea that the hydrocarbon reactions are very fast compared with Eqs. (169)–(171). Experimental measurements on flat flame burners indicate that the NO concentration is not zero at the flame front and that some NO is formed in the flame region. This NO is referred to as "prompt" NO [258].

The NO formation at the flame front can be explained by the following reactions:

$$HC + N_2 \rightleftarrows HCN + N \tag{187}$$

$$C_2 + N_2 \rightleftarrows 2CN \tag{188}$$

which generate N atoms, which in turn produce NO through the Zel'dovich mechanism; CN could also yield NO by O_2 and O attack.

The results of Hayhurst and Vince [258] indicate that NO concentration increases with those of nitrogen and hydrocarbons and that a molecule of C_2H_2 produces more NO than a molecule of CH_4. They also observed that the following chemical reactions are important in the formation of NO:

$$HC + N_2 \rightleftarrows HCN + N \tag{189}$$

$$CH_2 + N_2 \rightleftarrows HCN + NH \tag{190}$$

where HCN, N, and NH can be oxidized by O and O_2 to produce NO.

Multizone thermodynamic models can account for prompt NO by means of corrections to the extended Zel'dovich mechanism.

Pundir et al. [259] used the model proposed by Blumberg and Kummer [5] and the extended Zel'dovich mechanism for NO formation. They showed that the mixture preparation affects the NO emissions and that the degree of non-uniformity of the temperature and composition but not their scale affects the NO emissions.

4.7.2 Carbon Monoxide Emissions

The CO and unburnt HC emissions increase with the mixture equivalence ratio and are expected to be large for rich mixtures. The unburnt HC emissions also increase as the mixture equivalence ratio is decreased because of the slower flame propagation in lean mixtures.

The CO formed in the combustion process is oxidized to CO_2 at a rate that is relatively slow compared to the CO formation rate. The principal oxidation reactions considered important are

$$CO + OH \underset{k_2}{\overset{k_1}{\rightleftarrows}} CO_2 + H \tag{191}$$

$$CO_2 + O \underset{k_4}{\overset{k_3}{\rightleftarrows}} CO + O_2 \tag{192}$$

although the second reaction is relatively slow.

If O, OH, O_2, and CO_2 are assumed to be in equilibrium, the formation rate of CO can be expressed as

$$\frac{d[CO]}{dt} = C - D[CO] \tag{193}$$

where

$$C = R_1 + R_2 \tag{194}$$

$$D = \frac{R_1 + R_2}{[CO]_e} \tag{195}$$

$$R_1 = k_1[CO]_e[OH]_e = k_2[CO_2]_e[H]_e \tag{196}$$

$$R_2 = k_3[CO_2]_e[O]_e = k_4[CO]_e[O_2]_e \tag{197}$$

The integration of Eq. (193) yields the CO concentration, provided that the initial CO concentration is specified. In multizone models the CO concentration is assumed to be zero in the first cycle whereas in subsequent cycles it takes the value of the CO concentration in the residual gas.

An alternative approach to Eqs. (191)–(197) is to solve the rate equations for detailed chemical kinetics mechanisms such as in the model proposed by Zeleknik and McBride [15]. However, the combustion mechanisms for most fuels are not known with sufficient detail; methane, ethane, and propane are exceptions.

Baruah [260] used a two-zone thermodynamic model, the extended Zel'dovich mechanism for thermal NO formation, and the following equilibrium reaction:

$$CO + OH \rightleftarrows CO_2 + H \tag{198}$$

His calculations were based on the following flame speed ratio:

$$\frac{S_T}{S_L} = \left(1 + \frac{\epsilon_q}{\nu}\right)^{1/2} \tag{199}$$

where ϵ_q is the eddy diffusivity.

Baruah [260] showed that Eq. (198) does not predict the experimental CO measurements and proposed an empirical method to calculate the CO emissions.

Figure 4.40 shows that the CO emissions increase with the mixture equivalence ratio, whereas the NO emissions peak on the lean side of stoichiometry [69]. Figures 4.41–4.43 illustrate the effects of the ignition timing, equivalence ratio, and engine speed on the CO and NO emissions. An ignition timing of 40°BTDC (before top dead center) produces more NO and fewer CO emissions

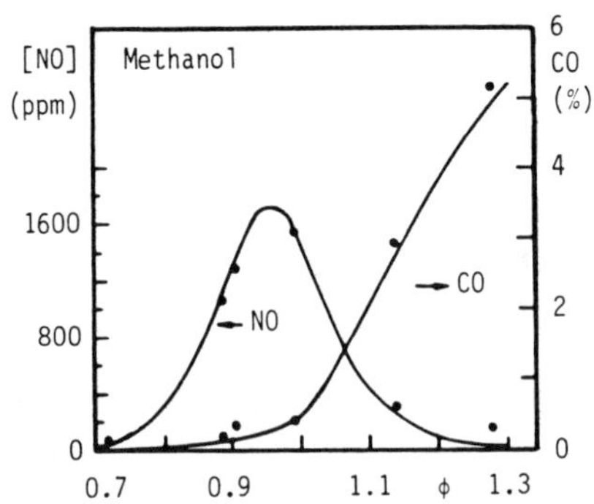

Figure 4.40 Nitric oxide and carbon monoxide emissions as a function of the mixture equivalence ratio [69].

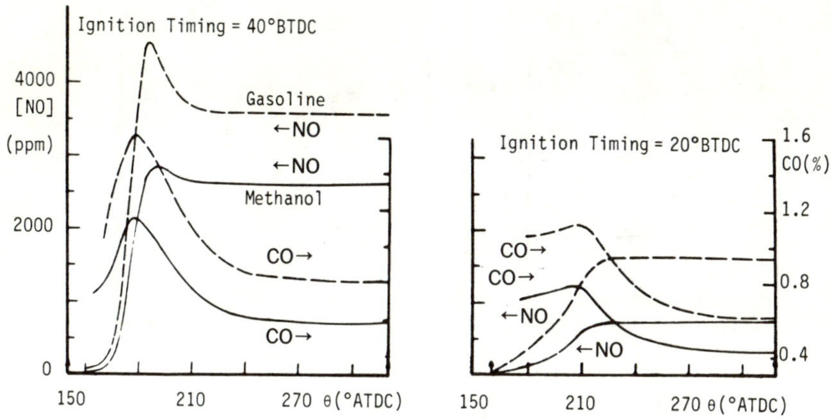

Figure 4.41 Nitric oxide and carbon monoxide emissions as a function of the ignition timing [69]. Solid lines correspond to a methanol-fueled engine; dashed lines correspond to a gasoline-fueled engine.

than an ignition timing of 20°BTDC (Fig. 4.41). The CO and NO emissions increase and decrease respectively as the mixture equivalence ratio is increased beyond stoichiometry (Fig. 4.42). The CO emissions are almost independent of the engine rpm, whereas the NO emissions increase as the engine speed is decreased (Fig. 4.43).

4.8 STRATIFIED–CHARGE SPARK–IGNITION ENGINES

Stratified-charge spark-ignition engines may replace current gasoline and diesel engines because of their performance, lower emissions, and possible multifuel

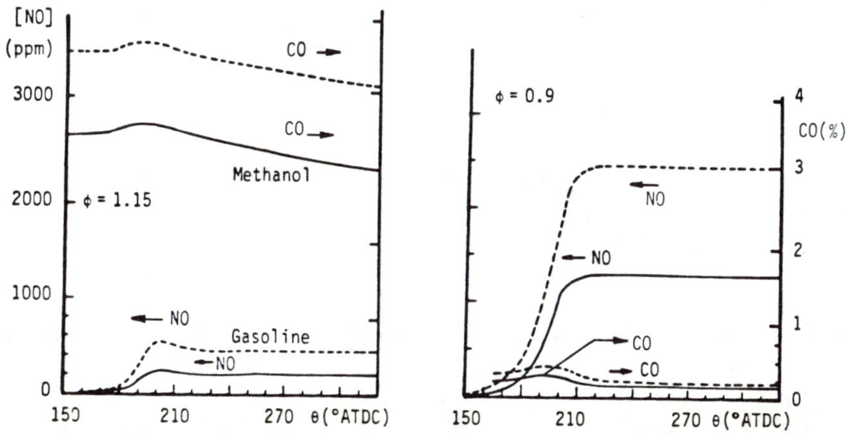

Figure 4.42 Nitric oxide and carbon monoxide emissions as a function of the mixture equivalence ratio [69]. Solid lines correspond to a methanol-fueled engine; dashed lines correspond to a gasoline-fueled engine.

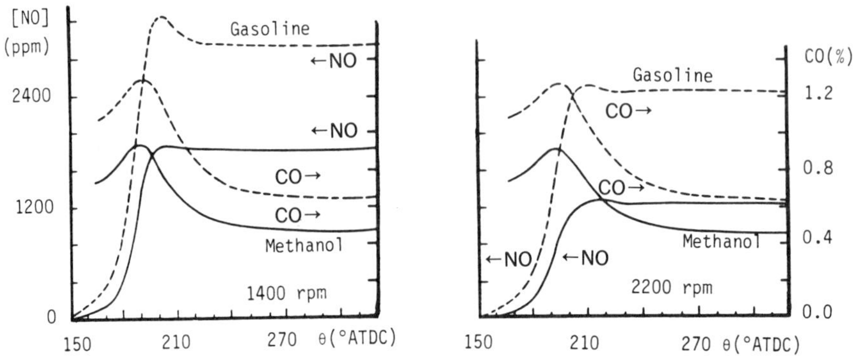

Figure 4.43 Nitric oxide and carbon monoxide emissions as a function of the engine speed [69]. Solid lines correspond to a methanol-fueled engine; dashed lines correspond to a gasoline-fueled engine.

capability. However, accurate control of the combustion process is critical to the performance of stratified-charge engines, and maintaining optimum stratification under a wide range of speed, load, temperature, and injector conditions has proved extremely difficult [261].

Studies on the now-defunct Texaco engine showed that gasoline-injected engines are efficient, clean, and practical power plants with multifuel capabilities [262]. Mizutani and Matsushita [263] injected light diesel fuel into the cylinders of spark-ignition engines and observed improved engine performance and reduced NO and CO emissions.

Al-Mamar et al. [264] observed faster combustion in dual-chamber than in open-chamber spark-ignition engines. They also observed that the rate of combustion increases with the flow velocity at the orifice due to orifice-generated turbulence. El-Messiri and Newhall [265] used a two-zone thermodynamic model to analyze combustion in divided-chamber engines. The primary chamber contained fuel and air, whereas the secondary chamber contained air only. El-Messiri and Newhall [265] showed that in their dual-chamber engine model, the hot products were quenched, NO was affected by the primary chamber volume and orifice area, and NO formation was substantially reduced by the fuel stratification.

Rychter et al. [266] used a catalyst in the prechamber of a spark-ignition engine and showed that the catalyst facilitates ignition, reduces the ignition delay time due to changes in the combustion chemistry, strengthens the flame in its initial stages, reduces the cycle-to-cycle variations, and extends the lean misfire limit.

The lean operating limit of spark-ignition engines can be extended by adding di-*t*-butyl peroxide to the combustion chamber. This peroxide dissociates and forms tertiary butoxy radicals, which react with unleaded gasoline and increases the fuel oxidation rate [267]. The addition of oxygen to the intake air flow increases the flame temperature and/or flame speed, decreases the emissions of unburnt HC, decreases the engine thermal efficiency, increases the NO emissions, and permits leaner fuel operation [268].

Rauckis and McLean [269] used a two-zone thermodynamic model and Woschni's heat transfer correlation to analyze the effects of hydrogen addition on flame propagation in spark-ignition engines. Their results show that H_2 addition reduces the ignition delay or induction time, particularly in lean mixtures, improves the engine efficiency, and results in fewer cycle-to-cycle variations. Rauckis and McLean [269] also observed that the induction time is dominated by chemical effects, whereas the turbulent burning period is controlled by turbulence transport.

Cassidy [270] studied the performance of gasoline- and gasoline-hydrogen-fueled multicylinder spark-ignition engines and showed that (1) the addition of small amounts of hydrogen to the gasoline results in efficient, lean operation, due to increased turbulent flame speed and reduced ignition delay time; (2) gasoline-hydrogen-fueled engines produce more NO_x emissions than do gasoline-fueled engines because of their higher peak combustion temperatures; (3) the unburnt HC emissions from gasoline-hydrogen-fueled engines, as compared with those from gasoline-fueled engines, are lower at equivalence ratios above 0.80 and higher at minimum-energy-consumption equivalence ratios; and (4) gasoline-fueled engines produce more CO emissions than do gasoline-hydrogen-fueled engines.

Krieger and Davis [271] examined the effects of enriching the prechamber mixture supply at a fixed overall stoichiometry in a single-cylinder stratified-charge engine. Their results show that (1) the combustion and emissions formation are determined by the jet plume issuing from the prechamber and the gases surrounding the jet plume; (2) the large increase in CO emissions with increased stratification appears to come from the prechamber gases; (3) the jet plume is the predominant source of NO emissions when the engine is operated with a strongly stratified air-fuel mixture supply; and (4) the decrease of HC emissions in jet-ignition engines with increased stratification is partly traceable to the higher exhaust temperatures.

Figure 4.44 shows that stratified jet-ignition engines produce lower unburnt HC emissions than do conventional engines operating at the same load [271]; and Fig. 4.45 indicates that the Ford PROCO stratified-charge engine produces

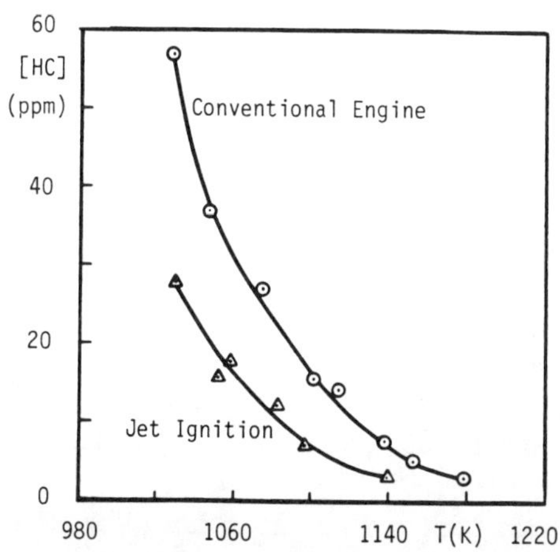

Figure 4.44 Exhaust hydrocarbon emissions as a function of the exhaust temperature in conventional engines and in strongly stratified jet-ignition engines at the same load [271].

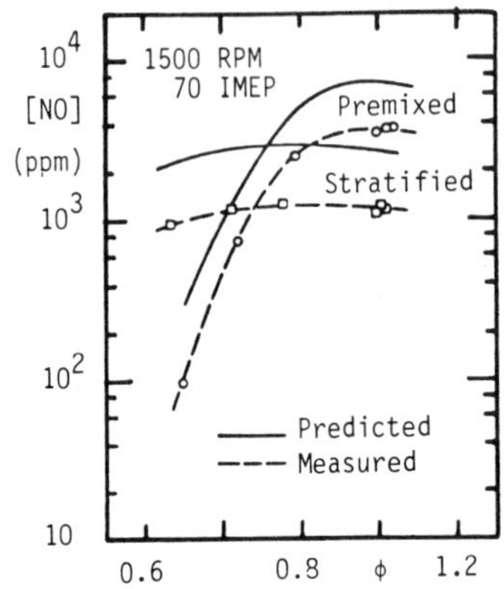

Figure 4.45 Exhaust NO concentrations for the Ford PROCO stratified-charge engine as a function of the overall equivalence ratio for normal and for premixed engine operation [247].

lower NO emissions than does a premixed-charge engine for equivalence ratios greater than about 0.7 [247].

Giovanetti et al. [272] showed that DI spark-ignition engines have a potential to tolerate a wide range of fuel types and equal the efficiency of diesel engines; however, they have poor combustion efficiency and produce large amounts of unburnt HC at light loads. Giovanetti et al. [272] also showed that fuel was located away from the primary combustion zone and that the lack of secondary burnup contributed to the unburnt HC emissions. An increase in the cylinder temperature results in lower HC emissions; such an increase can perhaps be achieved if ceramics are used in the combustion chamber.

Najit and Foster [273] used experimental heat release data and mass and energy balance equations for the fuel and burnt products. They showed that oxidation kinetics can be controlled to operate a spark-ignition engine in a compression-ignition mode.

Turbocharging may result in improved fuel economy in small-displacement spark-ignition engines [274]. Turbocharged engines produce high burning velocities and temperatures and may yield large amounts of NO emissions.

Accurate predictions of combustion phenomena in stratified-charge engines and important fluid motion such as squish and swirl can be obtained only with multidimensional models [275].

4.9 ENGINE KNOCK

In homogeneous-charge spark-ignition engines, large-amplitude pressure waves may develop rapidly under certain conditions (Fig. 4.46). This phenomenon is called knock and is detrimental to the engine efficiency, engine life, and automobile drivability [276]. The details of the onset of knock are not well known. Some researchers believe that knock is a detonation wave triggered by shock waves emanating from the flame front that result in an abnormally fast acceleration of the flame [277].

In this section we first describe some characteristics of hydrocarbon explosions and then present some of the models that have been used to predict engine knock.

4.9.1 Hydrocarbon Explosions

Hydrocarbons exhibit certain combustion charactersitics that are consistent with explosion limit curves. In particular, hydrocarbons (1) exhibit induction times that are followed by a very rapid reaction rate, (2) form aldehydes that shorten the ignition delay time, (3) (except for methane and ethane) exhibit cool flame phenomena, (4) exhibit negative temperature coefficients, (5) exhibit two-stage ignition phenomena, and (6) cause explosions without appreciable self-heating, which seem to be due to branched chain reactions [278].

The induction period of hydrocarbons can be explained by assuming that there are long-lived species that form as intermediates and then undergo different reaction paths according to the temperature. These reaction paths are of two types: non-chain-branching and chain-branching. At high temperatures the non-chain-branching path is faster than the chain-branching path. The negative coefficient of the reaction rate occurs when the temperature is increased and the reaction path changes from the chain-branching to the non-chain-branching type.

Knock is currently believed to be the result of the spontaneous thermal ignition of a certain amount of fuel-air mixture in the combustion chamber before it can be consumed by a flame propagating through the cylinder charge (Fig. 4.47). Engine knock provides a limit to the improvements in combustion efficiency, which would otherwise be achieved by increasing the compression ratio.

Engine knock can be related to cool flames and two-stage ignition phenomena at low temperatures and to a single-stage ignition process at high temperatures. A thorough understanding of the thermokinetic process of autoignition of hydrocarbon fuel-air mixtures under engine conditions may contribute to ameliorating engine knock. At moderate compression ratios the temperature of the end gas may be 600 K–800 K, and knock may be related to two-stage ignition and cool flame phenomena [278].

Gray et al. [279] showed that the combustion of acetaldehyde exhibits cool

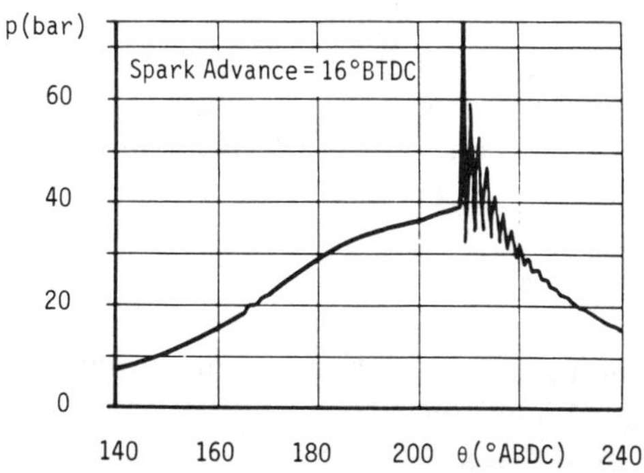

Figure 4.46 Schematic of the pressure diagram in a knocking engine.

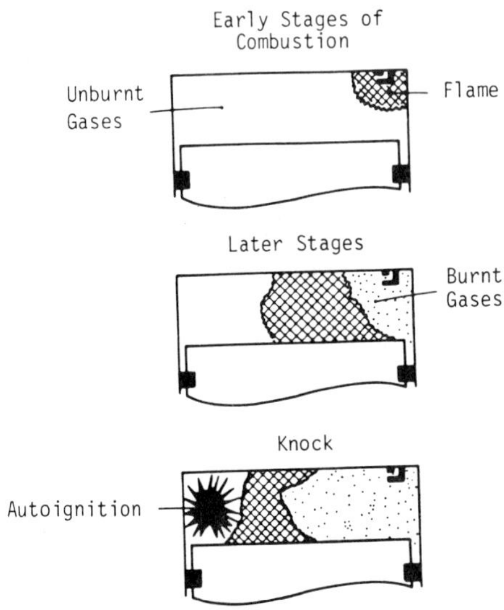

Figure 4.47 Graphic depiction of engine knock.

flames and ignition phenomena. In cool flames the maximum temperature rise is less than 200 K, intermediates such as hydrogen peroxide, methanol, and formaldehyde are formed, and the hydrocarbon oxidation is never completed; in ignition phenomena, however, the maximum temperature rise is greater than 200 K, molecular intermediates are destroyed, and the reaction is completed.

Kono et al. [280] used a fast Fourier transform to analyze pressure records from knocking engines. They showed that the gas vibrational energy originates from the difference between the displacement work of two types of burnt gases that are produced by normal flame propagation and spontaneous ignition. Cheng and Oppenheim [281] showed that strong ignition in methane-hydrogen mixtures depends on the sensitivity of the induction time to rate of temperature change.

Ferraro et al. [282] developed a knock limit detector with a knock intensity meter, and Patel and Henein [282] studied the knocking characteristics of methanol-indolene air mixtures in a CFR engine. Their studies showed that as the methanol concentration is increased, the ignition delay, the combustion duration, and the minimum advance for mean best torque (MBT) decrease, but the flame speed, the burning velocity, and the knock-limited compression ratio increase.

Czerwinski [284] showed that combustion oscillations in divided-chamber spark-ignition engines are caused by the turbulence generated in the prechamber nozzle due to rapid combustion. He also showed that oscillatory combustion and knock are different phenomena. Oscillatory combustion is caused by rapid inflammation and combustion rate due to the interactions between the flame front, pressure waves, and combustion chamber walls; the pressure waves can be amplified by resonance, high combustion rates, rapid inflammation, and turbulence. Oscillatory combustion is more intense in the prechamber and may occur in open-chamber engines at high combustion rates.

The results of Czerwinski [284] showed that knock occurs in the main chamber and is independent of the oscillatory combustion. They also showed that the influence of turbulence on knock cannot be defined universally.

Girard et al. [285] demonstrated that a cavity of small volume in the wall of a cylindrical combustion chamber is responsible for the occurrence of knock during the last phase of flame propagation. Their schlieren photographs and pressure records indicate that the fast end-gas burning is triggered by the turbulence generated by the unsteady jet issuing from the small cavity as the flame front approaches the combustion chamber wall. Girard et al. [285] found that the main factors that influence knock are the cavity diameter, the location of the cavity with respect to the spark plug, and the ratio of specific heats.

Gupta and Bracco [286] showed numerically that if the prechamber pressure in a divided-chamber engine increases over a time that is short compared with the acoustic time in the main chamber, a pressure imbalance results, which causes pressure oscillations. The decay of these oscillations depends on the throat Mach number and on the strength and extent of the first compression wave in the main chamber. The amplitude of the oscillations increases with the engine rpm and combustion rates in the prechamber and with decreasing throat area.

4.9.2 Phenomenological Models

The chemical reactions that lead to autoignition require a finite time to occur. This time is a complicated function of the mixture pressure, temperature, and equivalence ratio and of the fuel type. If the physical state of the mixture is constant, this time is called ignition delay. Rapid-compression machines can approximate closely the condition of constant physical state and can thus be used to measure ignition delay. However, the finite compression rate of the unburnt gases in a spark-ignition engine places an inherent limitation on the application of ignition delay formulas to prediction of engine knock.

Empirical relations for ignition delay time can be used to correlate autoignition-induced knock as follows. A global reaction is assumed to produce the chemical species necessary for autoignition. The rate of this reaction can be expressed as

$$\frac{d[x]}{dt} = f(T, p, \phi, \text{fuel}) \tag{200}$$

where $[x]$ denotes the concentration of the species necessary for autoignition.

If it is assumed that Eq. (200) produces a sudden transition to a process with a much faster reaction rate that rapidly completes combustion and that such a transition occurs at a critical concentration $[x_c]$, one can write

$$\frac{d([x]/[x_c])}{dt} = g\left(\frac{t}{\tau}\right) \tag{201}$$

where $[x_c]$ is constant in rapid-compression-machine experiments for a given fuel-air ratio and τ is the ignition delay time.

Equation (201) can be integrated to calculate the time t_k at which autoignition occurs and $[x] = [x_c]$, that is,

$$\int_{t(0)}^{t_k} g\left(\frac{t}{\tau}\right) dt = 1 \tag{202}$$

where $t_k - t(0)$ is the ignition delay time.

The function g cannot be obtained from ignition delay data [287]. However, if it is assumed that the reaction rate [Eq. (201)] does not change with time during a fixed-state process, then

$$g\left(\frac{t}{\tau}\right) = \frac{1}{\tau} \quad \text{and} \quad \int_{t(0)}^{t_k} \frac{dt}{\tau} = 1 \tag{203}$$

Equation (203) can be used to correlate engine knock with end-gas pressures and temperatures; it can also be used to predict the ignition delay in diesel engines.

Some of the induction time relationships used in the literature are [288–290]:

$$\tau = 0.017095\, p^{-1.49} \exp\left(\frac{7457}{T}\right) \qquad (7 \text{ atm} \leq p \leq 34 \text{ atm},\ 640 \text{ K} \leq T \leq 860 \text{ K}) \tag{204}$$

$$\tau = 0.916 \times 10^{-12}\, \phi^{-1.04}\, p^{-2.57} \exp\left(\frac{32{,}400}{T}\right) \qquad (1.25 \text{ atm} \leq p \leq 2 \text{ atm},\ 850 \text{ K} \leq T \leq 900 \text{ K}) \tag{205}$$

$$\tau = 10^{-3}\left(\frac{p}{\tilde{R}T}\right)^{-0.72} \exp\left(-9.387 + \frac{16{,}310}{T}\right) \qquad (1 \text{ atm} \leq p \leq 4 \text{ atm},\ 1200 \text{ K} \leq T \leq 1700 \text{ K}) \tag{206}$$

$$\tau = 19.75\left(\frac{\text{ON}}{100}\right)^{3.4107} p^{-1.7} \exp\left(\frac{3800}{T}\right) \tag{207}$$

where ON is the octane number of the fuel.

By et al. [287] used the Blizard-Keck (BK) model and Woschni's heat transfer correlation (see Chapters 2 and 3) to predict the temperature of the burnt and unburnt gases as follows

$$p\, v_b = R_b\, T_b \qquad p\, v_u = R_u\, T_u \tag{208}$$

$$\frac{V}{m} = x\, v_b + (1 - x)\, v_u \qquad v_b = \frac{V_b}{m_b} \qquad v_u = \frac{V_u}{m_u} \tag{209}$$

$$\frac{dE_u}{dt} = h_u \frac{dm_u}{dt} - p \frac{dV_u}{dt} - \frac{dQ_u}{dt} \tag{210}$$

$$\frac{dE_b}{dt} = h_u \frac{dm_b}{dt} - p \frac{dV_b}{dt} - \frac{dQ_b}{dt} \tag{211}$$

where E is the internal energy and v and x denote the specific volume and mass fraction burned, respectively.

By et al. [287] showed that a thermodynamic two-zone model can be used to predict the time of knock if the charge fraction involved in knock is assumed to be related to the knock intensity. However, the utility of a pressure-temperature threshold for defining the knock limit for a given fuel from a simple global relationship [Eq. (202)] is limited by the assumption that knock can be predicted by the end-gas state. This assumption is in fact an oversimplification of the complex autoignition phenomena that occur in spark-ignition engines. Furthermore, By et al. [287] had to modify the ignition delay time expressions in order to get reasonable agreement with experimental data; the best overall agreement between predictions and experimental data was obtained with Eq. (207) [291] and bulk gas temperatures.

It must be pointed out that use of an induction time to predict autoignition [Eqs. (203)–(207)] is questionable. Autoignition depends on several reaction paths and the state of the cylinder end gas (i.e., pressure, temperature, and composition); it also depends on the compression rate and the way that ignition is achieved (e.g., shock waves). It should be expected that different compression methods (e.g., piston compression, shock waves, etc.) may involve different chemical paths and may result in different induction times.

Equation (203) has also been used to predict the ignition delay time in diesel engines and has been generalized by Walsh and Cheng [292], who studied the ignition delay time of dodecane heated beyond its critical temperature at a compression ratio of 18.2 : 1.

Walsh and Cheng [292] assumed that the radical concentration r can be expressed as

$$\frac{dr}{dt} = \frac{r_0}{\tau_0} + \frac{r}{\tau_b} \tag{212}$$

where r_0/τ_0 and r/τ_b denote radical generation and branching, respectively.

Equation (212) can be integrated, under the assumption that r_0, τ_0, and τ_b are constant, to yield

$$r = r^* \left[\exp\left(\frac{t}{\tau_b}\right) - 1 \right] \tag{213a}$$

$$r^* = r_0 \frac{\tau_b}{\tau_0} \tag{213b}$$

where r^* denotes a critical radical concentration that can be achieved in a time t_d given by

$$r^* = \int_0^{t_d} \frac{dr}{dt}\, dt \tag{214}$$

t_d is the ignition delay time, and dr/dt can be expressed as

$$\frac{dr}{dt} = K \exp\left(-\frac{E}{\tilde{R}T}\right) \tag{215}$$

The mixing of the ambient air with the vaporizing fuel can be used to calculate T by means of the following expression:

$$\frac{dT}{dt} = \frac{T_a - T}{\tau_m} \tag{216}$$

where τ_m is a mixing time.

Substitution of Eqs. (213*b*), (215), and (216) into Eq. (214) yields

$$1 = \int_0^{t_d} a \exp\left(-\frac{E}{\tilde{R}T}\right) dt \qquad a = \frac{K\tau_m}{r^*} \tag{217}$$

Equation (216) can also be integrated to yield

$$T = T_{\text{inj}} + (T_a - T_{\text{inj}})\left[1 - \exp\left(-\frac{t}{\tau_m}\right)\right] \tag{218}$$

where T_{inj} is the injection temperature and T_a is the temperature of the surrounding air. Equations (217) and (218) indicate that the ignition delay time decreases with the charge air temperature.

Walsh and Cheng [292] observed that if the induction time is small, the premixed burning in diesel engines could not be detected from the heat release data and that the mixing-controlled phase of combustion is independent of the fuel temperature because during that phase, combustion is not affected by turbulence. They also found that the mixing-controlled phase of combustion is not very sensitive to the ignition delay time.

Equations (204)–(207) do not account for the effects of combustion chamber deposits on knock. It is important to indicate the similarities between deposit-induced surface ignition and the mechanisms by which depsoits influence knock. In deposit-induced ignition the local thermal and chemical interactions between the deposits and the fuel-air mixture can lead to ignition at locations and times different from those of the spark. The reactions can be initiated at a short distance away from the deposit surface if oxygen is consumed by the deposits [293]. For inert deposits the mechanism is the same as for local hot spots such as exhaust valves and spark plugs in deposit-free engines, there is no consumption of oxygen, and the air-fuel mixture is uniform throughout the combustion chamber. Unlike autoignition, surface ignition leads to the development of a flame front as in ignition from spark plugs [294]. Surface ignition prior to spark, also called preignition, may lead to abnormally high pressures and temperatures, which in turn may lead to knock. Deposits also increase the octane requirement of an engine. The concept of octane requirement increase (ORI) can be employed as a measure of the increase in fuel octane rating required in an engine due to combustion chamber deposit effects on knock propensity.

Octane requirement increase in an engine is believed to be the result of three different mechanisms: cylinder volume reduction by deposits, cylinder thermal insulation, and chemical processes induced by the deposits. Cylinder volume reduction and thermal insulation of the walls lead to higher end-gas pressures and temperatures and make knock more likely. Dumont [295] concluded that the thermal insulating effect is most dominant, with perhaps 20% to 40% of the ORI attributable to the cylinder volume reduction effect.

Anderson and Prakash [296] developed a one-dimensional model to analyze heat transfer through deposits, using experimental temperature conditions and assuming that the thermal conductivity of the deposits is $\lambda = AT^B$, where A and B are constants. They investigated the temperature amplitude and time lag at the wall-deposit interfaces and showed that two thermal capacities (ρc, where c is the specific heat) are needed to accurately predict both the temperature at the wall-deposit interface and the time lag.

The deposit structure and the buildup rate are expected to be a function of the engine-operating conditions, gasoline composition, and lubricating oil composition. Aromatics are considered to be responsible for a deposition rate that is high compared with that of paraffins. High-boiling-point components build up deposits more quickly than lower-boiling-point HC, and the deposit thickness is a function of the fuel and lubricating additives and decreases with the engine load. Furthermore, the deposits degrade on the combustion chamber walls, and their molecular structure is not well known.

DeGregoria [297] used Woschni's heat transfer correlation, a cosine law for the mass fraction of burnt gases [Eq. (7)], a thermodynamic two-zone model, and an induction time correlation to analyze engine deposits and their influence on the ORI. His induction time can be expressed as

$$\tau = 0.01869 \left(\frac{\text{RON}}{100}\right)^{3.4017} p^{-1.7} \exp\left(\frac{3800}{T}\right) \tag{219}$$

where RON is the research octane number.

DeGregoria [297] assumed one-dimensional heat transfer through the deposit layer and cylinder wall. However, his analysis neglects the chemical effects of the deposits on the fuel-air mixture (these effects may include heterogeneous reactions), the possible motion of the fuel-air mixture through pores in the deposits, and the anisotropy of the deposits. These effects are poorly understood at the present time.

Bush et al. [298, 299] used a one-dimensional time-dependent approach to model end-gas knock in a rapid-compression machine. They considered the propagation of a nonisorbaric laminar flame propagating through a premixed combustible mixture in the asymptotic limit of large Arrhenius activation energies where a thin flame separates the burnt and unburnt gases. They showed that expansion of the hot burnt gases results in compression heating of the unburnt gases, which may undergo spatially homogeneous autoconversion to combustion products prior to arrival of the propagating flame. If such an explosion is too rapid for acoustic flow adjustments, large spatial differences in pressure arise, and the resulting nonlinear waves produce audible knock.

The formulation used by Bush et al. [298] is not adequate to account for acoustic phenomena because the pressure was assumed to be uniform throughout the combustion chamber. However, their formulation shows that fuel autoconversion can be inhibited by increasing the heat transfer from the end gas; too large heat transfer from the end gas could result in flame quenching. Another possible strategy for inhibiting engine knock is to use fuel additives that inhibit end-gas autoignition without impeding flame propagation. Tetraethyl lead inhibits engine knock but poisons the catalysts used to treat the exhaust gases.

Carrier et al. [300] developed a three-zone model of nonisobaric combustion in a variable-volume enclosure that accounts for burnt gases, unburnt gases in the normally cooled region, and unburnt gases in the augmented cooled region. Their one-dimensional unsteady model shows that augmented heat transfer in the end gas can prevent high-temperature autoignition. However, if the conversion of the end gas to combustion products is much faster than the acoustic-adjustment time, spatial pressure nonuniformities result, producing an audible knock.

Maly and Ziegler [301] predicted engine knock with a thermodynamic model based on an energy balance and a quasistationary flame. Their results show that feedback causes flame acceleration due to turbulence enhancement and temperature enhancement of the end gas. Tabaczynski [302] suggested that the end gas is a mixture of burnt and unburnt gases, the burning rate of which is controlled by diffusion with no entrainment. Compact combustion chambers increase the heat transfer from the end and produce less knock.

Ferguson et al. [303] used a thermodynamic two-zone model to predict the unburnt-gas temperature in knocking and nonknocking engines. The model accounts for the presence of boundary layers and uses algebraic instead of differential equations of state. The model predictions and unburnt-gas temperatures agree well with the temperatures measured during the compression stroke. During combustion the model underpredicts the unburnt-gas temperatures.

The model employed by Ferguson et al. [303] uses a cosine Wiebe function for the mass fraction of burnt gas [Eq. (7)], and the mixture is divided into burnt and unburnt zones, so that the following equations apply:

$$\frac{V}{m} = (1 - x)v_u + xv_b \tag{220}$$

$$\frac{E}{m} = e = (1 - x)e_u + xe_b \tag{221}$$

where

$$e_b = e_b(T_b, p) \qquad e_u = e_u(T_u, p) \qquad v_u = v_u(T_u, p) \qquad v_b = v_b(T_b, p) \tag{222}$$

and e is the specific internal energy.

The mass m at any crank shaft angle is calculated assuming that blowby is maximum at TDC and zero at bottom dead center (BDC). Ferguson et al. [303] used the following expression to calculate the blowby:

$$m_{bb} = m_0 \exp\left[-2B\left(1 - \cos\frac{\theta - \pi}{2}\right)\right] \tag{223}$$

where m_0 is the mass at BDC and B is a constant that was determined in engine tests as a function of the cycle-averaged blowby rate. Equation (223) can be differentiated with respect to time to calculate the blowby flow rate.

The enthalpy of the gases leaking past the first ring is composed of unburnt gases early in the combustion period and burnt gases later in the period; it can be written as

$$h_l = (1 - x^n)\, h_u + x^n h_b \tag{224}$$

where h_l denotes the enthalpy of the leaking gases and n depends on the engine and rings.

The burnt and unburnt gases lose heat to the walls. Calculation of these heat transfer losses depends on the areas wetted by the burnt and unburnt gases, which can be easily calculated in the case of spherical flame propagation. However, one can estimate the wetted areas of the burnt and unburnt gases as

$$A_b = Ax^{1/2} \qquad A_u = A(1 - x^{1/2}) \tag{225}$$

where A is the total wall area.

If one-dimensional steady-state heat transfer is assumed through the boundary layers and combustion chamber walls, the heat transfer rates in the burnt and unburnt gases can be calculated as

$$\dot{Q}_b = h_b A_b \frac{T_b - T_c}{1 + R_w h_b A_b} \tag{226}$$

$$\dot{Q}_u = h_u A_u \frac{T_u - T_c}{1 + R_w h_u A_u} \tag{227}$$

where the wall temperature is

$$T_w = T_c + R_w\,(\dot{Q}_b + \dot{Q}_u) \tag{228}$$

R_w is the conduction heat transfer resistance of the solid walls, and T_c is the coolant temperature.

The mass contained in the cylinder at BDC can be calculated as

$$m_u = \frac{\Gamma m_i}{1 - \alpha} \tag{229}$$

where m_i is the mass inducted into the cylinder, α is the recycle fraction, and Γ is the trapping efficiency.

Ferguson et al. [303] accounted for crevice quenching as follows. Early in the combustion mostly unburnt gases are pushed into the crevices, whereas later on, burnt gases are pushed into the crevices. The crevices have a large surface-to-volume ratio, and the gases pushed into them rapidly equilibrate thermally to the wall temperature. If the gases contained in the crevices react, they release no heat since heat is efficiently transferred to the walls. In automobile engines the crevice volume, typically 1 cm^3, is not critical to the energy balance, but may be an important source of unburnt HC emissions (see Section 4.6).

In the expansion stroke the gases in the crevices reemerge into the cylinder and a substantial amount is oxidized. However, the heat release from this process is small.

If the fraction of gases pushed into the crevice volume is estimated as the enthalpy of the blowby gases [Eq. (224)], the fraction of crevice gases that enters unburnt is given by

$$\beta = \frac{\int (1 - x^n)\dot{m}_{cv}\, dt}{\int \dot{m}_{cv}\, dt} \tag{230}$$

where

$$\dot{m}_{cv} = \frac{V_{cv}}{RT_{cv}} \frac{dp}{dt} \tag{231}$$

The subscript cv denotes the crevice volume, and $\dot{m}_{cv}$ is the mass flow rate into the crevices. Note that Eq. (230) can be used to estimate the fraction of the gases pushed into the crevice volume that are burned.

The fractional pressure rise due to combustion is nearly equal to the mass fraction burned, especially for fast-burn engines [304], that is,

$$\beta = 1 - \frac{1}{\Delta p} \int x^n\, dp = \frac{n}{n + 1} \tag{232}$$

so that the total mass of unburnt charge in the crevices at the end of combustion is

$$m_{cv,u} = m_{cv,0} + \beta\, \Delta m_{cv} \tag{233}$$

where Δm_{cv} can be calculated from Eq. (231) and $m_{cv,0}$ is the mass in the crevice volume at the start of combustion, which can be estimated by calculating the density of the gases in the crevice using the wall temperature.

Ferguson et al. [303] calculated the mean squared and the squared mean gas temperatures, assuming that the fluid is a core at uniform temperature surrounded by boundary layers in which the temperature rises according to a 1/7 power law with the distance from the wall. Ferguson et al. [303] expanded in Taylor series the specific internal energy and specific volume of the burnt gases about the mass-averaged temperatures of the burnt gases; they accounted for the boundary layer thickness and crevice volume. One of the major drawbacks of their formulation is that their estimates require the value of the boundary layer thickness; however, their formulation yields algebraic equations for the specific internal energy and specific volume of the burnt gases.

4.9.3 Shell Model

The Shell model of engine knock gives a clear indication of how common chemical mechanisms underlie alkane oxidation at low pressures in batch reactors and at the high pressures of rapid-compression machines. The model seeks a common explanation of the ignition phenomena in these two regimes. At low pressures in batch reactors the model accounts for (1) the autocatalytic nature of slow combustion, (2) the existence of a region of negative temperature coefficients for the maximum rate of the slow combustion phenomena, (3) cool flames, (4) the occurrence of hot ignition, and (5) the existence of well-defined limits in a $p-T$ diagram between these three phenomena, the shape and disposition of these limits, and their dependence on stoichiometry [305].

The underlying chemical reactions involved in the model reflect (1) the early importance of conjugate olefin formation; (2) the formation of ring ethers, aldehydes, and ketones, which is consistent with peroxyalkil radical isomerization and decomposition mechanisms; and (3) the unselective nature of radical attack on the parent fuel.

In enginelike conditions the model explains (1) the two-stage nature commonly observed in autoignition phenomena and (2) the dependence of the induction period on pressure, temperature, and mixture stoichiometry.

The Shell model is of a thermokinetic nature and includes a degenerated branched reaction [306], which does not necessarily correspond to a detailed chemical mechanism. The model is not based on detailed kinetic mechanisms that involve many elementary species and chemical reactions, but it does contain a simplified mechanism that retains the essential features of exothermic degenerated branch chain reactions in a "lumped" manner [307]. These degenerated branch chain reactions are responsible for autoignition in the end-gas and for knock.

The thermokinetic nature of autoignition phenomena has been established in continuous stirred-tank reactors (CSTRs) at atmospheric pressure by Lignola et al. [308] under fuel-rich conditions. These researchers used different temperatures and residence times in CSTRs and showed that chemical kinetics play a major role in autoignition. In spark-ignition engines, high-speed cinematography has shown a sharp rise in light emission from the gas behind the flame front in knocking engines [309]. High-speed samples of the end gas in a cycle support the thermokinetic theory of autoignition and indicate that the chemical reactions involved in cool flame oxidation play an important role in engine knock [310].

Gibson et al. [311] studied acetaldehyde oxidation with a 25-chemical reaction mechanism in a well-stirred flow reactor and showed the presence of cool flames and complex multistage ignition phenomena. Their model includes initiation and low-temperature degenerate branching reactions, and it shows that thermokinetic interactions are responsible for the overall negative temperature-coefficient rate and oscillatory cool flames. Their model also shows that chain branching and ignition instability are due to reactions of the cool flame products.

Gibson et al. [311] showed that the origins of cool flames and degenerate reactions are the methyl radical oxidation and methyl hydroperoxide, respectively, whereas the precursor of hot ignition is formaldehyde. Chain branching was augmented by hydrogen peroxide decomposition, whereas the heat release in later stages is due primarily to the reaction between CO and OH.

The Shell model was developed by Halstead et al. [312] for high pressures and temperatures. It simulates the two-stage ignition of alkanes, which is characterized by a well-defined cool flame that is often quenched rapidly and completely before the onset of a sharp ignition. The model is based on a degenerate-branched-chain mechanism and contains two termination steps, one linear and the other quadratic in the radical concentration. Halstead et al. [313] used the Shell model to predict the autoignition of HC fuels at high temperatures and pressures and fitted the model with experimental data. Their results suggest that a degenerate-branching agent is formed by a radical reaction whose kinetic competition with the alternate main chain propagation reaction exhibits a first-order dependence on oxygen concentration rather than on fuel concentration.

The Shell model was used by Natarajan and Bracco [314] in their autoignition studies of spatially uniform charges in constant-volume bombs and rapid-compression machines. The Shell model predicted the experimental trends observed in cool flames of isooctane-air and *n*-heptane-air mixtures in CSTRs.

Natarajan and Bracco [314] also used the Shell model to analyze autoignition in a two-dimensional engine model and found that the structure of the finite thickness flames determines whether the flame is accelerated by preflame reactions or not. They also obtained good agreement with experimental data when the flame was not accelerated by preflame reactions.

Natarajan and Bracco [314] concluded that single-stage ignition is sufficient to model strong knock; however, single-stage ignition is inadequate to model engine cycles of marginal knock. These cycles are the more important ones in practice and are characterized by cool flames that precede and delay the onset of knock.

The Shell model can be written as [313]

$$\text{Initiation: } RH + O_2 \rightarrow 2R' \qquad (K_q) \tag{234}$$

$$\text{Propagation: } R' \rightarrow R' + \text{products} + \text{heat} \qquad (K_p) \tag{235}$$

$$\text{Branching agent formation: } R' \rightarrow R' + B \qquad (F_1K_p) \tag{236}$$

$$\text{Branching agent formation: } R' + Q \rightarrow R' + B \qquad (F_2K_p) \tag{237}$$

$$\text{Linear termination: } R' \rightarrow \text{termination} \qquad (F_3K_p) \tag{238}$$

$$\text{Propagation forming } Q\text{: } R' \rightarrow R' + Q \qquad (F_4K_p) \tag{239}$$

$$\text{Quadratic termination: } R' \rightarrow \text{termination} \qquad (K_t) \tag{240}$$

$$\text{Branching: } B \rightarrow 2R' \qquad (K_b) \tag{241}$$

where RH is the hydrocarbon, R represents the radicals, B is the branching agent, Q is an autocatalytic product, and F represents fraction terms.

For an overall reaction of the form

$$\frac{1}{n+1} C_nH_{2n+2} + sO_2 \rightarrow \frac{zn}{n+1} CO + \frac{(1-z)n}{n+1} CO_2 + H_2O \tag{242}$$

where $s = [(3 - z)n + 1)]/2(n + 1)$ and $z = 0.67$, assuming that fuel is consumed primarily by chain propagation and that two atoms of hydrogen are abstracted per cycle in the formation of products. The heat release rate per unit volume can be expressed as:

$$\text{HRR} = K_p\, q[R] \tag{243}$$

where $[R]$ is the radical concentration and q is the heat release rate per cycle of the chain, which can be calculated from Eq. (242).

The concentration of radicals R and agents B and Q can be calculated from the following expressions [313]:

$$\frac{d[R]}{dt} = 2\{K_q[\text{RH}][\text{O}_2] + K_b[B] - K_t\,[R]^2\} - F_3K_p\,[R] \tag{244}$$

$$\frac{d[B]}{dt} = F_1K_p[R] + F_2K_p[Q][R] - K_b[B] \tag{245}$$

$$\frac{d[Q]}{dt} = F_4K_p[R] - F_2K_p[Q][R] \tag{246}$$

whereas those of oxygen and fuel can be calculated from the reaction rate for Eq. (242).

The chain propagation rate K_p can be calculated as

$$K_p = \left\{\frac{1}{K_{p_1}[\text{O}_2]} + \frac{1}{K_{p_2}} + \frac{1}{K_{p_3}[\text{RH}]}\right\}^{-1} \tag{247}$$

where

$$K_{p_i} = A_{p_i} \exp\left(-\frac{E_{p_i}}{\tilde{R}T}\right) \tag{248}$$

$$F_1 = F_{1,0}[\text{O}_2]^{x_1}[\text{RH}]^{y_2} \tag{249}$$

$$F_2 = F_{2,0} \tag{250}$$

$$F_3 = F_{3,0}[\text{O}_2]^{x_3}[\text{RH}]^{y_3} \tag{251}$$

$$F_4 = F_{4,0}[\text{O}_2]^{x_4}[\text{RH}]^{y_4} \tag{252}$$

$$F_{\text{j},0} = A_{f_{\text{j},0}} \exp\left(-\frac{E_{f_{\text{j},0}}}{\tilde{R}T}\right) \tag{253}$$

The values of the parameters used in the Shell model are shown in Table 4.1 [313].

Hirst and Kirsch [315] used a two-zone thermodynamic model where the mass burning rate was described empirically by the following equation:

$$\frac{dn_b}{dt} = \psi \cdot \text{BRFAC} \cdot \frac{r_p}{A_p} \cdot V_{\text{tot}} \cdot \sin(\pi x) \cdot \frac{n_u}{V_u} \tag{254}$$

Table 4.1 Parameters of the Shell model[a]

Parameters	Research octane number (RON)	
	90	100
A_{p_1}	1.0×10^{11}	1.0×10^{12}
E_{p_1}	0.0	0.0
A_{p_2}	1.0×10^{11}	1.0×10^{11}
E_{p_2}	1.5×10^4	1.5×10^4
A_{p_3}	1.0×10^{13}	1.0×10^{13}
E_{p_3}	8.5×10^2	8.5×10^2
A_q	1.2×10^{12}	3.96×10^{13}
E_q	3.5×10^4	4.0×10^4
A_b	4.4×10^{17}	6.512×10^{15}
E_b	4.5×10^4	4.0×10^4
A_t	3.0×10^{12}	3.51×10^{12}
E_t	0.0	0.0
$A_{f_{1,0}}$	7.3×10^{-4}	7.3×10^{-4}
E_{f_1}	-1.5×10^4	-1.5×10^4
$A_{f_{2,0}}$	1.8×10^2	1.8×10^2
E_{f_2}	-7.0×10^3	-7.0×10^3
$A_{f_{3,0}}$	1.47	2.205
E_{f_3}	1.0×10^4	1.0×10^4
$A_{f_{4,0}}$	1.88×10^4	1.7×10^4
E_{f_4}	3.0×10^4	3.0×10^4
x_1	1.0	1.0
y_1	0.0	0.0
x_3	0.0	0.0
y_3	0.0	0.0
x_4	−1.0	−1.0
y_4	0.35	0.35
n	7.9	8.0
m	8.9	9.0
q	9.4×10^4	9.4×10^4
p	1.0903	1.0911

[a] Preexponentials in centimeters moles, and seconds; activation energies in calories per mole.

where n_u and n_b are the number of moles of the unburnt and burnt gases, respectively; BRFAC is the burning rate coefficient; r_p is the piston radius; A_p is the piston area; V_{tot} is the combustion chamber volume; V_u is the volume occupied by the unburnt gases; $x = V_b/V_{tot}$; and V_b is the volume occupied by the burnt gases.

Hirst and Kirsch [315] used Annand's heat transfer correlation [316] to estimate the heat transfer losses in the burnt and unburnt gases as follows:

$$q_{w,u} = \alpha_u C_{p,u}\left(\text{rpm}\,\frac{n_u}{V_u}\right)^{0.7} \frac{V_u}{V_{tot}} A_T(T_u - T_w) \tag{255}$$

$$T_w = \frac{\pi r_p^2 (T_{ch} + T_p) + 2 V_{\text{tot}} T_{cw}/r_p}{A_T} \tag{256}$$

$$A_T = 2\pi r_p^2 + \frac{2 V_{\text{tot}}}{r_p} \tag{257}$$

$$q_{w,b} = \alpha_b \frac{V_b}{V_{\text{tot}}} A_T T_b^4 \tag{258}$$

where α_u and α_b are constants and the temperatures of the cylinder head, piston, and cylinder wall are $T_{ch} = T_p = 525$ K and $T_{cw} = 425$ K, respectively.

Hirst and Kirsch [315] used Eqs. (244)–(246) but with the left-hand sides replaced by $V^{-1}\, dn_R/dt$, $V^{-1}\, dn_B/dt$, and $V^{-1}\, dn_Q/dt$, respectively, and they calculated the concentrations of fuel and oxygen by means of the following expressions:

$$\frac{1}{V}\frac{dn_{\text{O}}}{dt} = -p\, K_p[R] \tag{259}$$

$$n_{RH} = \frac{n_{\text{O}} - n_{\text{O}}(t=0)}{pm} + n_{RH}\,(t=0) \tag{260}$$

where n_{O} and n_{RH} represent the number of moles of oxygen and of fuel, and the values of p and m are shown in Table 4.1 for 90- and 100-RON fuels. The values of n and m in Table 4.1 correspond to the number of carbon atoms in the fuel [Eq. (242)] and $m = n + 1$.

Schapertons and Lee [317] used the degenerate branch chain mechanism of the Shell model [313, 315] to analyze knock in a two-dimensional spark-ignition engine model, where the fuel consumption ratio can be expressed as

$$\frac{d[\text{RH}]}{dt} = -A \exp\left(-\frac{E}{\tilde{R}T}\right)[\text{RH}]^{0.5}[\text{O}_2]^{0.5} \tag{261}$$

and A is the preexponential factor.

The two-dimensional planar model proposed by Schapertons and Lee [317] was able to predict the knocking tendencies as a function of the bore-to-stroke ratio, spark position, swirl, and exhaust gas recirculation.

Cox and Cole [318] modified the Shell model to predict the dependence of the ignition delay time on the temperature, pressure, and oxygen concentration. Their model includes 10 species and 15 generalized chemical reactions, accounts for the isomerization of alkylperoxy radicals, and can be written as

$RH + O_2 = R + HO_2$	initiation	(262)
$OH + RH \rightarrow R + H_2O$	propagation	(263)
$R + O_2 \rightleftarrows RO_2$	propagation	(264)

$$RO_2 + RH \rightarrow RO_2H + H \quad \text{formation and decomposition of branching agent} \tag{265}$$

$$RO_2 + RCHO \rightarrow RO_2H + RCO \quad \text{intermediate} \tag{266}$$

$$R + R \rightarrow \text{nonradical} \quad \text{termination} \tag{267}$$

$$RO_2 + RO_2 \rightarrow ROH + RCHO + O_2 \quad \text{termination} \tag{268}$$

$$R + O_2 \rightarrow \text{olefin} + HO_2 \quad \text{termination} \tag{269}$$

$$RCH_3 \rightarrow \text{olefin} + CH_3 \quad \text{termination} \tag{270}$$

$$HO_2 + RH \rightarrow R + H_2O_2 \tag{271}$$

$$H_2O_2 + X \rightarrow 2OH + X \tag{272}$$

where X is a third body and Eqs. (271) and (272) involve hydrogen peroxide (H_2O_2), which is a promoter in the second-stage ignition.

4.9.4 Detailed Chemical Kinetics Modeling

Pitz and Westbrook [319] added a low-temperature mechanism to examine the importance of low-temperature reaction paths in the autoignition of *n*-butane, and they showed that the reactions that involve the production and consumption of HO_2 and H_2O_2 play a crucial role in high-pressure autoignition. The reaction mechanism used by Pitz and Westbrook [319] involves 144 high-temperature reactions and 82 low-temperature reactions, which were solved with the piston engine HCT code. Pitz and Westbrook [319] showed that antiknock additives that remove HO_2 and H_2O_2 inhibit autoignition.

Cernansky et al. [320] used detailed chemical kinetics schemes to study the oxidation of *n*-butane-air and isobutane-air mixtures preceding the end-gas autoignition, and they showed that the chemical differences between *n*-butane and isobutane can be traced to the distribution of radical species that results from the consumption of various types of butyl radicals. Cernansky et al. [320] obtained good agreement between the numerical results and experimental gas chromatography data. Cernansky et al. [320] analyzed end-gas samples before autoignition and showed that branched hydrocarbons are more resistant to autoignition than unbranched hydrocarbons.

Smith et al. [321] used Raman spectroscopy, laser-induced fluorescence, schlieren photography, and gas sampling to analyze the knock characteristics of a single-cylinder engine with a compression ratio equal to 5.2 fueled with propane, *n*-butane, and isobutane. In their experiments only *n*-butane and isobutane knocked. Smith et al. [321] observed that the preflame reactions involve OH and CH radicals and produce CO, formaldehyde, and a weak blue-light emission. Their detailed chemical kinetics mechanisms predicted the experimental trends, and the computed results showed little sensitivity to the low-temperature reactions.

Griffiths and Nimmo [332] also showed that *n*-butane and isobutane have different knocking characteristics and that isobutane is more resistant to knock than *n*-butane. They also showed that knock is a thermokinetic phenomenon that includes low-temperature (<900 K) reaction paths.

Cullis et al. [323] showed that organosilicon sompounds used as antiknock additives produce solid siliceous deposits, which lengthen the induction period. They also studied the lower two-state ignition of 2-methylpentane and showed that the two-stage ignition is related to the cetane number.

The importance of HO_2 and H_2O_2 in autoignition has also been demonstrated by Jachinowski [324], who used 83 chemical reactions and 27 species to study the combustion of propane in shock tubes and obtained good agreement with experimental data for temperatures in the range $1150 \text{ K} < T < 2600 \text{ K}$. Jachinowski [324] also showed that the nonlinearity of the ignition delay time with decreasing temperatures can be interpreted in terms of the increased importance of the reactions involving HO_2 and H_2O_2 at low temperatures.

Westbrook [325] studied the autoignition of methane-air and propane-air mixtures by means of detailed chemical kinetics mechanisms, which include low-temperature oxidation ($500 \text{ K} \leq T \leq 800 \text{ K}$), cool flames and negative temperature coefficients, intermediate temperature kinetics, and high-temperature kinetics. He also studied the autoignition of *n*-octane and isooctane, and he showed that the reactions that occur in the cool flame region ($T < 700$ K) may have a sensitizing influence on the subsequent reactive history of the end gases.

Kaiser et al. [326] studied experimentally and numerically the oxidation of acetaldehyde at temperatures of 533 K and 713 K in a low-pressure reactor. Their experimental results were used to construct and validate a detailed chemical kinetics mechanism, which was used to examine acetaldehyde oxidation in the negative temperature regime between 550 K and 900 K. The overall rate of reaction and the properties of the negative temperature coefficient were found to be sensitive to the competition between radical decomposition reactions and the addition of molecular oxygen and methyl radicals, including

$$CH_3 + O_2 + X = CH_3O_2 + X \tag{273}$$

$$CH_3CO + X = CH_3 + CO + X \tag{274}$$

$$CH_3CO + O_2 = CH_3CO_3 \tag{275}$$

The importance of the pressure-dependent reactions [Eqs. (273) and (275)] may have serious implications for kinetic modeling of engine knock where the pressure can approach $30 \text{ atm} < p < 40 \text{ atm}$. Such pressures can shift the negative temperature coefficient to temperatures higher than 850 K.

The experimental and numerical studies of Kaiser et al. [326] show that since the end gases subject to knock experience temperatures below 1000 K for a large fraction of time in a typical engine, it is necessary to include the competitive reaction paths [Eqs. (273)–(275)] in any kinetic-modeling study of engine knock.

Axelsson et al. [327] developed detailed chemical kinetics mechanisms for the oxidation of *n*-octane and isooctane; they showed that elements of particular importance include site-specific abstraction of H atoms, radical isomerization of alkyl radicals by internal H atom abstraction, and rapid β scissions of the alkyl radicals. Comparisons between numerical and experimental data showed that the current octane oxidation mechanisms need further improvements and refinements in their treatment of the C_2–C_4 olefins and other unsaturated hydrocarbon species.

Green et al. [328] studied the autoignition of isobutane, which is a relatively knock-resistant fuel, and showed that there is a significant amount of fuel breakdown in the end gas, with resulting heat release and temperature increases. Their numerical and experimental results indicate that a reasonable amount of low-temperature chemical activity precedes autoignition. These low-temperature reactions or cool flames may produce reactive intermediate chemical species (presensitization of the fuel-air mixture) or heat. Although relatively small, this heat results in a quicker attainment of high-temperature ignition.

The numerical and experimental results of Green et al. [328] indicate that low-temperature chemical activity is not important as a means of preconditioning the end gas for autoignition and that these chemical processes affect the onset of autoignition through heat release, which raises the end-gas temperature to levels that support high-temperature ignition. Chemical kinetics studies [328] have shown that autoignition is very sensitive to reactions involving HO_2 and that autoignition times are very sensitive to reaction paths resulting from the addition of O_2 to alkyl radicals, but only moderately sensitive to alkylperoxy isomerization reactions.

Green and Lucht [329] used spontaneous Raman scattering and coherent anti-Stokes Raman spectroscopy (CARS) to measure the end-gas temperature in a spark-ignition optical research engine. High swirl and multipoint ignition were used to confine the end gas in the center of the combustion chamber, whereas the manifold pressure and temperature were used to control the knock responses. Green and Lucht [329] showed that knock appears to be the result of a nearly homogeneous and simultaneous autoignition of the end gas. They also showed that higher gas-end temperatures under knocking conditions are due to higher wall temperatures.

Green and Smith [330] used Raman and emission spectroscopy, laser-induced fluorescence, schlieren photography, and gas sampling to study preflame conditions and reactions that lead to knock in a single-cylinder spark-ignition engine. The intake manifold pressure and temperature were used to control the fraction of fuel autoigniting in a 5.1-compression-ratio engine. Knock in *n*-butane-air and isobutane-air mixtures was found to be the result of a dramatic increase in the heat release caused by the nearly homogeneous autoignition of the compression-heated end-gas. Detonation waves could not be detected and do not appear to be necessary for the occurrence of knock. However, preflame reactions involving OH and CH radicals and producing carbon monoxide, formaldehyde, and faint blue light were observed.

Lucht et al. [331] obtained broadband CARS measurements of unburnt-gas temperatures in an internal combustion engine fueled with *n*-butane and air and operated under knocking and nonknocking conditions. Their experimental data agree well with the predictions of a heat release model until TDC, when the CARS-measured temperatures were higher than those predicted by the heat release model. These differences suggest the development of temperature gradients in the core gases during compression and are very significant because autoignition probably begins in the unburnt-gas region that is at the highest temperature. Multidimensional calculations [332] in motored engines have shown that the temperature of the mixture core is uniform during most of the compression stroke but near TDC, axial temperature gradients develop. This sudden development has not been observed in other multidimensional calculations and could be explained by accounting for the interactions of the thermal boundary layers. However, it is questionable that such interactions may suddenly develop temperature nonuniformities in the combustion chamber, and the numerical results presented in reference 332 may be an artifact of the numerical algorithm and/or lack of numerical convergence. Therefore, these numerical results should not be used to justify the small differences between the measured and predicted temperatures after ignition observed by Lucht et al. [331].

Leppard [333] used a detailed chemical kinetics mechanism for the combustion of ethane-air mixtures, and he used experimentally determined pressure and temperature profiles to drive the chemical reactions. His model predicted engine knock within 2°CA compared with available experimental data.

Figure 4.48 presents a comparison between CARS temperature measurements and the predictions of a heat release model [303] in knocking and nonknocking engines [331]. Figure 4.49 presents the end-gas temperatures measured at the center of the combustion chamber by spontaneous Raman spectroscopy in knocking and nonknocking engines operating with stoichiometric *n*-butane-air mixtures at 600 rpm [330]. Figure 4.50 illustrates the time histories of formaldehyde, acetaldehyde, and acetone near the autoignition limit for an engine operating with a stoichiometric isobutane-air mixture at 600 rpm [328].

4.9.5 Antiknock Compounds

Li and Simmons [334] analyzed the distribution of gaseous lead compounds in the exhaust of a petrol system and showed that Pb(*g*) is present as PbO(*g*) to at least 1200 K. In rich mixtures, lead is present as Pb(*g*) but is converted to PbO(*g*). Li and Simmons [334] also showed that the exhaust gas becomes supersaturated with PbO(*g*) before lead halide is formed, and solid deposits are composed of PbO(*s*).

Haskell and Kiovsky [335] showed that titanium-based antiknock compounds are effective only if they vaporize before reacting with water. Salooja [336] showed that the effect of lead monoxides is to destroy peroxides completely, to

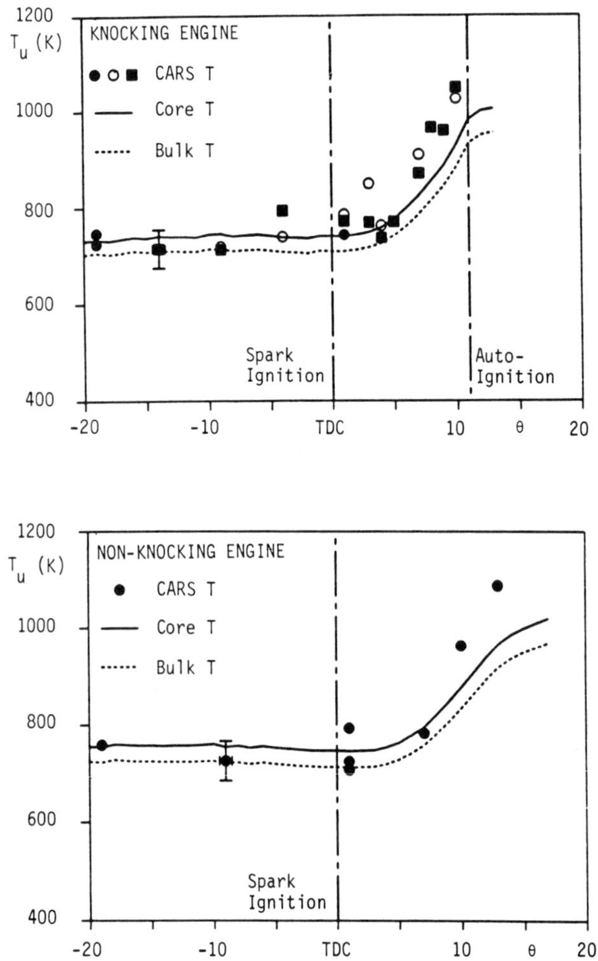

Figure 4.48 Comparisons of CARS temperature measurements and the predictions of a heat release model in knocking (top) and nonknocking (bottom) engines [331].

decrease the concentration of carbonylic compounds, and to increase that of the hydrocarbons.

4.10 IGNITION

The ignition of a combustible mixture by a spark discharge may be considered to proceed in three stages. In the first stage of the breakdown process, the electrical potential across the electrode gap is increased until a breakdown on the intervening mixture occurs. Ionizing streams then propagate from one electrode to the other. When the ionizing streams reach the opposite electrode, the impedance of the gap decreases drastically, and the electrical current through the gap increases rapidly. This results in a very large temperature rise and the formation of a shock wave. The first stage lasts about 10 ns.

The breakdown phase is then followed by the arc phase, which is characterized by a high electrical current flowing through the mixture while the plasma kernel inflates and endothermic reactions occur, followed by exothermic reactions. The arc phase lasts about 0.1 ms. In the third phase the flame propagation approaches a steady value while the flame thickness increases at a declining rate to reach a steady-state value at a later time. The third phase lasts about 0.2 ms [337].

In order to accurately resolve the ignition process it is necessary to calculate the breakdown phase, which involves the formation of a plasma and of shock waves. However, these calculations are difficult. Maly and Vogel [338] have shown that the breakdown phase is characterized by a very high peak value of voltage (~10 kV) and a resultant current of about 200 A for a duration of 1–10 ns. At a very early stage a cylindrical channel about 40 μm in diameter develops, together with a pressure jump and a rapid temperature rise. Maly and Vogel [338] showed that an increase in breakdown energy does not manifest itself as higher kernel temperatures; instead the channel diameter increases, causing a larger activated-gas volume. Sher and Refael [339] estimated the hot

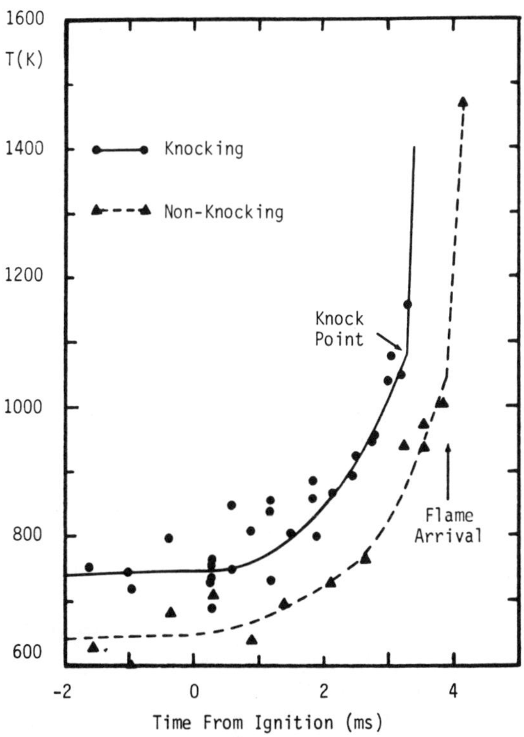

Figure 4.49 End-gas temperatures measured at the center of the combustion chamber by spontaneous Raman spectroscopy in knocking and nonknocking engines operating with stoichiometric *n*-butane-air mixtures [330].

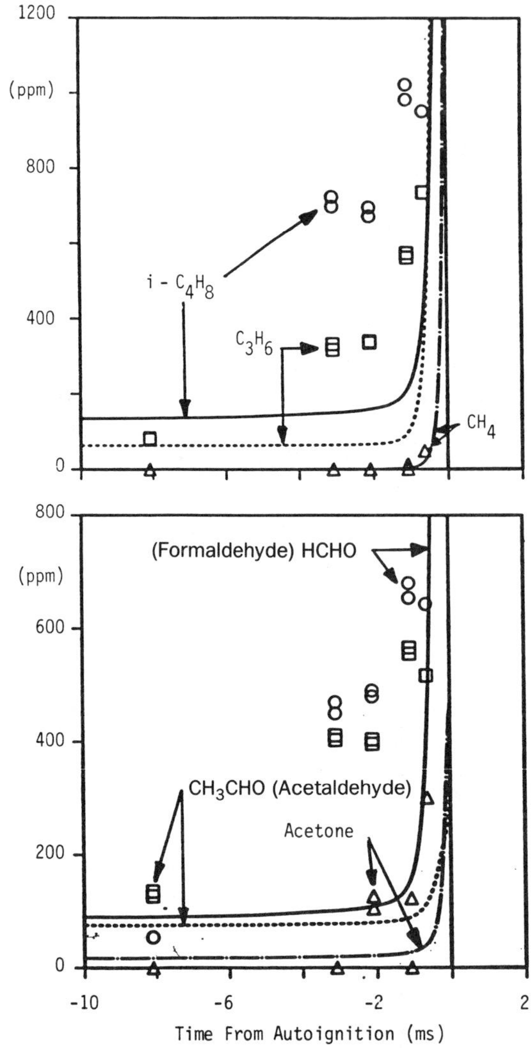

Figure 4.50 Predicted and measured concentration histories near the time of autoignition for formaldehyde, acetaldehyde, and acetone in an engine operating at 600 rpm and an equivalence ratio equal to 1 [328].

plasma at this early stage by assuming that the breakdown energy E_{bd} is introduced to the spark channel as a constant-volume process, during which no significant reactions occur, and that the high pressure that is produced results in the generation of a shock wave, which in turn reduces the pressure of the kernel to the ambient pressure and dissipates part of the energy to the surroundings. Since the ratio between the initial temperature of the mixture and the temperature of the spark channel is much smaller than unity, the diameter d of the cylindrical channel is given approximately by the following expression:

$$d = 2\left[(\gamma - 1)\frac{E_{bd}}{\gamma \pi g p}\right]^{1/2} \tag{276}$$

where $\gamma = C_p/C_v$, g is the electrode gap, p is the chamber pressure, the temperature of the cylindrical channel is T_i/γ, and T_i is the initial temperature of the channel during the breakdown period.

The driving force during the arc phase is the resistive heating of the plasma by the electric current. The rate of heat generation can be determined from Ohm's law by using the electrical conductivity of the mixture, which is a function of the density, temperature, and electron-ion cross section and includes dissociation and ionization processes [337].

In the next sections some of the phenomenological models and detailed models that have been used to study ignition are presented.

4.10.1 Phenomenological Models

Lim et al. [340] assumed that the kernel growth is initially controlled by the breakdown process and the subsequent electrical power input, and they used blast-wave theory to calculate the kernel radius R_0 as

$$R_0 = \left(\frac{E_0}{B\gamma p}\right)^{1/2} \tag{277}$$

where E_0 is the energy release and B is a function of the blast-wave shape and γ.

They also used a diffusion model to account for the rapid growth of the kernel:

$$\frac{dm}{dt} = \frac{d}{dt}\left(\frac{4}{3}\,\pi R^3 \rho\right) = 4\pi R^2 \rho_u c_s (S_L + c_d \rho_u u) \tag{278}$$

and assumed that some fraction of the input energy is converted into kinetic energy

$$\dot{E}_e = 4\pi R^2 \rho_u u^3 \tag{279}$$

where c_s and c_d are constants, u is the entrainment velocity, R is the kernel radius, and ρ_u is the density of the unburnt gases.

The energy equation applied to the kernel yields

$$\frac{dh}{dt} = \frac{1}{\rho}\frac{dp}{dt} - (h - h_u)\frac{1}{m}\frac{dm}{dt} + \frac{\dot{E}_e}{m} + \frac{\dot{m}}{m}Q_c \tag{280}$$

where $\dot{E}_e$ is the electric power, h is the enthalpy, and Q_c is the heat of combustion.

Equations (278)–(280) and the equation of state for the mixture can be used to predict the radius and temperature of the kernel.

Lim et al. [340] did not consider ionization and dissociation, and they showed that heat transfer to the electrodes is important and that combustion occurs rapidly after the electrical breakdown of the gap. They also showed that the flame kernel depends on how much energy is supplied and on the rate of application of the energy and the mode of discharge.

Ballal and Lefebvre [341] showed that the quenching distance and minimum ignition energy in flowing gaseous streams are a function of the turbulent length scale, fluctuating velocity, mean flow velocity, pressure, and equivalence ratio. Their model accounts for the heat generated by combustion and heat losses by means of the following expressions:

$$Q_c = \rho\, C_p\, \Delta T\, S_T \qquad Q_L = \lambda \frac{S_T}{S_L}\frac{dT}{dx} = \lambda \frac{\Delta T}{\delta_L}\frac{S_T}{S_L} \tag{281}$$

$$Q_D = \rho\, C_p \frac{\Delta T}{l} D \frac{S_T}{S_L} \qquad \frac{D}{u'l} = a \approx 0.16 \tag{282}$$

where Q_c, Q_L, and Q_D are the heat fluxes associated with combustion, heat losses, and heat diffusion, respectively; λ is the thermal conductivity; δ_L is the laminar flame thickness; S_T and S_L are the turbulent and laminar flame speeds, respectively; D is the eddy diffusivity; l is the integral length scale; and u' is the fluctuating velocity.

The balance between heat production and losses can be written as

$$Q_c = Q_D + Q_L \qquad \rho\, C_p\, S_T = \frac{\lambda}{\delta_L}\frac{S_T}{S_L} + 0.16\, \rho\, C_p\, u' \frac{S_T}{S_L} \tag{283}$$

and if the quenching distance d_q is assumed to be proportional to the laminar flame thickness, then

$$d_q = A\, \delta_L = \frac{A\lambda}{\rho\, C_p\, (S_L - 0.16u')} \tag{284}$$

where A is a constant.

Equation (284) has been used by Anderson and Lim [342] to calculate the minimum ignition energy as

$$E_{\min} = \frac{\pi}{6} d_q^3\, \rho\, C_p\, (T_{ad} - T_0) \tag{285}$$

where T_{ad} and T_0 are the adiabatic and initial temperatures, respectively. However, their model was unable to predict the effects of the engine rpm on misfire because the use of a fully developed value of the turbulent diffusivity [Eq. (282*b*)] during ignition in spark-ignition engines is not valid.

Ballal and Lefebvre [343] studied the effects of the breakdown voltage, spark duration, energy distribution, gap width, equivalence ratio, mean gas velocity, and electrode geometry and arrangement on the ignition of mixtures in a closed-circuit wind tunnel and showed that the optimum spark duration for minimum ignition energy is a function of the mixture pressure, equivalence ratio, and mean gas velocity. The energy released by the spark was found to be linearly proportional to the gap width and a function of the pressure and mean gas velocity. Ballal and Lefebvre [343] accounted for heat transfer by conduction, radiation, turbulent diffusion, and forced convection by means of the following expressions:

$$\mathrm{Nu}_g = 0.61\ \mathrm{Re}_g^{0.46} \qquad \mathrm{Re}_g = \rho\,\frac{Ug}{\mu} \tag{286}$$

where U is the mean gas velocity. They also consider the following expressions for the breakdown voltage and total spark energy:

$$V_{bd} = \frac{2.8 \times 10^5\ pg}{5.5 - \ln(pd)} \qquad E_s = \int_0^{t_s} VI\,dt \tag{287}$$

where t_s is the spark duration, V is the voltage, and I is the current.

Ballal and Lefebvre [344] studied the ignition of propane-air mixtures at subatmospheric pressures in a closed-circuit wind tunnel; they showed that both the quenching distance and the minimum ignition energy increase with departures from the mixture stoichiometry, as the pressure decreases, and with increases in the mean gas and fluctuating velocities. They also showed that the flame kernel thickness is a function of λ, ρ, C_p, S_L, u', U, δ_L, and l.

Ballal and Lefebvre [345] developed an ignition model for gaseous and liquid fuel-air mixtures, which accounts for the presence of fuel vapor and finite rate chemistry and results in good agreement with experimental data. Ballal and Lefebvre [345] calculated the quenching time as the ratio of the heat capacity of the flame kernel to the sum of the heat conduction and turbulent diffusion from the flame kernel, that is,

$$t_q = \frac{d^2}{8(\alpha + 0.08\ u'\ d)} \tag{288}$$

where d is the diameter of the flame kernel, u' is the rms of the fluctuating velocity, α is the molecular thermal diffusivity, and t_q is the quenching time. They also assumed that the quenching time is the sum of the evaporation and chemical kinetics times.

Maly [346] developed a thermal ignition model that predicts that the ignition time is a function of the mean flow velocity, rms of the fluctuating velocity, equivalence ratio, preheating, and mixture preparation. His model assumes that the ignition energy is used only to heat the outer shell of the plasma created by the spark discharge [337].

Kono et al. [347] developed the following phenomenological ignition model. The difference between the heat generated by combustion and the heat conducted away from the flame kernel represents the net heat increase in the flame kernel, that is,

$$\frac{dQ}{dt} = q_c\, v_u\, \rho_u\, S - \lambda\,\frac{T - T_0}{\delta_t}\, S \tag{289}$$

where q_c is the heat of reaction, v_u is the velocity, S is the flame surface area, T_0 is the ambient temperature, the subscript u denotes unburnt gases, and δ_t is the flame front thickness given by the following expression:

$$\delta_t = \left(\frac{\pi \lambda t}{\rho\, C_p}\right)^{1/2} \tag{290}$$

When ignition is achieved, there is a balance between the heat generated by combustion and the heat losses (i.e., $dQ/dt = 0$ at t_{ig}), and Eqs. (289) and (290) yield

$$t_{ig} = \left(\rho\, C_p\, \frac{\lambda}{\pi}\right)\left(\frac{T - T_0}{\rho\, v_u\, \rho_u}\right)^{1/2} \tag{291}$$

Kono et al. [348] used two successive sparks and showed that the discharge frequency rather than the spark duration is the most important factor for capacitance sparks. Kono et al. [348] studied the ignition of quiescent propane-air mixtures by means of a capacitance spark followed by DC (glow) or AC discharge, and they analyzed the effects of the gap width, electrode configuration, mixture strength, spark duration, and energy distribution. They showed that the minimum ignition energy is a function of the equivalence ratio and quenching distance. For DC discharges Kono et al. [348] showed that optimum ignition can be achieved with spark durations from 50 μs to 300 μs. Kono et al. [349] analyzed the effects of the electrodes, heat losses, radical species, and flame quenching by wake turbulence induced by the electrodes, and they showed that multiple sparks enhance the ignition ability due to their repetitive effect.

DeSoete [350] showed that the minimum ignition energy is a function of the spark duration, mean flow velocity, turbulence intensity, and nonadiabatic effects of the electrode walls. His results agreed with those of Ballal and Lefebvre [341, 343, 344].

Akindele et al. [51] and Bradley and Lung [351] developed an ignition model for turbulent gases that assumes that the flame kernel formation is spherical and that neglects chemical kinetics effects. In their model the pressure is assumed uniform and the velocity-temperature correlations are modeled according to Boussinesq. The one-dimensional continuity and energy equations in spherical coordinates can be written as

$$\frac{\partial \rho}{\partial t} + \frac{1}{r^2}\frac{\partial}{\partial r}(\rho r^2 u) = 0 \tag{292}$$

$$\rho C_p\left(\frac{\partial T}{\partial t} + u\,\frac{\partial T}{\partial r}\right) = \frac{1}{r^2}\frac{\partial}{\partial r}\left(r^2\, K\, \frac{\partial T}{\partial r}\right) \tag{293}$$

where

$$K = \lambda + \rho\, C_p\, \alpha \tag{294}$$

α is the eddy diffusivity, K is the effective thermal conductivity, and u is the mean gas velocity.

The eddy diffusivity can be calculated from the statistical theory of turbulence as [42, 352]

$$\alpha = u'^2 \int_0^t R(\xi)\, d\xi \tag{295}$$

where $R(\xi)$ is the Lagrangian correlation coefficient, which can be approximated by

$$R(\xi) = \exp\left(-\frac{\xi}{t_L}\right) \qquad t_L = \beta \frac{l}{u'} \qquad \beta \approx 0.44 \tag{296}$$

β is constant, l is the integral length scale, and u' is the turbulence intensity.

Substitution of Eq. (296) into Eq. (295) yields

$$\alpha = 0.44\, u'l \left[1 - \exp\left(-\frac{u't}{\beta l}\right)\right] \tag{297}$$

Ignition is achieved if

$$\int_0^t \frac{dt}{\tau} \geq 1 \tag{298}$$

where τ is the ignition delay time.

Experiments have shown that turbulence increases the spark energy required to achieve a given probability of ignition [353]. These increases can be correlated with the strain rate rms.

The electrodes of commercial spark plugs create their own flow field; downstream of the electrode, a wakelike flow is produced with a decreased mean velocity and an increased turbulence intensity. When the electrode distance is decreased, the lean ignition limit increases because of the larger quenching effect caused by the electrodes. For electrode gaps much smaller than the quenching distance, the transport of the flame kernel produces an optimum gas velocity. For electrode gaps of the same size as or larger than the quenching distance, the lean ignition limit increases with the velocity [354].

Akindele et al. [51] showed that in the early stages of ignition, molecular heat conduction is very important because of the high values of the thermal conductivity associated with the spark temperature. They also showed that local ignition must occur at a radius not less than the flame thickness and that the flame kernel moves away from the spark with a velocity approximately equal to u'. Akindele et al. [51] showed that low turbulence intensities wrinkle the flame front.

Bradley and Lung [351] used a model similar to that of Eqs. (292)–(298). They assumed that the gas is optically thick [355] and used Eq. (295), which comes from isotropic homogeneous turbulence [352]. They showed that there are three regimes in the ignition of gaseous mixtures: localized ignition, spark-assisted flame propagation, and normal flame propagation. They also showed that the minimum ignition energy increases as the aerodynamic and geometric flame stretchings are increased.

Adelman [356] developed a time-dependent theory of spark ignition, which accounts for the temporal dependence of the spark input, shock-wave formation, and distortion of the arc in moving media. His theory neglects transport effects and shows that the flame kernel must grow to a certain size before its expansion velocity falls to a critical value.

4.10.2 Detailed Modeling

Wu et al. [357] and Ramos [358] studied the ignition of one-dimensional, isobaric mixtures by burnt gases, sparks, and hot wires. They showed that the critical diameter for ignition and the minimum ignition time first decrease with the equivalence ratio and then increase in agreement with available experimental data [278]. Wu et al. [357] also showed that the critical diameter for ignition is about half the quenching distance.

Sibulkin and Siskind [359] studied the ignition of infinite, one-dimensional gaseous mixtures subjected to a sudden deposition of energy. They used a one-step chemical reaction and equations for the fuel mass fraction and temperature; they showed that ignition occurs if the radius of the hot gas kernel generated by the initial input of energy exceeds the thickness of the propagating flame. The flame speed at ignition was found to be a function of the Lewis number, kernel diameter, and ignition temperature.

Dixon-Lewis and Shepherd [360] used a detailed chemical kinetics mechanism to study the ignition of hydrogen-air mixtures. They showed that the minimum ignition energy depends on how the energy is supplied and the rate at which the energy is supplied.

Refael and Sher [361] used a high-temperature mechanism to study the ignition of methane-air mixtures. Their mechanism includes 18 chemical reactions and 14 species, realistic transport coefficients, and an equation of state for the plasma.

Refael and Sher [361] showed that radiation losses are important during the arc phase, where resistive and generated heats are produced at high rates and result in high temperatures. They decomposed the mixture into several zones, used an optically thin thickness approximation [355], and assumed negligible self-absorption in the zones. The zones were assumed to exchange heat with and absorb energy from the surroundings, but they did not absorb some of their emitted energy.

In cylindrical coordinates the radiative heat transfer can be calculated, for a zone of length L located at the distance r from the symmetry axis, as

$$\frac{dE_{\text{rad}}}{dt} = 2\pi rL\ \Delta r\ \sigma\ e\ (T^4 - T_0^4) \tag{299}$$

where Δr is the radial dimension of the zone, σ is the Stefan-Boltzmann constant,

e is the absorption coefficient, T is the zone's temperature, and T_0 is the ambient temperature.

During the arc phase, the resistive heating of the plasma by the electric current is most important and can be calculated from Ohm's law.

Refael and Sher [361] started their one-dimensional cylindrical computations after the start of the arc and used the following equation:

$$E_{bd} = E_{\text{ig}} + p_0 (V_{\text{ig}} - V_{bd}) \tag{300}$$

where E_{bd} and E_{ig} are the breakdown energy and the kernel energy after arc formation, respectively; V_{ig} and V_{bd} are the volumes of the kernel after arc formation and at breakdown, respectively; and p_0 is the ambient pressure.

The kernel energy can be calculated as

$$E = \frac{pV}{\gamma - 1}\left(1 - \frac{T_0}{T}\right) \tag{301}$$

where

$$\frac{p_0 V_{bd}}{T_0} = \frac{p(0)V(0)}{T(0)} = \frac{p_0 V_{\text{ig}}}{T_{\text{ig}}} \tag{302}$$

Equations (300)–(302) yield the following ignition temperature and kernel diameter:

$$T_{\text{ig}} = T_0 \left[\frac{1}{\gamma}\left(\frac{T(0)}{T_0} - 1\right) + 1\right] \tag{303}$$

$$d_{\text{ig}} = 2\left[\frac{\gamma - 1}{\gamma} \frac{E_{bd}}{p_0 \pi g(1 - T_0/T_{\text{ig}})}\right]^{1/2} \tag{304}$$

Equation (304) coincides with Eq. (276) if $T_{\text{ig}} >> T_0$.

Sher and Refael [339] also used a cylindrical model to study the early development of spark-ignition methane-air flames by means of a constant volumetric heat source. Sher and Keck [337] also used a one-dimensional model to analyze the ignition of methane-air mixtures and showed that (1) the initial stages of flame propagation depend strongly on the equivalence ratio, breakdown energy, and spark duration and, to a lesser extent, on the initial pressure, temperature, and residual gas fraction; (2) better sparks produce larger volumes of hot gases in short times, for which the probability of initiating a self-sustained combustion wave in nonhomogeneous mixtures is higher; and (3) cycle-to-cycle variations may be caused by the inhomogeneity of the trapped mixture.

Kailasanath el al. [362] studied the development of the temperature profile and the effect of the spark-channel radius on the minimum ignition energy of H_2-O_2-N_2 mixtures. They prescribed the time-dependent energy deposition in the spark channel from a similarity solution, which depends on the radius of the region in which the energy is deposited, the heating duration, and the total amount of energy deposited.

Sloane and Ratcliffe [363] used a detailed chemical kinetics model to analyze the chemistry of spark-ignition engines and obtained satisfactory agreement with experimental data.

4.10.3 Simulation of Ignition in Multidimensional Engine Models

Multidimensional models of reciprocating engines do not consider ignition phenomena in detail because of the high demands in terms of both computer time and storage that would be required in order to accurately resolve the breakdown and arc phases of ignition discharges and to resolve phenomena that occur in the electrode gap. The arc phase involves many chemical reactions and species.

In multidimensional models, ignition is frequently simulated by (1) adding heat to a few computational cells until a self-propagating flame kernel is established, (2) heating a localized region of the walls, or (3) consuming fuel in a few computational cells at a specified rate so that the heat released by the fuel consumption is used to heat the mixture in a few cells. The heat added to a few computational cells must be much smaller than the heat of combustion.

A fourth approach to simulating the ignition process uses a phenomenological model as follows. The heat rate supplied by the spark can be approximated as, for example,

$$q_{sp} = q_p \exp\left[-\left(\frac{t - t_{ig}}{\sigma}\right)^2\right] \tag{305}$$

where q_p is the maximum heat rate, t_{ig} corresponds to the time at which the spark plug is activated, and σ is the variance of the heat source distribution and represents a characteristic time of heat addition.

The total heat supplied by the spark can be obtained by integrating Eq. (305) from t_{ig} to t_{end} (i.e., the time at which the spark plug is deactivated) so that

$$q_T = \int_{t_{ig}}^{t_{end}} q_{sp}\, dt = \frac{2\sigma}{\pi^{1/2}} q_p \operatorname{erf}\left(\frac{t_{end} - t_{ig}}{\sigma}\right) \tag{306}$$

where erf is the error function.

In the absence of heat losses the energy required to raise the temperature of a small volume ΔT at constant pressure is

$$E = \rho\, C_p\, \Delta T \tag{307}$$

which must be equal to q_T, so that

$$q_p = \pi^{1/2} \rho\, C_p \frac{\Delta T}{2\sigma \operatorname{erf}[(t_{end} - t_{ig})/\sigma]} \tag{308}$$

If the temperature rise ΔT is specified as, for example, $\Delta T = 1700$ K, Eq. (308) provides the maximum heat rate to be supplied by the spark once t_{ig}, T_{end}, and σ are known. The value of σ must be smaller than the characteristic turbulent diffusion time; otherwise, the turbulence around the heat source will dissipate the steep temperature gradients generated by the spark plug, and no flame kernel would be formed. Thus,

$$\sigma < t_{\text{diff}} = \frac{\delta_T^2}{\nu_t} \tag{309}$$

where t_{diff} is the turbulence diffusion time, δ_T is the turbulent flame thickness, and ν_t is the eddy diffusivity. The value of ν_t can be calculated from a two-equation $k-\epsilon$ model of turbulence as (see Chapter 2)

$$\nu_t = C_\mu \frac{k^2}{\epsilon} \qquad C_\mu = 0.09 \tag{310}$$

The turbulent flame thickness depends on the engine rpm and can be estimated as 0.5 cm. Therefore, for an eddy diffusivity of 100 cm^2/s, the value of σ is about 2.5 ms [121, 177].

The model represented by Eqs. (305)–(310) does not account for the heat transfer losses by forced convection, conduction, and radiation from the flame kernel and neglects the heat transfer to the electrode gap. Therefore, it represents a highly idealized approximation to ignition by spark discharges and may require some tuning when the engine-operating conditions, engine geometry, etc., are varied.

4.10.4 Plasma Jet Ignition

The development of lean-charge, fast-burn spark-ignition engines depends on enhanced ignition, which provides the means for increasing the burning rate in mixtures characterized by low burning velocities. Enhanced ignition may involve high energies and long ignition durations and perhaps a wide dispersion of the ignition sources so that combustion occurs in as many sites throughout the charge as possible.

Systems for enhanced ignition can be grouped in the following categories: (1) high-energy spark plugs, (2) plasma jet igniters, (3) photochemical, laser, and microwave ignition systems, (4) torch cells, (5) divided-chamber stratified-charge engines, (6) flame jet igniters, (7) combustion jet ignition systems, and (8) EGR ignition systems. The first three ignition systems derive their power from electrical energy, whereas the rest are powered by exothermic chemical reactions that consume a small amount of fuel [364].

In plasma jet igniters the spark discharge is confined to a recessed cavity provided with a discharge orifice, while the electric power supply is augmented by the addition of a condenser, which discharges a relatively low voltage and a high current through the spark generated in a conventional manner by a high-voltage, low-current ignition system. The spark discharge produces a high-temperature plasma so rapidly that the cavity is pressurized, causing a plasma jet to be issued through the orifice and to penetrate into the charge. The plasma jet enters the combustion chamber as a turbulent plume imbedded in a blast wave, but the blast wave is dissipated by the time that combustion starts and ignition takes place in the turbulent plume. The penetration of the plasma jet depends on its initial velocity, energy deposited in the cavity, cavity size, and orifice size. Plasma jet ignition extends the lean operating limit of spark-ignition engines and may result in lower NO emissions.

Mittinti and Dabora [365] studied the plasma jet ignition of methane-air mixtures with an equivalence ratio of 0.7 and showed that CH_4 plasmas give better performance than do (CH_4 + $6O_2$ + NO_2) and (CH_4 + $7O_2$) plasmas. These studies show the importance of H radicals in plasma jet ignition, even though (CH_4 + $6O_2$ + NO_2) plasmas generate OH radicals. Mittinti and Dabora [365] showed that the plasma cavity, orifice size, spark duration, and aspect ratio have a great importance in plasma jet ignition.

Boston et al. [366] showed that surface-discharge ignition is superior to a two-electrode ignition system and that plasma jets with a wetted cavity torch from the prechamber create a critical volume of igniting hot gases. The turbulence generated by plasma jets decays sooner than that produced by surface-discharge ignition. Boston et al. [366] also showed that the flame speed is predominantly a function of the turbulence and that the ratio of the chemical conversion time to the eddy lifetime is a parameter that appears in the definition of the flame propagation.

Tozzi and Dabora [367] studied the plasma ignition of methane-air mixtures in a lean-burn CFR engine operating at 800 rpm; they showed that an increase in the plasma energy and a reduction in the energy deposition time result in improved engine performance. However, this improvement depends on the cavity size.

Orrin et al. [368] showed that high-speed jets of chemically active plasmas are very effective in igniting and promoting fast flame propagation in lean fuel-air mixtures due to the H radicals. They also found that flame convolutions are most effective in chemically inactive plasmas. Sloane et al. [369] used methane, acetylene, propane, and hydrogen plasma feeds to ignite methane-oxygen-argon mixtures and showed that a hydrogen plasma has the highest ignition efficiency in the leanest mixtures. The results of Sloane et al. [369] also indicate that local charge alteration may not have been eliminated or plasma products of the fuel are as effective in igniting lean mixtures as is a mixture enriched in the ignition region by the fuel itself.

Wiriyawit and Dabora [370] used an operator-splitting numerical method to study plasma ignition in one-dimensional enclosed geometries. They used the explicit predictor-corrector method of MacCormack [118], an Euler implicit method for the chemical reactions, and supplied H and OH radicals. Their calculations indicate that pressure waves are created during combustion; these waves result in a pulsating laminar flame. Wiriyawit and Dabora [370] indicated that the energy supplied as OH radicals is more efficient in igniting the mixture than the thermal energy and that the OH radicals must be allowed to spread out to obtain good ignition performance.

Fisher et al. [371] showed that puff-jet ignition decreases the ignition delay time and the burn duration and extends the lean misfire limit of internal combustion engines fueled with methane. They also showed that the power output of engines operating continuously with lean methane-air mixtures is also increased. The engine power output was almost the same as with a plasma jet igniter, and there was no electrode wear.

Dale et al. [372] showed that plasma jet ignition extends the lean operating limit and increases the engine performance, the specific fuel consumption, and the power output. However, it results in high NO emissions. Dale et al. [372] also found that the engine performance is quite insensitive to the ignition timing with plasma jet ignition.

Ridley et al. [373] showed that the increased rate of combustion with plasma jet ignition is directly related to the heat of combustion of the feed gas.

Pitt et al. [374] used puff jets as ignition sources in a single-cylinder, methane-fueled internal combustion engine. They showed that the ignition characteristics of puff jets are similar to those of plasma jets; both reduce the ignition delay and combustion times. Pitt and Clements [375] also showed that plasma jet ignition reduces the ignition delay times and extends the lean misfire operating limit. Plasma jet ignition does not alter the ignition timing when changing from one fuel to another in order to maintain maximum efficiency.

Behbahani et al. [376] showed that NO can be destroyed by nitrogen atoms from plasma jets. Behbahani et al. [377] used nitrogen atoms from a stream of molecular nitrogen and showed that plasma jet ignition heats a very small portion of the gas to exceedingly high temperatures. The thermal stratification produced by this localized heating can be used to reduce the NO emissions from reciprocating engines.

Grant et al. [378] showed that plasma jet ignition improves the lean operation of spark-ignition engines. They also showed that a balance between the plasma jet penetration and the jet mass flow must be achieved in order to produce optimal combustion enhancement.

Grant et al. [378] showed that the plasma jet performance increases as the jet orifice size is increased and that the most important factor that affects the plasma jet performance is the total energy input; the plasma current and voltage were found to be of secondary importance.

Plasma puff jets are pulsed plasma jets that discharge a fixed power pulse of electrical energy into the combustion chamber. Pitt et al. [379] used mechanical and electrical systems to produce a puff jet of combusting gases; they showed that the characteristics of plasma puff jets are very similar to those of plasma jet ignition (i.e., both reduce the ignition delay and combustion times). However, puff-jet ignition does not suffer from the same problems as plasma jet ignition (i.e., high energy consumption and high electrode erosion) and can be used to ignite ultralean mixtures.

Cavolowsky et al. [380] showed the existence of two distinct phases in the life of a puff: (1) initial acceleration associated with the process of formation and (2) a steady-state period. They also showed that a proper impedance match is required to achieve good jet mixing and to minimize the detrimental effects of energy dissipation.

Mittinti [381] showed that flame jet igniters are superior to plasma jet igniters due to the faster axial growth of the flame front. Mittinti [381] also showed that a great combustion improvement can be achieved if the initial flame kernel is sufficiently large and located away from solid walls, if the flame front is wrinkled, and if the initial kernel formation time is shorter than the ignition delay time.

Yamaguchi et al. [382] used torch jets to ignite mixtures in divided-chamber bombs with homogeneous and stratified charges; they measured ion currents, light and OH emissions, torch jet velocities, and main chamber pressures. They showed that the torch structure and ignition are a function of the nozzle diameter and volume ratio. Depending on the values of these parameters, four types of ignition were observed: (1) chemical chain ignition and well-dispersed burning, (2) composite ignition and well-dispersed burning followed by wrinkled laminar burning, (3) flame kernel torch ignition and wrinkled laminar flames, and (4) flame front torch ignition and wrinkled laminar flames. Yamaguchi et al. [382] found that type 2 is the most favorable for lean burning and that it is possible to burn lean mixtures outside their flammability limits by using a divided chamber.

Dale et al. [383] used conventional ignition systems and plasma jet igniters in single-cylinder engines fueled with gasoline or propane. They showed that plasma jet ignition extends the lean operating limit and engine operating performance (i.e., power output and specific fuel consumption). They also showed that NO emissions increase for constant timing whereas CO emissions are not affected by the ignition system.

Dale et al. [383] also found that the engine performance and pollutants are independent of the ignition system for MBT timing and air-fuel ratios between 10 and 17. For leaner mixtures the engine performance increases and the pollutant emissions decrease. They also found that the engine performance is insensitive to the ignition time with plasma jet igniters but very sensitive with conventional ignition systems.

Kono et al. [347] studied the ignition of gases by two successive sparks. They showed that the ignition capability of capacitance sparks increases as the frequency of the oscillatory discharge is decreased in propane-air mixtures; two sparks with 10 μs to 50 μs intervals have the highest ability to ignite propane-air mixtures with a frequency of 50 Hz to 100 Hz for the oscillatory discharge. Kono et al. [347] also showed that the frequency of the discharge rather than the spark duration is the predominant factor in ignition by capacitance sparks.

Dale et al. [384] used lasers to ignite mixtures in internal combustion engines. They showed that laser ignition is similar to plasma jet ignition.

Cote et al. [385] showed that plasma jets promote increased flame growth and extend the flammability limits of both rich and lean mixtures compared with commercial jets and spark gap igniters. Topham et al. [386] showed that the critical ignition of mixtures by pulsed plasma-produced elements depends on the heat content of the combustible mixture and on the energy-to-momentum ratio of the jet injected by the plasma source.

4.11 OTHER ENGINE MODELS

In this section we review some of the numerical methods that have been used to calculate the flow field in motored reciprocating engines. These methods can be classified into the following groups: inviscid, vortex, time linearization, and implicit iterative techniques.

Griffin and co-workers [179, 180, 387, 388] used MacCormack's explicit predictor-corrector method [118] to calculate the inviscid flow field in two- and three-dimensional engine geometries; they showed that a swirling flow pattern is created in the intake stroke and persists through the complete flow-stroke cycle. The calculations performed by Griffin et al. [179, 180, 387, 388] are unrealistic since the flow in an internal combustion engine is turbulent and viscous.

Vortex free-grid techniques have been used to calculate the flow field in planar and axisymmetric slightly viscous flows by means of vortex blobs or circular vortex filaments during the intake stroke in motored reciprocating engines [389–394]. Ashurst [393] showed that his vortex dynamics calculations agree quite well with the experimental data of Ekchian and Hoult [395], who demonstrated that the induction process is characterized by two ring vortices of concentrated vorticity, which are of the same scale as the engine cylinder. Ekchian and Hoult [395] also showed that the flow field is significantly affected by small changes in the engine geometry and that vortex breakdown is due to an inviscid instability, which is independent of the Reynolds number.

Ashurst [396] applied vortex dynamics to analyze the flow field in a motored four-stroke engine and showed that thin shear layers are created in the intake stroke and that the valve geometry has an important effect on determining the flow recirculation pattern within the engine cylinder. He also showed that the asymmetric flow patterns created by the intake and exhaust valves persist throughout the four-stroke cycle.

Martins and Ghoniem [397] used a vortex dynamics method to simulate the induction and compression strokes in an engine model. Their model is based on a zero-Mach-number approximation and accounts for the solenoidal and irrotational features of the flow. Recently, Oppenheim and Rotman [181] and Chau et al. [398] used a vortex dynamics model that accounts for volumetric expansion effects associated with combustion to predict the flow field and flame location in a variable-volume enclosure. Their numerical predictions are in good agreement with experimental pressure data and in qualitative agreement with schlieren photographs.

Some vortex dynamics calculations are presented in Figs. 4.51–4.54. Figure 4.51 shows the flow field in a planar geometry equipped with offset valves [393].

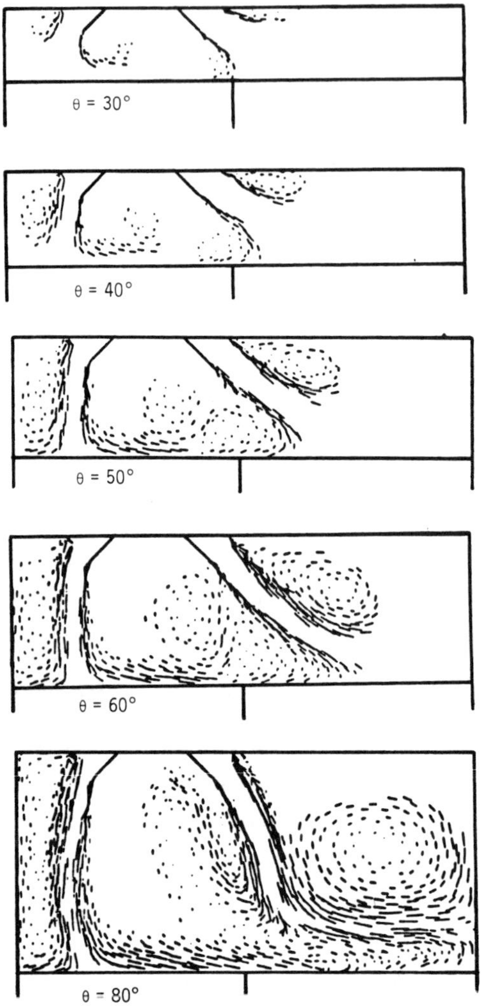

Figure 4.51 Vortex dynamics calculations in a planar geometry equipped with offset valves at selected crank shaft angles [393].

Figure 4.52 illustrates the velocity field prior to completion of the intake stroke and at the beginning of the compression stroke [392]. Figure 4.53 presents the flow field at the end of the compression stroke [397], and Fig. 4.54 shows the flame propagation as a function of the crank shaft angle [181].

Ghoniem used a vortex-finite element method to calculate the flow field during the induction stroke in ported engines. Some of his calculations are presented in Fig. 4.55, which also shows the numerical predictions obtained with multidimensional inviscid, algebraic turbulence and two-equation turbulence models. Figure 4.55 indicates that the results of the vortex and $k-\epsilon$ models agree well with experimental data, whereas the inviscid and algebraic models are unable to predict the in-cylinder flow field. In particular, the algebraic turbulence model does not predict the recirculation zone located near the cylinder liner and

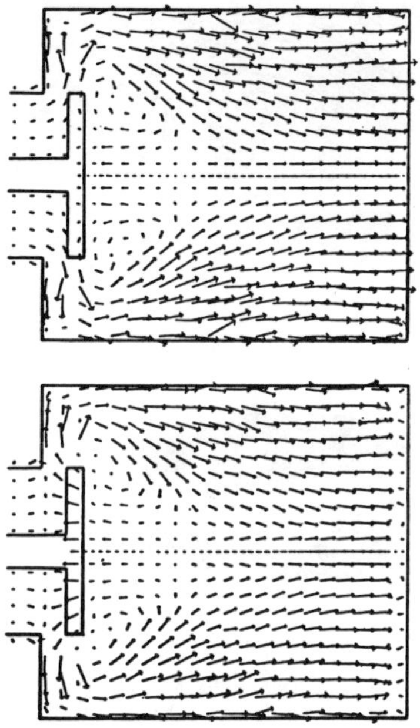

Figure 4.52 Velocity fields prior to completion of the intake stroke (top) and at the beginning of the compression stroke (bottom) [392].

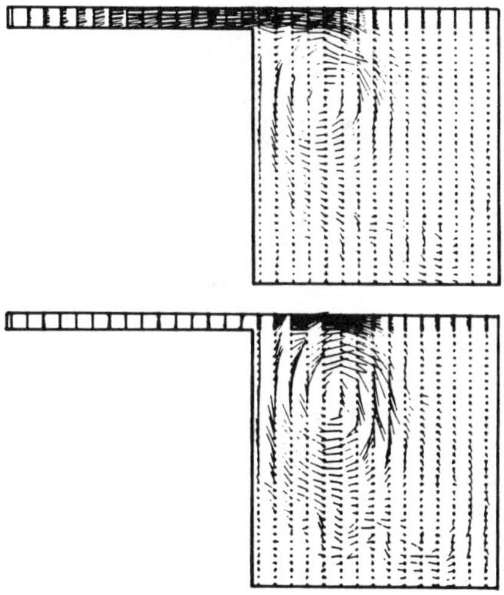

Figure 4.53 Flow field at the end of the compression stroke [397].

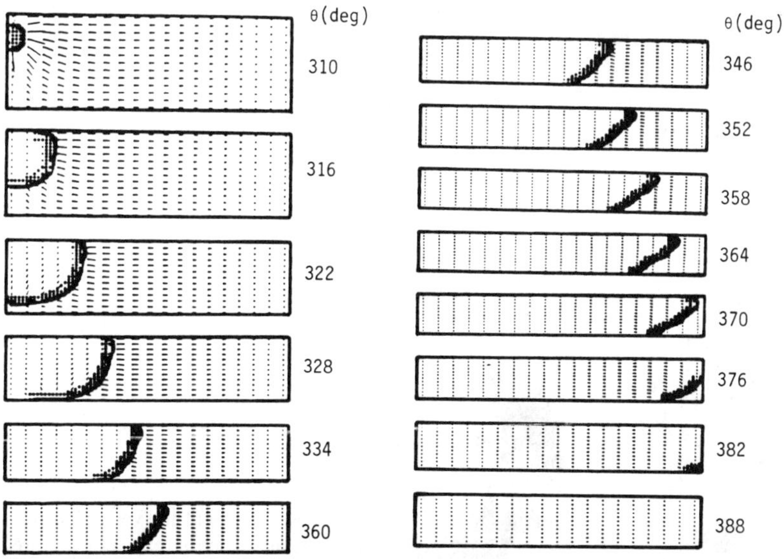

Figure 4.54 Flame location at selected crank shaft angles [181].

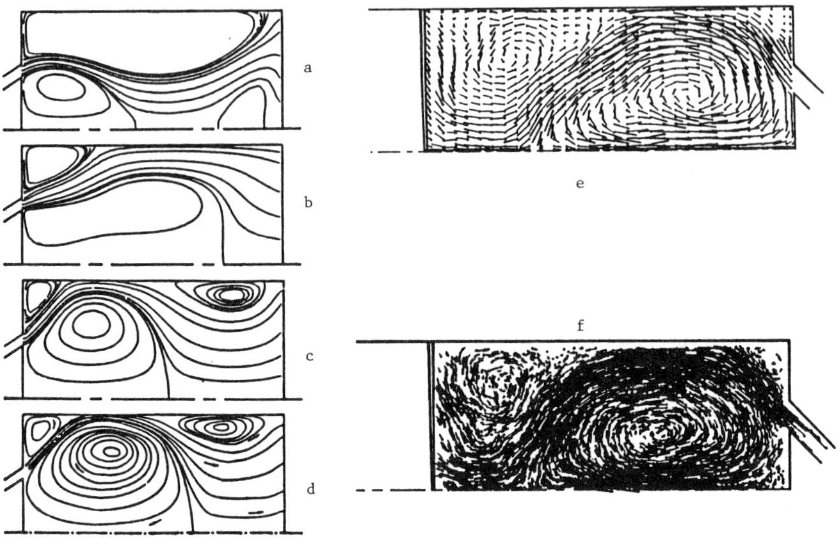

Figure 4.55 Flow field in the induction stroke predicted by (a) multidimensional inviscid; (b) algebraic turbulence; (c) two-equation turbulence; and (e) vortex-finite element methods; (f) location of the vortex elements; and (d) experimental data [399].

piston, whereas the inviscid model does not predict the length and width of the vortex located at the corner between the cylinder head and the cylinder liner.

Ghoniem also studied the flow field during the induction stroke in a single-cylinder engine equipped with a centrally located valve at the cylinder head and obtained good agreement with the experimental data of Wong and Hoult [108]. In Fig. 4.56 some sample results of these calculations are presented.

Barr and Witze [399] used a vortex method to analyze laminar, inviscid, adiabatic processes that occur at constant volume and showed that the flame slows down as it approaches the combustion chamber walls. They also showed

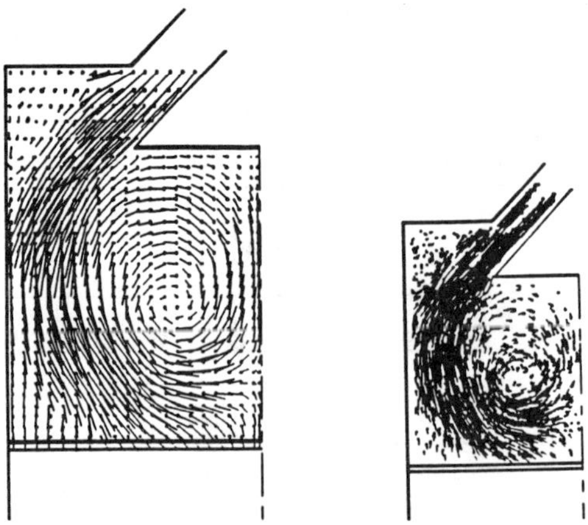

Figure 4.56 Flow field and vortex locations during the induction stroke in a motored single-cylinder engine equipped with a centrally located valve.

that the interactions between the flame and the combustion-generated flow field result in a highly curved flame of significantly greater surface area than the spherical flame approximation used in multizone models of spark-ignition engines. The flame curvature and distortion caused by the combustion-induced flow field strain the flame front and modify the local burning velocity.

The two-dimensional axisymmetric results of Barr and Witze [399] indicate that the assumption of a spherical flame with center at the ignition location can lead to a large underprediction of the flame surface area because the spherical flame is prematurely truncated by the combustion chamber walls and because the spherical flame assumption constrains the flame surface-to-volume ratio to be the absolute minimum.

Haselman [400] and Westbrook and Haselman [401] developed an explicit second-order accurate method for reciprocating engines that accounts for turbulence by means of a Boussinesq approximation. Westbrook and Haselman [401] calculated the flow field in a four-stroke reciprocating motored engine equipped with a centrally located valve and showed that vortex breakdown occurs in the intake stroke. During the compression stroke, vortex merging and complicated flow patterns were observed in the cylinder up to TDC. Vortex breakdown and merging were also observed in the turbulent calculations performed by Schock et al. [104, 105].

Shih et al. [402, 403] used a time linearization method and showed that vortex breakdown occurs in the intake stroke for bore-to-stroke ratios less than 1. If the vortices do not break down, they merge in the compression stroke. Shih et al. [403] used an eddy diffusivity equal to 100 times the laminar viscosity.

Henke and Hanel [404] used an implicit factorization method to solve the inviscid nonreactive flow field in planar or axisymmetric piston engine models. This method (see Chapter 2) reduces the coupled system of two-dimensional finite difference equations to a sequence of 2 one-dimensional linear problems that can be easily solved. Henke and Hanel [404] specified the intake density and flow angle, and the mass fluxes at the inlet and exhaust ports; they used linear extrapolation for the pressure. Their results show that the intake-generated vortices persist in the exhaust stroke and that the engine flow field depends on the boundary conditions at the inlet port. However, their inviscid calculations are not realistic; turbulent flow calculations and experimental data have shown that the intake-generated eddies may persist up to TDC but disappear in the expansion stroke.

Kondoh et al. [405] used an ICE-ALE (implicit continuous-fluid Eulerian-arbitrary Lagrangian-Eulerian) technique and a k–ϵ model of turbulence to analyze the compression and combustion strokes in a two-dimensional engine model provided with a bowl-in-piston combustion chamber, and they obtained good agreement with experimental data.

El Tahry [406] showed that the mean velocity field and the turbulence kinetic energy are sensitive to the initial swirl level and are influenced by the shape of the initial swirl-velocity profile. El Tahry [406] showed that the value of k (turbulence kinetic energy) at intake valve closing (IVC) has no effect on the value of k at TDC. However, in the absence of swirl, the initial value of k had a small effect on the value of k at TDC. Gosman et al. [407, 408] showed that a centrally located shrouded valve generates a long-lived vortex, which is sustained and amplified by compression and causes turbulence amplification at TDC (twice as much as with nonshrouded valves). Gosman et al. [408] obtained moderate agreement with experimental data for two engine configurations. Brandstatter et al. [409] performed theoretical-experimental work on the effects of a helical intake port on the flow field in motored reciprocating engines using a gas dynamics method to calculate the flow immediately upstream of the valve in order to determine the mass flow rate in the intake stroke (see Chapter 5). They also used quasisteady velocity profiles as boundary conditions. Brandstatter et al. [410] analyzed the three-dimensional flow field in the intake stroke and showed that the radial velocity exhibits large axial variations that generate swirl in the cylinder. Swirl in the valve gap was found to be independent of the valve lift and axial location within the valve gap. The intake port-generated swirl did not make any significant contribution to the in-cylinder swirl.

El Tahry [411] applied a Reynolds stress model and a high-order skew-upwind numerical scheme to an axisymmetry piston-cylinder configuration and obtained encouraging results for the mean flow. The discrepancies between the predicted and measured data were attributed to unknown boundary conditions and shortcomings in the Reynolds stress model. El Tahry [411] showed that

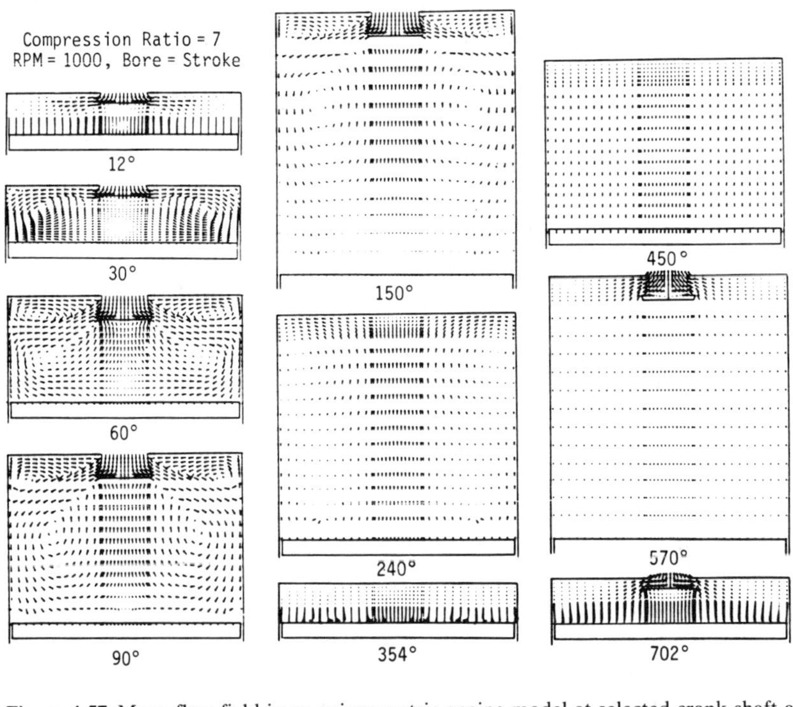

Figure 4.57 Mean flow field in an axisymmetric engine model at selected crank shaft angles [105].

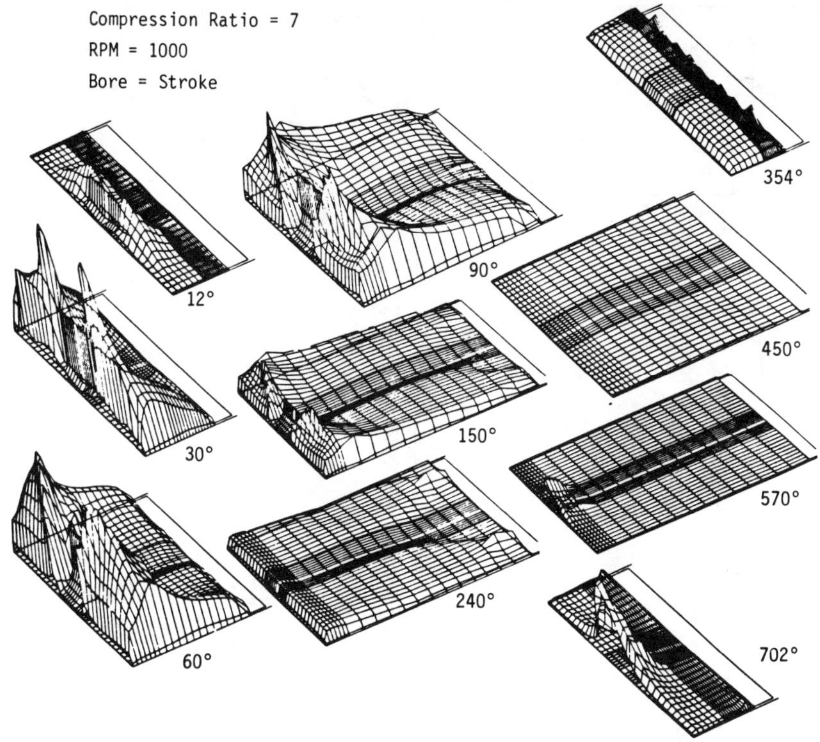

Figure 4.58 Normalized rms velocity profiles in an axisymmetric four-stroke motored engine model at selected crank shaft angles [105].

unbounded skew-upwind schemes may result in negative normal stresses; however, these schemes are more accurate than a first-order accurate upwind scheme. Gosman et al. [412] used a 9-point bounded-skew spatial-differencing scheme in a piston-cylinder configuration equipped with a centrally located valve; they compared their predictions with LDV data obtained in a motored engine having a compression ratio equal to 6.7 operated at 200 rpm. The predicted mean flow velocities and turbulent intensities were in moderate agreement with experimental data. Their 9-point scheme resulted in better agreement with experimental data than a 5-point hybrid central-upwind method.

Bernard [413] proposed a set of closed dynamical equations for the mean and variance of turbulent vorticity field, which were used to analyze the flow field in an internal combustion engine during the compression stroke. The flow was observed to be dominated by the development of large corner vortices; in their vicinity, intense production of turbulence occurs, which locally exceeds the compression-generated turbulence.

Some sample results illustrating the flow field in two- and three-dimensional engine models are presented in Figs. 4.57–4.62. Figures 4.57 and 4.58 illustrate the mean velocity and rms fields in an axisymmetric four-stroke motored engine

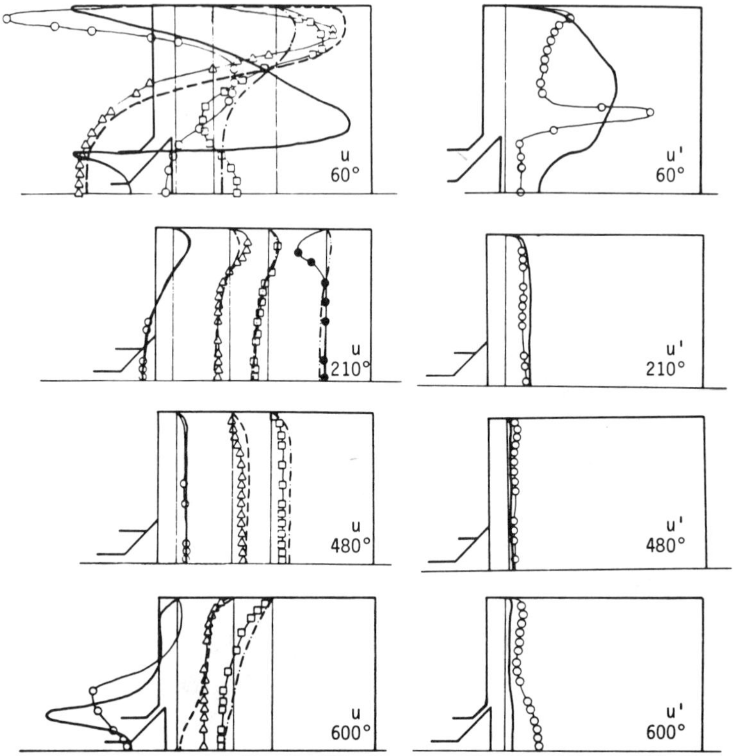

Figure 4.59 Comparisons between measured and predicted mean axial and turbulent rms velocities in a four-stroke motored engine at selected crank shaft angles [414] (u = mean axial velocity; u' = turbulent rms velocity).

[105]. Figure 4.59 shows a comparison between the measured and predicted mean velocity and turbulence intensity profiles in a four-stroke motored engine [415]. Figure 4.60 shows a schematic of a helical port, and Figs. 4.61 and 4.62 illustrate the mean velocity profiles at selected crank shaft angles in a motored four-stroke engine equipped with a helical intake port (Fig. 4.60) [409].

4.12 SUMMARY

In this chapter we have reviewed thermodynamic and multidimensional models of spark-ignition engines. Single-zone thermodynamic models provide a simple

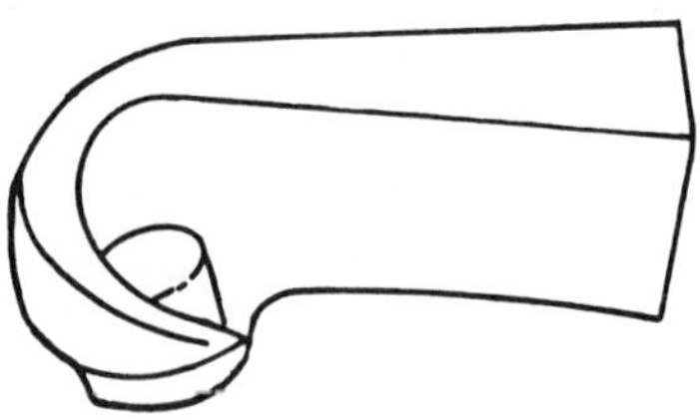

Figure 4.60 Schematic of a helical port [409].

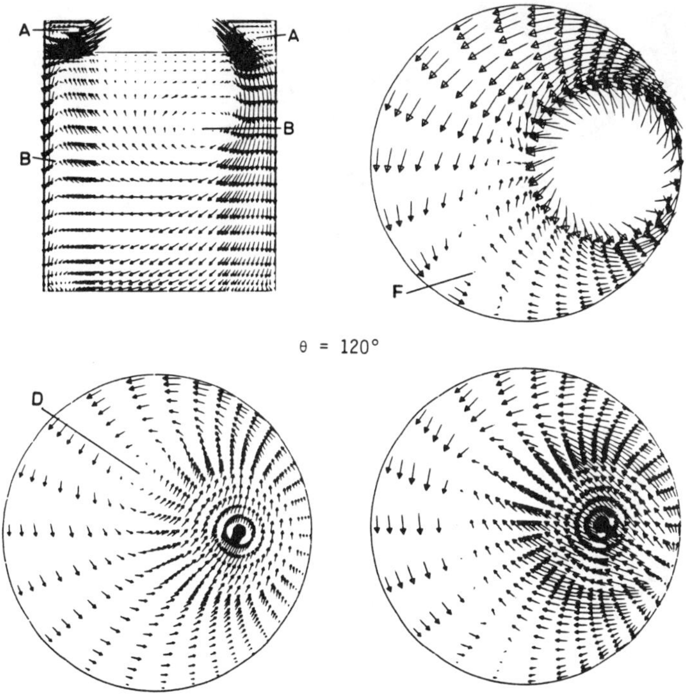

Figure 4.61 Velocity field at θ = 120° in an engine equipped with a helical port [409]. A and B are toroidal vortices in the axial-radial plane; D is the main swirl center; F is a plane vortex in the circumferential-radial plane.

way to evaluate the heat release rate when experimental pressure diagrams are used as an input in the first law of thermodynamics. Single-zone models can also be used as predictive tools if the heat release rate is specified as a function of the crank shaft angle. However, the heat release rate may depend on the combustion chamber geometry, engine-operating conditions, rpm, load, residual mass fraction, etc., and may have to be tuned up on a case-by-case basis.

Single-zone models ignore flame propagation phenomena and neglect the temperature and concentration gradients that exist throughout the charge. These gradients play an important role in determining the engine emissions, performance, heat transfer losses, etc.

Multizone thermodynamic models consider the flame propagation phenomena by dividing the cylinder charge into burnt and unburnt regions and quench layers. The burnt and unburnt regions are separated by the flame front that propagates throughout the combustion chamber at a turbulent flame speed. The flame speed or, alternatively, the mass burning rate can be specified as a function of the laminar flame speed, turbulence intensity, laminar flame thickness, and turbulence length scales. Depending on the values of these parameters, combustion may occur in the reaction-sheet regime or in the distributed-reaction regime.

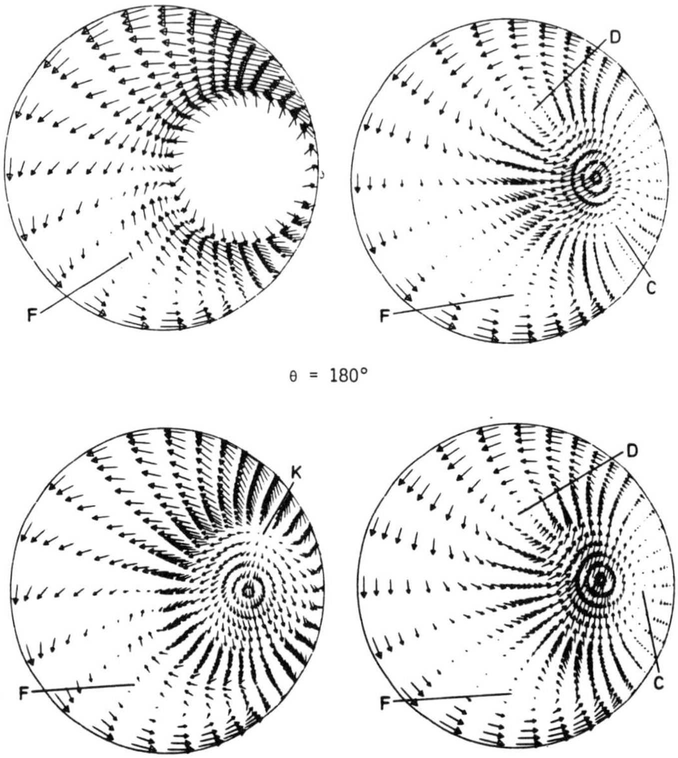

Figure 4.62 Velocity field at θ = 180° in an engine equipped with a helical port [409]. C and F are plane vortices in the circumferential-radial plane; D is the main swirl center.

Multizone models that calculate the mass burning rate from physical and phenomenological principles offer a great potential to evaluate the performance of spark-ignition engines. These models use equations for the turbulence kinetic energy and its dissipation rate as well as length scales that can be derived from isotropic turbulence. The validity of these equations and scale ratios in engine flows is a subject that deserves further research despite the agreement that has been obtained between numerical predictions and experimental data.

Multizone models can account for the spotty character of the dissipation rate of turbulence kinetic energy and thick flame fronts. However, the predictions of these models depend on the initial levels of turbulence in the combustion chamber. These levels cannot be accurately predicted by multizone models because the intake-generated turbulence is nonuniform; turbulence is generated at the shear layers of the incoming air flow in the intake stroke and at the solid walls. Turbulence is also generated by compression in the compression stroke but it is also affected by diffusion and dissipation, particularly near the combustion chamber walls. Nonuniform turbulence levels and mean velocity fields may be observed even beyond TDC, depending on the engine rpm, valve seat angle, valve lift, engine load, combustion chamber geometry, squish, swirl, etc.

It is doubtful that thermodynamic models can be used to accurately predict the in-cylinder turbulence levels in the intake stroke. However, some of these models have been used successfully to predict the engine performance and emissions under a variety of operating conditions.

The multidimensional models developed to analyze spark-ignition engine combustion use turbulence models that were developed for statistically stationary, incompressible, thin-shear flows. These models may not be valid for predicting unsteady, compressible, recirculating flows. Moreover, these models are based on time-averaged procedures that may not be applicable to engine flows, where the time scales of the energy-containing eddies are of the same order of magnitude as the time scales of the mean flow and where cycle times are short compared with relaxation times for important flow-adjustment phenomena.

The predictions of multidimensional models cannot be easily compared with LDV and cycle-resolved measurements. Any agreement between numerical and experimental data cannot be attributed to the accurate representation of the engine flow physics.

Multidimensional models have reached high levels of sophistication but are subject to several limitations: inaccurate knowledge of the boundary conditions at the inlet and exhaust ports, lack of accurate turbulence and combustion models, numerical resolution, etc. In addition, these models cannot as yet predict, for example, cycle-to-cycle variations.

Heat transfer losses in spark-ignition engines account for a significant fraction of the chemical energy. Multidimensional models calculate the heat transfer rates to the combustion chamber walls by means of the Reynolds analogy between heat and momentum transport. This analogy is strictly valid for statistically stationary, incompressible, thin shear flows with laminar and turbulent Prandtl numbers equal to 1 and with similar mean velocity and mean temperature profiles. These conditions may not be satisfied in engine flows, although the Reynolds analogy has resulted in good agreement between numerical predictions and experimental heat transfer data in some cases.

The main emissions from spark-ignition engines are unburnt HC, CO, and NO. The main sources of unburnt HC are the oil layers, crevices, quench layers, flame quenching, deposit layers, and bulk-flow quenching.

Nitric oxide formation occurs in the postflame burnt gases and is a very strong function of the temperature and concentration gradients throughout the combustion chamber. Therefore, in order to accurately predict the NO formation and decomposition processes, it is necessary to account for the history of the gas elements in the combustion chamber. Prompt NO may also be important in premixed gases.

The NO emissions have traditionally been calculated by means of the extended (thermal) Zel'dovich mechanism, with the atomic oxygen and hydrogen and hydroxyl radical assumed to be in chemical equilibrium. A steady-state assumption has often been used to calculate the atomic nitrogen concentration. However, during the expansion stroke, the atomic hydrogen and atomic oxygen and the hydroxyl radical may fail to recombine at rates sufficiently high to maintain shifting chemical equilibrium. The resulting persistence of excess atomic species and hydroxyl radicals may yield a hydrogen atom-to-hydroxyl radical ratio much greater than the ratio for chemical equilibrium.

The CO emissions are particularly important in fuel-rich mixtures and depend on the concentrations of atomic oxygen and hydrogen, molecular oxygen, hydroxyl radical, and carbon dioxide.

Engine knock is the result of the spontaneous ignition of a certain amount of unburnt fuel-air mixture in the combustion chamber before it can be consumed by a flame propagating through the cylinder charge; knock may be related to cool flames and two-stage ignition phenomena.

Thermodynamic and multidimensional models of spark-ignition engines have used the Shell model to predict knock. The Shell model includes a degenerated branched reaction; this reaction is responsible for autoignition in the end gas and for knock. Phenomenological models of engine knock have also been developed. Some of these models are based on ignition-delay time corrections.

Ignition phenomena occur in very short distances (e.g., the electrode gap) and cannot be adequately simulated by multidimensional models because of the required spatial and temporal resolution and the complex chemical reactions associated with ignition.

Future work on spark-ignition engine modeling will require improvements in the turbulence and combustion models, boundary conditions, and numerical methods as well as greater interaction between modelers, experimenters, and engine designers.

REFERENCES

1. Borman, G., and Nishiwaki, K., Internal-Combustion Engine Heat Transfer, *Progress in Energy and Combustion Science*, vol. 13, pp. 1–46, 1987.
2. Rao, V. K., and Bardon, M. F., Convective Heat Transfer in Reciprocating Engines, *Proceedings of the Institution of Mechanical Engineers*, vol. 199, no. D3, pp. 221–226, 1985.
3. Alkidas, A. C., and Myers, J. P., Transient Heat-Flux Measurements in the Combustion Chamber of a Spark-Ignition Engine, *ASME Journal of Heat Transfer*, vol. 104, no. 1, pp. 63–67, 1982.
4. Gatowski, J. A., Balles, E. N., Chun, K. M., Nelson, F. E., Ekchian, J. A., and Heywood, J. B., Heat Release Analysis of Engine Pressure Data, SAE International Congress and Exposition, Detroit, Mich., SAE Paper No. 841359, 1984.
5. Blumberg, P., and Kummer, J. K., Prediction of NO Formation in Spark-Ignition Engines—An Analysis of Methods of Control. *Combustion Science and Technology*, vol. 4, pp. 73–95, 1971.
6. Heywood, J. B., Engine Combustion Modeling—A Overview, in *Combustion Modeling in Reciprocating Engines*, eds. J. N. Mattavi and C. A. Amann, pp. 1–38, Plenum Press, New York, 1980.
7. Blumberg, P., Lavoie, G. A., and Tabaczynski, R. J., Phenomenological Models for Reciprocating Internal Combustion Engines, *Progress in Energy and Combustion Science*, vol. 5, pp. 123–167, 1979.
8. Caretto, L. S., Muzio, L. J., Sawyer, R. F., and Starkman, E. S., The Role of Kinetics in

Engine Emissions of Nitric Oxide, *Combustion Science and Technology*, vol. 3, pp. 53–61, 1971.

9. Heywood, J. B., Pollutant Formation and Control in Spark-Ignition Engines, in *Energy and Combustion Science*, ed. N. A. Chigier, student ed. 1, pp. 229–258, Pergamon Press, New York, 1979.
10. Henein, N. A., Analysis of Pollutant Formation and Control and Fuel Economy in Diesel Engines, in *Energy and Combustion Science*, ed. N. A. Chigier, student ed. 1, pp. 283–325, Pergamon Press, New York, 1979.
11. Tabaczynski, R. J., Turbulence and Turbulent Combustion in Spark-Ignition Engines, in *Energy and Combustion Science*, ed. N. A. Chigier, student ed. 1, pp. 259–281, Pergamon Press, New York, 1979.
12. Blizard, N. C., and Keck, J. C., Experimental and Theoretical Investigation of Turbulent Burning Model for Internal Combustion Engines, SAE International Congress and Exposition, Detroit, Mich., SAE Paper No. 740191, 1974.
13. Zeleknik, F. J., Combustion Modeling in Internal Combustion Engines, *Combustion Science and Technology*, vol. 12, pp. 159–164, 1976.
14. Zeleknik, F. J., and McBride, B. J., Modeling the Complete Otto Cycle—Preliminary Version, SAE International Congress and Exposition, Detroit, Mich., SAE Paper No. 770223, 1977.
15. Zeleknik, F. J., and McBride, B. J., Modeling the Internal Combustion Engine, NASA Reference Publication 1094, 1985.
16. Poulos, S. G., and Heywood, J. B., The Effect of Chamber Geometry on Spark-Ignition Engine Combustion, SAE International Congress and Exposition, Detroit, Mich., SAE Paper No. 830334, 1983.
17. Poulos, S. G., The Effect of Combustion Chamber Geometry on S.I. Engine Combustion Rates—A Modeling Study, M.S. thesis, Department of Mechanical Engineering, Massachusetts Institute of Technology, Cambridge, 1982.
18. Tabaczynski, R. J., Ferguson, C. R., and Radhakrishnan, K., A Turbulent Entrainment Model for Spark-Ignition Engine Combustion, SAE International Congress and Exposition, Detroit, Mich., SAE Paper No. 770647, 1977.
19. Belli, M., Bova, S. G., Danieli, G. A., Florio, G., and Leone, G., Predicting the Performance and Emission Characteristics of an Axisymmetric Spark Ignition Engine, SAE International Congress and Exposition, Detroit, Mich., SAE Paper No. 810152, 1981.
20. Ferguson, C. R., and Keck, J. C., On Laminar Flame Quenching and Its Application to Spark Ignition Engines, *Combustion and Flame*, vol. 28, pp. 197–205, 1977.
21. Goolsby, A. D., and Haskell, W. W., Flame-Quench Distance Measurements in a CFR Engine, *Combustion and Flame*, vol. 26, pp. 105–114, 1976.
22. Hires, S. D., Tabaczynski, R. J., and Novak, J. M., The Prediction of Ignition Delay and Combustion Intervals for a Homogeneous Charge, Spark Ignition Engine, SAE International Congress and Exposition, Detroit, Mich., SAE Paper No. 780232, 1978.
23. Lucas, G. G., and Brunt, M. F. J., The Effect of Combustion Chamber Shape on the Rate of Combustion in a Spark Ignition Engine, SAE International Congress and Exposition, Detroit, Mich., SAE Paper No. 820165, 1982.
24. Groff, E. G., and Matekunas, F. A., The Nature of Turbulent Flame Propagation in a Homogeneous Spark-ignited Engine, SAE International Congress and Exposition, Detroit, Mich., SAE Paper No. 800133, 1980.
25. Groff, E. G., An Experimental Evaluation of an Entrainment Flame-Propagation Model, *Combustion and Flame*, vol. 67, pp. 153–162, 1987.
26. Tabaczynski, R. J., Trinker, F. H., and Shannon, B. A. S., Further Refinement and Validation of a Turbulent Flame Propagation Model for Spark-Ignition Engines, *Combustion and Flame*, vol. 39, pp. 111–121, 1980.
27. Tagalian, J., and Heywood, J. B., Flame Initiation in a Spark-Ignition Engine, *Combustion and Flame*, vol. 64, pp. 243–246, 1986.
28. Kalghatgi, G. T., Early Flame Development in a Spark-Ignition Engine, *Combustion and Flame*, vol. 60, pp. 299–308, 1985.

29. Gatowski, J. A., Heywood, J. B., and Deleplace, C., Flame Photographs in a Spark-Ignition Engine, *Combustion and Flame*, vol. 56, pp. 71–81, 1984.
30. Hanson, R. J., and Thomas, A., Flame Development in Swirling Flows in Closed Vessels, *Combustion and Flame*, vol. 55, pp. 255–277, 1984.
31. Heywood, J. B., and Vilchis, F. R., Comparison of Flame Development in a Spark-Ignition Engine Fueled with Propane and Hydrogen, *Combustion Science and Technology*, vol. 38, pp. 313–324, 1984.
32. Zur Loye, A. O., and Bracco, F. V., Two-Dimensional Visualization of Ignition Kernels in an IC Engine, *Combustion and Flame*, vol. 69, pp. 59–69, 1987.
33. Keck, J. C., Turbulent Flame Structure and Speed in Spark-Ignition Engines, *Nineteenth Symposium (International) on Combustion*, pp. 1451–1466, The Combustion Institute, Pittsburgh, Pa., 1982.
34. DeSoete, G. G., Effects of Geometrical and Aerodynamic Induced Flame Stretch on the Propagation of Spark Fired Premixed Flames in Early Stages, *Twentieth Symposium (International) on Combustion*, pp. 161–168, The Combustion Institute, Pittsburgh, Pa., 1985.
35. Chomiak, J., and Jarosinski, J., Flame Quenching by Turbulence, *Combustion and Flame*, vol. 48, pp. 241–249, 1982.
36. Thomas, A., Flame Development in Spark-Ignition Engines, *Combustion and Flame*, vol. 50, pp. 305–322, 1983.
37. Beretta, G. P., Rashidi, M., and Keck, J. C., Turbulent Flame Propagation and Combustion in Spark Ignition Engines, *Combustion and Flame*, vol. 52, pp. 217–245, 1983.
38. Beretta, G. P., and Keck, J. C., Energy and Entropy Balances in a Combustion Chamber: Analytical Solution, *Combustion Science and Technology*, vol. 30, pp. 19–29, 1983.
39. Arici, O., Tabaczynski, R. J., and Arpaci, V. S., A Model for the Lean Misfire Limit in Spark-Ignition Engines, *Combustion Science and Technology*, vol. 30, pp. 31–45, 1983.
40. Anderson, R. W., and Lin, M. T., Investigation of Misfire in a Fast Burn Spark Ignition Engine, *Combustion Science and Technology*, vol. 43, pp. 183–196, 1985.
41. Tennekes, H., Simple Model for the Small-Scale Structure of Turbulence, *Physics of Fluids*, vol. 11, no. 3, pp. 669–671, 1968.
42. Tennekes, H., and Lumley, J. L., *A First Course in Turbulence*, MIT Press, Cambridge, Mass., 1972.
43. Annand, W. J. D., The Estimation of Flame Propagation Rates in Routine Computer Synthesis of Spark-Ignition Engine Combustion, *International Conference on Combustion Engineering*, vol. 1, pp. 125–134, I. Mech. E. Conference Publications 1983-3, London, 1983.
44. Tabaczynski, R. J., Turbulence Measurements and Modelling in Reciprocating Engines—An Overview, *International Conference on Combustion Engineering*, vol. 1, pp. 51–64, I. Mech. E. Conference Publications 1983-3, London, 1983.
45. Tabaczynski, R. J., Personal communication, Dubrovnik, Yugoslavia, August 25, 1987.
46. Daneshyar, H., and Hill, P. G., The Structure of Small-Scale Turbulence and Its Effect on Combustion in Spark Ignition Engines, *Progress in Energy and Combustion Science*, vol. 13, pp. 47–73, 1987.
47. Milane, R. E., Tabaczynski, R. J., and Arpaci, V. S., A Stochastic Model of Turbulent Mixing for the Prediction of Burn Rate in a Spark-Ignition Engine, *Combustion Science and Technology*, vol. 32, pp. 211–235, 1983.
48. Lavoie, G. A., Heywood, J. B., and Keck, J. C., Experimental and Theoretical Study of Nitric Oxide Formation in Internal Combustion Engines, *Combustion Science and Technology*, vol. 1, pp. 313–326, 1976.
49. Komiyama, K., and Heywood, J. B., Predicting NO_x Emissions and Effects of Exhaust Gas Recirculation in Spark-Ignition Engines, *SAE Transactions*, vol. 83, pp. 1458–1476, 1973.
50. Rashidi, M., The Nature of Cycle-by-Cycle Variations in the S.I. Engine from High Speed Photographs, *Combustion and Flame*, vol. 42, pp. 111–122, 1981.
51. Akindele, O. O., Bradley, D., Mak, P. W., and McMahon, M., Spark Ignition of Turbulent Gases, *Combustion and Flame*, vol. 47, pp. 129–155, 1982.

52. Hansel, J. G., Turbulent Combustion Model of Cycle-to-Cycle Combustion Variations in Spark-Ignition Engines, *Combustion Science and Technology*, vol. 2, pp. 223–225, 1970.
53. Tromans, P. S., The Interaction between Strain Fields and Flames—A Possible Source of Combustion Variations in Spark Ignition Engines, in *Fluid Mechanics of Combustion Systems*, eds. T. Morel, R. P. Lohmann, and J. M. Rackley, pp. 201–206, ASME, New York, 1981.
54. Chomiak, J., Flame Development from an Ignition Kernel in Laminar and Turbulent Homogeneous Mixtures, *Seventeenth Symposium (International) on Combustion*, pp. 255–263, The Combustion Institute, Pittsburgh, Pa., 1979.
55. Matekunas, F. A., A Schlieren Study of Combustion in a Rapid Compression Machine Simulating the Spark Ignition Engine, *Seventeenth Symposium (International) on Combustion*, pp. 1283–1294, The Combustion Institute, Pittsburgh, Pa., 1979.
56. Cole, J. B., and Swords, M. D., Optical Studies on Turbulence in an Internal Combustion Engine, *Seventeenth Symposium (International) on Combustion*, pp. 1295–1303, The Combustion Institute, Pittsburgh, Pa., 1979.
57. Gerber, A., Ecoulements Instationnaires a l'Interieur de Moteurs a Piston, Ph. D. thesis, École Centrale de Lyon, Lyon, France, 1982.
58. Cole, J. B., and Swords, M. D., On the Correlation between Gas Velocity and Combustion Pressure Fluctuations in a Spark-Ignition Engine, *Eighteenth Symposium (International) on Combustion*, pp. 1837–1846, The Combustion Institute, Pittsburgh, Pa., 1981.
59. Wakisaka, T., and Hamamoto, Y., Effects of Mixture Turbulence on the Limits of Flame Propagation, *Bulletin of the JSME*, vol. 22, no. 165, pp. 382–389, 1979.
60. Hamamoto, Y., Wakisaka, T., and Ohnishi, M., Cycle-to-Cycle Fluctuation of Lean Mixture Combustion in Spark-Ignition Engines, *Bulletin of the JSME*, vol. 25, no. 199, pp. 61–67, 1982.
61. Martin, J. K., Witze, P. O., and Borgnakke, C., Combustion Effects on the Preflame Flow Field in a Research Engine, SAE International Congress and Exposition, Detroit, Mich., SAE Paper No. 850122, 1985.
62. Davis, G. C., Tabaczynski, R. J., and Belaire, R. C., The Effect of Intake Valve Lift on Turbulence Intensity and Burn Rate in S.I. Engines—Model versus Experiment, SAE International Congress and Exposition, Detroit, Mich., SAE Paper No. 840030, 1984.
63. Kolmogorov, A. N., A Refinement of Previous Hypotheses Concerning the Local Structure of Turbulence in a Viscous Incompressible Fluid at High Reynolds Number, *Journal of Fluid Mechanics*, vol. 13, no. 1, pp. 82–85, 1962.
64. Belaire, R. C., Davis, G. C., Kent, J. C., and Tabaczynski, R. J., Combustion Chamber Effects on Burn Rates in a High Swirl Spark Ignition Engine, SAE International Congress and Exposition, Detroit, Mich., SAE Paper No. 830335, 1983.
65. Borgnakke, C., Arpaci, V. S., and Tabaczynski, R. J., A Model for the Instantaneous Heat Transfer and Turbulence in a Spark Ignition Engine, SAE International Congress and Exposition, Detroit, Mich., SAE Paper No. 800287, 1980.
66. Borgnakke, C., Davis, G. C., and Tabaczynski, R. J., Predictions of In-Cylinder Swirl Velocity and Turbulence Intensity for an Open Chamber Cup in Piston Engine, SAE International Congress and Exposition, Detroit, Mich., SAE Paper No. 810224, 1981.
67. Davis, G. C., and Borgnakke, C., The Effect of In-Cylinder Flow Processes (Swirl, Squish and Turbulence Intensity) on Engine Efficiency—Model Predictions, SAE International Congress and Exposition, Detroit, Mich., SAE Paper No. 820045, 1982.
68. Witze, P. O., Martin, J. K., and Borgnakke, C., Measurements and Predictions of the Precombustion Fluid Motion and Combustion Rates in a Spark Ignition Engine, SAE International Congress and Exposition, Detroit, Mich., SAE Paper No. 831697, 1983.
69. Mathur, H. B., Gajendra Babu, M. K., and Subba Reddi, K., A Thermodynamic Simulation Model for a Methanol Fueled Spark Ignition Engine, SAE International Congress and Exposition, Detroit, Mich., SAE Paper No. 830333, 1983.
70. Hiroyasu, H., and Kadota, T., Computer Simulation for Combustion and Exhaust Emissions on Spark-Ignition Engine, *Fifteenth Symposium (International) on Combustion*, pp. 1213–1223, The Combustion Institute, Pittsburgh, Pa., 1975.

71. Fagelson, J. J., McLean, W. J., and de Boer, P. C. T., Performance and NO_x Emissions of Spark Ignited Combustion Engines Using Alternative Fuels—Quasi One-Dimensional Modeling I. Hydrogen Fueled Engines, *Combustion Science and Technology*, vol. 18, pp. 47–57, 1978.
72. Rubin, M. B., and McLean, W. J., Performance and NO_x Emissions of Spark Ignited Combustion Engines Using Alternative Fuels—Quasi One-dimensional Modeling II. Methanol Fueled Engines, *Combustion Science and Technology*, vol. 18, pp. 199–206, 1978.
73. Andrews, G. E., Bradley, D., and Lwakabamba, S. B., Turbulence and Turbulent Flame Propagation—A Critical Appraisal, *Combustion and Flame*, vol. 24, pp. 285–304, 1975.
74. Haghgooie, M., Kent, J. C., and Tabaczynski, R. J., Turbulent Time-Scale Measurements in a Spark-Ignition Engine Using Hot Wire Anemometry and Fast Response Ion Probes, in *Flows in Internal Combustion Engines*, ed. T. Uzkan, pp. 1–7, ASME, New York, 1982.
75. Kido, H., Wakuri, Y., and Murase, E., Measurements of Spatial Scales and a Model for Small-Scale Structure of Turbulence in an Internal Combustion Engine, *ASME-JSME Thermal Engineering Joint Conference Proceedings, Honolulu, Hawaii*, vol. 4, pp. 191–198, ASME, New York, 1983.
76. Kneen, C., Woods, W. A., and Leung, K. K., Experiments on the Influence of an Intake Centre Body Port on the Turbulence and Flame Speed in a Spark Ignition Engine, *International Conference on Combustion Engineering*, vol. 1, pp. 143–151, I. Mech. E. Conference Publications 1983-3, London, 1983.
77. Nagao, A., and Tanaka, K., The Effect of Swirl Control on Combustion Improvement of Spark Ignition Engine, *International Conference on Combustion Engineering*, vol. 1, pp. 153–161, I. Mech. E. Conference Publications 1983-3, London, 1983.
78. Smith, J. R., The Influence of Turbulence on Flame Structure in an Engine, in *Flows in Internal Combustion Engines*, ed. T. Uzkan, pp. 67–72, ASME, New York, 1982.
79. Kalghatgi, G. T., and Swords, M. D., Flame-Speed Measurements in an Internal Combustion Engine, *Combustion and Flame*, vol. 49, pp. 163–169, 1983.
80. Bonini, J., Xia, L. Q., Chau, E., Hom, K., Stewart, H. E., Sawyer, R. F., Oppenheim, A. K., and Iiyama, A., Visualization of Flow and Combustion Processes in a Square Piston Engine Simulator, SAE International Congress and Exposition, Detroit, Mich., SAE Paper No. 870452, 1987.
81. Namazian, M., Hansen, S., Lyford-Pike, E., Sanchez-Barsse, J., Heywood, J., and Rife, J., Schlieren Visualization of the Flow and Density Fields in the Cylinder of a Spark-Ignition Engine, SAE International Congress and Exposition, Detroit, Mich., SAE Paper No. 800044, 1980.
82. Rassweiler, G. M., and Withrow, Ll., Motion Pictures of Engine Flames Correlated with Pressure Cards, SAE International Congress and Exposition, Detroit, Mich., SAE Paper No. 800131, 1980 (originally presented in 1938).
83. Metghalchi, M., and Keck, J. C., Laminar Burning Velocity of Propane-Air Mixtures at High Temperature and Pressure, *Combustion and Flame*, vol. 38, pp. 143–154, 1980.
84. Rhodes, D. B., Laminar Burning Speed Measurements of Indolene-Air-Diluent Mixtures at High Pressures and Temperatures, M. S. thesis, Department of Mechanical Engineering, Massachusetts Institute of Technology, Cambridge, 1981.
85. Lavoie, G. A., and Blumberg, P. N., Measurements of NO Emissions from a Stratified Charge Engine: Comparisons of Theory and Experiment, *Combustion Science and Technology*, vol. 8, pp. 25–37, 1973.
86. Mattavi, J. N., Groff, E. G., Lienesch, J. H., Matekunas, F. A., and Noyes, R. N., Engine Improvements through Combustion Modeling, in *Combustion Modeling in Reciprocating Engines*, eds. J. N. Mattavi and C. A. Amann, pp. 537–587, Plenum Press, New York, 1980.
87. Lavoie, G. A., and Blumberg, P. N., A Fundamental Model for Predicting Fuel Consumption, NO_x and HC Emissions of the Conventional Spark-Ignited Engine, *Combustion Science and Technology*, vol. 21, pp. 225–258, 1980.
88. Blumberg, P. N., Nitric Oxide Emissions from Stratified Charge Engines: Prediction and Control, *Combustion Science and Technology*, vol. 8, pp. 5–24, 1973.

89. Daneshyar, H., and Watfa, M., Prediction of Nitric Oxide (NO) Concentration in Spark-Ignition Engines, *Combustion and Flame*, vol. 21, pp. 395–398, 1973.
90. Hires, S. D., Ekchian, A., Heywood, J. B., Tabaczynski, R. J., and Wall, J. C., Performance and NO_x Emissions Modeling of a Jet Ignition Prechamber Stratified Charge Engine, SAE International Congress and Exposition, Detroit, Mich., SAE Paper No. 760161, 1976.
91. Heywood, J. B., and Watts, P. A., Parametric Studies of Fuel Consumption and NO Emissions of Dilute Spark-Ignition Engine Operation Using a Cycle Simulation, in *Fuel Economy and Emissions of Lean Burn Engines*, pp. 117–127, I. Mech. E. Conference Publications 1979-9, London, 1979.
92. Watts, P. A., Simulation Studies of the Effects of Lean Operation, Turbocharging and Heat Transfer on Spark Ignition Engines, M. S. thesis, Massachusetts Institute of Technology, Cambridge, 1979.
93. Watts, P. A., and Heywood, J. B., Simulation Studies of the Effects of Turbocharging and Reduced Heat Transfer on Spark-Ignition Engine Operation, SAE International Congress and Exposition, Detroit, Mich., SAE Paper No. 800289, 1980.
94. Sirignano, W. A., One-Dimensional Analysis of Combustion in a Spark-Ignition Engine, *Combustion Science and Technology*, vol. 7, pp. 99–108, 1973.
95. Payri, F., Desantes, J. M., Tinaut, F. V., and Ramos, J. I., Modelo Unidimensional de la Propagacion de la Llama en Motores de Encendido Provocado, *Anales de Ingenieria Mecanica*, vol. 3, no. 1, pp. 381–389, 1985.
96. Ramos, J. I., Comparisons between Thermodynamic and One-Dimensional Combustion Models of Spark-Ignition Engines, *Applied Mathematical Modelling*, vol. 10, no. 6, pp. 409–422, 1986.
97. Hayder, M. E., Varma, A. K., and Bracco, F. V., A Limit to TDC Turbulence Intensity in Internal Combustion Engines, *AIAA Journal of Propulsion and Power*, vol. 1, no. 4, pp. 300–308, 1985.
98. Grasso, F., and Bracco, F. V., Computed and Measured Turbulence in Axisymmetric Reciprocating Engines, *AIAA Journal*, vol. 21, no. 4, pp. 601–607, 1983.
99. Grasso, F., and Bracco, F. V., Sensitivity of Chamber Turbulence to Intake Flows in Axisymmetric Reciprocating Engines, *AIAA Journal*, vol. 21, no. 4, pp. 637–640, 1983.
100. Carpenter, M. H., and Ramos, J. I., Modelling of a Gasoline-Injected Two-Stroke Diesel Engine, SAE International Congress and Exposition, Detroit, Mich., SAE Paper No. 860167, 1986.
101. Nguyen, H. L., Carpenter, M. H, and Ramos, J. I., The Effects of Engine Speed and Injection Characteristics on the Flow Field and Fuel/Air Mixing in Motored Two-Stroke Diesel Engines, AIAA 25th Aerospace Sciences Meeting, Reno, Nev., Paper No. AIAA-87-0227, 1987.
102. Nguyen, H. L., Schock, H. J., Carpenter, M. H., Ramos, J. I., and Stegeman, J. D., Numerical Simulation of the Flow Field and Fuel Sprays in an IC Engine, SAE International Congress and Exposition, Detroit, Mich., SAE Paper No. 870599, 1987.
103. Carpenter, M. H., and Ramos, J. I., Mathematical Modelling of Uniflow-Scavenged Two-Stroke Diesel Engines, in *Modelado en Motores de Combustion Interna*, eds. F. Payri, J. M. Desantes, and F. Boada, pp. 321–332, Universidad Politecnica de Valencia, Valencia, Spain, 1987.
104. Schock, H. J., Sosoka, D. J., and Ramos, J. I., Numerical Studies of the Formation and Destruction of Vortices in a Motored Four-Stroke Piston-Cylinder Configuration, AIAA 21st Aerospace Sciences Meeting, Reno, Nev., Paper No. AIAA-83-0497, 1983.
105. Schock, H. J., Sosoka, D. J., and Ramos, J. I., Formation and Destruction of Vortices in a Motored Four-Stroke Piston-Cylinder Configuration, *AIAA Journal*, vol. 22, no. 7, pp. 948–949, 1984.
106. Carpenter, M. H., and Ramos, J. I., Spray Penetration, Vaporization and Mixing in Diesel Engines, XIX International Symposium on Heat and Mass Transfer in Gasoline and Diesel Engines, International Centre for Heat and Mass Transfer, Dubrovnik, Yugoslavia, 1987.
107. Reynolds, W. C., Modeling of Fluid Motions in Engines—An Introduction Overview, in *Combustion Modeling in Reciprocating Engines*, eds. J. N. Mattavi and C. A. Amann, pp. 41–68, Plenum Press, New York, 1980.

108. Wong, V. W., and Hoult, D. P., Rapid Distortion Theory Applied to Turbulent Combustion, SAE International Congress and Exposition, Detroit, Mich., SAE Paper No. 790357, 1979.
109. Hoult, D. P., and Wong, V. W., The Generation of Turbulence in an Internal Combustion Engine, in *Combustion Modeling in Reciprocating Engines*, eds. J. N. Mattavi and C. A. Amann, pp. 131–160, Plenum Press, New York, 1980.
110. Westbrook, C. K., Fuel Motion and Pollutant Formation in Stratified Charge Combustion, SAE International Congress and Exposition, Detroit, Mich., SAE Paper No. 790248, 1979.
111. Singh, T., and Surakomol, K., Mathematical Modeling of Combustion Process in a Spark-Ignition Engine, SAE International Congress and Exposition, Detroit, Mich., SAE Paper No. 790354, 1979.
112. Bellan, J. R., and Sirignano, W. A., A Theory of Turbulent Flame Development and Nitric Oxide Formation in Stratified Charge Internal Combustion Engines, *Combustion Science and Technology*, vol. 8, pp. 51–68, 1973.
113. Bellan, J. R., and Sirignano, W. A., Combustion and NO Formation in a Stratified-Charge Engine: A Two-Turbulent Equations Model, *Combustion Science and Technology*, vol. 12, pp. 75–104, 1976.
114. Westbrook, C. K., Propagation of a Flame through a Stratified Charge Combustion Chamber, *Acta Astronautica*, vol. 5, pp. 1185–1198, 1978.
115. Westbrook, C. K., A Generalized ICE Method for Chemically Reactive Flows in Combustion Systems, *Journal of Computational Physics*, vol. 29, no. 1, pp. 67–80, 1980.
116. O'Rourke, P. J., Personal communication, Dubrovnik, Yugoslavia, August 31, 1987.
117. Sorenson, S. C., and Pan, S. S., A One-Dimensional Combustion Model for a Dual Chamber Stratified Charge Spark Ignition Engine, SAE International Congress and Exposition, Detroit, Mich., SAE Paper No. 790355, 1979.
118. MacCormack, R. W., The Effect of Viscosity in Hypervelocity Impact Cratering, AIAA Hypervelocity Impact Conference, Cincinnati, Ohio, Paper No. AIAA-69-354, 1969.
119. Ramos, J. I., Numerical Methods for One-Dimensional Reaction-Diffusion Equations Arising in Combustion Theory, *Annual Review of Numerical Fluid Mechanics and Heat Transfer*, vol. 1, pp. 150–261, 1987.
120. Carpenter, M. H., and Ramos, J. I., Numerical Studies of the Combustion Process in a Four-Stroke Internal Combustion Engine, in *NUMETA 85, Numerical Methods in Engineering: Theory and Applications*, eds. J. Middleton and G. N. Pande, vol. 2, pp. 669–677, Balkema, Rotterdam, The Netherlands, 1985.
121. Ramos, J. I., and Sirignano, W. A., Turbulent Flow Field in Homogeneous-Charge, Spark-Ignition Engines, *Eighteenth Symposium (International) on Combustion*, pp. 1825–1835, The Combustion Institute, Pittsburgh, Pa., 1981.
122. Carpenter, M. H., and Ramos, J. I., Mathematical Modelling of Spark-Ignition Engines, *Applied Mathematical Modelling*, vol. 9, no. 1, pp. 40–52, 1985.
123. Ramos, J. I., Two-Dimensional Modelling of Spark-Ignition Engines, in *Modelado en Motores de Combustion Interna*, eds. F. Payri, J. M. Desantes, and F. Boada, pp. 217–231, Universidad Politecnica de Valencia, Valencia, Spain, 1987.
124. Ahmadi-Befrui, B., Gosman, A. D., Lockwood, F. C., and Watkins, A. P., Multidimensional Calculation of Combustion in an Idealised Homogeneous Charge Engine: A Progress Report, SAE International Congress and Exposition, Detroit, Mich., SAE Paper No. 810151, 1981.
125. Przekwas, A., Modeling of Fluid Flow, Heat Transfer and Flame Propagation in Reciprocating Engines, in *Flows in Internal Combustion Engines*, ed. T. Uzkan, pp. 73–79, ASME, New York, 1982.
126. Abraham, J., Bracco, F. V., and Reitz, R. D., Comparisons of Computed and Measured Premixed Charge Engine Combustion, *Combustion and Flame*, vol. 60, pp. 309–332, 1985.
127. Bray, K. N. C., Turbulent Flows with Premixed Reactants, in *Turbulent Reacting Flows*, eds. P. A. Libby and F. A. Williams, *Topics in Applied Physics*, vol. 44, pp. 115–183, Springer-Verlag, New York, 1980.
128. Gilaber, P., Validation of a Multidimensional Heat Transfer Model for S.I. Engines, XIX

International Symposium on Heat and Mass Transfer in Gasoline and Diesel Engines, International Centre for Heat and Mass Transfer, Dubrovnik, Yugoslovia, 1987.

129. Pinchon, P., and Baritaud, T., Modeling of Flow and Combustion in a Spark Ignition Engine with a Shrouded Valve: Comparison with Experiments, XIX International Symposium on Heat and Mass Transfer in Gasoline and Diesel Engines, International Centre for Heat and Mass Transfer, Dubrovnik, Yugoslavia, 1987.

130. Amsden, A. A., Ramshaw, J. D., O'Rourke, P. J., and Dukowicz, J. K., KIVA: A Computer Program for Two- and Three-Dimensional Fluid Flows with Chemical Reactions and Fuel Sprays, Report No. LA-10245-MS, Los Alamos National Laboratory, Los Alamos, N. M., 1985.

131. Rivard, W. C., Butler, T. D., and Farmer, O. A., The Transient Dynamics of Chemically Reactive Gaseous Mixtures, in *Proceedings of the Fourth International Conference on Numerical Methods in Fluid Dynamics*, ed. R. D. Richtmyer, *Lecture Notes in Physics*, vol. 35, pp. 334–340, Springer-Verlag, New York, 1975.

132. Butler, T. D., and O'Rourke, P. J., A Numerical Method for Two-Dimensional Unsteady Reacting Flows, *Sixteenth Symposium (International) on Combustion*, pp. 1503–1515, The Combustion Institute, Pittsburgh, Pa., 1977.

133. Bracco, F. V., Scaling of Some Flow Fields with Flames, in *Combustion Modeling in Reciprocating Engines*, eds. J. N. Mattavi and C. A. Amann, pp. 299–325, Plenum Press, New York, 1980.

134. Bracco, F. V., and O'Rourke, P. J., A Review of Initial Comparisons of Computed and Measured Two-Dimensional Unsteady Flame Fields, *Progress in Energy and Combustion Science*, vol. 7, pp. 103–124, 1981.

135. Diwakar, R., Assessment of the Ability of a Multidimensional Computer Code to Model Combustion in a Homogeneous-Charge Engine, SAE International Congress and Exposition, Detroit, Mich., SAE Paper No. 840230, 1984.

136. Spalding, D. B., Towards a Two-Fluid Model of Turbulent Combustion in Gases, with Special References to Spark-Ignition Engine, *International Conference on Combustion Engineering*, vol. 1, pp. 135–142, I. Mech. E. Conference Publications 1983-3, London, 1983.

137. Abraham, J., Williams, F. A, and Bracco, F. V., A Discussion of Turbulent Flame Structure in Premixed Charges, SAE International Congress and Exposition, Detroit, Mich., SAE Paper No. 850345, 1985.

138. Gosman, A. D., Aspects of the Simulation of Combustion in Reciprocating Engines, in *Numerical Simulation of Combustion Phenomena*, eds. R. Glowinski, B. Larrouturou, and R. Temam, *Lecture Notes in Physics*, vol. 241, pp. 46–73, Springer-Verlag, New York, 1985.

139. Smith, J. R., Turbulent Flame Structure in a Homogeneous-Charge Engine, SAE International Congress and Exposition, Detroit, Mich., SAE Paper No. 820043, 1982.

140. Hakberg, B., and Gosman, A. D., Analytical Determination of Turbulent Flame Speed from Combustion Models, *Twentieth Symposium (International) on Combustion*, pp. 225–232, The Combustion Institute, Pittsburgh, Pa., 1985.

141. Libby, P. A., Bray, K. N. C., and Moss, J. B., Effects of Finite Reaction Rate and Molecular Transport in Premixed Turbulent Combustion, *Combustion and Flame*, vol. 34, pp. 285–301, 1979.

142. Magnussen, B. F., and Hjertager, B. H., On Mathematical Modeling of Turbulent Combustion with Special Emphasis on Soot Formation, *Sixteenth Symposium (International) on Combustion*, pp. 719–729, The Combustion Institute, Pittsburgh, Pa., 1977.

143. Benjamin, S. F., Weaving, J. H., Glynn, D. R., Markatos, N. C., and Spalding, D. B., Development of a Mathematical Model of Flow, Heat Transfer and Combustion in a Stratified Charge Engine, *Stratified Charge Automotive Engines*, pp. 91–99, I. Mech. E. Conference Publications 1980-9, London, 1980.

144. Abdel-Gayed, R. G., Bradley, D., and McMahon, M., Turbulent Flame Propagation in Premixed Gases, *Seventeenth Symposium (International) on Combustion*, pp. 245–254, The Combustion Institute, Pittsburgh, Pa., 1979.

145. Abdel-Gayed, R. G., and Bradley, D., Dependence of Turbulent Burning Velocity on Turbulent Reynolds Number and Ratio of Laminar Burning Velocity to R.M.S. Turbulent Velocity, *Sixteenth Symposium (International) on Combustion*, pp. 1725–1735, The Combustion Institute, Pittsburgh, Pa., 1977.
146. Ballal, D. R., and Lefebvre, A. H., Turbulence Effects on Enclosed Flames, *Acta Astronautica*, vol. 1, pp. 471–483, 1974.
147. Ballal, D. R., An Experimental Study of Flame Turbulence, *Acta Astronautica*, vol. 5, pp. 1095–1112, 1978.
148. Chomiak, J., Dissipation Fluctuations and the Structure and Propagation of Turbulent Flames in Premixed Gases at High Reynolds Numbers, *Sixteenth Symposium (International) on Combustion*, pp. 1665–1673, The Combustion Institute, Pittsburgh, Pa., 1977.
149. Rask, R. B., Laser Doppler Anemometer Measurements in an Internal Combustion Engine, SAE International Congress and Exposition, Detroit, Mich., SAE Paper No. 790094, 1979.
150. Bray, K. N. C., Libby, P. A., Masuya, G., and Moss, J. B., Turbulence Production in Premixed Turbulent Flames, *Combustion Science and Technology*, vol. 25, pp. 127–140, 1981.
151. Lancaster, D. R., Effects of Engine Variables on Turbulence in a Spark-Ignition Engine, SAE International Congress and Exposition, Detroit, Mich., SAE Paper No. 760159, 1976.
152. Lancaster, D. R., Krieger, R. B., Sorenson, S. C., and Hull, W. L., Effects of Turbulence on Spark-Ignition Engine Combustion, SAE International Congress and Exposition, Detroit, Mich., SAE Paper No. 760160, 1976.
153. Witze, P. O., Hot-Wire Turbulence Measurements in a Motored Internal Combustion Engine, Report No. SAND-75-8641, Sandia National Laboratories, Livermore, Calif., 1975.
154. Witze, P. O., Hot-Wire Measurements of the Turbulence Structure in a Motored Spark-Ignition Engine, Report No. SAND-75-8697, Sandia National Laboratories, Livermore, Calif., 1975.
155. Witze, P. O., Measurements of the Spatial Distribution and Engine Dependence of Turbulent Air Motion in an I.C. Engine, SAE International Congress and Exposition, Detroit, Mich., SAE Paper No. 770220, 1977.
156. Witze, P. O., A Critical Comparison of Hot-Wire Anemometry and Laser Doppler Velocimetry for I.C. Engine Applications, SAE International Congress and Exposition, Detroit, Mich., SAE Paper No. 800132, 1980.
157. Witze, P. O., and Vilchis, F. R., Stroboscopic Laser Shadowgraph Study of the Effect of Swirl on Homogeneous Combustion in a Spark-Ignition Engine, SAE International Congress and Exposition, Detroit, Mich., SAE Paper No. 810226, 1981.
158. Witze, P. O., Martin, J. K., and Borgnakke, C., Conditionally-Sampled Velocity and Turbulence Measurements in a Spark Ignition Engine, *Combustion Science and Technology*, vol. 36, pp. 301–317, 1984.
159. Martin, J. K., Witze, P. O., and Borgnakke, C., Multiparameter Sampled Laser Velocimetry Measurements during Flame Propagation in a Spark-Ignition Engine, *Twentieth Symposium (International) on Combustion*, pp. 29–36, The Combustion Institute, Pittsburgh, Pa., 1985.
160. Spicher, U., and Velji, A., Measurements of Spatial Flame Propagation and Flow Velocities in a Spark-Ignition Engine, *Twentieth Symposium (International) on Combustion*, pp. 19–27, The Combustion Institute, Pittsburgh, Pa., 1985.
161. Wakisaka, T., Hamamoto, Y., Ohigashi, S., and Hashimoto, M., Measurements of Air Swirl and its Turbulence Characteristics in the Cylinder of an Internal Combustion Engine, in *Fuel Economy and Emissions of Lean Burn Engines*, pp. 51–62, I. Mech. E. Conference Publications 1979-9, London, 1979.
162. Wakisaka, T., Hamamoto, Y., and Shimamoto, Y., Turbulence Structure of Air Swirl in Reciprocating Engine Cylinders, ASME Winter Annual Meeting, Phoenix, Ariz., 1982.
163. Wakisaka, T., Hamamoto, Y., and Kinoshita, S., The Measurement of Turbulence in the Combustion Chamber of an I.C. Engine, *JSAE Review*, no. 3, pp. 9–16, 1980.
164. Fraser, R. A., Felton, P. G., and Bracco, F. V., Preliminary Turbulence Length Scale Measurements in a Motored IC Engine, SAE International Congress and Exposition, Detroit, Mich., SAE Paper No. 860021, 1986.
165. Liou, T.-M., Hall, M., Santavicca, D. A., and Bracco, F. V., Laser Doppler Velocimetry

Measurements in Valved and Ported Engines, SAE International Congress and Exposition, Detroit, Mich., SAE Paper No. 840375, 1986.

166. Liou, T.-M., and Santavicca, D. A., Cycle Resolved Turbulence Measurements in a Ported Engine with and without Swirl, SAE International Congress and Exposition, Detroit, Mich., SAE Paper No. 830419, 1983.
167. Liou, T.-M., and Santavicca, D. A., Cycle Resolved LDV Measurements in a Motored IC Engine, *ASME Journal of Fluids Engineering*, vol. 107, pp. 232–240, 1985.
168. James, E. H., and Lucas, G. G., Turbulent Flow in Spark Ignition Engine Combustion Chambers, SAE Congress and Exposition, Detroit, Mich., SAE Paper No. 750885, 1975.
169. Catania, A. E., and Mittica, A., Cycle-by-Cycle Correlation and Spectral Analysis of I.C. Engine Turbulence, in *International Symposium on Flows in Internal Combustion Engines—III*, eds. T. Uzkan, W. G. Tiederman, and J. M. Novak, pp. 39–46, ASME, New York, 1985.
170. Fraser, R. A., and Bracco, F. V., Measurements of Integral Length Scales in an I.C. Engine, in *Modelado en Motores de Combustion Interna*, eds. F. Payri, J. M. Desantes, and F. Boada, pp. 6–15, Universidad Politecnica de Valencia, Valencia, Spain, 1987.
171. Damkohler, G. Z., Elektrochemie Angewandte, *Zeitschrift für Elektrochemie*, vol. 46, p. 601, 1940; also NACA Technical Memorandum 1112, 1947.
172. Shchelkin, K. I., and Troshin, Y. K., *Gasdynamics of Combustion*, Monobook, Baltimore, Md., 1965.
173. Karlovitz, B., Denniston, D. W., Knapschaefer, D. H., and Wells, F. E., Studies on Turbulent Flames: A. Flame Propagation across Velocity Gradients. B. Turbulence Measurements in Flames, *Fourth Symposium (International) on Combustion*, pp. 613–620, Williams & Wilkins, Baltimore, Md., 1953.
174. Williams, F. A., *Combustion Theory*, 2nd ed., pp. 373–445, Benjamin-Cummings, Menlo Park, Calif., 1985.
175. Basso, A., Optimization of Combustion Chamber Design for Spark Ignition Engines, SAE International Congress and Exposition, Detroit Mich., SAE Paper No. 840231, 1984.
176. Basso, A., and Rinolfi, R., Two-Dimensional Computations of Engine Combustion: Comparison of Measurements and Predictions, SAE International Congress and Exposition, Detroit, Mich., SAE Paper No. 820519, 1982.
177. Gupta, H. C., REC: A Computer Program for the Computation of Combustion in Open and Divided Charge Engines with Uniform and Stratified Charges, AMS Report No. 1365, Princeton University, Princeton, N.J., 1978.
178. Markatos, N. C., and Mukerjee, T., Three-Dimensional Computer Analysis of Flow and Combustion in Automotive Internal Combustion Engines, *Mathematics and Computers in Simulation*, vol. 23, no. 4, pp. 354–366, 1981.
179. Griffin, M. D., Diwakar, R., Anderson, J. D., and Jones, E., Computational Fluid Dynamics Applied to the Flow in an Internal Combustion Engine, AIAA 16th Aerospace Sciences Meeting, Huntsville, Ala., Paper No. AIAA-78-57, 1978.
180. Griffin, M. D., Anderson, J. D., and Jones, E., Computational Fluid Dynamics Applied to Three-Dimensional Nonreacting Inviscid Flows in an Internal Combustion Engine, *ASME Journal of Fluids Engineering*, vol. 101, pp. 367–372, 1979.
181. Oppenheim, A. K., and Rotman, D. A., Fundamental Features of Ignition and Flame Propagation in Engines, Energy-Sources Technology Conference and Exhibition, Dallas, Tex., Paper No. 87-ICE-21, ASME, New York, 1987.
182. Adamczyk, A. A., Kaiser, E. W., Lavoie, G. A., and Isack, A. J., Single-Pulse Sampling Valve Measurements of Wall Layer Hydrocarbons in a Combustion Bomb, *Combustion and Flame*, vol. 52, pp. 1–9, 1983.
183. Sloane, T. M., and Schoene, A. Y., Computational Studies of End-Wall Quenching at Low Pressure: The Effects of Heterogeneous Radical Recombination and Crevices, *Combustion and Flame*, vol. 49, pp. 109–122, 1983.
184. Sloane, T. M., and Ratcliffe, J. W., A Molecular Beam Mass Spectrometer Study of Side-Wall Quenching at Low Pressure by Cooled Noncatalytic and Catalytic Surfaces, *Combustion and Flame*, vol. 47, pp. 83–92, 1982.

185. Westbrook, C. K., Adamczyk, A. A., and Lavoie, G. A., A Numerical Study of Laminar Flame Wall Quenching, *Combustion and Flame*, vol. 40, pp. 81–99, 1981.
186. Fendell, F., Fink, S., and Feldman, P., A Note on Unburned Hydrocarbon Emissions from Automotive Engines, *Combustion Science and Technology*, vol. 30, pp. 45–57, 1983.
187. Girard, A., and Leyer, J. C., Direct Measurements of the Head-on Flame Quenching Distance in Closed Chambers, in *Dynamics of Flames and Reactive Systems*, eds. J. R. Bowen, N. Manson, A. K. Oppenheim, and R. I. Soloukhin, *Progress in Astronautics and Aeronautics*, vol. 95, pp. 443–452, AIAA, New York, 1984.
188. LoRusso, J. A., Kaiser, E. W., and Lavoie, G. A., In-Cylinder Measurements of Wall Layer Hydrocarbons in a Spark Ignited Engine, *Combustion Science and Technology*, vol. 33, pp. 75–112, 1983.
189. Adamczyk, A. A., Kaiser, E. W., and Lavoie, G. A., A Combustion Bomb Study of the Hydrocarbon Emissions from Engine Crevices, *Combustion Science and Technology*, vol. 33, pp. 261–277, 1983.
190. LoRusso, J. A., Kaiser, E. W., and Lavoie, G. A., Quench Layer Contribution to Exhaust Hydrocarbons from a Spark-Ignited Engine, *Combustion Science and Technology*, vol. 25, pp. 121–125, 1981.
191. Kaiser, E. W., Rothschild, W. G., and Lavoie, G. A., Storage and Partial Oxidation of Unburned Hydrocarbons in Spark-Ignited Engines: Effect of Compression Ratio and Spark Timing, *Combustion Science and Technology*, vol. 36, pp. 171–189, 1984.
192. Adamczyk, A. A., Rothschild, W. G., and Kaiser, E. W., The Effect of Fuel and Oil Structure on Hydrocarbon Emissions from Oil Layers during Closed Vessel Combustion, *Combustion Science and Technology*, vol. 44, pp. 113–124, 1985.
193. Hocks, W., Peters, N., and Adomeit, G., Flame Quenching in Front of a Cold Wall under Two-Step Kinetics, *Combustion and Flame*, vol. 41, pp. 157–170, 1981.
194. Vosen, S. R., Greif, R., and Westbrook, C. K., Unsteady Heat Transfer during Laminar Flame Quenching, *Twentieth Symposium (International) on Combustion*, pp. 75–83, The Combustion Institute, Pittsburgh, Pa., 1985.
195. Adamczyk, A. A., Kaiser, E. W., Cavolowsky, J. A., and Lavoie, G. A., An Experimental Study of Hydrocarbon Emissions from Closed Vessel Explosions, *Eighteenth Symposium (International) on Combustion*, pp. 1695–1702, The Combustion Institute, Pittsburgh, Pa., 1981.
196. Clendening, C. W., Shackleford, W., and Hilyard, R., Raman Scattering Measurements in a Side-Wall Quench Layer, *Eighteenth Symposium (International) on Combustion*, pp. 1583–1590, The Combustion Institute, Pittsburgh, Pa., 1981.
197. Arpaci, V. S., and Tabaczynski, R. J., Radiation-Affected Laminar Flame Quenching, *Combustion and Flame*, vol. 57, pp. 169–178, 1984.
198. Carrier, G. F., Fendell, F. E., and Feldman, P. S., Interaction of a Planar Laminar Premixed Flame with a Perpendicular Wall, SAE International Congress and Exposition, Detroit, Mich., SAE Paper No. 800285, 1980.
199. Carrier, G. F., Fendell, F. E., Bush, W. B., and Feldman, P. S., Nonisenthalpic Interaction of a Planar Premixed Laminar Flame with a Parallel End Wall, SAE International Congress and Exposition, Detroit, Mich., SAE Paper No. 790245, 1979.
200. Carrier, G. F., Fendell, F. E., and Bush, W. B., Interaction of a Planar Premixed Flame with a Parallel Adiabatic End Wall, *Combustion Science and Technology*, vol. 20, pp. 195–207, 1979.
201. Bush, W. B., Fendell, F. E., and Fink, S. F., Effect of Boundary Thermal Constraint on Planar Premixed-Flame/Wall Interaction, *Combustion Science and Technology*, vol. 24, pp. 53–70, 1980.
202. Fink, S. F., Fendell, F. E., and Bush, W. B., Nonadiabatic Nonisobaric Propagation of a Planar Premixed Flame: Constant-Volume Enclosure, AIAA 21st Aerospace Sciences Meeting, Reno, Nev., Paper No. AIAA-83-0239, 1983.
203. Carrier, G. F., Fendell, F. E., and Feldman, P. S., Nonisobaric Flame Propagation, in *Dynamics and Modelling of Reactive Systems*, eds. W. E. Stewart, W. H. Ray, and C. C. Conley, pp. 333–351, Academic Press, New York, 1980.

204. Kooker, D. E., Numerical Study of a Confined Premixed Laminar Flame: Oscillatory Behavior and Wall Quenching, *Combustion and Flame*, vol. 49, pp. 141–149, 1983.
205. Goolsby, A. D., and Haskell, W. W., Flame-Quench Distance Measurements in a CFR Engine, *Combustion and Flame*, vol. 26, pp. 105–114, 1976.
206. Mori, Y., Miyauchi, T., and Hirano, M., Catalytic Combustion of Premixed Hydrogen-Air, *Combustion and Flame*, vol. 30, pp. 193–205, 1977.
207. Fakheri, A., and Buckius, R. O., The Effects of Radiation Heat Transfer on Catalytic Combustion, *Combustion and Flame*, vol. 57, pp. 275–282, 1984.
208. Jarosinski, J., Flame Quenching by a Cold Wall, *Combustion and Flame*, vol. 50, pp. 167–175, 1983.
209. Henningsen, S., A Theoretical Model for Propagating and Quenching of a One-dimensional, Laminar Two-Reaction Flame, SAE International Congress and Exposition, Detroit, Mich., SAE Paper No. 800105, 1980.
210. Shyy, W., and Adamson, T. C., Analysis of Hydrocarbon Emissions from Conventional Spark-Ignition Engines, *Combustion Science and Technology*, vol. 33, pp. 245–260, 1983.
211. Adamczyk, A. A., and Kach, R. A., The Effect of Oil Layers on Hydrocarbon Emissions: Low Solubility Oils, *Combustion Science and Technology*, vol. 26, pp. 227–234, 1984.
212. Kaiser, E. W., LoRusso, J. A., Lavoie, G. A., and Adamczyk, A. A., The Effect of Oil Layers on the Hydrocarbon Emissions from Spark-Ignition Engines, *Combustion Science and Technology*, vol. 28, pp. 69–73, 1982.
213. Carrier, G., Fendell, F., and Feldman, P., Cyclic Absorption/Desorption of Gas in a Liquid Wall Film, *Combustion Science and Technology*, vol. 25, pp. 9–19, 1981.
214. Kaiser, E. W., Adamczyk, A. A., and Lavoie, G. A., The Effect of Oil Layers on the Hydrocarbon Emissions Generated during Closed Vessel Combustion, *Eighteenth Symposium (International) on Combustion*, pp. 1881–1890, The Combustion Institute, Pittsburgh, Pa., 1981.
215. Adamczyk, A. A., and Kach, R. A., The Effect of Engine Deposit Layers on Hydrocarbon Emissions from Closed Vessel Combustion, *Combustion Science and Technology*, vol. 47, pp. 193–212, 1986.
216. Lavoie, G. A., Adamczyk, A. A., Kaiser, E. W., Cooper, J. W., and Rothschild, W. G., Engine HC Emissions Modeling: Partial Burn Effects, *Combustion Science and Technology*, vol. 49, pp. 99–105, 1986.
217. Wentworth, J. T., The Piston Crevice Volume Effect on Exhaust Hydrocarbon Emissions, *Combustion Science and Technology*, vol. 4, pp. 97–100, 1971.
218. Carrier, G. F., Fendell, F. E., and Feldman, P. S., Laminar Flame Propagation/Quench for a Parallel-Wall Duct, *Twentieth Symposium (International) on Combustion*, pp. 67–74, The Combustion Institute, Pittsburgh, Pa., 1985.
219. Fairchild, P. W., Fleeter, R. D., and Fendell, F. E., Raman Spectroscopy Measurements of Flame Quenching in a Duct-Type Crevice, *Twentieth Symposium (International) on Combustion*, pp. 85–90, The Combustion Institute, Pittsburgh, Pa., 1985.
220. Driscoll, J. F., Mirsky, W., Palaniswamy, S., and Liu, W., Experimental Study of the Flowfield Produced by a Wall Crevice in a Combustion Chamber, *Twentieth Symposium (International) on Combustion*, pp. 209–215, The Combustion Institute, Pittsburgh, Pa., 1985.
221. Lavoie, G. A., LoRusso, J. A., and Adamczyk, A. A., Hydrocarbon Emissions Modeling for Spark Ignition Engines, in *Combustion Modeling in Reciprocating Engines*, eds. J. N. Mattavi and C. A. Amann, pp. 409–445, Plenum Press, New York, 1980.
222. Herrin, R. J., Patterson, D. J., and Kadlec, R. H., Modeling Hydrocarbon Disappearance in Reciprocating Engine Exhaust, in *Combustion Modeling in Reciprocating Engines*, eds. J. N. Mattavi and C. A. Amann, pp. 447–482, Plenum Press, New York, 1980.
223. Caton, J. A., Heywood, J. B., and Mendillo, J. V., Hydrocarbon Oxidation in a Spark Ignition Exhaust Port, *Combustion Science and Technology*, vol. 37, pp. 153–169, 1984.
224. Mendillo, J. V., and Heywood, J. B., Hydrocarbon Oxidation in the Exhaust Port of a Spark Ignition Engine, SAE International Congress and Exposition, Detroit, Mich., SAE Paper No. 810019, 1981.

225. Caton, J. A., and Heywood, J. B., Models of Heat Transfer, Mixing and Hydrocarbon Oxidation in an Exhaust Port of a Spark-Ignited Engine, SAE International Congress and Exposition, Detroit, Mich., SAE Paper No. 800290, 1980.
226. Caton, J. A., and Heywood, J. B., An Experimental and Analytical Study of Heat Transfer in an Engine Exhaust Port, *International Journal of Heat and Mass Transfer*, vol. 24, pp. 581–595, 1981.
227. Kaiser, E. W., Rothschild, W. G., and Lavoie, G. A., The Effect of Fuel and Operating Variables on Hydrocarbon Species Distributions in the Exhaust from a Multicylinder Engine, *Combustion Science and Technology*, vol. 32, pp. 245–265, 1983.
228. Annand, W. J. D., Heat Transfer in the Cylinder and Porting, in *The Thermodynamics and Gas Dynamics of Internal Combustion Engines*, eds. J. H. Horlock and D. E. Winterbone, vol. 2, pp. 773–805, Clarendon Press, Oxford University Press, New York, 1986.
229. Newhall, H. K., Modelling of Engine Exhaust Emissions—An Introductory Overview, in *Combustion Modeling in Reciprocating Engines*, eds. J. N. Mattavi and C. A. Amann, pp. 331–343, Plenum Press, New York, 1980.
230. Daneshyar, H., Mendes-Lopez, J. M. C., Ludford, G. S. S., and Tromans, P. S., The Influence of Straining in a Premixed Flame and its Relevance to Combustion in IC Engines, in *International Conference on Combustion Engineering*, vol. I, pp. 191–199, I. Mech. E. Conference Publications 1983-3, London, 1983.
231. Al-Khishali, K. J., Boston, P. M., Bradley, D., Lawes, M., and Pegg, M. J., The Influence of Fluctuations in Turbulence upon Fluctuations in Turbulent Burning Velocity, in *International Conference on Combustion Engineering*, vol. I, pp. 175–180, I. Mech. E. Conference Publications 1983-3, London, 1983.
232. Strehlow, R., and Savage, L. D., The Concept of Flame Stretch, *Combustion and Flame*, vol. 31, pp. 209–211, 1978.
233. Borghi, R., and Escudie, D., Assessment of a Theoretical Model of Turbulent Combustion by Comparison with a Simple Experiment, *Combustion and Flame*, vol. 56, pp. 149–164, 1984.
234. Nohira, H., Konomi, T., and Ishiyama, T., Analysis on Cycle-by-Cycle Variation of Combustion from Pressure Indicator Data, *Japanèse Society of Automotive Engineers*, vol. 1, pp. 108–116, 1978.
235. Lancaster, D. R., Krieger, R. B., and Lienesch, J. H., Measurement and Analysis of Engine Pressure Data, SAE International Congress and Exposition, Detroit, Mich., SAE Paper No. 750026, 1975.
236. Smith, J. R., A Comparison of Ignition Locations in a High Swirl Engine, Report No. SAND79-8715, Sandia National Laboratories, Livermore, Calif., 1979.
237. Witze, P. O., The Effect of Spark Location on Combustion in a Variable-Swirl Engine, SAE International Congress and Exposition, Detroit, Mich., SAE Paper No. 820044, 1982.
238. Beretta, G. P., Rashidi, M., and Keck, J. C., Thermodynamic Analysis of Turbulent Combustion in a Spark Ignition Engine: Experimental Evidence, 1980 Western States Section of the Combustion Institute, Irvine, Calif., 1980.
239. Al-Alousi, Y. H., and Karim, Y. H., Some Considerations of Cyclic Variations in Spark Ignition Engines Fuelled with Gaseous and Liquid Fuels, SAE International Congress and Exposition, Detroit, Mich., SAE Paper No. 840232, 1984.
240. Lowry, S. O., and Devoto, R. S., Exhaust Emissions from a Single-Cylinder Engine Fueled with Gasoline, Methanol and Ethanol, *Combustion Science and Technology*, vol. 12, pp. 177–182, 1976.
241. Aiman, W. R., Extended Spark Duration Improves Engine Operation at High Exhaust Gas Recirculation Rates, *Combustion Science and Technology*, vol. 15, pp. 129–136, 1977.
242. Matsushita, S., Inoue, T., Nakanishi, K., Kato, K., and Kobayashi, N., Development of the Toyota Lean Combustion System, SAE International Congress and Exposition, Detroit, Mich., SAE Paper No. 850044, 1985.
243. Matsushita, S., Inoue, T., Nakanishi, K., Okumura, T., and Isogai, K., Effects of Helical Port with Swirl Control Valve on the Combustion and Performance of S.I. Engine, SAE International Congress and Exposition, Detroit, Mich., SAE Paper No. 850046, 1985.

244. Hirotomi, T., Nagayama, I., Kobayashi, S., and Yamamasu, M., Study of Induction Swirl in a Spark Ignition Engine, SAE International Congress and Exposition, Detroit, Mich., SAE Paper No. 810496, 1981.
245. Plee, S. L., A Computational Study of the Effects of Swirl Ratio, Squish Height and Bowl Geometry on the In-Cylinder Flow Field in Direct-Injection Diesel Engine, in *International Symposium on Flows in Internal Combustion Engines—III*, eds. T. Uzkan, W. G. Tiederman, and J. M. Novak, pp. 107–113, ASME, New York, 1985.
246. Yuen, W. W. and Servati, H., A Mathematical Engine Model Including the Effect of Engine Emissions, SAE International Congress and Exposition, Detroit, Mich., SAE Paper No. 840036, 1984.
247. Heywood, J. B., Pollutant Formation and Control in Spark Ignition Engines, *Fifteenth Symposium (International) on Combustion*, pp. 1191–1211, The Combustion Institute, Pittsburgh, Pa., 1975.
248. Heywood, J. B., Pollutant Formation and Control in Spark-Ignition Engines, in *Energy and Combustion Science*, ed. N. A. Chigier, Student ed. 1, pp. 229–258, Pergamon Press, New York, 1979.
249. Heywood, J. B., Fluid Motion within the Cylinder of Internal Combustion Engines—The 1986 Freeman Scholar Lecture, *ASME Journal of Fluids Engineering*, vol. 109, pp. 3–35, 1987.
250. Newhall, H. K., and Shahed, S. M., Kinetics of Nitric Oxide Formation in High-Pressure Flames, *Thirteenth Symposium (International) on Combustion*, pp. 381–389, The Combustion Institute, Pittsburgh, Pa., 1971.
251. Starkman, E. S., Experiment and Rationale in the Generation of Pollutants by Combustion, *Twelfth Symposium (International) on Combustion*, pp. 593–601, The Combustion Institute, Pittsburgh, Pa., 1969.
252. Newhall, H. K., Kinetics of Engine-Generated Nitrogen Oxides and Carbon Monoxide, *Twelfth Symposium (International) on Combustion*, pp. 603–613, The Combustion Institute, Pittsburgh, Pa., 1969.
253. James, E. H., Errors in NO Emission Prediction from Spark Ignition Engines, SAE International Congress and Exposition, Detroit, Mich., SAE Paper No. 820046, 1982.
254. Krieger, R. B., Applications of Engine Combustion Models—An Introductory Overview, in *Combustion Modeling in Reciprocating Engines*, eds. J. N. Mattavi and C. A. Amann, pp. 485–507, Plenum Press, New York, 1980.
255. Aiman, W. R., A Critical Test for Models of the Nitric Oxide Formation Process in Spark-Ignition Engines, *Fourteenth Symposium (International) on Combustion*, pp. 861–868, The Combustion Institute, Pittsburgh, Pa., 1973.
256. Brandstatter, W. R., and Decker, G., Fundamental Studies on the Volkswagen Stratified Charge Combustion Process, *Combustion and Flame*, vol. 25, pp. 15–23, 1975.
257. Fujimoto, S., Kaneko, Y., and Tsuruno, S., Possibility of Low-NO_x and High-Load Combustion in Pre-mixed Gases, *Twentieth Symposium (International) on Combustion*, pp. 61–66, The Combustion Institute, Pittsburgh, Pa., 1985.
258. Hayhurst, A. N., and Vince, J. M., The Origin and Nature of "Prompt" Nitric Oxide in Flames, *Combustion and Flame*, vol. 50, pp. 41–57, 1983.
259. Pundir, B. P., Zvonow, V. A., and Gupta, C. P., Nitric Oxide Formation in Spark-Ignition Engines with In-Cylinder Charge Non-homogeneity of a Random Nature, *Proceedings of the Institution of Mechanical Engineers*, vol. 199, no. D3, pp. 227–235, 1985.
260. Baruah, P. C., Combustion and Cycle Calculations in Spark Ignition Engines, in *The Thermodynamics and Gas Dynamics of Internal Combustion Engines*, eds. J. H. Horlock and D. E. Winterbone, vol. 2, pp. 823–865, Clarendon Press, Oxford University Press, New York, 1986.
261. Agnew, W. G., Room at the Piston Top: Contribution of Combustion Science to Engine Design, *Twentieth Symposium (International) on Combustion*, pp. 1–17, The Combustion Institute, Pittsburgh, Pa., 1985.
262. Mitchell, E., and Alperstein, M., Texaco Controlled-Combustion System—Multifuel, Efficient, Clean and Practical, *Combustion Science and Technology*, vol. 8, pp. 39–49, 1973.

263. Mizutani,Y., and Matsushita, S., Fuel Vapor-Spray-Air Mixture Operation of a Spark-Ignition Engine, *Combustion Science and Technology*, vol. 8, pp. 85–94, 1973.
264. Al-Mamar, F. W. A. A., Hynes, J., and Sheppard, C. G. W., Combustion Rate in a Dual Chamber S.I. Engine, *Combustion Science and Technology*, vol. 45, pp. 85–100, 1986.
265. El-Messiri, I. A., and Newhall, H. K., Analytical Characterization of the Divided-Chamber Combustion Process, *Fourteenth Symposium (International) on Combustion*, pp. 843–850, The Combustion Institute, Pittsburgh, Pa., 1973.
266. Rychter, T. J., Saragih, R., Lezanski, T., and Wojcicki, S., Catalytic Activation of a Charge in a Prechamber of a SI Lean-Burn Engine, *Eighteenth Symposium (International) on Combustion*, pp. 1815–1824, The Combustion Institute, Pittsburgh, Pa., 1981.
267. Harris, S. J., and Peters, B. D., Effect of Di-*t*-Butyl Peroxide on the Lean Operating Limit of a Spark-Ignition Engine, *Combustion Science and Technology*, vol. 29, pp. 293–298, 1982.
268. Quader, A. A., Exhaust Emissions and Performance of a Spark-Ignition Engine using Oxygen Enriched Intake Air, *Combustion Science and Technology*, vol. 19, pp. 81–86, 1978.
269. Rauckis, M. J., and McLean, W. J., The Effect of Hydrogen Addition on Ignition Delays and Flame Propagation in a Spark Ignition Engines, *Combustion Science and Technology*, vol. 19, pp. 207–216, 1979.
270. Cassidy, J. F., Emissions and Total Energy Consumption of a Multicylinder Piston Engine Running on Gasoline and a Hydrogen-Gasoline Mixture, NASA Technical Note D-8487, 1977.
271. Krieger, R. B., and Davis, G. C., The Influence of the Degree of Stratification on Jet-Ignition Engine Emissions and Fuel Consumption, *Stratified Charge Engines*, pp. 109–119, I. Mech. E. Conference Publications 1976-11, 1976.
272. Giovanetti, A. J., Ekchian, J. A., Heywood, J. B., and Fort, E. F., Analysis of Hydrocarbon Emissions in a Direct Injection Spark-Ignition Engine, SAE International Congress and Exposition, Detroit, Mich., SAE Paper No., 830587, 1983.
273. Najit, P. M., and Foster, E. M., Compression-Ignited Homogeneous Charge Combustion, SAE International Congress and Exposition, Detroit, Mich., SAE Paper No. 830264, 1983.
274. Emmenthal, K.-D., Hagemann, G., and Hucho, W.-H., Turbocharging Small Displacement Spark Ignition Engines for Improved Fuel Economy, SAE International Congress and Exposition, Detroit, Mich., SAE Paper No. 790311, 1979.
275. Borman, G. L., Modeling Flame Propagation and Heat Transfer Release in Engines—An Introductory Overview, in *Combustion Modeling in Reciprocating Engines*, eds. J. N. Mattavi and C. A. Amann, pp. 165–191, Plenum Press, New York, 1980.
276. Obert, E. F., *Internal Combustion Engines and Air Pollution*, Intext, New York, 1973.
277. Oppenheim, A. K., The Knock Syndrome—Its Cures and Its Victims, SAE International Congress and Exposition, Detroit, Mich., SAE Paper No. 841339, 1984.
278. Lewis, B., and von Elbe, G., *Combustion, Flames and Explosions of Gases*, 2nd ed., Academic Press, New York, 1961.
279. Gray, P., Griffiths, J. F., Hasko, S. M., and Lignola, P. G., Novel, Multi-Stage Ignition in the Spontaneous Combustion of Acetaldehyde, *Combustion and Flame*, vol. 43, pp. 175–186, 1981.
280. Kono, M., Shiga, S., Kumagai, S., and Iinuma, K., Thermodynamic and Experimental Observations of Knock Intensity by Using a Spark-ignited Rapid Compression Machine, *Combustion and Flame*, vol. 54, pp. 33–47, 1983.
281. Cheng, R. K., and Oppenheim, A. K., Autoignition in Methane-Hydrogen Mixtures, *Combustion and Flame*, vol. 58, pp. 125–139, 1984.
282. Ferraro, C. V., Marzano, M., and Nuzzio, P., Knock-Limit Measurement in High-Speed S.I. Engines, SAE International Congress and Exposition, Detroit, Mich., SAE Paper No. 850127, 1985.
283. Patel, K. S., and Henein, N. A., Burning Velocities in Methanol-Indolene-Air Mixtures in a CFR Engine, SAE International Congress and Exposition, Detroit, Mich., SAE Paper No. 850111, 1985.
284. Czerwinski, J., Oscillatory Combustion and Knocking in S.I. Engines with Divided Combustion Chambers, SAE International Congress and Exposition, Detroit, Mich., SAE Paper No. 850045, 1985.

285. Girard, A., Fisson, F., and Leyer, J. C., Vibratory Combustion Triggered by a Small Cavity in the Wall of a Constant Volume Combustion Chamber, in *Dynamics of Flames and Reactive Systems*, eds. J. R. Bowen, N. Manson, A. K. Oppenheim, and R. I. Soloukhin, *Progress in Astronautics and Aeronautics*, vol. 95, pp. 433–442, AIAA, New York, 1984.
286. Gupta, H. C., and Bracco, F. V., The Origin of Pressure Oscillations in Divided Chamber Engines, *Combustion and Flame*, vol. 48, pp. 33–49, 1982.
287. By, A., Kempinski, B., and Rife, J. M., Knock in Spark Ignition Engines, SAE International Congress and Exposition, Detroit, Mich., SAE Paper No. 810147, 1981.
288. Rifkin, E., and Walcutt, C., A Basis for Understanding Antiknock Action, *SAE Transactions*, vol. 65, pp. 552–566, 1957.
289. Burwell, W. G., and Olson, D. R., The Spontaneous Ignition of Isooctane-Air Mixtures under Steady Flow Conditions, SAE International Congress and Exposition, Detroit, Mich., SAE Paper No. 650520, 1975.
290. Vermeer, D. J., Meyer, J. W., and Oppenheim, A. K., Auto-Ignition of Hydrocarbons behind Reflected Shock Waves, *Combustion and Flame*, vol. 18, pp. 327–336, 1972.
291. Douad, A. M., and Eyzat, P., Four-Octane-Number Method for Predicting the Anti-knock Behavior of Fuels and Engines, SAE International Congress and Exposition, Detroit, Michigan, SAE Paper No. 780080, 1978.
292. Walsh, G. J., and Cheng, W. K., Effects of Highly-Heated Fuel on Diesel Combustion, SAE International Congress and Exposition, Detroit, Mich., SAE Paper No., 850088, 1985.
293. Bowditch, F. W., and Yu, T. C., A Consideration of the Deposit Ignition Mechanism, *SAE Transactions*, vol. 69, pp. 435–447, 1961.
294. Sokolik, A. S., *Self-Ignition, Flame and Detonation in Gases*, NASA TT F-125, Israel Program for Scientific Translation, Jerusalem, 1963.
295. Dumont, L. F., Possible Mechanisms by Which Combustion Chamber Deposits Accumulate and Influence Knock, *SAE Quarterly Transactions*, vol. 5, pp. 565–576, 1951.
296. Anderson, C. L., and Prakash, C., The Effects of Variable Conductivity on Unsteady Heat Transfer in Deposits, SAE International Congress and Exposition, Detroit, Mich., SAE Paper No. 850048, 1985.
297. DeGregoria, A. J., A Theoretical Study of Engine Deposit and Its Effect on Octane Requirement Using an Engine Simulation, SAE International Congress and Exposition, Detroit, Mich., SAE Paper No. 820072, 1982.
298. Bush, W. B., Fendell, F. E., and Fink, S. F., Modeling End-Gas Knock in a Rapid-Compression Machine, AIAA 22nd Aerospace Sciences Meeting, Reno, Nev., Paper No. AIAA-84-0208, 1984.
299. Bush, W. B., Fendell, F. E., and Fink, S. F., Modeling End-Gas Knock in a Rapid Compression Machine: Review and Extension, in *Reacting Flows: Combustion and Chemical Reactors, Part 1*, ed. G. S. S. Ludford, *Lectures in Applied Mathematics*, vol. 24, pp. 349–373, AMS, Providence, R.I., 1986.
300. Carrier, G., Fendell, F., Fink, S., and Feldman, P., Heat Transfer as a Determent of End-Gas Knock, *Combustion Science and Technology*, vol. 38, pp. 1–48, 1984.
301. Maly, R., and Ziegler, G., Thermal Combustion Modeling—Theoretical and Experimental Investigation of the Knocking Process, *SAE Transactions*, vol. 91, pp. 2569–2602, 1983.
302. Tabaczynski, R. J., Flame Propagation and Detonation in Gasoline Engines, XIX International Symposium on Heat and Mass Transfer in Gasoline and Diesel Engines, International Centre for Heat and Mass Transfer, Dubrovnik, Yugoslavia, 1987.
303. Ferguson, C. R., Green, R. M., and Lucht, R. P., Unburned Gas Temperatures in an Internal Combustion Engine: II. Heat Release Computations, Report No. SAND-86-8605, Sandia National Laboratories, Livermore, Calif., 1986.
304. Amann, C. A., Cylinder-Pressure Measurement and Its Use in Engine Research, SAE International Congress and Exposition, Detroit, Mich., SAE Paper 852067, 1985.
305. Kirsch, L. J., and Quinn, C. P., Progress towards a Comprehensive Model of Hydrocarbon Autoignition, *Journal de Chemie Physique*, vol. 82, no. 5, pp. 459–473, 1985.
306. Semenov, N. N., *Chemical Kinetics and Chain Reactions*, Oxford University Press, New York, 1935.

307. Kirsch, L. J., and Quinn, C. P., A Fundamentally Based Model of Knock in the Gasoline Engine, *Sixteenth Symposium (International) on Combustion*, pp. 233–246, The Combustion Institute, Pittsburgh, Pa., 1977.
308. Lignola, P. G., Reverehon, E., and Piro, R., Dynamics of Combustion Processes of *n*-Heptane, i-Octane and Their Mixtures and Knocking, *Twentieth Symposium (International) on Combustion*, pp. 123–131, The Combustion Institute, Pittsburgh, Pa., 1985.
309. Lightfoot, N. S., and Negus, C. R., Investigation of the "Knock" Phenomenon in an Optically Accessed Engine, *Twentieth Symposium (International) on Combustion*, pp. 111–122, The Combustion Institute, Pittsburgh, Pa., 1985.
310. Luck, C. J., Burgess, A. R., Desty, D. H., Whitehead, D. M., and Pratley, D., A Study of the Combustion of *n*-Heptane in an Engine Using a Novel High-Speed Sampling Technique, *Fourteenth Symposium (International) on Combustion*, pp. 501–512, The Combustion Institute, Pittsburgh, Pa., 1973.
311. Gibson, C., Gray, P., Griffiths, J. F., and Hasko, S. M., Spontaneous Ignition of Hydrocarbons and Related Fuels: A Fundamental Study of Thermokinetic Interactions, *Twentieth Symposium (International) on Combustion*, pp. 101–109, The Combustion Institute, Pittsburgh, Pa., 1985.
312. Halstead, M. P., Kirsch, L. J., Prothero, A., and Quinn, C. P., A Mathematical Model for Hydrocarbon Autoignition at High Pressures, *Proceedings of the Royal Society of London, Series A*, vol. 346, pp. 515–538, 1975.
313. Halstead, M. P., Kirsch, L. J., and Quinn, C. P., The Autoignition of Hydrocarbon Fuels at High Temperatures and Pressures—Fitting of a Mathematical Model, *Combustion and Flame*, vol. 30, pp. 45–60, 1977.
314. Natarajan, B., and Bracco, F. V., A Multidimensional Modeling of Auto-Ignition in Spark-Ignition Engines, *Combustion and Flame*, vol. 57, pp. 179–197, 1984.
315. Hirst, S. L., and Kirsch, L. J., The Application of a Hydrocarbon Autoignition Model in Simulating Knock and Other Engine Combustion Phenomena, in *Combustion Modeling in Reciprocating Engines*, eds. J. N. Mattavi and C. A. Amann, pp. 193–229, Plenum Press, New York, 1980.
316. Annand, W. J. D., Heat Transfer in the Cylinders of Reciprocating Internal Combustion Engines, *Proceedings of the Institution of Mechanical Engineers*, vol. 177, pp. 973–990, 1963.
317. Schapertons, H., and Lee, W., Multidimensional Modelling of Knocking Combustion in SI Engines, SAE International Congress and Exposition, Detroit, Mich., SAE Paper No. 850502, 1985.
318. Cox, R. A., and Cole, J. A., Chemical Aspects of Autoignition of Hydrocarbon-Air Mixtures, *Combustion and Flame*, vol. 60, pp. 109–123, 1985.
319. Pitz, W. J., and Westbrook, C. K., Chemical Kinetics of the High Pressure Oxidation of *n*-Butane and Its Relation to Engine Knock, *Combustion and Flame*, vol. 63, pp. 113–133, 1986.
320. Cernansky, N. P., Green, R. M., Pitz, W. J., and Westbrook, C. K., Chemistry of Fuel Oxidation Preceding End-Gas Autoignition, *Combustion Science and Technology*, vol. 50, pp. 3–25, 1986.
321. Smith, J. R., Green, R. M., Westbrook, C. K., and Pitz, W. J., An Experimental and Modeling Study of Engine Knock, *Twentieth Symposium (International) on Combustion*, pp. 91–100, The Combustion Institute, Pittsburgh, Pa., 1985.
322. Griffiths, J. F., and Nimmo, N., Spontaneous Ignition and Engine Knock under Rapid Compression, *Combustion and Flame*, vol. 60, pp. 215–218, 1985.
323. Cullis, C. F., Herron, D., and Hirschler, M. M., Organosilicon Compounds as Antiknock Additives, *Combustion and Flame*, vol. 59, pp. 151–161, 1985.
324. Jachinowski, C. J., Chemical Kinetic Reaction Mechanism for the Combustion of Propane, *Combustion and Flame*, vol. 55, pp. 213–224, 1984.
325. Westbrook, C. K., Chemical Kinetics and Engine Knock, in *Proceedings of the Workshop on the Gas Flame Structure, Part 1*, ed. V. K. Baev, pp. 206–230, Institute of Theoretical and Applied Mechanics, USSR Academy of Sciences, Siberian Division, Novosibirsk, U.S.S.R., 1988.
326. Kaiser, E. W., Westbrook, C. K., and Pitz, W. J., Acetaldehyde Oxidation in the Negative Temperature Coefficient Regime: Experimental and Modeling Results, *International Journal of Chemical Kinetics*, vol. 18, pp. 655–688, 1986.

327. Axelsson, E. I., Brezinsky, K., Dryer, F. L., Pitz, W. J., and Westbrook, C. K., Chemical Kinetic Modeling of the Oxidation of Large Alkane Fuels: *n*-Octane and Iso-Octane, *Twenty-first Symposium (International) on Combustion*, pp. 783–793, The Combustion Institute, Pittsburgh, Pa., 1988.
328. Green, R. M., Parker, C. D., Pitz, W. J., and Westbrook, C. K., The Autoignition of Isobutane in a Knocking Spark Ignition Engine, Internal Report, Sandia National Laboratories, Livermore, Calif., 1986.
329. Green, R. M., and Lucht, R. P., An Application of Laser Diagnostics to the Study of End-Gas Autoignition and Engine Knock, *International Symposium on Diagnostics and Modeling of Combustion in Reciprocating Engines*, pp. 259–266, JSME Tokyo, 1985.
330. Green, R. M., and Smith, J. R., An Experimental Study of Engine Knock, Spring Meeting of the Western States Section of the Combustion Institute, University of Colorado, Boulder, Paper No. WSS/CI 84-46, 1984.
331. Lucht, R. P., Green, R. M., Palmer, R. E., Teets, R. E., and Ferguson, C. R., Unburned Gas Temperatures in an Internal Combustion Engine: I. CARS Temperature Measurements, Report No. SAND-86-8749, Sandia National Laboratories, Livermore, Calif., 1986.
332. Ikegami, M., Kidoguchi, Y., and Nishiwaki, K., A Multidimensional Model Prediction of Heat Transfer in Non-Fired Engines, SAE International Congress and Exposition, Detroit, Mich., SAE Paper 860467, 1986.
333. Leppart, W. R., A Detailed Chemical Kinetics Simulation of Engine Knock, *Combustion Science and Technology*, vol. 43, pp. 1–20, 1985.
334. Li, T.-M., and Simmons, R. F., A Theoretical Examination of the Distribution of Gaseous Lead Compounds in the Exhaust from a Petrol Engine, *Combustion and Flame*, vol. 56, pp. 113–122, 1984.
335. Haskell, W. W., and Kiovsky, T. E., Antiknock Compounds for Unleaded Gasolines: Organic Titanates, *Combustion and Flame*, vol. 43, pp. 303–315, 1981.
336. Salooja, K. C., Studies Relating to the Mechanism of Antiknock Action of Tetraethyl Lead, *Combustion and Flame*, vol. 9, pp. 211–217, 1965.
337. Sher, E., and Keck, J. C., Spark Ignition of Combustible Gas Mixtures, *Combustion and Flame*, vol. 66, pp. 17–25, 1986.
338. Maly, R., and Vogel, M., Initiation and Propagation of Flame Fronts in Lean CH_4-Air Mixtures by the Three Modes of the Ignition Spark, *Seventeenth Symposium (International) on Combustion*, pp. 821–831, The Combustion Institute, Pittsburgh, Pa., 1979.
339. Sher, E., and Refael, S., Numerical Analysis of the Early Phase Development of Spark-ignited Flames in CH_4-Air Mixture, *Nineteenth Symposium (International) on Combustion*, pp. 251–257, The Combustion Institute, Pittsburgh, Pa., 1982.
340. Lim, M. T., Anderson, R. W., and Arpaci, V. S., Prediction of Spark Kernel Development in Constant Volume Combustion, *Combustion and Flame*, vol. 69, pp. 303–316, 1987.
341. Ballal, D. R., and Lefebvre, A. H., Ignition and Flame Quenching in Flowing Gaseous Streams, *Proceedings of the Royal Society of London, Series A*, vol. 357, pp. 163–181, 1977.
342. Anderson, R. W., and Lim, M. T., Investigation of Misfire in a Fast Burn Spark Ignition Engine, *Combustion Science and Technology*, vol. 43, pp. 183–196, 1985.
343. Ballal, D. R., and Lefebvre, A. H., The Influence of Spark Discharge Characteristics on Minimum Ignition Energy in Flowing Gases, *Combustion and Flame*, vol. 24, pp. 99–108, 1975.
344. Ballal, D. R., and Lefebvre, A. H., The Influence of Flow Parameters on Minimum Ignition Energy and Quenching Distance, *Fifteenth Symposium (International) on Combustion*, pp. 1473–1481, The Combustion Institute, Pittsburgh, Pa., 1975.
345. Ballal, D. R., and Lefebvre, A. H., A General Model of Spark Ignition for Gaseous and Liquid Fuel-Air Mixtures, *Eighteenth Symposium (International) on Combustion*, pp. 1737–1746, The Combustion Institute, Pittsburgh, Pa., 1981.
346. Maly, R., Ignition Model for Spark Discharges and the Early Phase of Flame Front Growth, *Eighteenth Symposium (International) on Combustion*, pp. 1747–1754, The Combustion Institute, Pittsburgh, Pa., 1981.
347. Kono, M., Kumagai, S., and Sakai, T., Ignition of Gases by Two Successive Sparks with

Reference to Frequency Effect of Capacitance Sparks, *Combustion and Flame*, vol. 27, pp. 85–98, 1976.
348. Kono, M., Kumagai, S., and Sakai, T., The Optimum Conditions for Ignition of Gases by Composite Sparks, *Sixteenth Symposium (International) on Combustion*, pp. 757–766, The Combustion Institute, Pittsburgh, Pa., 1977.
349. Kono, M., Hatori, J., and Iinuma, K., Investigation of Ignition Ability of Composite Sparks in Flowing Mixtures, *Twentieth Symposium (International) on Combustion*, pp. 133–140, The Combustion Institute, Pittsburgh, Pa., 1985.
350. DeSoete, G. C., Propagation Behavior of Spark Ignited Flames in Early Stages, *International Conference on Combustion Engineering*, vol. I, pp, 93–100, I. Mech. E. Conference Publications 1983-3, London, 1983.
351. Bradley, D., and Lung, F. K.-K., Spark Ignition and the Early Stages of Turbulent Flame Propagation, *Combustion and Flame*, vol. 69, pp. 71–93, 1987.
352. Hinze, J. O., *Turbulence*, 2nd ed., McGraw-Hill, New York, 1975.
353. Tromans, P. S., and O'Connor, S. J., Influence of Turbulent Motion on Spark Ignition, in *Dynamics of Flames and Reactive Systems*, eds. J. R. Bowen, N. Mason, A. K. Oppenheim, and R. I. Soloukhin, *Progress in Astronautics and Aeronautics*, vol. 95, pp. 421–432, AIAA, New York, 1984.
354. Bjørge, T., Influence of Electrode Orientation, Temperature and Pressure on Lean Ignition Limits of Methane, XIX International Symposium on Heat and Mass Transfer in Gasoline and Diesel Engines, International Centre for Heat and Mass Transfer, Dubrovnik, Yugoslavia, 1987.
355. Siegel, R., and Howell, J. R., *Thermal Radiation Heat Transfer*, 2nd ed., Hemisphere, Washington, D.C., 1981.
356. Adelman, H. G., A Time Dependent Theory of Spark Ignition, *Eighteenth Symposium (International) on Combustion*, pp. 1333–1342, The Combustion Institute, Pittsburgh, Pa., 1981.
357. Wu, Y., Huang, Z., and Xiong, T., The Numerical Solution of Governing Equations for Ignition of a Sphere of Burned Gas or Spark, *Twentieth Symposium (International) on Combustion*, pp. 151–159, The Combustion Institute, Pittsburgh, Pa., 1985.
358. Ramos, J. I., Ignition of Confined Gaseous Mixtures by Hot Surfaces and Hot Wires, AIAA 21st Aerospace Sciences Meeting, Reno, Nev., Paper No. AIAA-83-0240, 1983.
359. Sibulkin, M., and Siskind, K. S., Numerical Study of a Combustion Wave by an Ignition Kernel, *Combustion and Flame*, vol. 69, pp. 49–57, 1987.
360. Dixon-Lewis, G., and Shepherd, I. G., Some Aspects of Ignition by Localized Sources, and of Cylindrical and Spherical Flames, *Fifteenth Symposium (International) on Combustion*, pp. 1483–1491, The Combustion Institute, Pittsburgh, Pa., 1975.
361. Refael, S., and Sher, E., A Theoretical Study of the Ignition of a Reactive Medium by Means of an Electrical Discharge, *Combustion and Flame*, vol. 59, pp. 17–30, 1985.
362. Kailasanath, K., Oran, E., and Boris, J., A Theoretical Study of the Ignition of Premixed Gases, *Combustion and Flame*, vol. 47, pp. 173–190, 1982.
363. Sloane, T. M., and Ratcliffe, J. W., Chemistry of Spark Ignition: An Experimental and Computational Study, *American Chemical Society Symposium Series*, no. 249, pp. 205–222, 1984.
364. Dale, J. D., and Oppenheim, A. K., Enhanced Ignition for I.C. Engines with Premixed Gases, SAE International Congress and Exposition, Detroit, Mich., SAE Paper No. 810146, 1981.
365. Mittinti, D. N. R., and Dabora, E. K., Plasma Jet Ignition Studies, *Twentieth Symposium (International) on Combustion*, pp. 169–177, The Combustion Institute, Pittsburgh, Pa., 1985.
366. Boston, P. M., Bradley, D., Lung, F. K.-K., Vince, I. M., and Weinberg, F. J., Flame Initiation in Lean, Quiescent Mixtures with Various Igniters, *Twentieth Symposium (International) on Combustion*, pp. 141–149, The Combustion Insitute, Pittsburgh, Pa., 1985.
367. Tozzi, L., and Dabora, E. K., Plasma Jet Ignition in a Lean-Burn CFR Engine, *Nineteenth Symposium (International) on Combustion*, pp. 1467–1474, The Combustion Institute, Pittsburgh, Pa., 1982.
368. Orrin, J. E., Vince, I. M., and Weinberg, F. J., A Study of Plasma Jet Ignition Mechanisms, *Eighteenth Symposium (International) on Combustion*, pp. 1755–1765, The Combustion Institute, Pittsburgh, Pa., 1981.

369. Sloane, T. M., Ratcliffe, J. W., and Dobosiu-Halalai, I., Effects of Plasma Chemistry and Charge Stratification in Plasma Jet Ignition, *Combustion Science and Technology*, vol. 50, nos. 1–3, pp. 61–77, 1986.
370. Wiriyawit, S., and Dabora, E. K., Modeling the Chemical Effects of Plasma Ignition in One-Dimensional Chamber, *Twentieth Symposium (International) on Combustion*, pp. 179–186, The Combustion Institute, Pittsburgh, Pa., 1985.
371. Fisher, P. D., Pitt, P. L., Ridley, J. D., and Clements, R. M., Puff-Jet Ignition of Methane in an Internal Combustion Engine, *Combustion Science and Technology*, vol. 46, pp. 137–143, 1986.
372. Dale, J. D., Smy, P. R., and Clements, R. M., The Effects of a Coaxial Spark Igniter on the Performance and Emissions from an Internal Combustion Engine, *Combustion and Flame*, vol. 31, pp. 173–185, 1978.
373. Ridley, J. D., Pitt, P. L., Clements, R. M., and Smy, P. R., The Effect of Feed Gases on Plasma Jet Ignition, *Combustion Science and Technology*, vol. 43, pp. 39–47, 1985.
374. Pitt, P. L., Ridley, J. D., and Clements, R. M., Low Energy Ignition of an Ultra Lean Methane Fueled Internal Combustion Engine, *Combustion Science and Technology*, vol. 38, pp. 217–225, 1984.
375. Pitt, P. L., and Clements, R. M., The Effects of Plasma Jet Ignition on a Methane Fueled Internal Combustion Engine, *Combustion Science and Technology*, vol. 30, pp. 327–333, 1983.
376. Behbahani, H. F., Warris, A.-M., and Weinberg, F. J., The Destruction of Nitrogen Oxide by Nitrogen Atoms from Plasma Jets: Designing for Thermal Stratification, *Combustion Science and Technology*, vol. 30, pp. 289–302, 1983.
377. Behbahani, H. F., Fontijn, A., Muller-Dethlefs, K., and Weinberg, F. J., The Destruction of Nitrogen Oxide by Nitrogen Atoms from Plasma Jets, *Combustion Science and Technology*, vol. 27, pp. 123–132, 1982.
378. Grant, J. F., Marran, E. P., and McIlwain, M. E., Optimizaton of Plasma Jet Ignition Properties: Ignition of Lean-Quiescent Mixtures of Propane, *Combustion Science and Technology*, vol. 30, pp. 171–184, 1983.
379. Pitt, P. L., Ridley, J. D., and Clements, R. M., An Ignition System for Ultra Lean Mixtures, *Combustion Science and Technology*, vol. 35, pp. 277–285, 1984.
380. Cavolowsky, J. A., Faris, D. W., Smy, P. R., and Oppenheim, A. K., Formation of a Plasma Puff, SAE International Congress and Exposition, Detroit, Mich., SAE Paper No. 870609, 1987.
381. Mittinti, D. N. R., Flame Dynamics of Plasma Jet and Flame Jet Igniters, XIX International Symposium on Heat and Mass Transfer in Gasoline and Diesel Engines, International Centre for Heat and Mass Transfer, Dubrovnik, Yugoslavia, 1987.
382. Yamaguchi, S., Ohiwa, N., and Hasegawa, T., Ignition and Burning Process in a Divided Chamber Bomb, *Combustion and Flame*, vol. 59, pp. 177–187, 1985.
383. Dale, J. D., Smy, P. R., and Clements, R. M., The Effects of a Coaxial Spark Igniter on the Performance of and the Emissions from an Internal Combustion Engine, *Combustion and Flame*, vol. 31, pp. 173–185, 1978.
384. Dale, J. D., Smy, P. R., Way-Nee, D., and Clements, R. M., Laser-Ignited Internal Combustion Engines, *Combustion and Flame*, vol. 30, pp. 319–320, 1977.
385. Cote, T., Ridley, J. D., Clements, R. M., and Smy, P. R., The Ignition Characteristics of Igniters at Sub-Atmospheric Pressures, *Combustion Science and Technology*, vol. 48, pp. 151–162, 1986.
386. Topham, D. R., Clements, R. M., Ridley, J. D., and Smy, P. R., The Initiation of Combustion by a Hot Turbulent Mixing Element, *Combustion Science and Technology*, vol. 50, pp. 41–60, 1986.
387. Diwakar, R., Anderson, J. D., Griffin, M. D., and Jones, E., Inviscid Solutions for the Flow Field in an Internal Combustion Engine, *AIAA Journal*, vol. 14, no. 12, pp. 1667–1668, 1976.
388. Griffin, M. D., Anderson, J. D., and Diwakar, R., Navier-Stokes Solutions for the Flow Field in an Internal Combustion Engine, *AIAA Journal*, vol. 14, no. 12, pp. 1665–1666, 1976.
389. Sod, G. A., A Generalized Hybrid Random Choice Method with Application to Internal

Combustion Engines, in *Seventh International Conference on Numerical Methods in Fluid Dynamics*, eds. W. C. Reynolds and R. W. MacCormack, *Lecture Notes in Physics*, vol. 141, pp. 387–390, Springer-Verlag, New York, 1981.

390. Sod, G. A., A Hybrid Random Choice Method with Application to Internal Combustion Engines, in *Sixth International Conference on Numerical Methods in Fluid Dynamics*, eds. H. Cabannes, M. Holt, and V. Rusanov, *Lecture Notes in Physics*, vol., 60, pp. 492–501, Springer-Verlag, New York, 1978.
391. Sod, G. A., Automotive Engine Modeling with a Hybrid Random Choice Method, SAE International Congress and Exposition, Detroit, Mich., SAE Paper No. 790242, 1979.
392. Sod, G. A., Automotive Engine Modeling with a Hybrid Random Choice Method, II, SAE International Congress and Exposition, Detroit, Mich., SAE Paper No. 800288, 1980.
393. Ashurst, W. T., Vortex Dynamic Calculation of Fluid Motion in a Four Stroke Piston "Cylinder"—Planar and Axisymmetric Geometry, Report No. SAND-78-8229, Sandia National Laboratories, Livermore, Calif., 1978.
394. Witze, P. O., Comparisons between Measurement and Analysis of Fluid Motion in Internal Combustion Engines, Report No. SAND-81-8242, Sandia National Laboratories, Livermore, Calif., 1981.
395. Ekchian, A., and Hoult, D. P., Flow Visualization Study of the Intake Process of an Internal Combustion Engine, SAE International Congress and Exposition, Detroit, Mich., SAE Paper No. 790095, 1979.
396. Ashurst, W. T., Piston-Cylinder Fluid Motion via Vortex Dynamics, in *Fluid Mechanics of Combustion Systems*, eds. T. Morel, R. P. Lohmann, and J. M. Rackley, pp. 191–199, ASME, New York, 1981.
397. Martins, L.-F., and Ghoniem, A. F., Vortex Simulation of the Induction and Compression Processes in an Engine Model, SAE International Fuels and Lubricants Meeting and Exposition, Toronto, Ont., Canada, SAE Paper No. 872102, 1987.
398. Chau, E. L., Rotman, D. A., Schipperijn, F. W., Oppenheim, A. K., Sawyer, R. F., Iiyama, A., and Xia, L. Q., Experimental and Numerical Modeling Study of Engine Combustion, SAE International Fuels and Lubricants Meeting and Exposition, Toronto, Ont., Canada, SAE Paper No. 872103, 1987.
399. Barr, P. K., and Witze, P. O., Some Limitations of the Spherical Flame Assumption used in Phenomenological Engine Models, SAE International Congress and Exposition, Detroit, Mich., SAE Paper No. 880129, 1988.
400. Haselman, L. C., TDC—A Computer Code for Calculating Chemically Reacting Hydrodynamic Flows in Two-Dimensions, Report No. UCRL-52931, Lawrence Livermore National Laboratory, Livermore, Calif., 1980.
401. Westbrook, C. K., and Haselman, L. C., Computer Modeling of Automotive Engine Combustion, Report No. UCRL-80451, Lawrence Livermore Natonal Laboratory, Livermore, Calif., 1977.
402. Shih, T. I-P., Smith, G. E., Springer, G. S., and Rimon, Y., Boundary Conditions for the Solution of the Compressible Navier-Stokes Equations by an Implicit Factored Method, *Journal of Computational Physics*, vol. 52, pp. 54–79, 1983.
403. Shih, T. I-P., Smith, G. E., and Springer, G. S., Vortex Motion in Axisymmetric Piston-Cylinder Configurations, *AIAA Journal*, vol. 22, pp. 512–513, 1984.
404. Henke, H., and Hanel, D., Numerical Simulation of Gas Motion in Piston Engines, in *Ninth International Conference on Numerical Methods in Fluid Dynamics*, eds. Soubbaramayer and J. P. Boujot, *Lecture Notes in Physics*, vol. 218, pp. 267–271, Springer-Verlag, New York, 1985.
405. Kondoh, T., Fukumoto, A., Ohsawa, K., and Ohkubo, Y., An Assessment of a Multidimensional Numerical Method to Predict the Flow in Internal Combustion Engines, SAE International Congress and Exposition, Detroit, Mich., SAE Paper No. 850500, 1985.
406. El Tahry, S. H., A Numerical Study on the Effects of Fluid Motion at Inlet-Valve Closure on Subsequent Fluid Motion in a Motored Engine, SAE International Congress and Exposition, Detroit, Mich., SAE Paper No. 820035, 1982.

407. Gosman, A. D., Tsui, Y. Y., and Vafidis, C., Flow in a Model Engine with a Shrouded Valve—A Combined Experimental and Computational Study, SAE International Congress and Exposition, Detroit, Mich., SAE Paper No. 850498, 1985.
408. Gosman, A. D., Tsui, Y. Y., and Watkins, A. P., Calculation of Three-Dimensional Air Motion in Model Engines, SAE International Congress and Exposition, Detroit, Mich., SAE Paper No. 840229, 1984.
409. Brandstatter, W., Johns, R. J. R., and Wigley, G., The Effect of Inlet Port Geometry on In-Cylinder Flow Structure, SAE International Congress and Exposition, Detroit, Mich., SAE Paper No. 850499, 1985.
410. Brandstatter, W., Johns, R. J. R., and Wigley, G., Calculation of the Flow Produced by a Tangential Inlet Port, in *International Symposium of Flows in Internal Combustion Engines—III*, eds. T. Uzkan, W. G. Tiederman, and J. M. Novak, pp. 135–148, ASME, New York, 1985.
411. El Tahry, S., Application of a Reynolds Stress Model to Engine-Like Flows, *ASME Journal of Fluids Engineering*, vol. 107, pp. 444–450, 1985.
412. Gosman, A. D., Jahanbakhsh, A., and Watkins, A. P., Evaluation of Multidimensional Predictions of Flow in Piston-Bowl Configurations, in *International Symposium on Flows in Internal Combustion Engines—III*, eds. T. Uzkan, W. G. Tiederman, and J. M. Novak, pp. 115–124, ASME, New York, 1985.
413. Bernard, P. S., Computation of the Turbulent Flow in an Internal Combustion Engine during Compression, *ASME Journal of Fluids Engineering*, vol. 103, pp. 75–81, 1981.
414. Ramos, J. I., Gany, A., and Sirignano, W. A., Study of Turbulence in a Motored Four-Stroke Internal Combustion Engine, *AIAA Journal*, vol. 19, no. 5, pp. 595–600, 1981.

CHAPTER

FIVE

MATHEMATICAL MODELS OF GAS EXCHANGE PROCESSES AND TWO-STROKE ENGINES

5.1 INTRODUCTION

The objective of this chapter is to present and review some of the models that have been used to analyze flows in manifolds, gas exchange processes, and two-stroke engines. Flows in manifolds have been traditionally studied by means of the one-dimensional gas dynamics equations including the effects of heat transfer. Recently, however, multidimensional models have been used to analyze the flow through the intake valve and flows in manifolds.

The combustion phenomena in two-stroke engines can be simulated by models similar to those presented in Chapters 2, 3, and 4 (i.e., single-zone, multizone, and multidimensional models).

To accurately predict the in-cylinder flow field and combustion, it is necessary to know the boundary conditions at the inlet and exhaust ports. These boundary conditions depend on the gas exchange process, the cylinder pressure and temperature, the geometry of the ports and valves, etc. These boundary conditions are provided by the methods presented in this chapter, which deals with flows outside the engine cylinder (as compared with Chapters 2, 3, and 4, which deal with in-cylinder flows).

The indicated power of an internal combustion engine at a given speed is proportional to the mass of air contained in the cylinder. Thus, inducting the maximum air mass at wide throttle or full load and retaining that mass within the cylinder is one of the major goals of the gas exchange processes.

The purpose of the inlet and exhaust processes in four-stroke cycle engines or of the scavenging process in two-stroke cycle engines is to remove the burnt gases at the end of the power stroke and admit the fresh charge for the next cycle. An additional goal of the intake process is to set up the appropriate flow field within the cylinder. The effectiveness of these exchange processes is defined by overall parameters such as the volumetric efficiency for four-stroke cycle engines and the scavenging efficiency and the trapping efficiency for two-stroke cycle engines.

In the following sections, calculation methods for flows in manifolds, scavenging processes, and two-stroke engines are described. Some of these methods are based on the multidimensional conservation equations presented in Chapter 2; those equations are not repeated here.

5.2 FLOWS IN ENGINE MANIFOLDS

The intake and exhaust flows of reciprocating internal combustion engines play an important role in determining the engine performance, efficiency, and emissions. In this section we describe some of the models that have been used to analyze the flow fields in engine manifolds. These models can be classified into two groups: one- and multidimensional models.

5.2.1 One-Dimensional Models

The flow field through the intake and exhaust manifolds has frequently been treated as one-dimensional. Benson [1] used the method of characteristics to solve the governing equations of one-dimensional, unsteady, compressible flows in ducts. These equations can be written as

$$\frac{\partial}{\partial t}(\rho A) + \frac{\partial}{\partial x}(\rho u A) = 0 \qquad A = \frac{\pi}{4}d^2 \tag{1}$$

$$\rho\left(\frac{\partial u}{\partial t} + u\frac{\partial u}{\partial x}\right) = -\frac{\partial p}{\partial x} + \rho\frac{u|u|}{2}\frac{4f}{d} \tag{2}$$

$$\frac{\partial}{\partial t}\left[A\rho\left(e + \frac{1}{2}u^2\right)\right] + \frac{\partial}{\partial x}\left[A\rho u\left(e + \frac{p}{\rho} + \frac{1}{2}u^2\right)\right] = q\pi d \tag{3}$$

where t is time, x is the distance along the pipe, u is the fluid velocity, ρ is the density, f is the friction coefficient, d is the pipe diameter, A is the cross-sectional area of the pipe, q is the heat transfer rate per unit area, and e is the internal energy, that is,

$$e = e^0 + \int_{T_0}^{T} C_v\, dT \tag{4}$$

e^0 is the internal energy of formation, T_0 is a reference temperature, and C_v is the specific heat at constant volume.

If the gases flowing through the intake and exhaust manifolds are assumed to be ideal, the fluid pressure, density, and temperature are related by the following equation:

$$p = \rho RT \tag{5}$$

where R is the gas constant.

If diffusion is negligible (high-Reynolds-number flow), the mixture fraction obeys the following equation:

$$\frac{\partial Y}{\partial t} + u \frac{\partial Y}{\partial x} = 0 \tag{6}$$

The friction coefficient is frequently assumed to be that of a fully developed, steady-state, turbulent flow [2]

$$f = 0.316 \text{ Re}^{-1/4} \tag{7}$$

where the Reynolds number is defined as

$$\text{Re} = \rho |u| \frac{d}{\mu} \tag{8}$$

For fully developed, turbulent pipe flow, the shear stress at the wall is defined as

$$\tau_{\omega} = \frac{f}{8} \rho u^2 \tag{9}$$

and the film heat transfer coefficient can be calculated by using the Reynolds-Colburn analogy [2] as

$$\text{St Pr}^{2/3} = \frac{f}{8} \tag{10}$$

where St is the Stanton number (St = Nu/Re Pr), Nu is the Nusselt number, and Pr is the Prandtl number. Substitution of Eq. (7) into Eq. (10) yields

$$\text{Nu} = \frac{hd}{\lambda} = 0.0395 \text{ Re}^{3/4} \text{ Pr}^{1/3} \tag{11}$$

where h is the film heat transfer coefficient and λ is the thermal conductivity.

For constant-temperature walls, the heat transfer rate per unit area can be expressed as

$$q = h (T_w - T) = 0.0395 \frac{\lambda}{d} \text{Re}^{3/4} \text{ Pr}^{1/3} (T_w - T) \tag{12}$$

where T_w is the wall temperature.

Equations (1)–(3) and (6) can be written as

$$\frac{\partial U}{\partial t} + \frac{\partial F}{\partial x} = G \tag{13}$$

where

$$U = \left(\rho,\ \rho u,\ \frac{p}{\gamma - 1} + \frac{1}{2}\rho u^2,\ \rho Y\right)^T \tag{14}$$

$$F = \left[\rho u,\ \rho u^2 + p,\ \left(\frac{1}{2}\rho u^2 + \frac{\gamma}{\gamma - 1}p\right)u,\ \rho Y u\right]^T \tag{15}$$

$$G = \left[-\frac{\rho u}{A}\frac{dA}{dx},\ -\frac{\rho u^2}{A}\frac{dA}{dx} - \rho\frac{4f}{d}\frac{u|u|}{2},\ \frac{q\pi d}{A} - \frac{1}{A}\frac{dA}{dx}u\left(\frac{1}{2}\rho u^2 + \frac{\gamma}{\gamma - 1}p\right),\ -\frac{\rho u Y}{A}\frac{dA}{dx}\right]^T \tag{16}$$

the superscript T denotes transpose, $\gamma = C_p/C_v$ is the ratio of specific heats, and C_p is the specific heat at constant pressure.

For a calorically perfect gas (C_v = constant, C_p = constant)

$$\rho e = C_v T\rho = \frac{C_v}{R}p = \frac{1}{\gamma - 1}p \tag{17}$$

where we have used Eq. (4) and $R = C_p - C_v$.

Equation (13) can be solved by a variety of numerical schemes. Benson [1] used the method of characteristics, and Takizawa et l. [3] employed the two-step Lax-Wendroff method, which can be written as [4]

$$U_{i+1/2}^{n+1/2} = \frac{1}{2}(U_i^n + U_{i+1}^n) - \frac{\Delta t}{\Delta x}[F(U_{i+1}^n) - F(U_i^n)] + \frac{\Delta t}{2}[G(U_i^n) + G(U_{i+1}^n)] \tag{18}$$

$$U_i^{n+1} = U_i^n - \frac{\Delta t}{\Delta x}[F(U_{i+1/2}^{n+1/2}) - F(U_{i-1/2}^{n+1/2})] + \frac{\Delta t}{2}[G(U_{i+1/2}^{n+1/2}) + G(U_{i-1/2}^{n+1/2})] \tag{19}$$

where n denotes the time ($t^n = n\,\Delta t$), Δt is the time step, Δx is the grid spacing, and $x_i = i\,\Delta x$.

Equations (18) and (19) are second-order accurate in both space and time, that is, the truncation errors are $O(\Delta t^2, \Delta x^2)$. The two-step Lax-Wendroff method is stable, provided that the Courant-Friedrichs-Lewy condition is satisfied, that is,

$$(|u| + c)\frac{\Delta t}{\Delta x} \le 1 \tag{20}$$

where c is the sound speed defined as

$$c^2 = \left(\frac{\partial p}{\partial \rho}\right)_s = \gamma RT = \frac{\gamma p}{\rho} \tag{21}$$

and s denotes entropy.

Equation (13) can also be solved by flux-corrected techniques [5, 6] and other finite-difference methods [7]. However, most studies have used the method of characteristics [8, 9].

Desantes et al. [10] used the method of characteristics to design the exhaust pipe system and optimize the scavenging process. Winterbone et al. [11] used a wave-action method to evaluate the transmission of energy from the cylinder through the exhaust valve via the exhaust pipes and junctions to the turbine of the turbocharger. Winterbone et al. [12] used the nonhomentropic method of characteristics to analyze the flow field in manifolds, and they showed that compact manifolds can lead to an improved transmission efficiency of energy from the cylinder to the turbine of the turbocharger despite the pressure losses at junctions.

Newlyn and Few [13] used the method of characteristics and resonance equations to predict the engine volumetric efficiency, and they showed that improvements in the volumetric efficiency can be achieved by increasing the bore-to-stroke ratio at high engine speeds. (The volumetric efficiency is defined as the ratio of the air charge in the cylinder to the mass of air that would fill the cylinder at inlet pressure and temperature [14].) Vorum [15] also used resonance equations to optimize the primary and secondary pipe velocities and to provide maximum boost in volumetric efficiency in a naturally aspirated engine.

Payri et al. [16, 17] showed that variable-valve timing systems can reduce the pumping losses in spark-ignition engines and improve the efficiency of engines at partial loads.

The use of heat transfer correlations for steady-state, fully developed turbulent flows in Eqs. (3) and (11) is questionable. Flows in inlet and exhaust manifolds are unsteady, have entrance effects, and involve variations in the flow rates. Kalinin and Dreitser [18], Dreitser [19], Dreitser et al. [20, 21], and Galitseiskii et al. [22] showed that the heat transfer coefficient in unsteady turbulent flows is determined not by the time-dependent boundary conditions but only by the rates of their changes, and that unsteady heat transfer calculations based on steady-state correlations result in large errors. Kalinin and Dreitser [18] proposed a variety of heat transfer correlations as a function of the wall heating and cooling, Reynolds and Prandtl numbers, and rate of change of the wall temperature.

The friction coefficient [Eq. (7)] may not be valid in manifolds due to flow reversals, resonance, and unsteady effects. As the gas in the manifolds is heated, its dynamic viscosity and the boundary layer thickness increase, and the friction coefficients decrease. Conversely, as the gas in the manifolds is cooled, its dynamic viscosity and the boundary layer thickness decrease, and the friction coefficients increase. Furthermore, in naturally aspirated spark-ignition engines, fuel droplets may deposit on the pipe walls and form liquid films. The fuel droplet evaporation and acceleration may affect the heat transfer and friction [23].

Mahdi Ali [24] studied the motion of fuel droplets in a gasoline engine manifold by means of a steady-state flow model that neglects the interaction between droplets and droplet vaporization. His studies show that (1) the fuel distribution in the engine manifold can be improved by using high air velocities, (2) the fuel distribution is more important than the air distribution in controlling the fuel-air ratio, and (3) there is a critical droplet diameter that should not be exceeded in order to prevent the droplets from impinging on the manifold walls.

Milton and Behnia [25] investigated the fuel vaporization in an unbranched manifold under steady-state conditions. Their one-dimensional model does not account for the flow pulsations induced by the cylinder but provides the velocity and thickness of the vapor film at the manifold wall. The wall film is of primary importance in determining the lag and in-cylinder fuel composition during acceleration-deceleration transients.

The fuel droplets in a manifold can be reentrained from the liquid film; however, this process is difficult to model accurately.

Several models have been developed to study the hydrodynamics and heat transfer characteristics of two-phase annular flows [26]. Based on experimental observations, these analyses divide the liquid film into a continuous liquid layer adjacent to the channel surface and a wavy layer close to the gas-liquid interface. The continuous liquid layer is assumed to have a structure similar to that of a single-phase turbulent pipe flow, whereas a mixing length concept is used in the wavy layer region of the film.

Unsteady one-dimensional equations such as Eq. (13) have been used to predict the flow field in pipe junctions and branches [3]. One-dimensional models of flows in inlet and exhaust manifolds can be used in conjunction with the combustion models presented in Chapters 2, 3, and 4 to predict the performance of spark-ignition and diesel engines [27].

Some sample results that illustrate the pressure as a function of the crank shaft angle at different locations in the intake and exhaust manifolds are shown in Fig. 5.1 [3].

Bingham and Blair [28] used the method of characteristics and pressure loss coefficients from steady-state tests to predict the flow field in branched pipes of multicylinder engines. Their method agrees better with experimental data than a constant-pressure approach, where the pressure is assumed to be the same at the pipe junction. It also eliminates the complexity of the techniques based on momentum theory. However, the flows in pipe junctions and branches are not one-dimensional and can be accurately predicted only by using two- and three-dimensional models.

5.2.2 Multidimensional Models

The governing equations of multidimensional flows have been described in Chapter 2 and are not repeated here.

Chapman [29] used a two-dimensional model to analyze the flow field in the inlet manifold of a four-cylinder internal combustion engine. His model can be used to develop an optimal fuel-injection timing strategy at various loads and under various operating conditions and to examine the effects of the plenum chamber geometry on wave interactions in the manifold. Tosa et al. [30] used

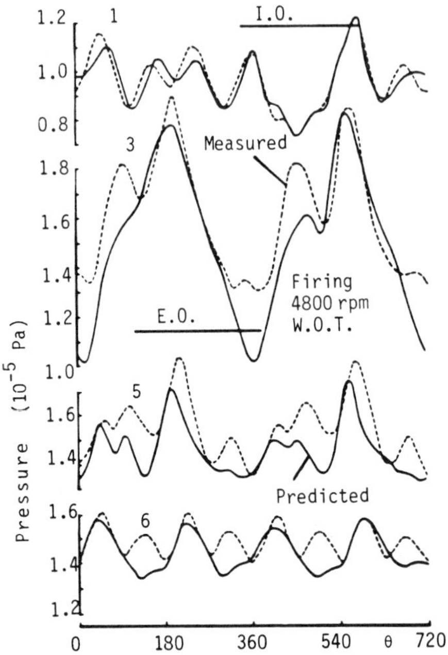

Figure 5.1 Pressure as a function of the crank shaft angle at different locations in the intake and exhaust manifolds [3] (I.O. = intake valve opening; E.O. = exhaust valve opening.).

a two-dimensional fluid-in-cell method to study the flow field in pipe branches and confluences and a one-dimensional method to analyze the flow in the pipes. Dimitriadis et al. [31] used a three-dimensional model of the control-volume variety to analyze the flow field in pipe junctions.

Seppen and Janssen [32] developed a three-dimensional model to optimize the design of intake and exhaust manifolds based on the PHOENICS code [33].

Some sample results that illustrate the flow field in a four-stroke engine manifold are shown in Fig. 5.2 [29].

5.3 FLOWS AROUND VALVES

The purpose of the exhaust and inlet processes in four-stroke cycle engines or of the scavenging processes in two-stroke cycle engines is to remove the burnt gases at the end of the power (expansion) stroke and admit the fresh charge for the next cycle. The indicated power, performance, emissions, and efficiency of an internal combustion engine at a given speed are proportional to the air mass flow rate. Thus, inducting the maximum air mass at wide-open throttle (WOT) or full load and retaining that mass within the cylinder are the major goals of the gas exchange processes. The effectiveness of these gas exchange processes is defined by the volumetric efficiency for four-stroke engines and by the scavenging efficiency and trapping efficiency for two-stroke engines. These efficiencies depend on the flow field in the intake and exhaust manifolds (see Section 5.2) and especially on the valves and ports.

Early intake valve closing (IVC) results in lower pumping losses, lower specific fuel consumption, lower cylinder gas temperatures, lower heat transfer, lower NO_x emissions, and higher hydrocarbon (HC) emissions than conventionally throttled engines operated at part load [34]. The rate of exhaust valve opening (EVO) and the timing have a significant influence on the performance and gas exchange work of turbocharged truck engines; the earlier the timing, the more efficient is the gas exchange. In contrast, the later the timing, the greater is the available energy for the turbocharger turbine [35].

The valves or the ports are the most important restrictions in the intake and exhaust systems. In addition, the intake valve sets up the flow within the cylinder. The instantaneous valve flow area depends on the valve lift and the geometric details of the valve head, seat, and stem. The effect of the valve geometry and timing on air flow can be illustrated by defining a pseudo-flow-velocity v_{ps} as

$$v_{ps} = \frac{1}{A_m}\frac{dV}{d\theta} \tag{22}$$

where A_m is the valve area, V is the cylinder volume, and θ is the crank shaft angle. The cylinder volume is given by the following expression:

$$V = \frac{\pi B^2}{4}\left(c + s\,\frac{1 - \cos\theta}{2} + L\left\{1 - \left[1 - \left(\frac{s}{2L}\sin\theta\right)^2\right]^{1/2}\right\}\right) \tag{23}$$

where B is the cylinder bore, c is the clearance, $s/2$ is the crank radius, L is the connecting rod length, and θ is the crank shaft angle measured with respect to top dead center (TDC), and s is the stroke.

Low pseudo-velocities early in the exhaust stroke and late in the intake stroke indicate that phenomena other than quasisteady flow govern the flow rate [36]. These are the periods when exhaust blowdown as well as ram and tuning effects in the intake are the most important.

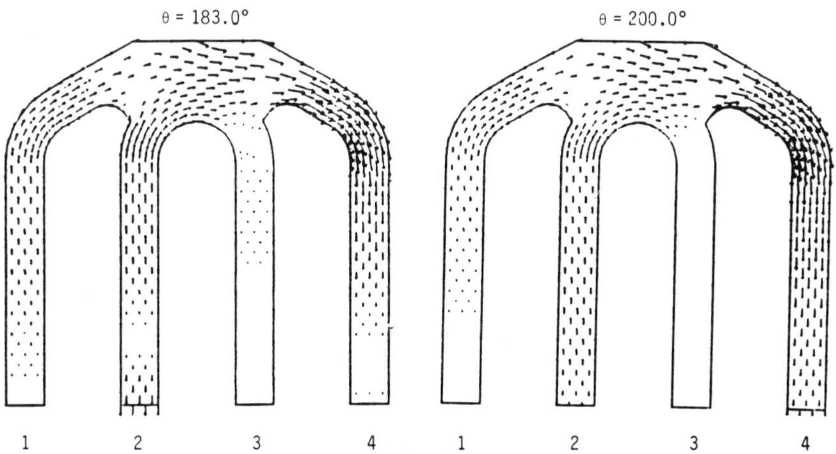

Figure 5.2 Two-dimensional flow fields in a four-stroke engine manifold [29].

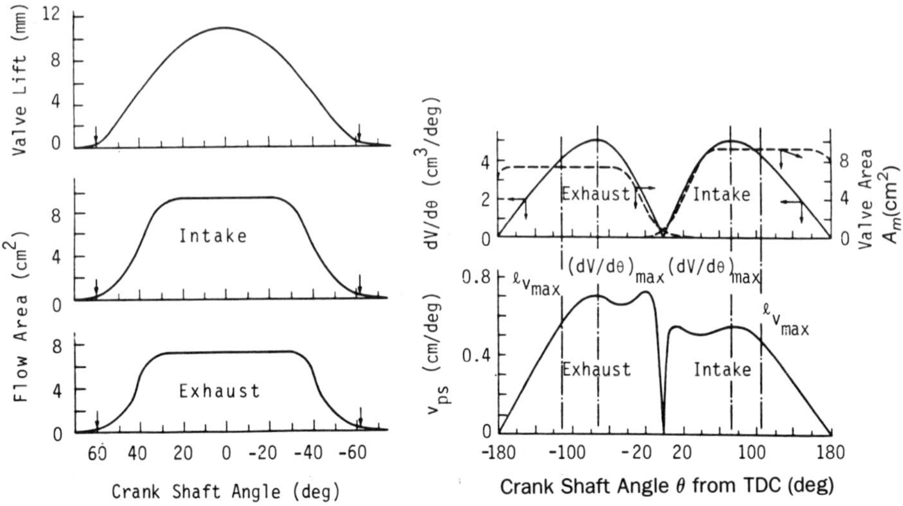

Figure 5.3 Valve lift, flow areas of the intake and exhaust valves, rate of change of the cylinder volume $dV/d\theta$, valve minimum flow area A_m, and pseudo-flow-velocity v_{ps} as a function of the crank shaft angle θ for typical exhaust and intake valves [36].

Figure 5.3 shows the valve lift, the flow areas of the intake and exhaust valves, the rate of change of the cylinder volume $dV/d\theta$, the valve minimum flow area, and the pseudo-flow-velocity as a function of the crank shaft angle for typical exhaust and intake valves [36].

In the next sections we describe some of the models that have been used to calculate the air flow rate through the valves.

5.3.1 One-Dimensional Models

The mass flow rate through the valves is usually described by one-dimensional isentropic analyses through a flow restriction, and irreversibilities and real gas flow effects are included by means of an experimentally determined discharge coefficient. The air flow rate is related to the upstream stagnation pressure p_0 and temperature T_0, static pressure just downstream of the flow restriction (assumed to be equal to the pressure at the restriction p_R), and a characteristic area of the valve design A_v as

$$\dot{m} = c_D A_v p_0 (RT_0)^{-1/2} \left(\frac{p_R}{p_0}\right)^{1/\gamma} \left\{ \frac{2\gamma}{\gamma - 1} \left[1 - \left(\frac{p_R}{p_0}\right)^{(\gamma-1)/\gamma} \right] \right\}^{1/2} \tag{24}$$

when the flow is choked

$$\frac{p_R}{p_0} \leq \left(\frac{2}{\gamma + 1}\right)^{\gamma/(\gamma-1)} \tag{25}$$

$$\dot{m} = c_D A_v p_0 \gamma^{1/2} (RT_0)^{-1/2} \left(\frac{2}{\gamma + 1} \right)^{(\gamma+1)/2(\gamma-1)} \tag{26}$$

For flow into the cylinder through an intake valve, p_0 is the intake manifold pressure, and p_R is the cylinder pressure. For flow out of the cylinder through an exhaust valve, p_0 is the cylinder pressure, and p_R is the exhaust manifold pressure. The valve area can be estimated as

$$A_v = \pi d_v l_v \tag{27}$$

where d_v is the valve head diameter and l_v is the valve lift.

The discharge coefficient based on Eq. (27) is a discontinuous function of the valve lift. At very low lifts the flow remains attached to the valve head and seat, giving high values for the discharge coefficient. At intermediate lifts the flow separates from the valve head at the inner edge of the valve seat, and the discharge coefficient decreases abruptly. The discharge coefficient then increases with increasing lift since the size of the separated flow region remains approximately constant while the minimum flow area is increasing. At high lifts the flow separates from the inner edge of the valve seat, and the discharge coefficient decreases.

Figure 5.4 shows the discharge coefficient of a typical inlet poppet valve as a function of the valve lift [36].

Experiments with compressed air indicate that the discharge coefficient is a function of the rate of volumetric expansion and of length-to-diameter ratio and geometry. If the downstream-to-upstream pressure ratio is kept constant, the discharge coefficient decreases as the volume expansion or the piston speed is increased [37]. The valve discharge coefficient can be expressed as

$$C_D = C_{Dss} + C_{Dt} \tag{28}$$

where the steady C_{Dss} and transient C_{Dt} discharge coefficients are a function of the Mach number and the engine-operating conditions [38, 39].

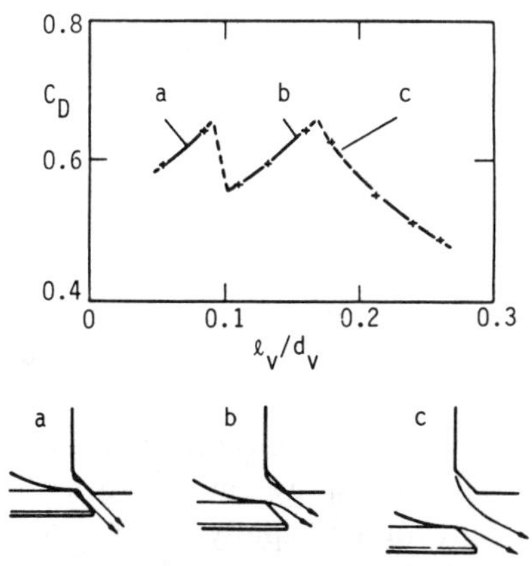

Figure 5.4 Discharge coefficient of a typical inlet poppet valve as a function of the valve lift [36].

The flow through and the boundary conditions at the intake valve have a profound effect on the combustion characteristics of spark-ignition engines. Haghgooie et al. [40] showed that the combustion rate correlates with the kinetic energy and angular momentum fluxes measured at the valve as well as with the isentropic flow velocity computed from the measured pressure drop.

The effects of the flow and boundary conditions on the flow field in motored engines have been investigated experimentally by Bicen et al. [41], Gany et al. [42], and Arcoumanis et al. [43]. The laser Doppler velocimetry (LDV) data of Arcoumanis et al. [43] show that the valve seat angle has a great influence on the engine flow field. Their studies indicate that a valve seat angle of 30° produces an intake jet that adheres to the cylinder head. A valve seat angle of 60° produces vortices at the cylinder wall, piston, and cylinder head, increases the turbulence levels, and results in longer living vortices when the incoming air flow strikes the cylinder wall. Arcoumanis et al. [43] also observed that longer strokes produce longer vortices and earlier decay. If the clearance is very small, the intake jet may strike on the piston and generate wall turbulence.

Bicen et al. [41] showed that in engines with open annular ports, the intake-generated vortex structures are destroyed before the piston reaches TDC.

Arcoumanis et al. [44] and Arcoumanis and Whitelaw [45] showed that the flow around the intake valve is not uniform; the axial velocity is more sensitive to the valve lift than is the radial velocity. Bicen et al. [46] observed that the flow regimes within the engine cylinder are affected by the valve seat angle and cylinder confinement, but are insensitive to the valve operation; valve seat angles of 45° and 60° produce four and three flow regimes, respectively.

The importance of the inlet boundary conditions on engine modeling has been demonstrated by Gosman et al. [47], who showed that the mean velocity field is more strongly affected by the engine geometry than by the speed. These investigators also showed that their two-dimensional model predicts the main features of the flow, but there are discrepancies in the small scale due to inaccurate simulation of the boundary conditions.

The performance of the inlet valve depends on the valve lift, valve seat width, valve seat angle, port design, cylinder head shape, and rounding of the seat corners. In many designs the port and valve assembly are used to generate a rotational motion (swirl) inside the engine during the induction process, or the cylinder is shaped to restrict the flow through one side of the valve open area to generate swirl. Swirl reduces the valve and port flow discharge coefficients.

Elkotb et al. [48] showed that the port area has a greater effect on the tangential velocity and velocity rms than the engine rpm.

At high speeds, unless the inlet valve is of sufficient size, the inlet flow during the induction process can become choked at the minimum valve flow area. Choking substantially reduces the volumetric efficiency.

When the exhaust valve opens toward the end of the expansion stroke, the cylinder pressure is substantially higher than the exhaust system pressure. A blowdown process results as burnt gases exit the cylinder rapidly. A vigorous

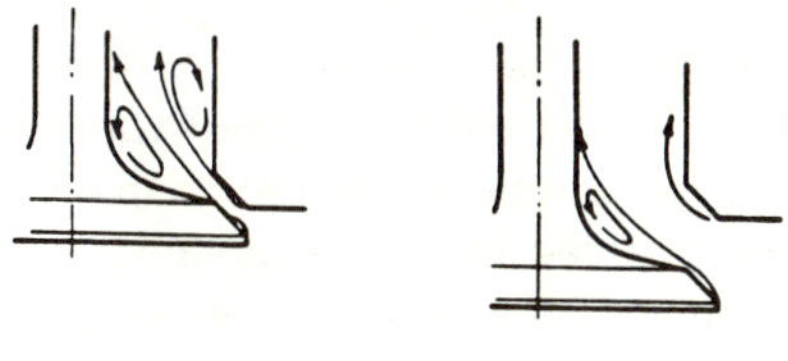

Figure 5.5 Flow pattern through the exhaust valve at low and high lifts [36].

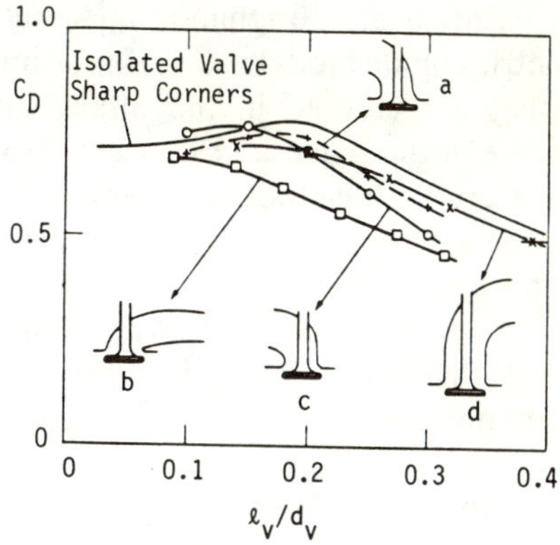

Figure 5.6 Discharge coefficient as a function of the valve lift for several exhaust valves and port designs [36].

in-cylinder gas flow toward the exhaust valve occurs, generating in-cylinder velocities up to several times the mean piston speed, and a recirculation zone is created beneath the open exhaust valve.

The discharge coefficient through the exhaust depends on the valve lift. At low lifts the flow separates from the valve seat and port, whereas at high lifts the flow separates from the valve seat, and the discharge coefficient decreases.

Figure 5.5 shows the flow patterns through the exhaust valve at low and high lifts [36]. Figure 5.6 shows the discharge coefficient as a function of the valve lift for several exhaust valves and port designs [36].

Because the flows through the intake and exhaust valves are three-dimensional, realistic predictions of the flow field in the port can be achieved only with multidimensional computer codes. Multidimensional predictions of the flow field around the intake and exhaust valves are described in the next section.

5.3.2 Multidimensional Models

Multidimensional models for analyzing the flow around the intake and exhaust valves are based on the equations presented in Chapter 2.

In the first multidimensional studies, the flow around the valve was not considered, and boundary conditions were imposed along a cylindrical surface from the valve seat diameter to the cylinder head [49, 50]. Those studies did not represent the inlet and exhaust processes accurately, and they required specification of the time-dependent boundary conditions along the cylindrical surface located between the valve seat and the cylinder head. More recently, Pinchon and Baritaud [51] used a similar simulation of the inlet and exhaust boundary conditions to analyze the effects of a shrouded valve on the flow field and combustion in spark-ignition engines. Their simulations used experimentally determined inlet boundary conditions.

Ramos and Sirignano [52] simulated the intake-exhaust valve in an axisymmetric engine model as an infinitesimally thin disk; they showed that recirculation zones are created in the intake and exhaust strokes beneath the valve and between the valve and its stem. However, because this valve representation is very crude, it cannot accurately predict the flow around the intake and exhaust valves or the flow in the engine cylinder, which is governed by that through the intake valve.

Schock et al. [53, 54] also simulated the intake-exhaust valve as an infinitesimally thin disk, and they showed that the flow field within the engine cylinder depends strongly on the valve seat angle. For valve seat angles of less than 45°, the in-cylinder flow field consists of two vortices located beneath the valve and near the corner between the cylinder head and the cylinder wall. For valve seat angles of about 45°, three elongated vortices exist in the cylinder and persist into the early parts of the compression stroke. These vortices break up and merge in the compression stroke. Different flow regions as a function of the valve seat angle have also been observed experimentally by Bicen et al. [46].

Figures 5.7 and 5.8 show the flow field in the engine cylinder at different crank shaft angles for different valve seat angles [53, 54].

The flow upstream of a valve was analyzed by means of a steady-state model by Demirdzic et al. [55], who used an arbitrary mesh finite volume method,

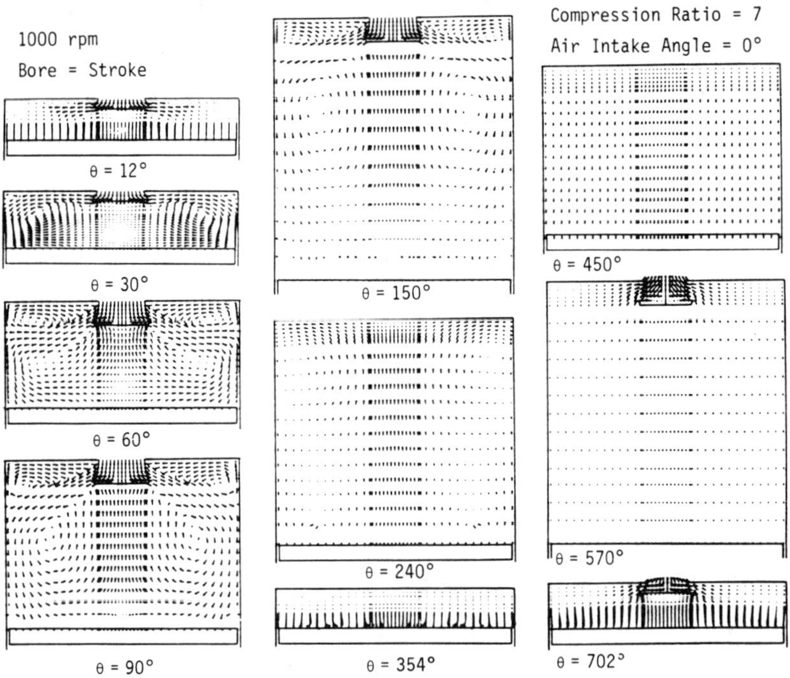

Figure 5.7 Flow field in the engine cylinder at selected crank angles and at air intake angle = 0° [53, 54].

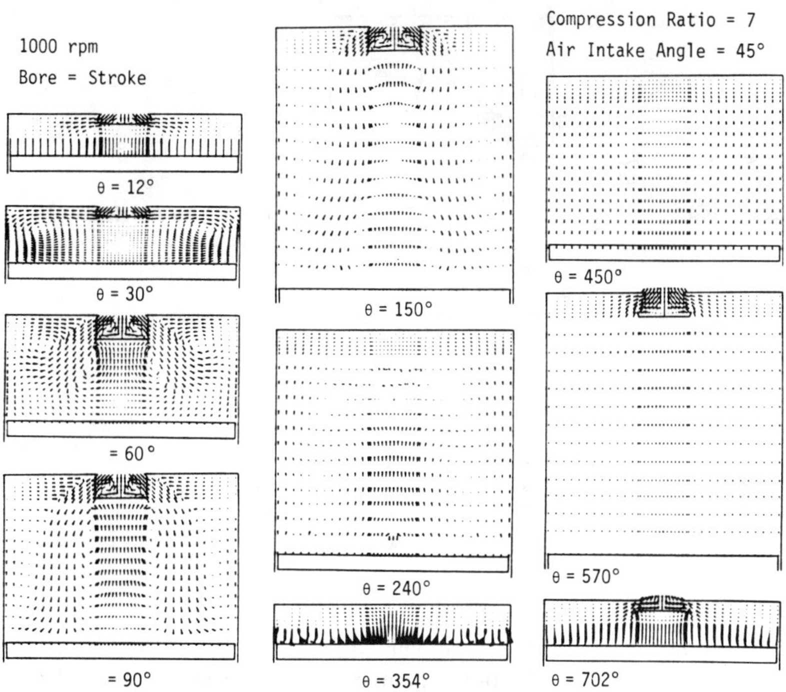

Figure 5.8 Flow field in the engine cylinder at selected crank angles and at air intake angle = 45° [53, 54].

contravariant velocities, and a semistrong conservation formulation that minimizes interpolation and numerical diffusion and eases the imposition of the boundary conditions, but does not involve cross derivatives in the divergence operator. Contravariant formulations can also be used to analyze the flow field around the intake and exhaust valves and in the piston bowl [56, 57].

Sanatian and Moss [58] used a multidimensional model to analyze the steady flow field through alternative induction system throttling devices (i.e., a butterfly valve and a variable geometry ramp restriction). Their flow field predictions agree satisfactorily with flow visualization studies. They also studied the flow field around the intake valve, which was simulated as a finite-width disk. A more realistic valve representation was used by Demirdzic et al. [59], who compared their numerical predictions with LDV data and showed that the predicted mean velocities are in good agreement with experimental data, except in the region of very steep gradients located between the valve seat and head. Their studies also indicate that the numerical predictions of the turbulence kinetic energy show less agreement with the LDV data in the vicinity of the cylinder wall. This disagreement was attributed to numerical inaccuracies, deficiencies in the k–ϵ model, and measurement uncertainties.

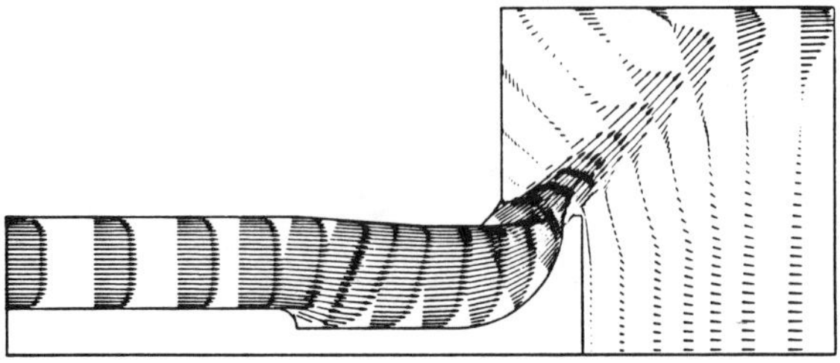

Figure 5.9 Mean velocity vectors around an inlet valve [59].

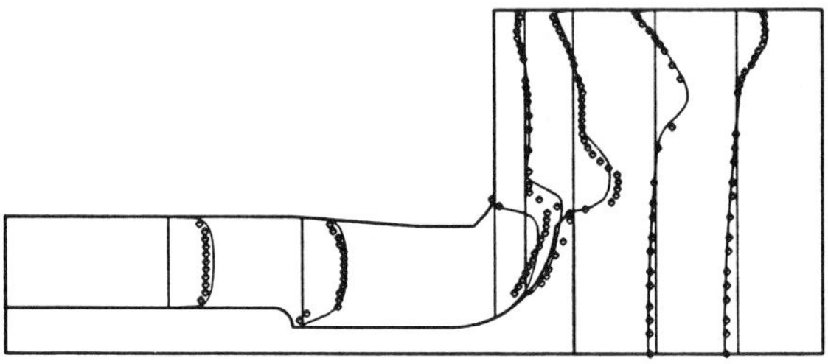

Figure 5.10 Comparisons between measured and predicted mean axial velocity profiles [59].

Figure 5.9 shows the predicted mean velocity vectors around an inlet valve, whereas Figs. 5.10–5.12 present comparisons between the measured and predicted mean axial velocity, mean radial velocity, and turbulence intensity profiles [59].

Figure 5.13 shows the velocity vectors around a butterfly valve at setting angles of 30° and 50° [58].

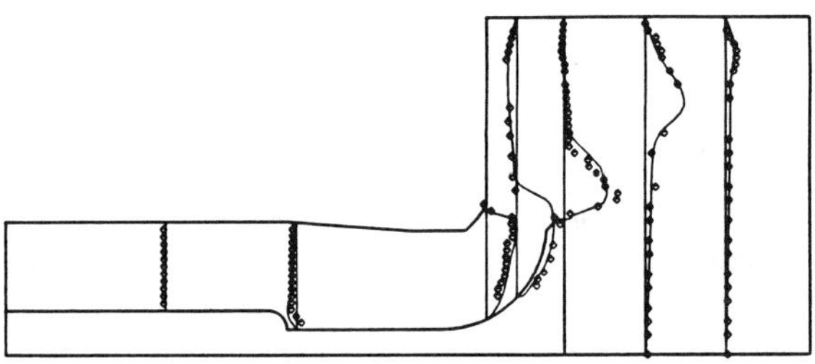

Figure 5.11 Comparisons between measured and predicted mean radial velocity profiles [59].

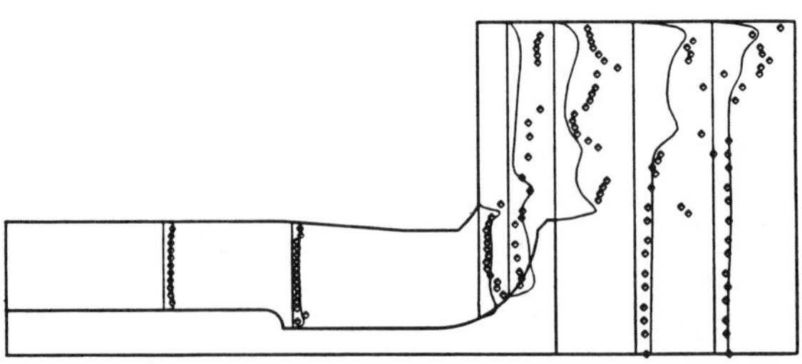

Figure 5.12 Comparisons between measured and predicted turbulence intensity profiles [59].

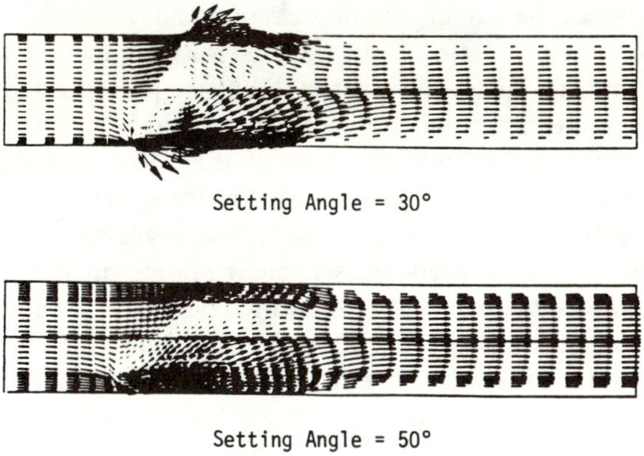

Figure 5.13 Velocity vectors around a butterfly valve at setting angles of 30° and 50° [58].

5.4 TWO-STROKE ENGINES AND SCAVENGING PROCESSES

In two-stroke cycle engines, each outward stroke of the piston is a power stroke. To achieve this operating cycle, the fresh charge must be supplied to the engine cylinder at a pressure high enough to displace the burnt gases from the previous cycle. The pressure of the intake mixture may be raised in a separate pump or compressor. The operation of clearing the cylinder of burnt gases and filling it with fresh mixture is called scavenging.

Scavenging flow arrangements are classified into three groups: cross-scavenged, loop-scavenged, and uniflow-scavenged (Fig. 5.14). Cross- and loop-scavenged systems use exhaust and inlet ports in the cylinder wall, which are uncovered by the piston as it approaches bottom dead center (BDC). The uniflow-scavenged system may use inlet ports with exhaust valves in the cylinder head, or it may use inlet and exhaust ports with opposed pistons. The location and orientation of the scavenging ports control the scavenging process; despite the different flow patterns obtained with each scavenging system, however, the general operating principles are the same (i.e., air in a diesel engine or a fuel-air mixture in a spark-ignition engine must be supplied to the inlet ports at a pressure higher than that of the exhaust system).

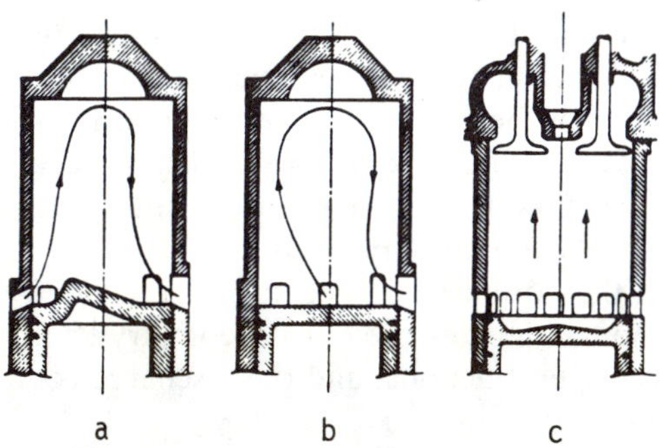

Figure 5.14 Two-stroke cycle flow configurations: (a) cross-scavenged; (b) loop-scavenged; and (c) uniflow-scavenged [36].

For a uniflow-scavenged engine, the scavenging process proceeds as follows. Early in the second half of the expansion stroke, the exhaust valve opens, and a blowdown discharge process begins. (Blowdown is the discharge period up to the time of scavenging port opening.) Initially, the pressure ratio across the exhaust valve may exceed a critical value, and the flow at the valve may be sonic. As the cylinder pressure decreases, the pressure ratio drops below the critical value. The scavenging ports open 45–70 crank shaft angle degrees before BDC, when the cylinder pressure slightly exceeds the scavenging pump pressure. Because the burnt gases flow toward the exhaust valves, which now have a large open area, the exhaust flow continues, and no backflow occurs. When the cylinder pressure falls below the inlet pressure, air enters the cylinder, and the scavenging process starts. This flow continues as long as the inlet ports are open and the inlet pressure exceeds the cylinder pressure. As the cylinder pressure rises above the exhaust pressure, the fresh charge flowing into the cylinder discharges the burnt gases, and part of the fresh mixture mixes with the burnt gases and is expelled with them. The exhaust valves close at the same time as, or before, the inlet ports close. Since the exhaust valve open area is less than that of the scavenging ports, additional charging can be obtained [36].

The scavenging process can be evaluated in terms of the scavenging efficiency (SE) and scavenging ratio (SR). Scavenging efficiency is defined as the ratio between the volume of the fresh charge trapped and the total volume of the trapped charge, whereas scavenging ratio is defined as the ratio between the volume of the fresh charge supplied to the cylinder volume at trapping. Trapping efficiency (TE) is defined as the product of the scavenging efficiency and the scavenging ratio, that is, TE = SE · SR [60].

If the fresh mixture mixes perfectly with the burnt gases,

$$\mathrm{SE} = 1 - \exp(-\mathrm{SR}) \tag{29}$$

However, the scavenging process includes a short-circuit proportion y and a perfect displacement proportion p of the scavenging ratio, so that the following empirical expressions hold:

$$\text{If} \quad (1-y)p < \mathrm{SR} \quad \text{then} \quad \mathrm{SE} = (1-y)p \tag{30}$$

$$\text{If} \quad (1-y)p > \mathrm{SR} \quad \text{then} \quad \mathrm{SE} = 1 - (1-\mathrm{SR})\exp[\mathrm{SR} - (1-y)p] \tag{31}$$

Scavenging flow is affected by the crank shaft angle at which the inlet ports open, the size, number, geometry, and location of the ports around the cylinder periphery, and the direction and velocity of the jets issuing from the ports into the cylinder. For small openings, the flow is attached to the port walls; however, for fully open ports with sharp corners, the flow detaches at the upstream corners. Rounded entries help prevent flow detachment within the port.

The discharge coefficient through the ports depends on the port geometry and operating conditions and is a function of the pressure ratio. Geometry effects are most significant at small and large, open fractions, and the discharge coefficient increases as the pressure ratio is increased.

Tangentially inclined inlet ports are used when swirl is desired to improve scavenging. However, the discharge coefficient decreases as the jet tangential inclination increases. The jet angle and the port angle can deviate significantly from each other, depending on the details of the port design and the open fraction. The discharge coefficient through the exhaust valve depends on the valve geometry, open fraction, and pressure ratio across the exhaust port.

Unsteady gas flow calculations in two-stroke cycle engines contain two models: combustion and scavenging. These models vary in their level of sophistication. For example, combustion can be modeled by means of simple heat-release-rate, two-zone thermodynamic, or multidimensional models (see Chapters 2, 3, and 4). The scavenging process can be modeled by means of mixing [61], one-dimensional, and multidimensional models. The combustion model, the unsteady gas dynamics calculations, and the scavenging model provide the information for the closed-cycle process (i.e., trapped charge mass, state conditions, etc.). The overall accuracy of the entire calculation depends on the combustion and scavenging models as well as on the accuracy of the gas dynamics calculations. Of the two models, the weaker one is the scavenging model, which is dealt with in the next sections.

5.4.1 Semiempirical Models

Sher [62] developed semiempirical models to represent the gas exchange processes in cross-, loop- and uniflow-scavenged engines. In these models the mass fraction of fresh air contained in the gas passing through the exhaust port,

$$\beta = \frac{dm_{ae}/dt}{dm_e/dt} \tag{32}$$

where dm_e/dt is the mass flow rate of the mixture emerging from the cylinder through the exhaust ports and dm_{ae}/dt is the fresh charge mean flow rate contained in the mixture, is represented by the following expression:

$$\beta = 1 - \exp\left[-c\,\lambda\left(\frac{\theta - \theta_{IO}}{\theta_{IC} - \theta_{IO}}\right)^b\right] \tag{33}$$

where θ is the crank angle, b and c are form and shape factors, λ is the delivery ratio, and the subscripts IO and IC denote inlet port opening and closing, respectively. The delivery ratio is defined as the ratio between the volume of the air delivered to the cylinder and the displaced volume. The charging efficiency is defined as the ratio between the mass of delivered charge retained m_{ac} and the mass of pure air that could be trapped in the cylinder at atmospheric conditions m_0, that is,

$$\eta_c = \frac{m_{ac}}{m_0} \tag{34}$$

Conservation of mass states that

$$\frac{dm_{ac}}{dt} = \frac{dm_{as}}{dt} - \frac{dm_{ae}}{dt} \tag{35}$$

where dm_{as}/dt is the mean flow rate of the fresh charge entering through the inlet ports, which is related to the delivery ratio by the following equation:

$$\frac{dm_{as}}{dt} = m_0 \frac{d\lambda}{dt} \tag{36}$$

If the scavenging process occurs at constant volume and pressure and the gases are calorically perfect (constant specific heats),

$$pV = mRT = \text{constant} \qquad \frac{dm}{dt} = -\frac{m}{T}\frac{dT}{dt} \tag{37}$$

In the absence of heat losses, the energy equation can be expressed as

$$\frac{d}{dt}(mC_vT) = -\beta\frac{dm_e}{dt}C_pT_{as} - \frac{dm_e}{dt}(1-\beta)C_pT_{ub} + \frac{dm_{as}}{dt}C_pT_{as} = 0 \tag{38}$$

where p, V, and T are the pressure, volume, and temperature, respectively; m is the mass; R is the gas constant; C_p is the specific heat at constant pressure; and T_{ub} is the temperature of the unmixed burnt gases.

Equation (38) can be written as

$$\frac{dm_e}{dt} = T_{as}\frac{dm_{as}/dt}{\beta\, T_{as} + (1-\beta)\, T_{ub}} \tag{39}$$

Substitution of Eqs. (32) and (35)–(39) into Eq. (34) yields

$$\eta_c = \lambda - \int_0^t \beta \frac{d\lambda/dt}{\beta + (1-\beta)\, T_{ub}/T_{as}}\, dt \tag{40}$$

which can be integrated numerically.

Sher [62] suggested $c = 1.7$ and $b = 2$ for loop-scavenged engines and showed that b and c do not depend on the engine speed.

5.4.2 Isothermal Pure Mixing Model

Pure mixing models assumed that the scavenging process occurs at constant pressure and volume and that the fresh charge and residual gas are calorically perfect with identical constant specific heats. These models can be obtained from Eqs. (37)–(39) by setting

$$T = \beta\, T_{as} + (1-\beta)\, T_{ub} \tag{41}$$

Conservation of mass implies that

$$\frac{dm}{dt} = \frac{dm_{as}}{dt} - \frac{dm_e}{dt} = -\frac{T}{m}\frac{dT}{dt} \tag{42}$$

Substitution of Eqs. (38) and (39) into Eq. (42) yields

$$\frac{dm_{as}}{dt} = -\frac{m}{T - T_{as}}\frac{dT}{dt} = -\frac{mT}{T(T - T_{as})}\frac{dT}{dt} = -\frac{C}{T(T - T_{as})}\frac{dT}{dt} \tag{43}$$

where

$$C = mT = \text{constant}. \tag{44}$$

If the inlet conditions are equal to the standard conditions,

$$mT = m_0 T_{as} \tag{45}$$

Substitution of Eq. (45) into Eq. (43) and integration of the resulting equation yields

$$\lambda m_0 = -m_0 T_{as}\int_{T_{be}}^{T_{se}} \frac{dT}{T(T - T_{as})} \quad \text{or} \quad T_{se} = \frac{T_{as}}{1 - (1 - T_{as}/T_{be})\exp(-\lambda)} \tag{46}$$

where T_{be} and T_{se} are the cylinder temperatures at the end of the blowdown period and at the end of the scavenging process, respectively.

The temperature at the end of the scavenging process can be calculated as

$$T_{se} = \eta\, T_{as} + (1-\eta)\, T_{be} \tag{47}$$

where the scavenging efficiency η is here defined as the ratio of the mass of delivered charge retained in the cylinder to the mass contained in the cylinder.

Substitution of Eq. (46) into Eq. (47) yields

$$\eta = 1 - \frac{T_{se}}{T_{be}}\exp(-\lambda) \tag{48}$$

If it is further assumed that there is perfect mixing between the residual gases and the fresh charge,

$$\beta = \eta \tag{49}$$

The charging efficiency η_c is related to the scavenging efficiency as

$$\eta_c = \frac{T_{as}}{T_{se}}\eta \tag{50}$$

Substitution of Eqs. (46)–(48) into Eq. (50) yields

$$\eta_c = 1 - \exp(-\lambda) \tag{51}$$

which implies that the charging efficiency is independent of the temperature ratio between the burnt gases and the fresh charge. By way of contrast, Eq. (40) depends on that temperature ratio.

5.4.3 Pure Displacement Model

The pure displacement model assumes a constant-pressure process, in which the fresh charge entering the cylinder pushes out the burnt gases by a perfect displacement mechanism, during which no mass or heat transfer is allowed at their interface. This yields [61]

$$\eta_c = \lambda \quad \text{for} \quad \lambda < 1 \tag{52}$$

$$\eta_c = 1 \quad \text{for} \quad \lambda > 1 \tag{53}$$

5.4.4 Nonisothermal Mixing Model

In this model a temperature difference exists between the cylinder charge and the fresh charge entering the cylinder. However, this model yields the same charging efficiency as the isothermal pure mixing model [Eq. (51)]. Equation (51) does not take into account the method of scavenging and the combustion chamber geometry.

5.4.5 Benson's Model

Benson [63] developed a thermodynamic model of the scavenging process and divided the cylinder charge into three zones: an air zone, a gas zone, and a mixing zone. In the mixing zone three different phases were considered: displacement phase, short-circuiting phase, and mixing phase. Two differential equations for each zone were used to account for the mass and energy exchanges, and two functions were prescribed to specify the flow rates of air and burnt gases into the mixing zone. This model accounts for the combustion chamber geometry and the method of scavenging by means of mixing rates and phase timings.

Figure 5.15 shows the effects of the delivery ratio on the charging efficiency for the pure displacement model, mixing model, and Sher's model [62].

5.4.6 Other Models

Krieger et al. [64] divided a two-stroke engine into three thermodynamic systems: the cylinder gases, the crankcase gases, and the inlet system gases. Energy and

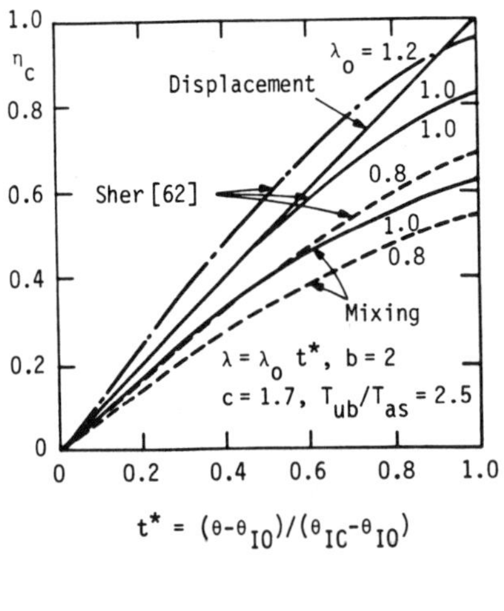

Figure 5.15 Effect of the overall delivery ratio λ on the charging efficiency η_c [62].

mass balances, the ideal gas law, and thermodynamic properties were combined to obtain a set of ordinary differential equations that described the thermodynamic state of each system.

Krieger et al. [64] used a two-zone thermodynamic model that includes the effects of dissociation of the combustion gases through empirical curve fits for the equilibrium thermodynamic properties and calculated the mass flow rate through the engine by means of one-dimensional unsteady equations using discharge coefficients from steady-state flow experiments. They also assumed perfect mixing between the residual gases and the fresh charge and used a theoretical expression that relates the scavenging efficiency and the delivery ratio. Krieger et al. [64] determined the heat transfer losses through the engine walls by means of Eichelberg's heat transfer correlation [65].

The scavenging efficiency, defined as the ratio of the fresh air-vapor mass in the cylinder to the total mass in the cylinder, was expressed as

$$\eta_c = 1 - \exp(-\lambda R) \tag{54}$$

where R is a constant, $R = 1$ corresponds to a perfect mixing model, and $R = \infty$ corresponds to perfect scavenging.

Muñoz et al. [66] used one-dimensional gas dynamics equations and discharge coefficients to analyze the gas exchange process in two-stroke engines. Numora et al. [67] used a perfect mixing model, a global reaction mechanism for the oxidation of the fuel to carbon dioxide and water vapor, specific heats at constant pressure as a function of temperature, and exhaust timings that depend on the engine load and rpm to analyze the fuel consumption in two-stroke gasoline-fueled engines.

Numora et al. [67] observed that variable exhaust timings reduce short circuiting and improve engine thermal efficiency. Nuti and Martorano [68] showed that short circuiting correlates with the short circuit ratio, and unburnt HC are generated by short circuiting. Ishihara [69] demonstrated that a U-cylinder uniflow-scavenged engine yields low unburnt HC emissions and is one of the best methods for engine scavenging.

Onishi et al. [70] proposed a new method to control the scavenging flow by separating the mixture generated by the carburetor into rich and lean mixtures between the inlet manifold and the scavenging port. This scavenging control improves the engine thermal efficiency and the unburnt HC emissions without sacrificing the break specific power output and mechanical simplicity.

Hashimoto et al. [71] estimated the performance of two-stroke engines from the trapping efficiency that was obtained from measurements of the oxygen concentration, air-fuel ratio, ignition timing, and gas-sampling location.

5.4.7 Multidimensional Models

Sher [72] studied the effects of atmospheric conditions on the performance of a two-stroke crankcase-scavenged spark-ignition engine by means of a two-dimensional model. His studies include a Wiebe function (see Chapters 2, 3, and

4) and Annand's correlation to analyze combustion and heat transfer, respectively.

Sher [73] developed a two-dimensional model to analyze the gas exchange process in two-stroke, cross-, or loop-scavenged engines. His calculations indicate that the mixing process, but not the shear stresses, plays an important role in the scavenging process, and the predicted instantaneous interface between the fresh charge and the burnt gases fits well the observed profile of a layer that contains a mixture of residuals and fresh charge.

Figure 5.16 shows the interface between the fresh charge and the burnt gases in a two-stroke cycle engine at different crank angles [73]. Figure 5.17 shows the scavenging efficiency as a function of the scavenging ratio for three models and experimental results.

Awn and Spalding [74] used a Lagrangian-Eulerian formulation to predict two-dimensional unsteady flows in loop-scavenged engines and obtained qualitative agreement with experimental data. Their model does not include the influence of mixing on the scavenging efficiency but uses a Lagrangian technique to predict the behavior of the fresh mixture-residual gases interface and an Eulerian finite difference algorithm to iteratively solve the conservation equations of mass and linear momentum.

Diwakar [75] analyzed the gas exchange process in a uniflow-scavenged two-stroke diesel engine. The flow was assumed to be axisymmetric, and a single annular slot on the cylinder head was used to simulate the exhaust valves. His results show that during the gas exchange process, two recirculation zones are in the cylinder, one near the cylinder axis and the other near the cylinder wall just above the inlet ports.

Diwakar [75] observed that inlet port flow angle and associated swirl motion determine the shape and size of the recirculation zones that trap residual gases and influence the scavenging efficiency. His studies indicate that the recirculation zone located near the cylinder wall is more important than the one near the cylinder axis and that both the scavenging and the trapping efficiencies increase with the intake flow angle (swirl). Increasing the inlet flow angle decreases the

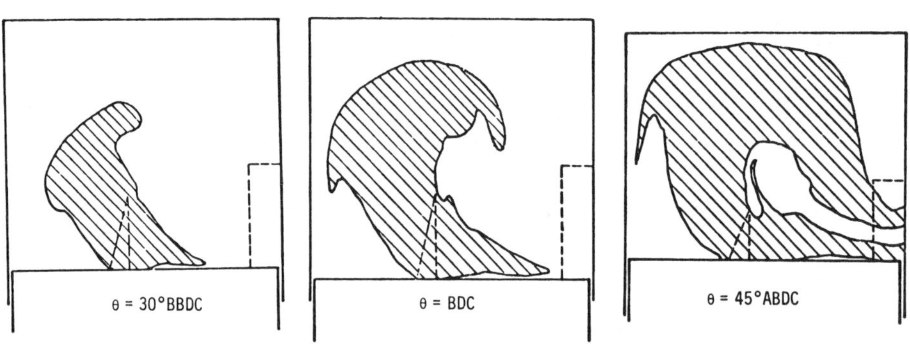

Figure 5.16 Interface between the fresh charge and burnt gases at different crank shaft angles [73].

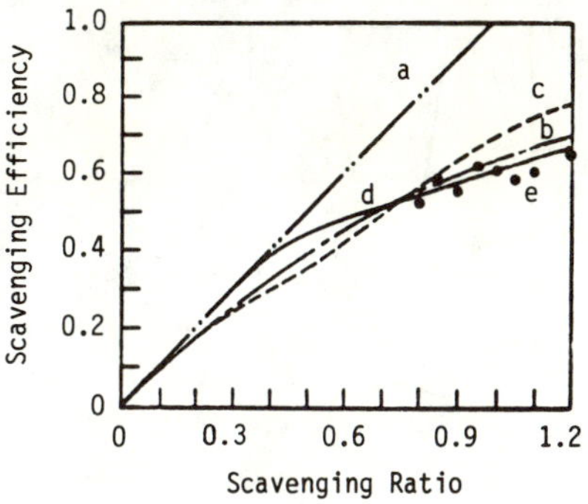

Figure 5.17 Scavenging efficiency as a function of the scavenging ratio for a (a) pure displacement model; (b) perfect mixing model; (c) two-dimensional compressible fluid model; (d) two-dimensional incompressible fluid model; and (e) experiments (solid circles) [73].

size of the recirculation zone located near the cylinder wall and increases the size of the recirculation zone near the cylinder axis. Similar conclusions were reached in the three-dimensional calculations of Diwakar [76], which indicate that the three-dimensional model predicts lower trapping and higher scavenging efficiencies than the two-dimensional model. The three-dimensional calculations also predict less charge stratification than the two-dimensional ones.

Figure 5.18 shows the air purity contours at 191° and 121° BTDC (before top dead center) for an intake port angle of 20° [75]. Figure 5.19 shows the extent of the fresh charge inside the cylinder at various crank shaft angles for three intake port angles [75].

Uzkan and Hazelton [77] used a two-dimensional computer code to study the influence of swirl (inlet port angle) on the fresh charge stratification in a uniflow-scavenged two-stroke engine. They simulated the inlet ports as an annular opening around the liner and the exhaust valves as an annular slot in the cylinder head. Their results indicate that two vortices are present in the cylinder when the intake port and exhaust valve are open. The first vortex is located near the cylinder liner, and the second one is located near the cylinder axis. The incoming fresh charge flows toward the cylinder head between the two vortices with reverse flows on the liner wall and near the symmetry axis. Increasing the swirl (inlet port angle) increases the strength and the region covered by the

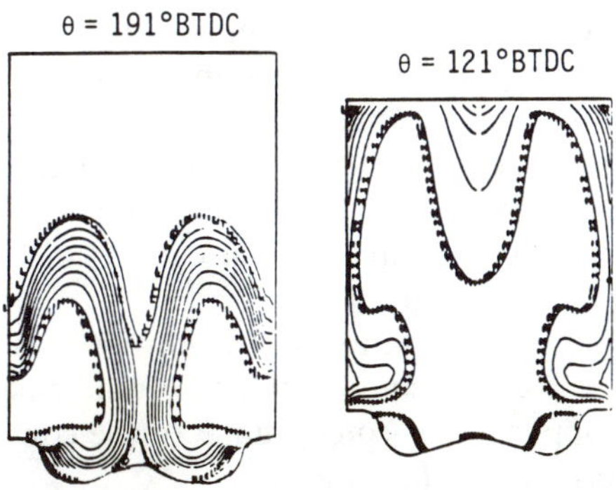

Figure 5.18 Air purity contours at 191° and 121° BTDC for an intake port angle of 20° [75].

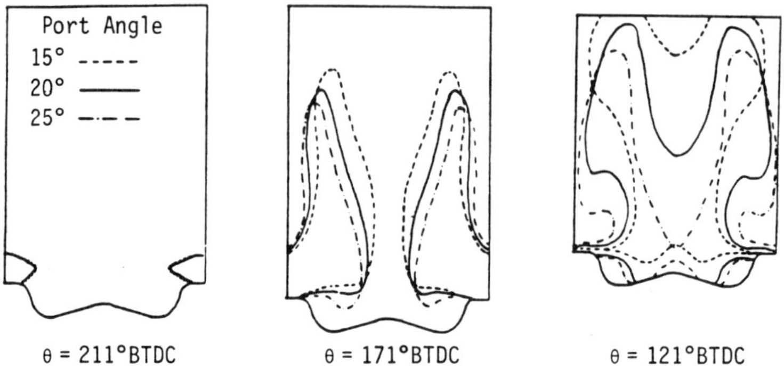

Figure 5.19 Extent of the fresh charge inside the cylinder at various crank angles for three intake port angles [75].

backflow vortex, redirects the upward fresh charge to a larger radius, and restricts the short circuiting of the fresh air to the exhaust valve. This increases the fresh charge trapping and the scavenging efficiency.

Uzkan and Hazelton [77] showed that the gases do not mix completely with the fresh charge during scavenging and compression and that there is considerable stratification at TDC. The inlet port angle affects the mixing of the fresh charge and residual gases and the stratification of the fresh charge at TDC. In order to enhance mixing of the fuel spray with the fresh charge, inlet port angles of about 20° are preferable. However, the mixing of residual gases with the fresh charge is better at low swirl numbers, whereas the trapping and scavenging efficiencies increase with the inlet port angle.

Blair [60] used the PHOENICS computer code [33] to determine the flow field and scavenging processes in two-stroke engines. Sweeney et al. [78] showed that the PHOENICS code was capable of (1) correctly predicting the magnitude and ranking of the scavenging efficiency of five test engine cylinders; (2) giving a superior set of data on the variation of the cylinder purity, short circuit proportion, and displacement proportion; and (3) explaining the reasons for the different scavenging behavior of different cylinders.

Blair [60] suggested that the design of two-stroke engines may be achieved by computer codes and may eliminate the necessity of experiments. He also suggested that perhaps someday, two-stroke engines will be designed with unsteady gas dynamics models with built-in multidimensional models for the scavenging part of the cycle and combustion. This suggestion seems reachable in the near future. However, improvements on turbulence, combustion, and numerical techniques will be needed if multidimensional models are to be used in engine design.

At present, multidimensional models do not seem to have been used for more than evaluation of the performance of combustion chamber geometries under motored conditions. Even in these applications, some tuning of the constants that appear in the turbulence models may be necessary.

Multidimensional models may never eliminate the need for experimental work. They may, however, be used to select the most efficient combustion chambers, which can then be tested experimentally. A close collaboration among theoreticians, experimenters, and design engineers is needed to develop truly predictive multidimensional models and to reduce the costs associated with new engine developments.

Carpenter and Ramos [79, 80] studied the flow field in uniflow-scavenged gasoline-injected two-stroke engine by means of a two-dimensional axisymmetric model. In these simulations as well as in those of Nguyen et al. [81, 82] and Carpenter and Ramos [83], the inlet port was simulated as an annular orifice located in the cylinder liner, while the exhaust valve was simulated as an infinitesimally thin disk that penetrates into the cylinder.

Carpenter and Ramos [79] showed that during the scavenging process two eddies appear in the engine cylinder. The first one is located near the cylinder liner, and the second one is located near the cylinder symmetry axis. The injection of gaseous fuel was observed to reinforce the eddy located near the symmetry axis.

Carpenter and Ramos [80] developed an axisymmetric, uniflow-scavenged two-stroke engine model and simulated combustion by means of a mean-reaction-rate model of the Arrhenius type (see Chapter 2). Their results show that complex flow patterns exist within the engine cylinder due to the interactions among swirl, combustion, and gaseous fuel injection. Carpenter and Ramos [80] also showed that turbulence is generated at the shear layers of the incoming annular air jet but becomes uniform in the compression and expansion strokes. Turbulence is also generated at the flame front during combustion; however, the flame-generated turbulence levels are smaller than those generated by the intake process.

The gaseous fuel injected into the cylinder as a hollow cone displaces the oxidizer and burns as in a diffusion-type flame. The burning rate is controlled by the slow diffusion of fuel and oxidizer and might be increased by the presence of large-scale convective motions such as those induced by squish. For a flat piston, however, the large-scale motions generated during the scavenging process are damped away in the combustion stroke, and poor fuel-air mixing and combustion result.

Nguyen et al. [81, 82] studied the spray penetration, vaporization, and mixing in motored two-stroke diesel engines for different inlet port angles (swirl levels), rpm, droplet number distributions at the injector nozzle, and spray cone angles. Their results show that the spray penetration and vaporization rate increase with the inlet port angle due to the large gas velocities and turbulence levels generated by swirl. The spray penetration and vaporization also increase with the spray cone angle up to a limit. Beyond a critical spray cone angle, some droplets evaporate in the injection-generated eddy located near the cylinder centerline, and others evaporate slowly at the cold cylinder wall. Early injection results in larger spray penetrations because of the low cylinder temperatures and small vaporization cooling. Late injection results in fast vaporization near the injector, where the droplets are trapped in the injection-generated eddies. The turbulence levels and mean gas-phase velocities increase as the engine speed (rpm) is increased. This results in strong deceleration of the injected droplets and fast vaporization rates near the injector and cylinder centerline.

Nguyen et al. [81] showed numerically that there are optimum conditions for injection. These conditions depend on the droplet distribution function at the injector nozzle exit, swirl (inlet port angle), spray cone angle, rpm, and injection timing. Nguyen et al. [82] also found that swirl increases the gas-phase turbulence levels and the vaporization rate. Large swirl levels result in large spray penetration and dispersion; they may have detrimental effects, however, because the large droplets may be thrown against the cold cylinder walls, where they may form a liquid film. This liquid film vaporizes slowly. Under firing conditions the high temperature and pressure in the cylinder increases the vaporization rate, and large swirl numbers may result in good fuel-air mixing throughout the combustion chamber. However, the pyrolysis of the liquid fuel located at the solid walls is a source of unburnt HC.

Carpenter and Ramos [83] analyzed the flow field in a two-stroke uniflow-scavenged engine by means of a mean-reaction-rate model and showed that combustion occurs at the spray envelope, where the fuel-air mixture is within the flammability limits. Their results indicate that most fuel droplets vaporize and burn between the cylinder axis and midcylinder. However, for open combustion chambers without bowl in the piston, the absence of large-scale motions, such as those generation by squish, results in large temperature nonuniformities and poor spray penetration.

Adachi et al. [84] studied uniflow-scavenged engines by means of a multi-dimensional model that uses an eddy diffusivity proportional to the mesh spacing and turbulence dissipation rate (see Chapter 2); they obtained good agreement with experimental data.

Nishimoto and Kamimoto [85] studied the gas exchange process in a uniflow-scavenged two-stroke engine. Initially the cylinder contained cool air, and hot air was forced into the cylinder through the inlet port. The arrival of the hot air was detected with a thermocouple, and the interface between the cool air and the hot air charge was determined from measurements at 20 different locations within the combustion chamber.

Nishimoto and Kamimoto [85] used swirl in their experiments and assumed that the flow field in the cylinder has characteristics similar to vortex breakdown [86, 87] in the sense that an increase in the swirl number increases the depression in the middle portion at the fresh-charge-cool-air interface. However, numerical calculations [75] do not support the analogy between vortex breakdown and the flow during the gas exchange process.

The term vortex breakdown refers to a disturbance characterized by the formation of an internal stagnation point on the vortex axis, followed by a reversed flow in a region of limited axial extent [87]. Vortex breakdown has been observed in steady-state, concentric, swirling flows as the magnitude of the swirl number (i.e., the ratio between the axial flux of angular momentum to the axial flux of axial linear momentum) is increased [88–93].

Nguyen et al. [81, 82] observed complex recirculation zones and stagnation regions in two-stroke diesel engines as the swirl angle at the inlet port was increased. In some cases, vortex breakdown and merging were observed in the calculations.

5.4.8 Experimental Data

Mirko and Radislav [94] performed experiments with liquids and used similarity concepts [95] to determine the scavenging efficiency and delivery ratio in two-stroke engines as a function of the Strouhal and Euler numbers, pressure and density ratios, and engine geometry.

Phatak [96] used water, air, and smoke as working media to visualize the scavenging processes in two-stroke engines. Sher [97] devised the following nondimensional (similarity) groups from the governing equations:

$$\text{Reynolds number:} \quad \mathrm{Re} = \frac{ul}{\nu} \tag{55}$$

$$\text{Prandtl number:} \quad \mathrm{Pr} = \frac{\nu}{\alpha} \tag{56}$$

$$\text{Schmidt number:} \quad \mathrm{Sc} = \frac{\nu}{D} \tag{57}$$

$$\text{Eckert number:} \quad \mathrm{Ec} = \frac{u^2}{h} \tag{58}$$

$$\text{Euler number:} \quad \mathrm{Eu} = \frac{\Delta p}{\rho u^2} \tag{59}$$

where u is the mean piston velocity, l is the stroke, Δp is the pressure drop across the inlet ports, ρ is the density, ν is the kinematic viscosity, α is the thermal diffusivity, D is the mass diffusivity, and h is the enthalpy.

Sher [97] used liquids and a pH indicator to determine the flow patterns and isoconcentration profiles in the engine as a function of time for values of the nondimensional groups typical of two-stroke engines. His experimental data agreed satisfactorily with the results of a perfect mixing model and with those of Eq. (40).

The use of liquids to analyze the scavenging efficiency has the following advantages compared to experiments with gases [98]: simpler measurements, simpler determination of the concentrations in the cylinder and in the exhaust, smaller velocities, and simpler visualization.

5.5 SUMMARY

In this chapter we have described some of the mathematical models that have been used to predict the flow field in engine manifolds and around the intake and exhaust valves in four-stroke engines. Flows in engine manifolds can be calculated by means of the one-dimensional gas dynamics equations accounting for friction and heat transfer. These equations are believed to be approximately valid in the engine manifolds except at pipe branches and junctions, where two- and three-dimensional flow effects are important. These effects can be incorporated into one-dimensional models through appropriate discharge coefficients.

One-dimensional flow models can also be used to predict the intake and blowdown periods, provided that the discharge coefficients through the intake and exhaust valves are known as a function of the valve geometry and lift, valve seat angle and diameter, Reynolds and Mach numbers, etc. Frequently, these discharge coefficients are taken from steady-state flow experiments.

The flows through valves and in pipe branches and junctions can also be calculated by means of multidimensional models. These models are more expensive, however, but provide more details about flows in manifolds than do one-dimensional models.

Models of two-stroke spark-ignition or diesel engines can be classified into three main groups: single-zone, multizone, and multidimensional models. Single-zone models consider that the cylinder charge is uniform in both composition and temperature, and they account for heat transfer by means of experimental correlations. Single-zone models can be used as predictive tools if the heat release rate or the mass burning rate are specified as a function of the engine geometry and operating conditions. They can also be used in heat release analyses if experimentally determined pressure diagrams are used as an input in the first law of thermodynamics.

In multizone models of spark-ignition engines, the cylinder charge is divided into several regions: the burnt gases, the unburnt gases, and the quench layers. The burnt gases are in chemical equilibrium, whereas the unburnt gases are assumed to be frozen. The burnt and unburnt gases are separated by a flame front having a velocity that may be specified as a function of the turbulence intensity, turbulence length scales, laminar flame thickness, and laminar flame speed. Alternatively, the mass burning rate may be specified as a function of the crank shaft angle or calculated from entrainment models that account for energy cascade processes.

In multizone models of diesel engines, the spray injected into the cylinder is divided into several parcels that exchange mass, momentum, and energy with adjacent parcels.

In multidimensional models, the time-averaged conservation equations of mass, linear momentum, and energy are solved at a finite number of grid points. These models may include turbulence submodels, different combustion models, and stochastic representations of sprays; they can be used to calculate the scavenging process in two-stroke diesel engines.

Phenomenological models of scavenging phenomena include short circuiting of the fresh charge from the inlet to the exhaust port, displacement of the burnt or residual gases by the fresh charge, mixing of the burnt gases and residual charge, and blowdown (flows through the exhaust port) processes.

It is believed that with future computers, development of comprehensive multidimensional models of the gas exchange process, scavenging, and combustion would be possible. However, in order to achieve this goal and obtain truly predicted models of engine flows, improvements in turbulence and combustion modeling and in numerical methods are needed.

Global engine models may be used to determine the performance of new engine designs. However, it is unlikely that engine models will entirely eliminate the need for experimental tests. On the contrary, it is believed that a greater interaction among engine modelers, experimenters, and designers is needed to obtain truly predictive engine models and to reduce the time and expense associated with current design techniques.

REFERENCES

1. Benson, R. S., *The Thermodynamics and Gas Dynamics of Internal Combustion Engines*, vol. 1, Oxford University Press, New York, 1982.
2. Holman, J. P., *Heat Transfer*, 6th ed., McGraw-Hill, New York, 1986.
3. Takizawa, M., Uno, T., Oue, T., and Yura, T., A Study of Gas Exchange Process Simulation of an Automotive Multi-Cylinder Internal Combustion Engine, SAE International Congress and Exposition, Detroit, Mich., SAE Paper No. 820410, 1982.
4. Anderson, D. A., Tannehill, J. C., and Pletcher, R. H., *Computational Fluid Mechanics and Heat Transfer*, Hemisphere, Washington, D.C., 1984.
5. Bulaty, T., and Niessner, H., Calculation of 1-D Unsteady Flows in Pipe Systems of I.C. Engines, *ASME Journal of Fluids Engineering*, vol. 107, pp. 407–412, 1985.
6. Zalesak, S. T., Fully Multidimensional Flux-Corrected Transport Algorithms for Fluids, *Journal of Computational Physics*, vol. 31, pp. 335–362, 1979.
7. Chapman, M., Novak, J. M., and Stein, R. A., Numerical Modeling of Inlet and Exhaust Flows in Multi-Cylinder Internal Combustion Engines, in *Flows in Internal Combustion Engines*, ed. T. Uzkan, pp. 9–19, ASME, New York, 1982.
8. Glennie, A. E., Solution by Characteristics of the Equations of One-Dimensional Unsteady Flow, in *Numerical Solution of Ordinary and Partial Differential Equations*, ed. L. Fox, pp. 325–365, Pergamon Press, New York, 1962.
9. Hoskin, N. E., Solution by Characteristics of the Equations of One-Dimensional Unsteady Flow, in *Methods in Computational Physics*, eds. B. Alder, S. Fernbach, and M. Rotenberg, vol. 3, pp. 265–293, Academic Press, New York, 1964.
10. Desantes, J. M., Boada, F., and Corberan, J. M., Exhaust Pipe Design Method for the Optimization of the Scavenging Process, SAE International Congress and Exposition, Detroit, Mich., SAE Paper No. 850083, 1985.
11. Winterbone, D. E., Nichols, J. R., and Alexander, G. I., Efficiency of the Manifolds of Turbocharged Engines, *Proceedings of the Institution of Mechanical Engineers*, vol. 199, no. D2, pp. 137–149, 1985.
12. Winterbone, D. E., Alexander, G. I., and Nichols, J. R., Developments in Methods of Considering Wave Action in Pipes Connected to I.C. Engines, in *International Symposium on Flows in Internal Combustion Engines—III*, eds. T. Uzkan, W. G. Tiederman, and J. M. Novak, pp. 71–77, ASME, New York, 1985.
13. Newlyn, H. A., and Few, P. C., Effects of Engine Geometry on the Air Consumption of a Single Cylinder Internal Combustion Engine, SAE International Congress and Exposition, Detroit, Mich., SAE Paper No. 840424, 1984.

14. Obert, E. F., *Internal Combustion Engines and Air Pollution*, Intext, New York, 1973.
15. Vorum, P. C., Short Pipe Manifold Design for Four-Stroke Engines, Paper No. 76-WA/DGP-4, ASME, New York, 1976.
16. Payri, F., Boada, F., and Macian, V., Reduction of Pumping Losses by the Use of a Variable Valve Timing System, *Proceedings of the Institution of Mechanical Engineers*, vol. 198, no. D15, pp. 295–300, 1984.
17. Payri, F., Desantes, J. M., and Boada, F., Calculation Model for the Optimization of a Variable Distribution (Reduction of the Pump Losses), SAE Special Publication No. 143, pp. 3.1–3.9, Society of Automotive Engineers, Warrendale, Pa., 1984.
18. Kalinin, E. K., and Dreitser, G. A., Unsteady Convective Heat Transfer for Turbulent Flows of Gases and Liquids in Tubes, *International Journal of Heat and Mass Transfer*, vol. 28, no. 2, pp. 361–369, 1985.
19. Dreitser, G. A., Limits of Applicability of Quasistationary Values of Heat Transfer Coefficients in Calculating Real Nonstationary Thermal Processes, *Inzhenerno-Fizicheskii Zhurnal*, vol. 36, no. 5, pp. 814–820, 1979.
20. Dreitser, G. A., Kalinin, E. K., and Kuz'minov, V. A., Transient Convective Heat Transfer during Various Modes of Cooling Hot Gases in Pipes, *Inzhenerno-Fizicheskii Zhurnal*, vol. 25, no. 2, pp. 208–216, 1973.
21. Dreitser, G. A., Evdokimov, V. D., and Kalinin, E. K., Unsteady Convective Heat Transfer in the Heating of a Liquid in a Pipe by Variable Heat Flux, *Inzhenerno-Fizicheskii Zhurnal*, vol. 31, no. 1, pp. 5–12, 1976.
22. Galitseiskii, B. M., Nozdrin, A. A., Ryzhov, Yu. A, and Yakush, E. V., Investigation of Local Heat-Transfer Coefficients under Conditions of the Resonance Oscillations of a Gas in Channels, *Inzhenerno-Fizicheskii Zhurnal*, vol. 31, no. 2, pp. 197–201, 1976.
23. Servati, H. B., and Yuen, W. W., Deposition of Fuel Droplets in Horizontal Intake Manifolds and the Behavior of Fuel Film Flow on Its Walls, SAE International Congress and Exposition, Detroit, Mich., SAE Paper No. 840239, 1984.
24. Mahdi Ali, E., Flow and Distribution Patterns within the Gasoline Engine Manifold, XIX International Symposium on Heat and Mass Transfer in Gasoline and Diesel Engines, International Centre for Heat and Mass Transfer, Dubrovnik, Yugoslavia, 1987.
25. Milton, B. E., and Behnia, M., A Numerical Study of the Interchanging Vapour, Droplet and Film Flows in a Gasoline Engine Manifold, XIX International Symposium on Heat and Mass Transfer in Gasoline and Diesel Engines, International Centre for Heat and Mass Transfer, Dubrovnik, Yugoslavia, 1987.
26. Dobran, F., Hydrodynamic and Heat Transfer Analysis of Two-Phase Annular Flow with a New Liquid Film Model of Turbulence, *International Journal of Heat and Mass Transfer*, vol. 26, no. 8, pp. 1159–1171, 1983.
27. Pucher, H., Internal Combustion Engine Cycle Simulation Methods Aid Engine Development, *Journal of Non-Equilibrium Thermodynamics*, vol. 11, pp. 123–142, 1986.
28. Bingham, J. F., and Blair, G. P., An Improved Branched Pipe Model for Multi-cylinder Automotive Engine Calculations, *Proceedings of the Institution of Mechanical Engineers*, vol. 199, no. D1, pp. 63–77, 1985.
29. Chapman, M., Two Dimensional Numerical Simulation of Inlet Manifold Flow in a Four Cylinder Internal Combustion Engine, SAE Congress and Exposition, Detroit, Mich., SAE Paper No. 790244, 1979.
30. Tosa, Y., Shimoda, K., and Oikawa, H., Calculation of 2-Dimensional Unsteady Flows in Inlet Pipe Systems and Its Application for V-8 Resonant Intake System, in *International Symposium on Flows in Internal Combustion Engines—III*, eds. T. Uzkan, W. G. Tiederman, and J. M. Novak, pp. 63–70, ASME, New York, 1985.
31. Dimitriadis, C., Leschziner, M. A., Winterbone, D. E., Alexander, G. I., and Sierens, R., Computation of Three-dimensional Flow in Manifold-Type Junctions, in *International Symposium on Flows in Internal Combustion Engines—III*, eds. T. Uzkan, W. G. Tiederman, and J. M. Novak, pp. 57–62, ASME, New York, 1985.
32. Seppen, J. J., and Janssen, G. A. H. F., Development and Verification of a Numerical Flow

Package for Manifolds in Combustion Engines, XIX International Symposium on Heat and Mass Transfer in Gasoline and Diesel Engines, International Centre for Heat and Mass Transfer, Dubrovnik, Yugoslavia, 1987.

33. Spalding, D. B., The PHOENICS Flow-Simulation System, Advanced Course on Computer Simulation of Fluid Flow, Heat and Mass Transfer and Combustion in Reciprocating Engines, International Centre for Heat and Mass Transfer, Dubrovnik, Yugoslavia, 1987.
34. Tuttle, J. H., Controlling Engine Load by Means of Early Intake-Valve Closing, *SAE Transactions*, vol. 91, pp. 1648–1662, 1983.
35. Charlton, S. J., and Pappas, P., A Parametric Study of the Exhaust Valve Requirements of High Output Truck Engines, SAE International Congress and Exposition, Detroit, Mich., SAE Paper No. 850243, 1985.
36. Heywood, J. B., Fluid Motion within the Cylinder of Internal Combustion Engines—The 1986 Freeman Scholar Lecture, *ASME Journal of Fluids Engineering*, vol. 109, pp. 3–35, 1987.
37. Deckker, B. E. L., and Hung, S. C., Transient Discharge from a Cylinder at High Rates of Volume Expansion, SAE International Congress and Exposition, Detroit, Mich., SAE Paper No. 850082, 1985.
38. Rabbitt, R. D., Fundamentals of Reciprocating Engine Airflow: Part I. Valve Discharge and Combustion Chamber Effects, SAE International Congress and Exposition, Detroit, Mich., SAE Paper No. 840337, 1984.
39. Tabaczynski, R. J., The Role of Fluid Mechanics in Spark-Ignition Engine Design, in *International Symposium on Flows in Internal Combustion Engines—III*, eds. T. Uzkan, W. G. Tiederman, and J. M. Novak, pp. 149–157, ASME, New York, 1985.
40. Haghgooie, M., Kent, J. C., and Tabaczynski, R. J., Intake Valve Cylinder Boundary Flow Characteristics in an Internal Combustion Engine, *Combustion Science and Technology*, vol. 38, pp. 49–57, 1984.
41. Bicen, A. F., Vlachos, N. S, Whitelaw, J. H., The Creation and Destruction of Vortices in Unsteady Flows, *Letters in Heat and Mass Transfer*, vol. 7, pp. 77–82, 1980.
42. Gany, A., Larrea, J.-J., and Sirignano, W. A., Laser-Doppler Velocimetry Measurements in a Motored IC Engine Simulator, AIAA 1980 Aerospace Sciences Meeting, Pasadena, Calif., Paper No. AIAA-80-0079, 1980.
43. Arcoumanis, C., Bicen, A. F., Vlachos, N. S., and Whitelaw, J. H., Effects of Flow and Geometry Boundary Conditions on Fluid Motion in a Motored IC Model Engine, *Proceedings of the Institution of Mechanical Engineers*, vol. 196, pp. 1–10, 1982.
44. Arcoumanis, C., Vafidis, C., and Whitelaw, J. H., Valve and In-Cylinder Flow Field Generated by a Helical Port in a Production Diesel Engine, in *International Symposium on Flows in Internal Combustion Engines—III*, eds. T. Uzkan, W. G. Tiederman, and J. M. Novak, pp. 29–37, ASME, New York, 1985.
45. Arcoumanis, C., and Whitelaw, J. H., Fluid Mechanics of Internal Combustion Engines—A Review, *Proceedings of the Institution of Mechanical Engineers*, vol. 201, no. C1, pp. 57–74, 1987.
46. Bicen, A. F., Vafidis, C., and Whitelaw, J. H., Steady and Unsteady Airflow through the Intake Valve of a Reciprocating Engine, *ASME Journal of Fluids Engineering*, vol. 107, pp. 413–420, 1985.
47. Gosman, A. D., Melling, A., Whitelaw, J. H., and Watkins, P., Axisymmetric Flow Model in a Motored Reciprocating Engine, *Proceedings of the Institution of Mechanical Engineers*, vol. 192, no. 11, pp. 213–223, 1978.
48. Elkotb, M. M., Abou-Ellail, M. M. M., and Salah-Salem, I., Influence of Swirl Chamber Geometry on Turbulence in Multifuel Engines, *International Conference on Combustion Engineering*, vol. 1, pp. 101–112, I. Mech. E. Conference Publications 1983-3, London, 1983.
49. Gosman, A. D., Flow Processes in Cylinders, in *The Thermodynamics and Gas Dynamics of Internal Combustion Engines*, eds. J. H. Horlock, and D. E. Winterbone, vol. 2, pp. 616–772, Clarendon Press, Oxford University Press, New York, 1986.
50. Gosman, A. D., Multidimensional Modeling of Cold Flows and Turbulence in Reciprocating Engines, SAE International Congress and Exposition, Detroit, Mich., SAE Paper No. 850344, 1985.

51. Pinchon, P., and Baritaud, T., Modeling of Flow and Combustion in a Spark Ignition Engine with a Shrouded Valve: Comparison with Experiments, XIX International Symposium on Heat and Mass Transfer in Gasoline and Diesel Engines, International Centre for Heat and Mass Transfer, Dubrovnik, Yugoslavia, 1987.
52. Ramos, J. I., and Sirignano, W. A., Axisymmetric Flow Model in a Piston-Cylinder Arrangement with Detailed Analysis of the Valve Region, SAE International Congress and Exposition, Detroit, Mich., SAE Paper No. 800286, 1980.
53. Schock, H. J., Sosoka, D, J., and Ramos, J. I., Numerical Studies of the Formation and Destruction of Vortices in a Motored Four-Stroke Piston-Cylinder Configuration, AIAA 21st Aerospace Sciences Meeting, Reno, Nev., Paper No. AIAA-83-0497, 1983.
54. Schock, H. J., Sosoka, D. J., and Ramos, J. I., Formation and Destruction of Vortices in a Motored Four-Stroke Piston-Cylinder Configuration, *AIAA Journal*, vol. 22, no. 7, pp. 948–949, 1984.
55. Demirdzic, I., Gosman, A. D., and Issa, R. I., A Finite-Volume Method for the Prediction of Turbulent Flow in Arbitrary Geometries, in *Seventh International Conference on Numerical Methods in Fluid Dynamics*, eds. W. C. Reynolds and, R. W. MacCormack, *Lecture Notes in Physics*, vol. 141, pp. 144–150, Springer-Verlag, New York, 1981.
56. Jovanovic, Z., and Petrovic, S., The Modeling of Combustion in a Wedge-Type Combustion Chamber and Its Comparison with Experiments, Congress on Internal Combustion Engine Modeling, Universidad Politecnica de Valencia, Valencia, Spain, 1987.
57. Jovanovic, Z., Some Solutions to the Problems That Result from the Complex Geometries of Combustion Chambers in Multidimensional Numerical Modeling of Combustion Processes in Otto Engines, *Fuels and Combustion in Otto Engines*, pp. 141–163, Kragujevac, Yugoslavia, 1986 (in Serbo-Croatian).
58. Sanatian, R., and Moss, J. B., Computer Simulation of Induction Flows in Spark Ignition Engines, XIX International Symposium on Heat and Mass Transfer in Gasoline and Diesel Engines, International Centre for Heat and Mass Transfer, Dubrovnik, Yugoslavia, 1987.
59. Demirdzic, I., Peric, M., and Yianneskis, M., Numerical Predictions of the Mean Flow and Turbulence in an Axisymmetric Port and Their Assessment against Experimental Data, XIX International Symposium on Heat and Mass Transfer in Gasoline and Diesel Engines, International Centre for Heat and Mass Transfer, Dubrovnik, Yugoslavia, 1987.
60. Blair, G. P., Motion and Mixing in the Cylinder of a Two-Stroke Engine prior to Combustion, Advanced Course on Computer Simulation of Fluid Flow, Heat and Mass Transfer and Combustion in Reciprocating Engines, International Centre for Heat and Mass Transfer, Dubrovnik, Yugoslavia, 1987.
61. Benson, R. S., and Whitehouse, N. D., *Internal Combustion Engines*, Pergamon Press, New York, 1979.
62. Sher, E., A New Practical Model for the Scavenging Process in a Two-Stroke Cycle Engine, SAE International Congress and Exposition, Detroit, Mich., SAE Paper No. 850085, 1985.
63. Benson, R. S., A New Gas Dynamic Method for the Gas Exchange Process in Two-Stroke Loop and Cross Scavenged Engines, *International Journal of Mechanical Sciences*, vol. 19, pp. 693–711, 1977.
64. Krieger, R. B., Booy, R. R., Myers, P. S., and Uyehara, O. A., Simulation of a Crankcase Scavenged, Two-Stroke, SI Engine and Comparisons with Experimental Data, SAE Automotive Engineering Congress, Detroit, Mich., SAE Paper No. 690135, 1969.
65. Eichelberg, G., Some New Investigations of Old Internal Combustion Engine Problems, *Engineering*, vol. 149, pp. 297–299, 1939.
66. Muñoz, M., Bethencourt, A. N., and Salazar-Simpson, J. R., Two-Stroke Engines—Models for Charge Renovation, SAE International Congress and Exposition, Detroit, Mich., SAE Paper No. 850084, 1985.
67. Numora, K., Hirano, S., Gotoh, T., and Motoyama, Y., Improvement of Fuel Consumption with Variable Exhaust Port Timing in a Two-Stroke Gasoline Engine, SAE International Congress and Exposition, Detroit, Mich., SAE Paper No. 850183, 1985.
68. Nuti, M., and Martorano, L., Short-Circuit Ratio Evaluation in the Scavenging of Two-Stroke

S.I. Engines, SAE International Congress and Exposition, Detroit, Mich., SAE Paper No. 850177, 1985.
69. Ishihara, S., An Experimental Development of a New U-Cylinder Uniflow Scavenged Engine, SAE International Congress and Exposition, Detroit, Mich., SAE Paper No. 850181, 1985.
70. Onishi, S., Jo, S. H., Jo, P. D., and Kato, S., Multi-Layer Stratified Scavenging (MULS)—A New Scavenging Method for Two-Stroke Engine, SAE International Congress and Exposition, Detroit, Mich., SAE Paper No. 840420, 1984.
71. Hashimoto, E., Tottori, T., and Terata, S., Scavenging Performance Measurements of High-Speed Two-Stroke Engines, SAE International Congress and Exposition, Detroit, Mich., SAE Paper No. 850182, 1985.
72. Sher, E., The Effect of Atmospheric Conditions on the Performance of an Air-Borne Two-Stroke Spark-Ignition Engine, *Proceedings of the Institution of Mechanical Engineers*, vol. 198D, no. 15, pp. 239–251, 1984.
73. Sher, E., Prediction of the Gas Exchange Performance in a Two-Stroke Cycle Engine, SAE International Congress and Exposition, Detroit, Mich., SAE Paper No. 850086, 1985.
74. Awn, A. G., and Spalding, D. B., Fluid Flow within Reciprocating-Engine Cylinders, in *Numerical Methods in Thermal Problems*, eds. R. W. Lewis, K. Morgan, and B. A. Schreffler, vol. 2, pp. 1189–1203, Pineridge Press, Swansea, U.K., 1981.
75. Diwakar, R., Multidimensional Modeling of the Gas Exchange Processes in a Uniflow-Scavenged Two-Stroke Diesel Engine, in *International Symposium on Flows in Internal Combustion Engines—III*, eds. T. Uzkan, W. G. Tiederman, and J. M. Novak, pp. 125–134, ASME, New York, 1985.
76. Diwakar, R., Three-Dimensional Modeling of the In-Cylinder Gas Exchange Processes in a Uniflow-Scavenged Two-Stroke Engine, SAE International Congress and Exposition, Detroit, Mich., SAE Paper No. 870596, 1987.
77. Uzkan, T., and Hazelton, J. R., The Influence of Swirl on the Fresh Charge Stratification in an IC Engine Combustion Chamber, SAE International Congress and Exposition, Detroit, Mich., SAE Paper No. 860466, 1986.
78. Sweeney, M. E. G., Kenny, R. G., Swann, G. B. G, and Blair, G. P., Computational Fluid Dynamics Applied to Two-Stroke Engine Scavenging, SAE International Off-Highway Vehicle Meeting, Milwaukee, Wis., SAE Paper No. 851519, 1985.
79. Carpenter, M. H., and Ramos, J. I., The Numerical Solution of the Flowfield in an Internal Combustion Engine, in *Advances in Computer Methods for Partial Differential Equations—V*, eds. R. Vichnevetsky, and R. S. Stepleman, pp. 24–31, IMACS, Rutgers University, New Brunswick N.J., 1984.
80. Carpenter, M. H., and Ramos, J. I., Modeling of a Gasoline-Injected Two-Stroke Cycle Engine, SAE International Congress and Exposition, Detroit, Mich., SAE Paper No. 860167, 1986.
81. Nguyen, H. L., Carpenter, M. H., and Ramos, J. I., The Effects of Engine Speed and Injection Characteristics on the Flow Field and Fuel/Air Mixing in Motored Two-Stroke Diesel Engines, AIAA 25th Aerospace Sciences Meeting, Reno, Nev., Paper No. AIAA-87-0227, 1987.
82. Nguyen, H. L., Schock, H. J., Carpenter, M. H., Ramos, J. I., and Stegeman, J. D., Numerical Simulation of the Flow Field and Fuel Sprays in an IC Engine, SAE International Congress and Exposition, Detroit, Mich., SAE Paper No. 870599, 1987.
83. Carpenter, M. H., and Ramos, J. I., Mathematical Modelling of Uniflow-Scavenged Two-Stroke Diesel Engines, in *Modelado en Motores de Combustion Interna*, eds. F. Payri, J. M. Desantes, and F. Boada, pp. 217–231, Universidad Politecnica de Valencia, Valencia, Spain, 1987.
84. Adachi, T., Kato, S., and Tsujimura, H., Numerical Analysis of Scavenging Flow in Two-Cycle Internal Combustion Engines, in *Flows in Internal Combustion Engines*, ed. T. Uzkan, pp. 29–34, ASME, New York, 1982.
85. Nishimoto, K., and Kamimoto, T., A Study on the Influence of Inlet Angle and Reynolds Number on the Flow-Pattern of Uniflow Scavenging Air, SAE International Congress and Exposition, Detroit, Mich., SAE Paper No. 841056, 1984.
86. Hall, M. G., Vortex Breakdown, *Annual Review of Fluid Mechanics*, vol. 4, pp. 195–218, 1972.

87. Leibovich, S., The Structure of Vortex Breakdown, *Annual Review of Fluid Mechanics*, vol. 10, pp. 221–246, 1978.
88. Ramos, J. I., Incompressible Swirling Flows, *Engineering Computations*, vol. 3, no. 1, pp. 53–63, 1986.
89. Ramos, J. I., The Numerical Solution of Non-premixed Reactive Flows in a Swirl Combustor Model, *Engineering Computations*, vol. 1, no. 2, pp. 173–182, 1984.
90. Ramos, J. I., A Numerical Study of a Swirl Stabilized Combustor, *Journal of Non-Equilibrium Thermodynamics*, vol. 10, pp. 263–286, 1985.
91. Ramos, J. I., Turbulent Non-Reacting Swirling Flows, *AIAA Journal*, vol. 22, no. 6, pp. 846–848, 1984.
92. Ramos, J. I., and Sommer, H. T., Swirling Flow in a Research Combustor, *AIAA Journal*, vol. 23, no. 2, pp. 241–248, 1985.
93. Vu, B. T., and Gouldin, F. C., Flow Measurements in a Model Swirl Combustor, *AIAA Journal*, vol. 20, no. 5, pp. 642–651, 1982.
94. Mirko, C., and Radislav, P., A Model Method for the Evaluation of the Scavenging System in a Two-Stroke Engine, SAE International Congress and Exposition, Detroit, Mich., SAE Paper No. 850176, 1985.
95. Birkhoff, G., *Hydrodynamics: A Study in Logic, Fact, and Similitude*, Dover, New York, 1955.
96. Phatak, R. G., A New Method of Analyzing Two-Stroke Cycle Engine Gas Flow Patterns, SAE Congress and Exposition, Detroit, Mich., SAE Paper No. 790487, 1979.
97. Sher, E., Investigating the Gas Exchange Process of a Two-Stroke Cycle Engine with a Flow Visualization Rig, *Israel Journal of Technology*, vol. 20, pp. 127–136, 1982.
98. Sanborn, D. S., and Roeder, W. M., Single Cycle Simulation Simplifies Scavenging Study, SAE International Congress and Exposition, Detroit, Mich., SAE Paper No. 850175, 1985.

INDEX